DESIGN FOR SUSTAINABILITY

DESIGN FOR SUSTAINABILITY

GREEN MATERIALS AND PROCESSES

Edited by

S.M. Sapuan

Advanced Engineering Materials and Composites Research Centre (AEMC), Department of Mechanical and Manufacturing Engineering, Faculty of Engineering, Universiti Putra Malaysia, Selangor, Malaysia; Laboratory of Biocomposite Technology, Institute of Tropical Forestry and Forest Products (INTROP), Universiti Putra Malaysia, Selangor, Malaysia

Muhd Ridzuan Mansor

Faculty of Mechanical Engineering, Universiti Teknikal Malaysia Melaka, Melaka, Malaysia

ELSEVIER

Elsevier
Radarweg 29, PO Box 211, 1000 AE Amsterdam, Netherlands
The Boulevard, Langford Lane, Kidlington, Oxford OX5 1GB, United Kingdom
50 Hampshire Street, 5th Floor, Cambridge, MA 02139, United States

British Library Cataloguing-in-Publication Data

A catalogue record for this book is available from the British Library

Library of Congress Cataloging-in-Publication Data

A catalog record for this book is available from the Library of Congress

ISBN: 978-0-12-819482-9

For Information on all Elsevier publications
visit our website at https://www.elsevier.com/books-and-journals

Publisher: Matthew Deans
Acquisitions Editor: Christina Gifford
Editorial Project Manager: Emily Thomson
Production Project Manager: Kamesh Ramajogi
Cover Designer: Greg Harris

Typeset by MPS Limited, Chennai, India

Contents

Part I

Design methods

3. Sustainable materials selection: principles and applications

M.A. Shaharuzaman, S.M. Sapuan and Muhd Ridzuan Mansor

4. Implementation of design for sustainability in developing trophy plaque using green kenaf polymer composites

S.M. Sapuan, G. Hemapriya, R.A. Ilyas, M.S.N. Atikah, M.R.M. Asyraf and Muhd Ridzuan Mansor

5. Dynamic analysis of a laminated rubber–metal spring vibration isolator for sustainable design

Mohd Azli Salim, Azma Putra, Mohammed Hussin Ahmed Al-Mola, Norbazlan Mohd Yusof and Muhd Ridzuan Mansor

6. Design for sustainability integration in education

Ainur Zaireen Zainudin, Norhidayah Md Yunus, Siti Radiaton Adawiyah Zakaria and Aminah Mohsin

Part II

Green materials

7. A review of natural fiber reinforced recycled thermoplastic polymer composites

M.J. Taufiq, Muhd Ridzuan Mansor and Zaleha Mustafa

8. Degradable composites: processes and applications

Zaleha Mustafa, Siang Yee Chang, Siti Hajar Sheikh Md Fadzullah, Azlin Fazlina Osman and Ismail Ibrahim

9. Thermoplastic starch as renewable plastics

R. Jumaidin and Z.A.S. Saidi

13. Recent development in kenaf (*Hibiscus cannabinus*)-based biocomposites and their potential industrial applications: A review

M.D. Hazrol, S.M. Sapuan, M.Y.M. Zuhri, E.S. Zainudin, N.I.A. Wahab and R.A. Ilyas

Part III

Manufacturing processes

14. Recent progress on fused filament fabrication research: sustainable materials and processing parameters

S.N.M Farhan Han, M.T. Mastura and Muhd Ridzuan Mansor

15. Out-of-autoclave as a sustainable composites manufacturing process for aerospace applications

Nadlene Razali, Muhd Ridzuan Mansor, Ghazali Omar, Syed Ahmad Faiz Syed Kamarulzaman, Mohd Hanafee Zin and Nadia Razali

List of contributors

A.F. Ab Ghani Faculty of Mechanical and Manufacturing Engineering Technology, Universiti Teknikal Malaysia Melaka, Melaka, Malaysia

A.B. Abdullah School of Mechanical Engineering, Universiti Sains Malaysia (Engineering Campus), Penang, Malaysia

N. Abdullah UTM International, Level 8, Menara Razak, Universiti Teknologi Malaysia, Jalan Sultan Yahya Petra, Kuala Lumpur, Malaysia

I. Ahmad Algae and Biomass Research Laboratory, Malaysia-Japan International Institute of Technology, Universiti Teknologi Malaysia, Jalan Sultan Yahya Petra, Kuala Lumpur, Malaysia

M.D. Ahmad Department of Post-Harvest Engineering and Technology, Aligarh Muslim University, Aligarh, Uttar Pradesh, India

M.R. Alkahari Faculty of Mechanical Engineering, Universiti Teknikal Malaysia Melaka, Melaka, Malaysia; Advanced Manufacturing Centre, Universiti Teknikal Malaysia Melaka, Melaka, Malaysia

Mohammed Hussin Ahmed Al-Mola Faculty of Petroleum and Mining Engineering, University of Mosul, Mosul, Iraq

M.R.M. Asyraf Department of Aerospace Engineering, Universiti Putra Malaysia, Selangor, Malaysia

M.S.N. Atikah Department of Chemical and Environmental Engineering, Universiti Putra Malaysia, Serdang, Selangor, Malaysia

Siang Yee Chang Faculty of Manufacturing Engineering, Universiti Teknikal Malaysia Melaka, Melaka, Malaysia

M. Fadzil School of Mechanical Engineering, Universiti Sains Malaysia (Engineering Campus), Penang, Malaysia; Centre of Advanced Manufacturing and Material Processing (AMMP Centre), University of Malaya, Kuala Lumpur, Malaysia

Siti Hajar Sheikh Md Fadzullah Centre for Advanced Research on Energy, Fakulti Kejuruteraan Mekanikal, Universiti Teknikal Malaysia Melaka, Melaka, Malaysia

S.N.M Farhan Han Faculty of Mechanical Engineering, Universiti Teknikal Malaysia Melaka, Melaka, Malaysia

M.D. Hazrol Advanced Engineering Materials and Composites Research Centre (AEMC), Department of Mechanical and Manufacturing Engineering, Faculty of Engineering, Universiti Putra Malaysia, Selangor, Malaysia

G. Hemapriya Advanced Engineering Materials and Composites Research Centre (AEMC), Department of Mechanical and Manufacturing Engineering, Faculty of Engineering, Universiti Putra Malaysia, Selangor, Malaysia

Ismail Ibrahim Center of Excellence Geopolymer and Green Technology (CEGeoGTech), Faculty of Chemical Engineering Technology, Universiti Malaysia Perlis (UniMAP), Perlis, Malaysia

R.A. Ilyas School of Chemical and Energy Engineering, Faculty of Engineering, Universiti Teknologi Malaysia, Johor, Malaysia; Centre for Advanced Composite Materials, Universiti Teknologi Malaysia, Johor, Malaysia

N.M. Ishak Faculty of Mechanical Engineering, Universiti Teknikal Malaysia Melaka, Melaka, Malaysia

S. Mohd Izwan Advanced Engineering Materials and Composites Research Centre (AEMC), Department of Mechanical and Manufacturing Engineering, Faculty of Engineering, Universiti Putra Malaysia, Selangor, Malaysia

J. Jaafar Advanced Membrane Technology Research Centre (AMTEC), Universiti Teknologi Malaysia, Johor, Malaysia

R. Jumaidin Fakulti Teknologi Kejuruteraan Mekanikal dan Pembuatan, Universiti Teknikal Malaysia Melaka, Hang Tuah Jaya, Durian Tunggal, Melaka, Malaysia

Syed Ahmad Faiz Syed Kamarulzaman Section of Environmental Engineering Technology, Universiti Kuala Lumpur—Malaysian Institute of Chemical and Bioengineering Technology, Melaka, Malaysia

Muhd Ridzuan Mansor Faculty of Mechanical Engineering, Universiti Teknikal Malaysia Melaka, Melaka, Malaysia

Y.H.P. Manurung Smart Manufacturing Research Institute (SMRI), Faculty of Mechanical Engineering Universiti Teknologi MARA, Selangor, Malaysia

M.T. Mastura Faculty of Mechanical and Manufacturing Engineering Technology, Universiti Teknikal Malaysia Melaka, Melaka, Malaysia; Centre for Advanced Research on Energy, Universiti Teknikal Malaysia Melaka, Melaka, Malaysia

A.R. Mohamed Department of Manufacturing and Material Engineering, Kulliyyah of Engineering, International Islamic University Malaysia, Kuala Lumpur, Malaysia

Aminah Mohsin Faculty of Built Environment and Surveying, Universiti Teknologi Malaysia, Johor Bahru, Malaysia

N.M. Mokhtar Faculty of Civil Engineering Technology, Universiti Malaysia Pahang, Pahang, Malaysia

Zaleha Mustafa Faculty of Manufacturing Engineering, Universiti Teknikal Malaysia Melaka, Melaka, Malaysia

N.A.H. Nordin Chemical Engineering Department, Universiti Teknologi Petronas, Perak, Malaysia

A.H. Nurfaizey Faculty of Mechanical Engineering, Universiti Teknikal Malaysia Melaka, Melaka, Malaysia; Centre for Advanced Research on Energy, Universiti Teknikal Malaysia Melaka, Melaka, Malaysia

Ghazali Omar Faculty of Mechanical Engineering, Universiti Teknikal Malaysia Melaka, Melaka, Malaysia; Aerospace Malaysia Innovation Centre (AMIC), Kompleks Teknologi 3, German Malaysian Institute, Jalan Ilmiah, Taman Universiti, Selangor, Malaysia

Azlin Fazlina Osman Center of Excellence Geopolymer and Green Technology (CEGeoGTech), Faculty of Chemical Engineering Technology, Universiti Malaysia Perlis (UniMAP), Perlis, Malaysia

Azma Putra Faculty of Mechanical Engineering, Universiti Teknikal Malaysia Melaka, Melaka, Malaysia

Nadia Razali Centre for Advanced Research on Energy (CARe), Universiti Teknikal Malaysia Melaka, Melaka, Malaysia

Nadlene Razali Faculty of Mechanical Engineering, Universiti Teknikal Malaysia Melaka, Melaka, Malaysia; Aerospace Malaysia Innovation Centre (AMIC), Kompleks Teknologi 3, German Malaysian Institute, Jalan Ilmiah, Taman Universiti, Selangor, Malaysia

Z.A.S. Saidi Fakulti Teknologi Kejuruteraan Mekanikal dan Pembuatan, Universiti Teknikal Malaysia Melaka, Hang Tuah Jaya, Durian Tunggal, Melaka, Malaysia

Mohd Azli Salim Advanced Manufacturing Centre, Universiti Teknikal Malaysia Melaka, Melaka, Malaysia; Faculty of Mechanical Engineering, Universiti Teknikal Malaysia Melaka, Melaka, Malaysia

Z. Samad School of Mechanical Engineering, Universiti Sains Malaysia (Engineering Campus), Penang, Malaysia

S.M. Sapuan Advanced Engineering Materials and Composites Research Centre (AEMC), Department of Mechanical and Manufacturing Engineering, Faculty of Engineering, Universiti Putra Malaysia, Selangor, Malaysia; Laboratory of Biocomposite Technology, Institute of Tropical Forestry and Forest Products (INTROP), Universiti Putra Malaysia, Selangor, Malaysia

M.A. Shaharuzaman Faculty of Mechanical Engineering, Universiti Teknikal Malaysia Melaka, Melaka, Malaysia

A.K. Syahibudil Ikhwan Faculty of Mechanical and Manufacturing Engineering Technology, Universiti Teknikal Malaysia Melaka, Melaka, Malaysia; Centre for Advanced Research on Energy, Universiti Teknikal Malaysia Melaka, Melaka, Malaysia

M.J. Taufiq Faculty of Mechanical Engineering, Universiti Teknikal Malaysia Melaka, Melaka, Malaysia

N.I.A. Wahab Department of Electrical and Electronic Engineering, Universiti Putra Malaysia, Selangor, Malaysia

Norhidayah Md Yunus Faculty of Built Environment and Surveying, Universiti Teknologi Malaysia, Johor Bahru, Malaysia

F. Yusof Centre of Advanced Manufacturing and Material Processing (AMMP Centre), University of Malaya, Kuala Lumpur, Malaysia; Department of Mechanical Engineering, Faculty of Engineering, University of Malaya, Kuala Lumpur, Malaysia

Norbazlan Mohd Yusof Centre of Excellence, Projek Lebuhraya Usahasama Berhad, Menara Korporat, Selangor, Malaysia

Ainur Zaireen Zainudin Faculty of Built Environment and Surveying, Universiti Teknologi Malaysia, Johor Bahru, Malaysia

E.S. Zainudin Advanced Engineering Materials and Composites Research Centre (AEMC), Department of Mechanical and Manufacturing Engineering, Faculty of Engineering, Universiti Putra Malaysia, Selangor, Malaysia

Siti Radiaton Adawiyah Zakaria Faculty of Built Environment and Surveying, Universiti Teknologi Malaysia, Johor Bahru, Malaysia

Mohd Hanafee Zin Section of Environmental Engineering Technology, Universiti Kuala Lumpur—Malaysian Institute of Chemical and Bioengineering Technology, Melaka, Malaysia

M.Y.M. Zuhri Advanced Engineering Materials and Composites Research Centre (AEMC), Department of Mechanical and Manufacturing Engineering, Faculty of Engineering, Universiti Putra Malaysia, Selangor, Malaysia; Laboratory of Biocomposite Technology, Institute of Tropical Forestry and Forest Products (INTROP), Universiti Putra Malaysia, Selangor, Malaysia

About the editors

Professor **S.M. Sapuan** is a professor of composite materials at Universiti Putra Malaysia (UPM). He earned his Bachelor of Engineering degree in Mechanical Engineering from University of Newcastle, Australia, in 1990, MSc from Loughborough University, United Kingdom, in 1994 and PhD from De Montfort University, United Kingdom, in 1998. His research interests include natural fiber composites, materials selection, and concurrent engineering. To date, he has authored or coauthored more than 1521 publications [730 papers published/accepted in national and international journals, 16 authored books, 25 edited books, 153 chapters in books, and 597 conference proceedings/seminar papers/presentation (26 of which are plenary and keynote lectures and 66 of which are invited lectures)]. He was the recipient of the Rotary Research Gold Medal Award 2012, The Alumni Medal for Professional Excellence Finalist, 2012 Alumni Awards, University of Newcastle, NSW, Australia, Khwarizmi International Award (KIA). In 2013 he was awarded the 5 Star Role Model Supervisor award by UPM. He has been awarded "Outstanding Reviewer" by Elsevier for his contribution in reviewing journal papers. He received Best Technical Paper Award in UMIMAS STEM International Engineering Conference in Kuching, Sarawak, Malaysia. He was recognized as the first Malaysian to be conferred Fellowship by the US-based Society of Automotive Engineers International in 2015. He was the 2015/2016 recipient of the Southeast Asian Regional Center for Graduate Study and Research in Agriculture (SEARCA) Regional Professorial Chair. In the 2016 ranking of UPM researchers based on the number of citations and h-index by SCOPUS, he was ranked sixth of 100 researchers. In 2017 he was awarded the IOP Outstanding Reviewer Award by the Institute of Physics, United Kingdom, National Book Award, The Best Journal Paper Award, UPM, Outstanding Technical Paper Award, Society of Automotive Engineers (SAE) International, Malaysia, and Outstanding Researcher Award, UPM. He also received in 2017, the Citation of Excellence Award from Emerald, United Kingdom, SAE Malaysia the Best Journal Paper Award, IEEE/Teertenkar Mahivel University (TMU) Endeavour Research Promotion Award, Best Paper Award by Chinese Defence Ordnance, and Malaysia's Research Star Award, from Elsevier. In addition, in 2019, he was awarded the Top Research Scientist Malaysia in 2019 and Professor of Eminence Award from AMU, India.

M.R. Mansor (also known as **Muhd Ridzuan Mansor**) holds a PhD in Mechanical Engineering with specialization in concurrent design of biocomposites product. He is currently a senior lecturer at the Faculty of Mechanical Engineering, Universiti Teknikal Malaysia Melaka, Melaka, Malaysia. To date, he has published more than 10 book chapters, 50 journals, and 50 technical proceedings as the main and coauthor, mostly focusing on the areas of product design, materials selection, life cycle assessment, biocomposite materials, and materials characterization. He has also published one book project on the concurrent design of natural fiber composite products with Springer.

Preface

In product development, traditionally designers consider all downstream activities during the outset of the design process. These activities include maintainability, manufacturability, assemblability, safety, cost, recycling, and other issues. These are done with improved product quality, time to market, and cost in mind. This concurrent engineering approach of product development has been successfully implemented in different industries worldwide since the late 1980s. A few years later, suppliers were also involved, and they were also involved in the concurrent engineering team. However, one important aspect, which is often neglected in product development is design for environment and sustainability. Nowadays, the aspects of the environment and sustainability have been included as important agendas in the product development process. With this idea in mind, this book is presented to five details on different activities in design for the environment and sustainability. It focuses on design for sustainability (DfS) by taking materials and processes as central topics. The chapters in this book are broadly divided into three different categories, that is, concurrent "green" design, "green" materials, and "green" processing. Chapter 1, 'Green' Conceptual Design Toward Design for the Environmental Sustainability, by S.M. Sapuan, introduces design for environmental sustainability and taking the conceptual design of biocomposite products as case studies. Mansor et al. (Chapter 2: Conceptual Design Development and Selection of Green Product) and Shaharuzaman et al. (Chapter 3: Sustainable Materials Selection: Principles and Applications) explore conceptual design development and materials selection of "green" composites as important activities in DfS. In Chapter 4, Implementation of Design for Sustainability in Developing Trophy Plaque Using Green Kenaf Polymer Composites, a case study of DfS in the development of a biocomposite product (i.e., a trophy plaque) is presented. It explores different concept generation and evaluation techniques such as brainstorming and Pugh evaluation techniques in the conceptual design stage of the sustainable product development process. The development of a sustainable laminated rubber–metal spring vibration isolator is presented by Salim et al. in Chapter 5, Dynamic Analysis on Laminated Rubber-Metal Spring Vibration Isolator for Sustainable Design. In an attempt to educate future generations on the importance of DfS, a chapter by Zaireen et al. is devoted to DfS in education. Seven chapters focus on review and characterization of different types of "green" biocomposites utilizing different natural fibers and biopolymers. Different "green" materials' performance determinations are highlighted, all with the aim of producing sustainable products. The major advantages of these materials include being environmentally friendly, low cost, comparable specific strength and stiffness performance to synthetic materials, renewability, abundantly available, low-energy consumption, and not posing an health hazard. In the third section of the book, four chapters are devoted to sustainable manufacturing processes in the context of DfS. Life cycle assessment analysis as well as the utilization of sustainable and low-energy processing such as fused filament fabrication, out-of-autoclave, and tailor welded blanks welding methods are the topics covered. It is hoped that the publication of this book will contribute to the new knowledge in DfS by focusing on sustainable design, materials, and processing.

Muhd Ridzuan Mansor

Faculty of Mechanical Engineering, Universiti Teknikal Malaysia Melaka, Melaka, Malaysia

S.M. Sapuan

Advanced Engineering Materials and Composites Research Centre (AEMC), Department of Mechanical and Manufacturing Engineering, Faculty of Engineering, Universiti Putra Malaysia, Selangor, Malaysia; Laboratory of Biocomposite Technology, Institute of Tropical Forestry and Forest Products (INTROP), Universiti Putra Malaysia, Selangor, Malaysia

June 2020

Acknowledgments

Alhamdulillah—all praises to almighty Allah who has made it possible for the Editors to complete this book.

The editors are highly honored to dedicate this book to Research Collaboration between Universiti Putra Malaysia [Laboratory of Biocomposite Technology (INTROP) and Advanced Engineering Materials and Composites Research Centre (AEMC)] and Universiti Teknikal Malaysia Melaka, Melaka, Malaysia (Faculty of Mechanical Engineering).

They are also very thankful to all authors for their highly valuable and insightful contributions in this book toward advancement in the field of Design for Sustainability.

Last but not least, the editors also wish to express their highest gratitude to the Publishers and their family members for the overwhelming support throughout the completion of this book.

PART I

Design methods

CHAPTER

1

"Green" conceptual design toward design for environmental sustainability

S.M. Sapuan[1,2]

[1]Laboratory of Biocomposite Technology, Institute of Tropical Forestry and Forest Products (INTROP), Universiti Putra Malaysia, Selangor, Malaysia, [2]Advanced Engineering Materials and Composites Research Centre (AEMC), Department of Mechanical and Manufacturing Engineering, Faculty of Engineering, Universiti Putra Malaysia, Selangor, Malaysia

1.1 Introduction

Design for sustainability is a design approach that aims to reduce the negative impact of design on the environment. According to Ashby (2005), design for the environment is the activity normally carried out by design engineers that considers environmental degradation of the products in their current design practice for the next 10 years of the product (the average expected life duration of the product). Design for sustainability goes beyond the average expected life of the product and any designs currently developed should consider the well-being of future generations, so that their needs are not sacrificed. In discussing the term design for sustainability, it is generally related to sustainability and sustainable development. There are three pillars of sustainability proposed by lawmakers—social, environment, and economy—also known as by the 3Ps, people, planet, and profits, respectively. This chapter focuses mainly on the environmental aspect of sustainability or environmental sustainability. Environmental sustainability requires the implementation of new and improved use of materials and energy, in

Design for Sustainability
DOI: https://doi.org/10.1016/B978-0-12-819482-9.00008-3

order to carry out effective sustainable design. Sustainability considers peoples' current needs without compromising future generations' ability to fulfill their needs (Vallero & Brasier, 2008).

In the past, design engineers performed the design of a product in isolation; without concern for colleagues who would fabricate the product at the manufacturing site. However, this issue was addressed sufficiently with the introduction of concurrent engineering in the late 1980s. With concurrent engineering (Sapuan, 2017), not only manufacturing issues, but also other related functions within the organization such as maintenance, planning, sales, marketing, and finance, as well as top-level management were in communication with design engineers. The problems associated with quality, time to market, and cost were greatly resolved (Sapuan & Mansor, 2014). As a step beyond that, external functions, particularly suppliers and vendors, were involved at the beginning of the product development process and the improvements were further enhanced.

After some time, it was realized that, in the existing product development practice, the issue of environmental sustainability was not considered to be an important agenda compared with cost, time, and product quality (Gutowski et al., 2005). The principles and methods related to the environment were not fully utilized in the product development process. Therefore, there was a need to cater for this in a more formal way; and the terms design for the environment and environmental sustainability became major talking points in more recently.

In this connection, Pahl, Beitz, Feldhusen, and Grote (2007) stated that in product design, environmental safety along with other safety measures in design such as operator safety and operational safety should be taken into consideration. Environmental safety is limitation of damage to the environment in which a product/system is used. Industries are gearing toward that goal; for instance, algae are considered for aircraft fuels, the need to use "green" materials in automobiles, laws passed prohibiting the use of nonbiodegradable plastics, and so on. This chapter focuses on the conceptual design stage of the product development process and the emphasis is on consideration of the environmental and sustainability issues during the outset of the product development stage.

Conceptual design is defined by Pahl et al. (2007) as:

"the part of the design process – where by identifying the essential problems through abstraction, establishing function structures, searching for appropriate working principles and combining these into a working structure – the basic solution path is laid down through the elaboration of a solution principle."

Pugh (1991) emphasized the importance of carrying out (1) concept generation and (2) concept evaluation during the conceptual design stage of product development. In this chapter, these principles are followed. Concept generation is aided by various idea generation techniques and concept evaluation is performed using different evaluation techniques. Ashby (2005) stated that during the conceptual design stage, designers should be in the position to define the specifications, determine function structure, seek working principles, and evaluate and select the concepts.

1.2 Conceptual design and design for environmental sustainability

Many conceptual design works in the past did not consider environmental strategies as major important considerations (Mansor, Sapuan, Zainudin, Nuraini, & Hambali, 2014; Noor Azammi, Sapuan, Ishak, & Sultan, 2018) and the focus was mainly on cost, quality, and other technical requirements. Concerns over environmental damage triggered the need to develop environmentally friendly products, which led to the term design for environmental sustainability or green design. Design for environmental sustainability considers the product life cycle (Fig. 1.1) during the conceptual design stage of product development processes to ensure the later stage of product development, which ensures the manufacturing process will be more environmentally friendly.

Some examples of the implementation of design for environmental sustainability are presented here. Ulrich and Eppinger (2004) reported that the Lakes concept team from Xerox had strived toward the concept of design for environmental sustainability. The major aim was that none of their product should be disposed to landfill but instead go for

FIGURE 1.1 Product life cycle consideration in design for environmental sustainability.

remanufacturing and recycling. This was not only their mission statement, but had also been implemented during the conceptual design stage and eventually during the production process.

Design for environment sustainability, which emphasizes the achievement of environmental sustainability, is a design concept that encourages environmental considerations such as safety and health, life-cycle environmental analysis, and pollution prevention in the product development process (Ertas & Jones, 1996). Fairs (2009) provided an example for environmental consideration in conceptual design stage, that is, a design concept that catered for the environment, which was translated into the production of effective products, water containers (jugs and vases). He stated that the design of these water containers was creative, sustainable, and made from natural materials, recycled and unwanted materials, which were less wasteful and low cost. These water containers were made from waste/shredded newspaper pulp mixed with wallpaper glue and ink and lined with polyurethane/epoxy to make them waterproof. There are many more examples of such practices, which can be found in Fair (2009), Guidice, La Rosa, and Risitano (2006), Vallero and Brasier (2008), Vezzoli and Manzini (2008), and other publications.

1.2.1 Proposed guidelines and strategies toward design for environmental sustainability to be implemented in conceptual design

Guidice et al. (2006) highlighted that during the conceptual design stage, it is important to consider environmental aspects, and designer engineers should define and introduce environmental strategies during this stage of the design process. The strategies can broadly be divided into two design levels—system and component levels. Some of these strategies are illustrated in Fig. 1.2, where it is shown that one of these strategies is the use of biodegradable and low-impact materials.

Dieter and Schmidt (2004) stated that in the engineering design practice, in order to protect the environment, several measures should be adopted. In line with the strategies provided by Guidice et al. (2006), the use of biodegradable and recycled materials is also proposed (Fig. 1.3).

Vallero and Brasier (2008), in discussing sustainable designs, have listed guidelines or principles of the green engineering program. These principles appear to be more relevant in the field of chemistry and chemical engineering design but one important point relevant to this chapter is on the use of renewable materials. These principles are shown in Fig. 1.4.

In another related publication, Vezzoli and Manzini (2008) considered the selection of renewable and biocompatible materials as important criteria in design for sustainability, under the topic of selection of

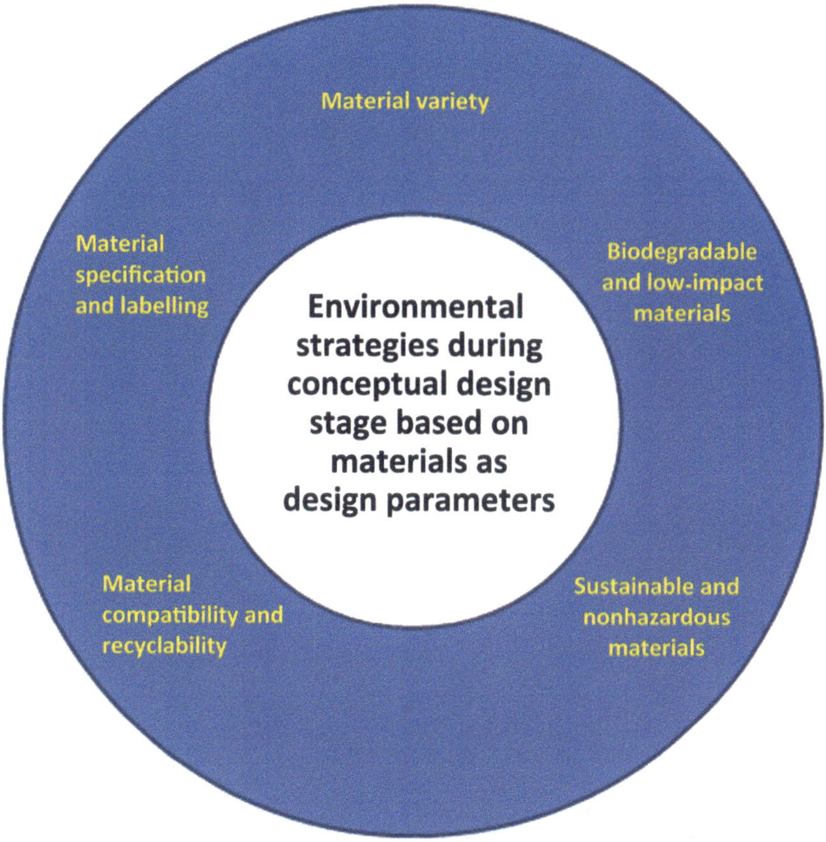

FIGURE 1.2 Selected environmental strategies during the conceptual design stage.

low-impact resources and processes. The guidelines for the selection of renewable and biocompatible materials are as shown in Fig. 1.5.

Based on the strategies and guidelines outline in Figs. 1.2–1.5, which emphasized the importance of the use of biocompatible, renewable, recyclable, and biodegradable materials, in the context of design for the environmental sustainability, it is decided to focus the remainder of this chapter on the use of one class of such materials, that is, biocomposite materials.

1.3 Renewable materials: Biocomposites

This section presents a discussion on one of the renewable materials, that is, natural fiber composites or biocomposites, in the context of design for environmental sustainability. In the discussion on natural fiber composites,

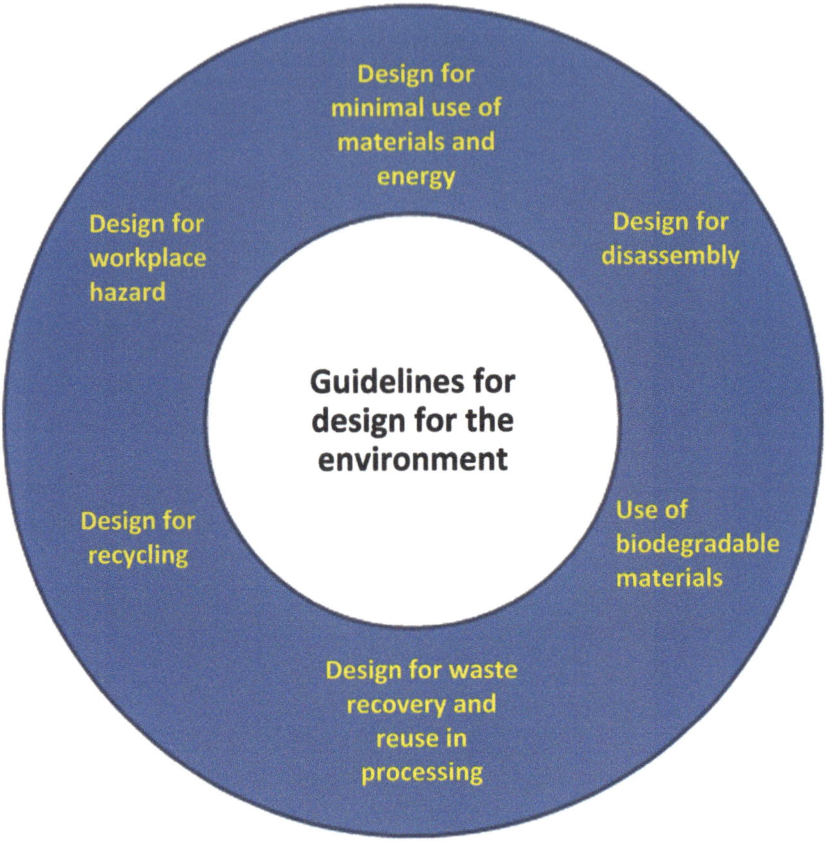

FIGURE 1.3 Guidelines for design for the environment.

two different types of materials are often referred to: (1) natural fiber reinforced synthetic polymer composites and (2) natural fiber reinforced biopolymer composites. The latter is discussed in the next subsection.

Purists in green design often argued that biocomposites are not fully degradable due to the presence of synthetic polymers in the composite system. However, natural fibers from plant or animal sources are normally used as fiber reinforcements. An indication of how "green" a material is is determined by the carbon footprint, and the carbon footprints of these materials are better than many synthetic composites and metals. The conceptual design of biocomposites is an area of research that is not widely carried out by the researchers because the biocomposite technology itself is not an established technology. There are many companies that have invested and manufactured products from biocomposites but their major focus is more on the final products irrespective of how the conceptual design is performed.

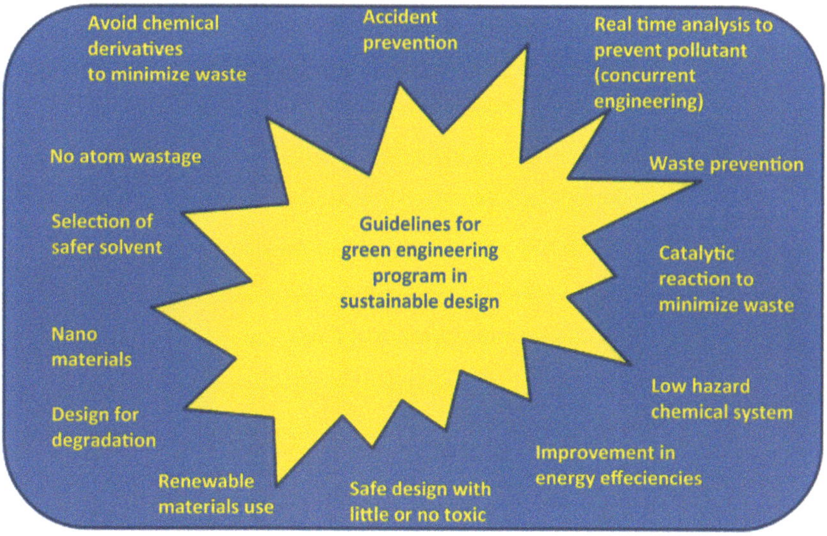

FIGURE 1.4 Guidelines for the green engineering program.

Generally, product design is considered to be "materials independent" (Pugh, 1991). No matter what product is developed, the design concepts remain the same, particularly for homogeneous materials like metals, plastics, and ceramics. However, for composites, design consideration is completely different (Mayer, 1993); but design routes of conventional (synthetic fiber) composites and natural fiber composites do much differ very much. However, research on conceptual design for biocomposites will become important when the biocomposite technology itself becomes established in the near future.

Natural fibers that were reported as being used as reinforcements in synthetic or biopolymer composites are numerous, depending on the functional requirements of the products. Examples include oil palm, pineapple leaf, sugar palm (Fig. 1.6), kenaf, abaca, roselle, hemp, jute, sisal, henequen, sugarcane bagasse, coconut coir and shell, betel nut, arrowroot, cassava bagasse, cornstalk, date palm, rice husk, durian skin, banana pseudo-stem, mengkuang, silk, wool, luffa, rattan, bamboo, and rambutan skin. Major advantages of natural fibers over synthetic fibers include lighter weight, renewable, recyclable, abundance, no health hazard, and lower cost. Natural fibers have successfully been used as reinforcements and fillers in polymer composites and the composites are used in different industries, ranging from the mundane to the most sophisticated applications. In fact, among the first manmade composites were natural fiber composites, that is, the use of wheat straws to reinforce clay to make bricks for building materials during the era of ancient Egypt. Books related to natural fiber

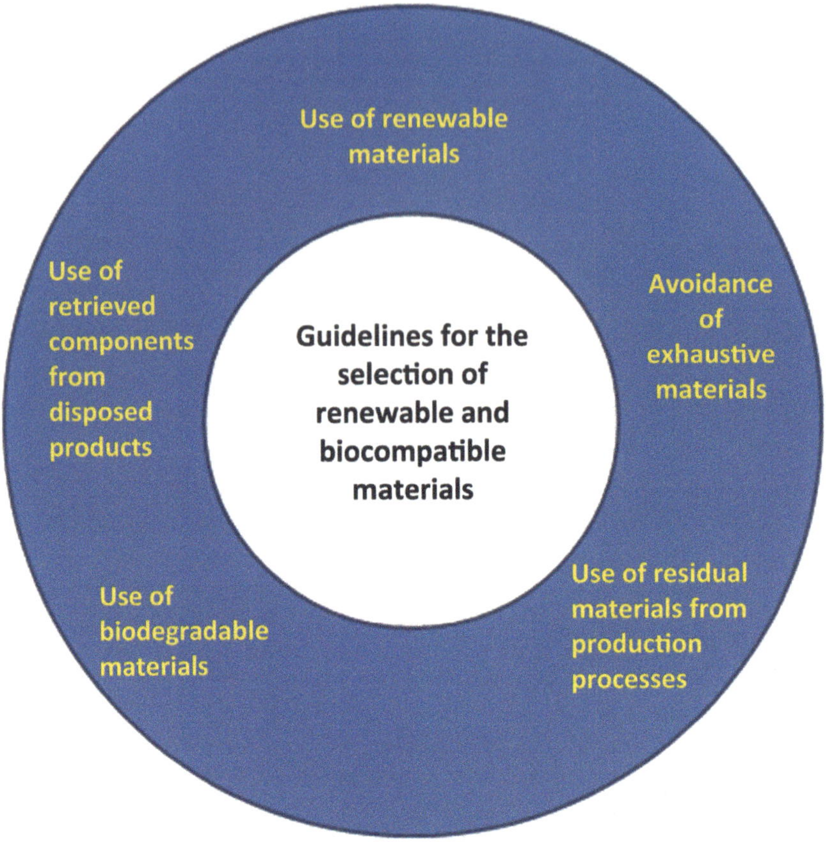

FIGURE 1.5 Guidelines for the selection of renewable and biocompatible materials.

FIGURE 1.6 Roof of the palace of Istana Pagar Ruyung, West Sumatera, Indonesia, that is made of sugar palm fibers.

composites are available in large collections and interested readers can read these books to gain a better understanding on the topics (Al-Oqla & Sapuan, 2017; Fan & Fu, 2017; Jawaid, Sapuan, & Alothman, 2017a, 2017b; Kheng, Aswathi, De Silva, & Thomas, 2020; Mastura & Sapuan, 2020; Sapuan, 2009, 2014a; Sapuan, Ismail, & Zainudin, 2018; Sapuan, Sahari, Ishak, & Sanyang, 2019; Visakh & Yu, 2016).

Malaysia is blessed with an abundance of natural resources, and natural fibers can be obtained from these resources, such as oil palm, sugar palm, pineapple leaf, coconut coir and shell, roselle, and sugarcane bagasse, and these materials are generally regarded as agricultural residues or even wastes. All these fibers are suitable candidates for reinforcements in polymer composites.

The Laboratory of Biocomposite Technology, INTROP, Universiti Putra Malaysia (UPM) is one of the institutions that anchors research in tropical wood and fibers in Malaysia and has been recognized since 2016 as a Higher Institution Centre of Excellence (HICoE) in those fields. Some of the products from natural fiber reinforced synthetic polymer composites are shown in Fig. 1.7. One of the products, called *PutraFrame*, was successfully commercialized recently (bottom left). It is made of kenaf fiber reinforced high-density polypropylene composites. The conceptual design was carried out by a group of undergraduate students in product

FIGURE 1.7 Biocomposite products developed at INTROP, UPM.

design at UPM and a small design competition was organized. The products were fabricated using injection molding and the fabrication was performed by a group of UPM students under the program of *Asnaf*.

1.3.1 Renewable and biodegradable materials: Fully degradable biopolymer composites

As many researchers have emphasized the use of renewable and biodegradable materials, this section is particularly devoted to this topic. Henry Ford developed cars and was a pioneer of the automotive industry. He was considered a concurrent engineer (Donovan, 1989), because during his time, the factory was small with few staff, and he was involved in the design as well as the manufacturing of cars. It is believed that he considered all manufacturing-related activities at the very early stage of the product development process, which is the essence of concurrent engineering (Sapuan, 2017). On top of that, it is also believed that Henry Ford performed design for sustainability in an informal way, in the development (including perhaps conceptual design) of one of the models he manufactured. In the 1930s, he and his team developed a car body panel from natural fiber reinforced soya bean biopolymer composites (Hogg, 2004). However, Hogg (2004) also states that the primary reason for choosing natural fiber reinforced biopolymer composites by Henry Ford was for economic and not environmental reasons. The project was not taken up as the price of synthetic polymers was then reduced.

Wool and Sun (2005) reported on the development of a fully biodegradable biopolymer composite chair made from soya biopolymer and natural fiber composites. Marques, Esteves, Viana, Loureiro, and Arteiro (2012) reported the use of similar materials for city bus bodies, where the materials used were cork fiber reinforced poly(lactic acid) (PLA)/ polyhydroxyalkanoate (PHA) biopolymer composites.

Fuad-Luke (2009) introduced an ecologically sound design in his book and provided an excellent guide to renewable, recyclable, and energy-efficient products. One particular product class of interest is biocomposites. These materials have no impact on the environment, and are easily recycled. An example is EnvironTM, a biocomposite made from recycled paper and soya flour-filled plant-based resin biopolymer composites. The material is renewable and recyclable and has been mass produced by Environ Biocomposites, United States. The material is harder than oak wood and can be used as an interior decoration and furniture. However, biopolymer composites require little processing and are low-embodied energy materials (low energy required to produce), which have a shorter product life compared to high-embodied energy materials such as metals, due to durability and extended product life.

Design for environmental sustainability has been addressed by Sapuan and Yusoff (2015), regarding its relevance to natural fiber composite products. Fifteen questions related to natural fiber reinforced biopolymer composites in meeting the criteria of design for environmental sustainability have been answered, which confirmed that natural fiber reinforced biopolymer composite is one of the solutions to design for environmental sustainability issues. Selected questions and answers are highlighted in Fig. 1.8.

The major difference between natural fiber reinforced synthetic polymer composites and fully biodegradable biopolymer composites is in the use of matrices, as was highlighted earlier. The latter used biobased polymers. The use of commercial biopolymers such as PLA and PHA has already been established. Attempts have been made at UPM to utilize different types of plants as the sources of starches to be used as raw materials to make biopolymers. In place of synthetic polymers, these biopolymers were used as matrices together with natural fibers to form fully biodegradable biopolymer composites, as shown in Fig. 1.9.

In recent years, at UPM, starches were extracted from different types of plants such as *Dioscorea hispida* (locally known as *ubi gadong*), (Zatil, Sapuan, & Jumaidin, 2020) arrowroot (locally known *as ubi bemban*) (Tarique & Sapuan, 2020), sago (Halimatul, Sapuan, Jawaid, Ishak, & Ilyas, 2019), cassava (Edhirej, Sapuan, Jawaid, & Zahari, 2017), corn (Ibrahim, Sapuan, Zainudin, & Zuhri, 2019), and sugar palm (Ilyas, Sapuan, Ishak, & Zainudin, 2018).

In the work of Zatil et al. (2020) (Fig. 1.10), *Dioscorea hispida* starch was extracted from *Dioscorea hispida* tuber. The tuber was sliced into small pieces and soaked in sodium chloride/distilled water to remove unwanted

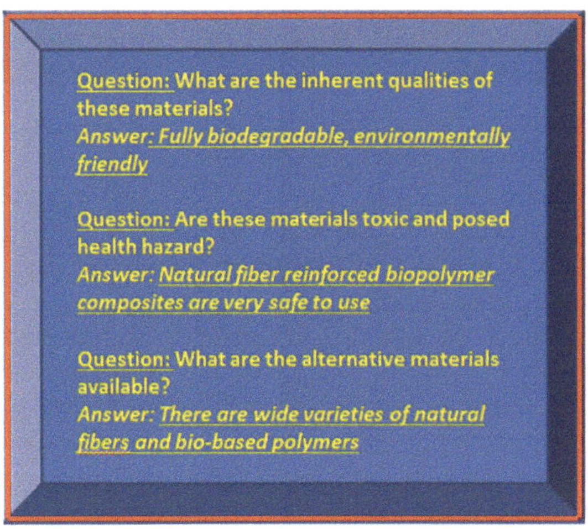

FIGURE 1.8 Selected elements of design for environmental sustainability fulfilled by fully biodegradable biopolymer composites.

FIGURE 1.9 Constituents of fully biodegradable polymer composites.

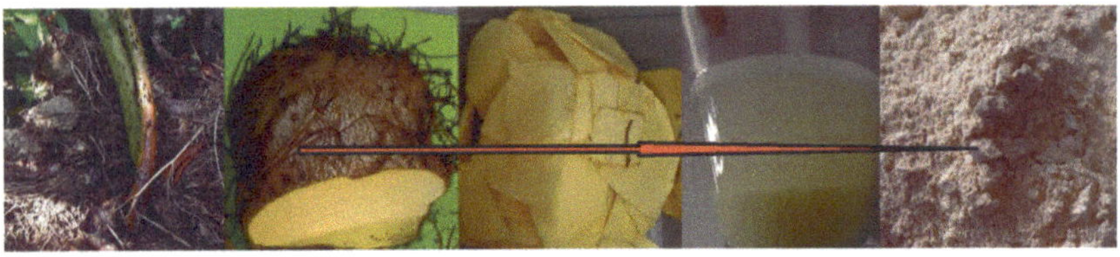

FIGURE 1.10 Process of making *Dioscorea daemona* starch biopolymers; from tubers, cutting into slices, and undergoing sedimentation to form starch.

alkaloid discorine. Thereafter, the tuber slices were crushed into powder form using a food processor and soaked in fresh water. This was followed by squeezing in order to dissolve the starch into the water. Water containing starch was filtered to separate the starch from unwanted items. The dewatering process then took place for starch sedimentation followed by drying up the sedimentation to produce *Dioscorea hispida* starch. In order to produce biopolymers, plasticizers (sorbitol and glycerol) were added to provide plasticity or flexibility and low brittleness. The starch biopolymers from *Dioscorea hispida* were mixed with *Dioscorea hispida* bagasse fibers to produce fully biodegradable biopolymer composites targeted mainly for food packaging applications.

1.4 Biocomposites: "Green" concept generation and evaluation

Concept generation is a design activity to generate design concepts. For the design of products from biocomposites, in the past, concept generation was performed using any available idea generation techniques, which were used by experts in different fields. So far, there is no single method that can be claimed to specifically belong to concept generation, neither in tradition composites nor in biocomposites. So far, a very limited number of researchers have been focusing their research on concept generation in traditional composites as well as in biocomposites.

Mayer (1993) published a book in which much valuable information on design guides during conceptual design for synthetic-based composites was discussed. The design guides are generally very useful for those who are involved in conceptual design of biocomposites as well. Mayer (1993) proposed a conceptual design framework based on some particular materials (in this case composites). In the conceptual design of components or products, design concepts must be ascertained with any essential requirements, codes, or standards for the products. Fig. 1.11 shows the design requirements for a component from composites proposed by Mayer (1993).

Design engineers, in carrying out design concept development, have to consider these principal design considerations. In Fig. 1.11, although the considerations are meant for the requirements of conventional composites, it is noted that the environmental impact and operating environment are included, and they can be taken as strategies for design for environmental sustainability. In fact, in the context of design for environmental sustainability, the issues of environment are considered paramount and they focus on life cycle analysis. Mayer (1993) also explained the conceptual design measures for the product under discussion, such as the possibility of reuse, refurbish, and recycle of the component, corrosion resistance, matching manufacturing cost with production volume, and consideration for simple changes in styling, texture, or coloring at low cost. Although Mayer (1993) was trying to describe conceptual design of the composite component, none of the concept generation technique was introduced. It can be assumed that the published literature related to conceptual design for composites is really very limited and this has opened up the huge scope of research in this field.

Sapuan (2017) focused on concurrent engineering (design for manufacture) for composites. He covered new topics on conceptual design of composites, materials selection for composites, and design for environmental sustainability for traditional composites, with special attention being paid to biocomposite product development as well. His co-workers were also involved in similar research in tackling different problems in the area of biocomposites (e.g., Mahmood, 2019; Mansor et al., 2014; Mastura, Sapuan, Mansor, & Nuraini, 2017; Noor Azammi et al.,

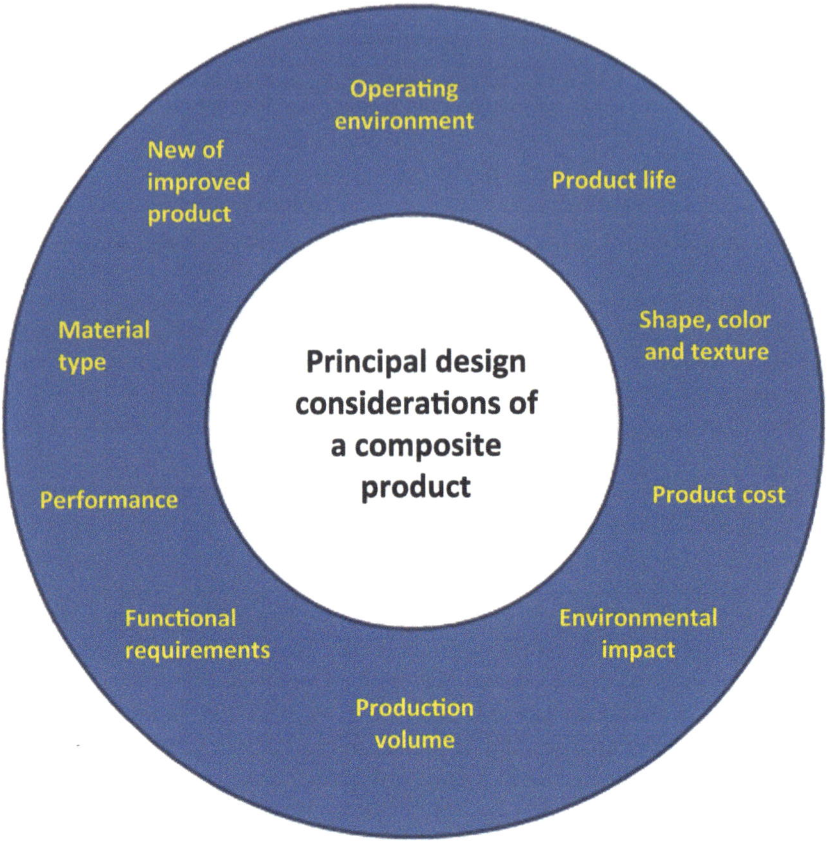

FIGURE 1.11 Principal design considerations for a composite product.

2018; Salwa, Sapuan, Mastura, & Zuhri, 2020; Shaharuzaman, Sapuan, Mansor, & Zuhri, 2020; Yusof, Sapuan, Sultan, & Jawaid, 2020).

These authors have invented new or adopted different design concept generation techniques, such as Theory of Inventive Problem Solving (TRIZ), mind mapping, VIseKriterijumska Optimizacija I Kompromisno Resenje (VIKOR), function-oriented search (FOS), function analysis diagram (FAD), six sigma, blue ocean strategy (BOS), quality function deployment for environment (QFDE), fuzzy analysis hierarchy process with extent analysis, conceptual design with input from anthropometric data (AD), biomimicry or biomimetic, cause-and-effect chain analysis (CECA), and morphology chart in the development of biocomposite products. All these activities were performed to explore new and adopted methods within the scope of the conceptual design of biocomposite products, and to contribute new knowledge in this "green" area of research. These authors believed that these developments had shaped a new dimension in biocomposite research, which was otherwise mainly dominated by the domain of characterization and determination of different properties of biocomposite materials.

In addition, in line with the move to use renewable and biodegradable materials as one of the strategies in the design for sustainability or environmental sustainability proposed by earlier experts, the current research endeavors are looking at the issues from a completely different prospective, that is, providing solutions to the current environmental problems with fully biodegradable biopolymer composites in the conceptual design stage and the earlier work was initiated by this author (Sapuan, 2014b). The most recent work in this area by his research group is on the design for environmental sustainability of fully biodegradable biopolymer composites for takeaway food packaging (Salwa et al., 2020).

The basic design concept evaluation for biocomposite products is a straightforward process. There is no specific method prescribed for such a task. The two main concept evaluation methods commonly used and reported in the literature are the weighted objective method and the Pugh evaluation method (Sapuan & Mansor, 2014). Sapuan (2017) also reported the use of a less conventional technique, that is, a direct comparison method for a simple problem, and Mansor et al. (2014) used an analytic hierarchy process for concept evaluation in a more advanced design problem in biocomposites. These are exactly the same methods used in concept evaluation unconventional (e.g., glass and carbon fibers) composites. They are also the same methods used in almost all engineering product designs. The methods did not even consider materials as major determining factors in the design of any products.

However, it must be remembered that the basic principles of design with composites are unique, unlike homogeneous materials, because, in composites, the products and materials are completely different due to the fact that composites are generally regarded as nonhomogeneous and anisotropic materials. Design with composites requires the consideration of many different factors, such as fiber orientation, arrangement, types and length, fabrication techniques, variation in matrices used, and other factors.

Mahmood (2019) carried out the conceptual design of a natural fiber composite chair and table for students at Malaysian polytechnics. The inputs in the product design specification and design concept generation were based on anthropometric data, in addition to the requirements for biocomposite material design. The descriptive statistics of anthropometric data (dimensions) for Malaysian males and females, which include the mean, standard deviation, standard error of mean, coefficient of variation, minimum, maximum, 1st, 50th, 95th, and 99th percentiles were determined and used as inputs in the conceptual design stage of natural fiber composite products.

Kim and Mauborgne (2005) pioneered important scholarly and ground-breaking research, development, and implementation of a successful business strategy called the blue ocean strategy (BOS), and this approach to business practice is widely accepted worldwide with those who have implemented it reaping tremendous benefits. BOS seeks to achieve the creation of uncontested market space and make the competition irrelevant. Among the major tools proposed include the strategy canvas and the four actions framework in the quest for value creation. BOS concept and activities have been effectively adapted in the conceptual design of a biocomposite automotive antiroll bar by Mastura et al. (2017). These tools were used to compare the performance of a biocomposite automotive antiroll bar against its metal counterparts. In addition, Mastura et al. (2017) used FAD, TRIZ-BOS, CECA, and FBD in the concept generation stage. In order to cater for design for environmental sustainability, the QFDE method was also incorporated. Finally, the concept evaluation process was accomplished using the AHP method.

Sapuan and Mansor (2016) presented a case study of conceptual design of a natural fiber reinforced composite laptop case using the TRIZ method. Design concepts were generated for a new laptop case with an embedded external cooling fan. Design concept generation was performed by clarification of design brief, specific problem definition using engineering contradiction with TRIZ, conversion of the specific problem into TRIZ general problem, identification of general solutions using TRIZ contradiction matrix, and finally generating specific solutions for the new laptop case design. Four TRIZ general inventive principles were used: local quality, composite materials, segmentation, and dynamization. The TRIZ method was also used in concept generation of other biocomposite products such as a shelf bracket (Mansor & Sapuan, 2018a), automotive engine rubber mounting (Noor Azammi et al., 2018), and ironing board leg (Mansor & Sapuan, 2018b).

Conceptual design works for biocomposite products were also carried out for different products such as a portable laptop table (Sapuan, Purushothman, Sanyang, & Mansor, 2018), shoe shelf (Mazani, Sapuan, Sanyang, Atiqah, & Ilyas, 2019), automotive crash box (Yusof et al., 2020), and automotive side-door impact beam (Shaharuzaman et al., 2020). Mazani et al. (2019) used TRIZ, morphology chart, and mind mapping in their concept generation and weighted objectives method in the concept evaluation of a biocomposite shoe shelf, while Sapuan et al. (2018) used the Pugh evaluation method in the design of a biocomposite portable laptop table.

In the recent work of Shaharuzaman et al. (2020), the conceptual design of an automotive side-door impact beam from natural fiber composites was carried out. Concept generation techniques used were TRIZ, FOS, biomimicry, or biomimetic (Table 1.1) and finite element analysis was used for design analysis. Biomimetic was used to generate design concepts

TABLE 1.1 Idea generation using the theory of inventive problem solving, function-oriented search-biomimetics method (Shaharuzaman et al., 2020).

Biomimetic strategies	Design	Name
A 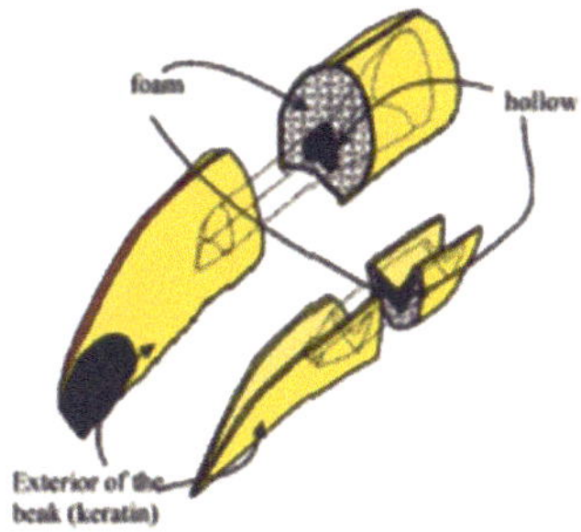 Toucan beak		A-01
	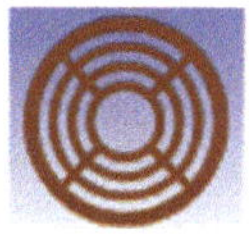	A-02
		A-03
B 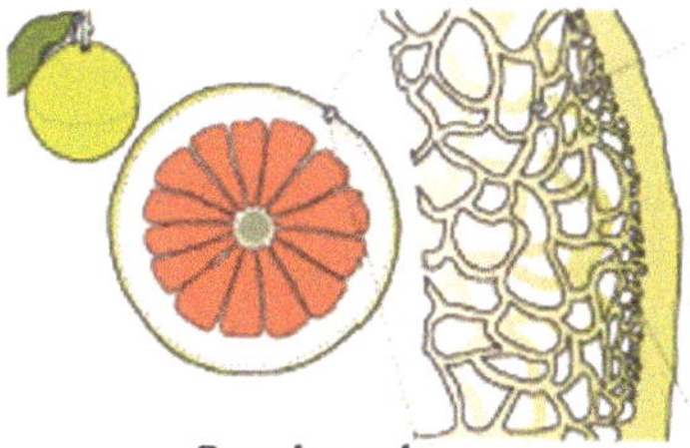 Pomelo peel	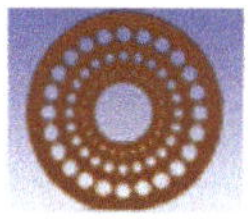	B-01
		B-02
	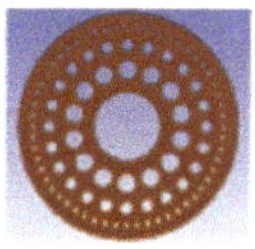	B-03
C 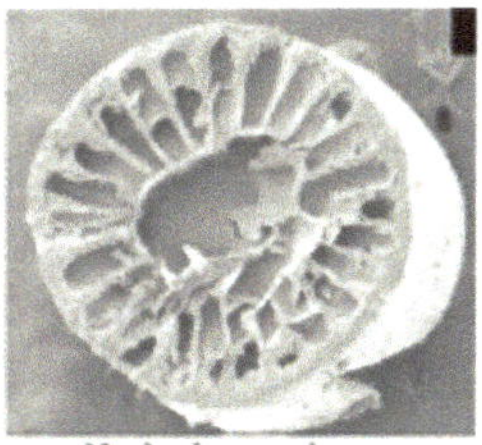 Hedgehog spine	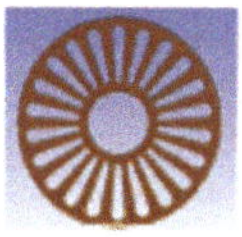	C-01
		C-02

inspired by nature. Concept evaluation to select the best design concept for the composite side-door impact beam was performed using the VIKOR method. Yusof et al. (2020) performed the conceptual design of a biocomposite automotive crash box (ACB). TRIZ and biomimetic methods were also used to generate the concepts against material characteristics, function specifications, force identification, root cause analysis, and geometry profile. The concept evaluation was performed using the AHP method.

The work of Salwa et al. (2020) is related to design concept generation and evaluation of fully biodegradable biopolymer composites for takeaway food containers. It is the first attempt by the research group at UPM to carry out the conceptual design of fully biodegradable biopolymer composite products, in an effort to create fully environmentally sustainable products.

1.5 Conclusions

The major lesson learnt from this review is that renewable and biodegradable materials play important roles in research into designs for environmental sustainability. Biodegradable materials were mainly represented by biopolymers and natural fiber-based biopolymer composites in this chapter, with researchers at UPM are actively involved in research in this field. In the conceptual design stage of product development, activities such as design concept generation and concept evaluation are two major elements that were performed in biocomposite product development process while keeping design for environmental sustainability in mind. This concurrent engineering approach has enabled the product development team to consider not only technical issues such as performance, cost, and reliability, but also environmental sustainability, that is, catering to the needs and well-being of future generations.

Acknowledgments

The author wishes to thank UPM for providing facilities to carry out this research work. The patience, perseverance, and support from Ms Nadiah Zainal Abidin and Ms Qurratu Aini Mohd Sapuan, during the course of preparation of this chapter as well as editing of the entire book, during the Movement Control Order (MCO) in Malaysia due to the COVID-19 pandemic in March–May 2020, are greatly appreciated. It was a once in a life time experience and will always be remembered.

References

Al-Oqla, F., & Sapuan, S. M. (2017). *Materials selection for natural fiber composites*. Duxford: Woodhead Publishing.

Ashby, M. F. (2005). *Materials selection in mechanical design* (3rd ed.). Amsterdam: Butterworth Heinemann.

Dieter, G. E., & Schmidth, L. C. (2004). *Engineering design*. Boston: McGraw Hill.

Donovan, J. R. (1989). Training for design for manufacture. *Proceedings in the Institution of Mechanical Engineers, Part D, 203*(1), 47–53.

Edhirej, A., Sapuan, S. M., Jawaid, M., & Zahari, N. I. (2017). Preparation and characterization of cassava bagasse reinforced thermoplastic cassava starch. *Fibers and Polymers, 18*(1), 162–171.

Ertas, A., & Jones, J. C. (1996). *The engineering design process* (2nd ed.). New York: John Wiley & Sons, Inc.

Fair, M. (2009). *Green design: creative sustainable designs for the twenty-first century*. London: Carlton Books Limited.

Fan, M., & Fu, F. (2017). *Advanced high strength natural fibre composites in construction*. Duxford: Woodhead Publishing.

Fuad-Luke, A. (2009). *EcoDesign, the source book* (3rd ed.). San Francisco, CA: Chronicle Books LLC.

Guidice, F., La Rosa, G., & Risitano, A. (2006). *Product design for the environment: a life cycle approach*. Boca Raton, FL: CRC Press.

Gutowski, T., Murphy, C., Allen, D., Bauer, D., Bert Bras, B., Piwonka, T., et al. (2005). Environmentally benign manufacturing: Observations from Japan, Europe and the United States. *Journal of Cleaner Production, 13*, 17.

Halimatul, M. J., Sapuan, S. M., Jawaid, M., Ishak, M. R., & Ilyas, R. A. (2019). Effect of sago starch and plasticizer contents on thermoplastic sago films: Mechanical and repeated soaking and drying test. *Polimery, 64*(6), 32–41.

Hogg, P.J. (2004). The role of materials in creating a sustainable economy, In *The 4th International Materials Technology Conference & Exhibition*, Kuala Lumpur, Malaysia, 23–25 March, p. 19.

Ibrahim, M. I. J., Sapuan, S. M., Zainudin, E. S., & Zuhri, M. Y. M. (2019). Physical, thermal, morphological, and tensile properties of corn starch-based films as affected by different plasticizers. *International Journal of Food Properties, 22*(1), 925–941.

Ilyas, R. A., Sapuan, S. M., Ishak, M. R., & Zainudin, E. S. (2018). Development and characterization of sugar palm nanocrystalline cellulose reinforced sugar palm starch bionanocomposites. *Carbohydrate Polymers, 202*, 186–202.

Jawaid, M., Sapuan, S. M., & Alothman, O. Y. (2017a). *Green biocomposites: design and applications*. Cham: Springer International Publishing AG.

Jawaid, M., Sapuan, S. M., & Alothman, O. Y. (2017b). *Green biocomposites: manufacturing and properties*. Cham: Springer International Publishing AG.

Kheng, L. G., Aswathi, M. K., De Silva, R. T., & Thomas, S. (2020). *Interfaces in particle reinforced composites: current perspective from polymer, ceramic, metal and extracellular matrices*. Duxford: Woodhead Publishing.

Kim, W. C., & Mauborgne, R. (2005). *Blue ocean strategy: how to create uncontested market space and make the competition irrelevant*. Boston, MA: Harvard Business School Press.

Mahmood, A. (2019). Conceptual design of bio-composite chair and desk based on anthropometric data, PhD Thesis, Universiti Putra Malaysia.

Mansor, M. R., & Sapuan, S. M. (2018a). Concurrent design of kenaf composite products. In S. M. Sapuan, J. Sahari, M. R. Ishak, & M. L. Sanyang (Eds.), *Kenaf fibres and composites*. Boca Raton, FL: CRC Press, pp. 205–226.

Mansor, M. R., & Sapuan, S. M. (2018b). *Concurrent conceptual design and materials selection of natural fiber composite products*. Singapore: Springer Nature Singapore Pte Ltd.

Mansor, M. R., Sapuan, S. M., Zainudin, E. S., Nuraini, A. A., & Hambali, A. (2014). Conceptual design of kenaf fiber polymer composite automotive parking brake lever using integrated TRIZ-morphological chart-AHP Method. *Materials & Design, 54*, 473–482.

Marques, A.T., Esteves, J.L., Viana, J., Loureiro, H., & Arteiro, A. (2012). Design for sustainability with composite systems, In The 15th International Conference on Experimental Mechanics, Porto, 22–27 July, p. 2.

Mastura, M. T., & Sapuan, S. M. (2020). *Evaluating the implementation and performance of green materials in technology development, emerging research and opportunities*. Hershey, PA: IGI Global.

Mastura, M. T., Sapuan, S. M., Mansor, M. R., & Nuraini, A. A. (2017). Conceptual design of a natural fibre reinforced composite automotive anti-roll bar using a hybrid approach. *The International Journal of Advanced Manufacturing Technology, 91*(5-8), 2031–2048.

Mayer, R. M. (1993). *Design with reinforced plastics: a guide for engineers and designers*. London: The Design Council.

Mazani, N., Sapuan, S. M., Sanyang, M. L., Atiqah, A., & Ilyas, R. A. (2019). Design and fabrication of a shoe shelf from kenaf fiber reinforced unsaturated polyester composites. In H. Ariffin, S. M. Sapuan, & M. A. Hassan (Eds.), *Lignocellulose for future bioeconomy*. Amsterdam: Elsevier, pp. 315–332.

Noor Azammi, A. M., Sapuan, S. M., Ishak, M. R., & Sultan, M. T. H. (2018). Conceptual design of automobile engine rubber mounting composite using TRIZ-Morphological chart-analytic network process technique. *Defence Technology, 14*(4), 268–277.

Pahl, G., Beitz, W., Feldhusen, J., & Grote, K. H. (2007). *Engineering design: a systematic approach* (3rd ed). London: Springer-Verlag London Limited.

Pugh, S. (1991). *Total design: integrated methods for successful product engineering*. Wesley Publishing Company.

Salwa, N. H., Sapuan, S. M., Mastura, M. T., & Zuhri, M. M. Y. (2020). Post life cycle processing of reinforced thermoplastic polymer composites. In P. Kumar Bajpai, & I. Singh (Eds.), *Reinforced polymer composites: processing, characterization and post life cycle assessment*. Weinheim: Wiley- VCH Verlag GmbH & Co., pp. 225–244

Sapuan, S. M. (2009). *Research on natural fibre reinforced polymer composites*. Serdang, Selangor, Malaysia: UPM Press.

Sapuan, S. M. (2014a). *Tropical natural fibre composites: properties, manufacture and applications*. Singapore: Springer.

Sapuan, S. M. (2014b). Concurrent engineering for natural fibre composites, In The 3rd International Conference on Design and Concurrent Engineering (iDECON 2014), Melaka, Malaysia, 22–23 September, pp. 38–39.

Sapuan, S. M. (2017). *Composites materials: concurrent engineering approach*. Oxford: Butterworth Heinemann.

Sapuan, S. M., Ismail, H., & Zainudin, E. S. (2018). *Natural fiber reinforce vinyl ester and vinyl polymer composites: development, characterization and applications*. Duxford: Woodhead Publishing.

Sapuan, S. M., & Mansor, M. R. (2014). Concurrent engineering approach in the development of composite products: a review. *Materials & Design, 58*, 161–167.

Sapuan, S. M., & Mansor, M. R. (2016). Design of natural fiber-reinforced composite structures. In R. D. S. G. Campilho (Ed.), *Natural fiber composites*. Boca Raton, FL: CRC Press pp. 255–278.

Sapuan, S. M., Purushothman, K. R., Sanyang, M. L., & Mansor, M. R. (2018). Design and fabrication of kenaf fibre reinforced polymer composites for portable laptop table. In S. Kalia (Ed.), *Lignocellulosic composite materials*. Cham, Switzerland: Springer International Publishing AG, pp. 232–356.

Sapuan, S. M., Sahari, J., Ishak, M. R., & Sanyang, M. L. (2019). *Sugar palm, biofibers, biopolymers, and biocomposites*. Boca Raton, FL: CRC Press.

Sapuan, S. M., & Yusoff, N. B. (2015). The relationship between manufacturing and design for manufacturing in product development of natural fibre composites. In M. S. Sapuan, M. Jawaid, N. Bin Yusoff, & M. Enamul Hoque (Eds.), Manufacturing of natural fibre reinforced polymer composites. Cham, Switzerland: Springer International Publishing AG, pp. 1–15.

Shaharuzaman, M. A., Sapuan, S. M., Mansor, M. R., & Zuhri, M. Y. M. (2020). Conceptual design of natural fiber composites as a side-door impact beam using hybrid approach. *Journal of Renewable Materials, 8*(5), 549–563.

Tarique, J.& Sapuan, S.M. (2020). Review of arrowroot fibre reinforced arrowroot biopolymer composites, Presented at Materials Lecture Competition 2020 (MLC 2020), 11 March, Faculty of Engineering, UPM, Serdang, Selangor, Malaysia.

Ulrich, K. T., & Eppinger, S. D. (2004). *Product design and development* (3rd ed.). Boston: McGraw Hill Irwin.

Vallero, D., & Brasier, C. (2008). *Sustainable design: the science of sustainability and green engineering*. Hoboken, NJ: John Wiley & Sons, Inc.

Vezzoli, C., & Manzini, E. (2008). *Design for environmental sustainability*. London: Springer Verlag London Limited.

Visakh, P. M., & Yu, L. (2016). *Starch-based blends, composites and nanocomposites*. Cambridge.: Royal Society of Chemistry.

Wool, R. P., & Sun, X. S. (2005). *Bio-based polymers and composites*. Burlington: Elsevier.

Yusof, N. S. B., Sapuan, S. M., Sultan, M. T. H., & Jawaid, M. (2020). Conceptual design of oil palm fibre reinforced polymer hybrid composite automotive crash box using integrated approach. *The Journal of Central South University, 27*(1), 64–75.

Zatil, H., Sapuan, S. M., & Jumaidin, R. (2020). Ubi gadongas potential biocomposite material: A comprehensive review. In S. M. Sapuan, R. Jumaidin, & I. Hanafi (Eds.), *Biofiller-reinforced biodegradable polymer composites*. Boca Raton, FL: CRC Press.

CHAPTER

2

Conceptual design development and selection of green product

N.M. Ishak[1], *Muhd Ridzuan Mansor*[1] *and A.F. Ab Ghani*[2]

[1]Faculty of Mechanical Engineering, Universiti Teknikal Malaysia Melaka, Melaka, Malaysia, [2]Faculty of Mechanical and Manufacturing Engineering Technology, Universiti Teknikal Malaysia Melaka, Melaka, Malaysia

2.1 Introduction to conceptual design in design for sustainability

In 2015, the United Nations launched a holistic target toward future sustainability, called the Sustainable Development Goals (SDG), to tackle climate change and environmental protection. Within similar SDG, other areas related to sustainability were also included which encompass areas related to economic and social needs. The SDG is aimed at promoting prosperity while protecting the planet. The SDG were divided into 17 goals, including sustainable industry, innovation and infrastructure, affordable and clean energy, consumption and production, and climate change, as shown in Fig. 2.1 (United Nations, 2020). The launch of the SDG amplified the drive toward sustainability activities around the world. From a product and process development perspective, there are goals included in the SDG which have a direct impact such as responsible consumption and production, which relates to the efficient use of natural resources, reducing waste, and adopting sustainable practices.

Similarly, the concept of sustainability in design also seeks to reduce negative impacts on the environment while at the same time creating healthy and comfortable quality of life for all people. The United States General Service Administration (GSA) further elaborated that the basic

Design for Sustainability
DOI: https://doi.org/10.1016/B978-0-12-819482-9.00012-5

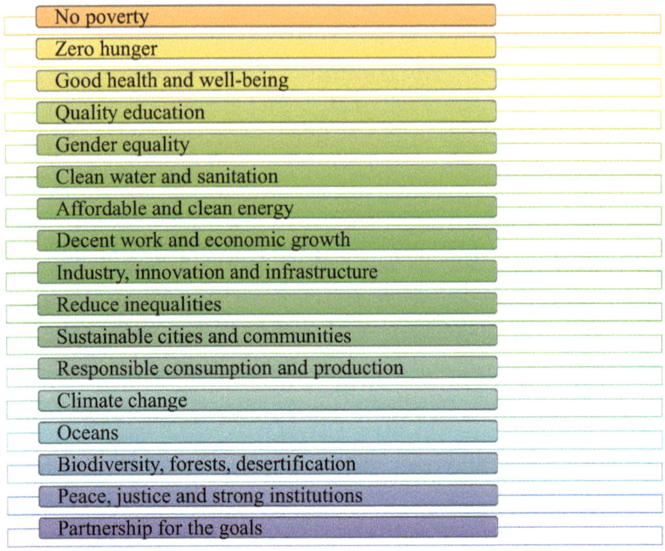

FIGURE 2.1 Summary of the 17 goals in the Sustainable Development Goals by the United Nations (United Nations, 2020).

objectives of sustainability are to reduce consumption of nonrenewable resources, minimize waste, and create healthy and productive environments (US GSA, 2020).

Based on the various descriptions of sustainability as aforementioned, it can be noted that a sustainable product or processes should basically incorporate optimum quantity of resources and use of environmentally friendly materials as the substances, without compromising the product functionality and safety requirements. Hence, the use of green materials in new product design and development projects are strongly aligned to achieve the sustainability goals. Moreover, clever planning of space and creating innovative design features also help green products to be developed that successfully meet customer requirements, while minimizing waste and product cost to perform a similar function. With this basic understanding, green products can be developed successfully to compete with conventional products, and appeal to the targeted end user.

Meanwhile, in some cased related to the use of green materials such as natural fiber composites for developing green products, designers also require an understanding of the material properties, manufacturing process, and design features concurrently. This further makes the product development process of using such green material more complex compared to using traditional engineering materials (Mansor et al., 2016). It has also been reported that up to 70 percent of the total product cost is decided during the earliest stage of the product development process, which involves activities related to product conceptual development and selection (Sapuan & Mansor, 2014). In general, the conceptual design is the part of the design process which involves the process of identifying the essential problems through abstraction, establishing function structures, searching for appropriate working principles, and combining these into a working structure to generate the basic solution path as laid down through the elaboration of a solution principle (Pahl, Beitz, Feldhusen, & Grote, 2007). Therefore acquiring proper product design knowledge and implementing it at the earliest stage in product development is essential and valuable to product designers in developing their green product successfully (such as quickest development time, lowest cost, and highest quality).

In this chapter, the conceptual design development and selection for producing green products are discussed to help product designers perform effective actions during the early stage of the product development process. Among the contents are an explanation of the overall framework for developing conceptual design and selection of green products, process in conceptual design development to generate creative and innovative designs solutions, and a conceptual design selection process which deals with the method to systematically perform a multicriteria decision-making process in selecting the best concept design among the generated solutions. Furthermore, a case study on a regular household item is also included to demonstrate how conceptual design development and selection for green products may be performed successfully to achieve the targeted design requirements. Finally, the overall conclusions to this chapter are described together with some remarks on future directions which may be taken related to the conceptual design process of green products.

2.2 Framework of conceptual design and selection process in green product development

The overall framework of the conceptual design and selection process proposed for green product development is illustrated in Fig. 2.2. Similar to conventional products, green products are also developed to satisfy various customer requirements such as functionality, cost, attractiveness, and user friendliness, as well as being environmentally friendly. For product manufacturers, green products also need to be produced in high quality, with a fast development time to market, and low cost. Hence, the framework for the conceptual design and selection process of green product development is recommended to start with identifying customer requirements to determine the design intent, followed by formalizing the design intent into a reference document called the product design specifications (PDS). Based on the PDS, product designers later work to produce initial ideas and solutions to meet the

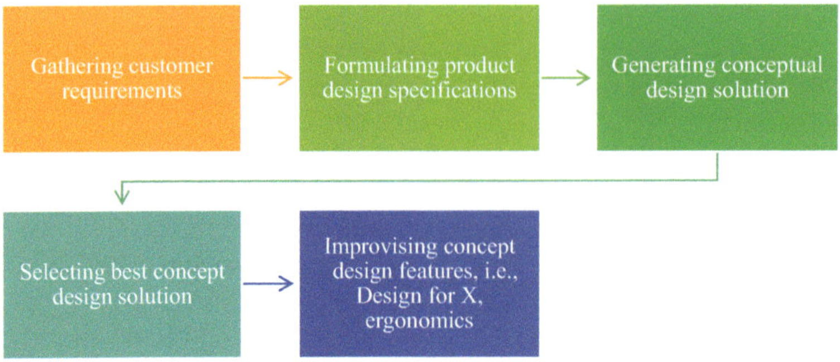

FIGURE 2.2 Overall framework in the conceptual design process for green products.

design intent, which is called the conceptual designs. At the end of this stage, usually several conceptual solutions which adhere to the product design requirements are generated.

Based on the various conceptual design solutions generated, the green product development process is later carried out by selecting the best conceptual design solution. In this stage, the multicriteria decision-making method (MCDM) is often applied for the decision-making process by all stakeholders involved in the product development. MDCM methods are applied due to their capability to perform systematic and justified screening and ranking analysis involving multiple solutions with varying attributes based on varying selection criteria, simultaneously. This capability really helps all decision makers involved to make the best judgment for the best solution. Furthermore, it is also recommended that green product designers improvise and further optimize the initial selected concept design feature. For example, using a Design for X (DfX) approach to improve the assembly and disassembly performance, manufacturability, cost, and recyclability performance (Maxwell & Van der Vorst, 2003). Improvements may also be enabled by enhancing the ergonomic features in the final selected conceptual design. This enable better integration between the user and the product in terms of comfort and safety, hence increasing user acceptance for the newly introduced green products in the market (Sun, Houssin, Renaud, & Gardoni, 2019). Once the development process enhanced version of the final conceptual design is completed, it is sent for the next detailed design stage in the product development process.

2.3 Conceptual design development process

The conceptual design stage is one of the crucial steps in any product development as well as the development of products that use green materials. This process is carried out after PDS and is developed based on the information obtained from the market investigation. At the concept design stage, functional models are developed and these models describe the inputs and outputs of the product (Sapuan, 2010). There are many methods involved in the conceptual design development process, such as mindmapping, morphological chart, problem decomposition, analysis of existing technical system, quality function deployment, failure mode and effect analysis, and many more. In general, designing green products at the conceptual design stage is similar to the development of other products. Many researchers have explored conceptual design development on products using green materials. Hambali, Sapuan, Ismail, and Nukman (2009) generated conceptual designs for an automotive composite bumper beam. This case study accommodates that the materials applied must be recyclable at the end of their useful life. Brainstorming was involved at the concept generation stage, where eight design concepts for automotive bumper beams were carried out, as depicted in Fig. 2.3. Brainstorming is regarded as the most accepted and widely used technique for idea generation. This technique allows ideas to be shared in a free-thinking environment and thus expands the solution space. In designing the automotive bumper beams, all the proposed design concepts are assumed to curve flat-faced since the shape

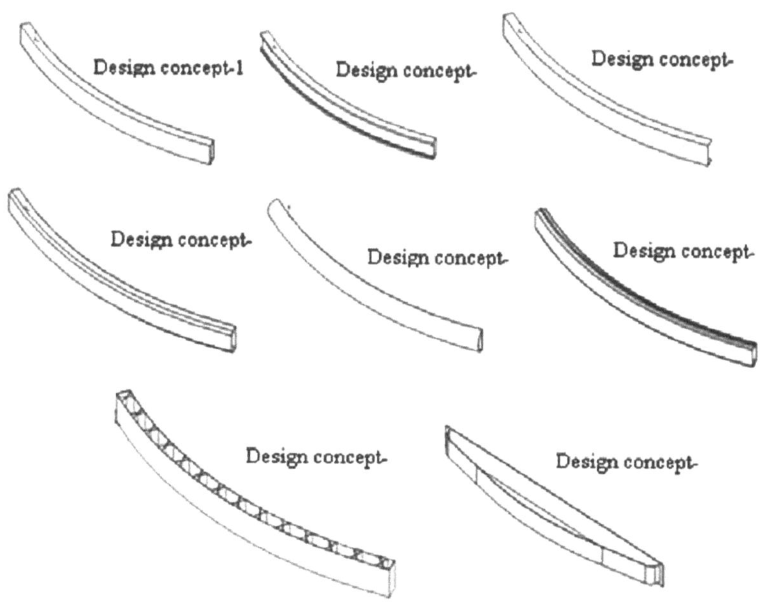

FIGURE 2.3 Conceptual designs developed for the automotive composite bumper beam (Hambali et al., 2009).

is complex. The degree of curvature, depth of draw, and standard dimensions were considered during the brainstorming process.

Mansor, Sapuan, Zainudin, Nuraini, and Hambali (2014) developed the conceptual design of kenaf fiber polymer composite for an automotive parking brake lever using integrated Theory of Inventive Problem Solving (TRIZ) and a morphological chart to explore the suitable solutions for the issue: lower mechanical strength during the conceptual design stage for the development of an automotive parking brake lever. The TRIZ method was applied as the solution tool for idea generation, while the morphological chart was applied as an idea refinement tool based on the TRIZ recommended solutions for concept design developments. Based on the TRIZ solution approach, the improving and worsening parameters were determined and matched with lists of 39 engineering parameters. The process is continued by selecting the appropriate TRIZ 40 inventive principles which matched the system parameters using the contradiction matrix. The appropriate TRIZ solution principles were then further refined into relevant alternative system elements using the morphological chart method. The chart provides the range of elements, components, or subsolutions that can be combined to form solutions. Fig. 2.4 shows the morphological chart matrix for the TRIZ solution principles and the specific functional design features relevant to the solution principles of the automotive parking brake lever, while Fig. 2.5 shows the conceptual designs of the automotive parking brake lever that have been developed based on the integration of two methods. Besides integration of TRIZ and the morphological chart, stress distribution also has been analyzed for

TRIZ solution principles and design strategy	Design features	Solutions				
		1	2	3	4	5
#1. Segmentation i) Product the component in different sections and the sections should be asymmetric as possible to simplify and ease the design and manufacturing process ii) Joining the sections using pin-and-boss method for easy assembly and disassembly process	**Cross-section profile**	Double-T	Square	U-type		
	No. of sections	2	3	4		
	Sectioning type	Symmetrical	Non-symmetrical			
	Part section assembly method	Screw-boss	Circular pin-boss	Square pin-boss	Adhesive	
#3. Local quality i) Vary the thickness of the component according to the stress concentration value. Thicker component at higher stress location points. ii) Brake lever body casing designed with ribs to reinforce and strengthened the structure as well as with pin-and-boss features to provide quick and easy assembly method. Joint different functioning feature together to the same component	**Body type**	Solid	Shell			
	Ribbing pattern	I	V	X	I+V	I+X

Note: *Example for Concept Design 1 = Square + 2-sections + symmetrical + circular pin-boss + shell + I-rib*

FIGURE 2.4 Morphological chart matrix for the TRIZ solution principles of the automotive parking brake lever (Mansor et al., 2014).

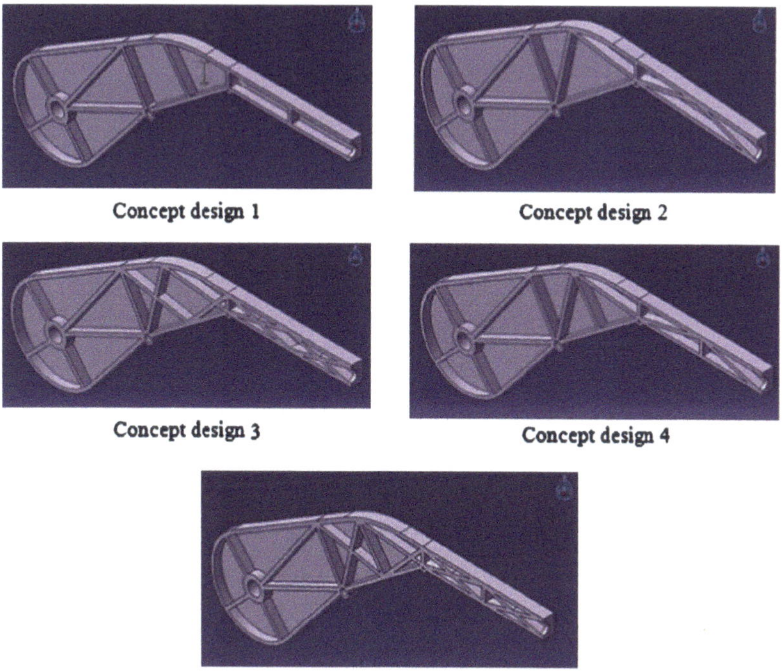

FIGURE 2.5 Conceptual designs of the automotive parking brake lever developed using integrated TRIZ and the morphological chart (Mansor et al., 2014).

each of the conceptual designs to predict their structural performance based on the generated stress and deformation values.

The integration of TRIZ and the morphological chart are also practiced in the development of the green product—kenaf polymer composites automotive spoiler (Mansor, Sapuan, Hambali, Zainudin, & Nuraini, 2015). The idea generation process started with the identification of the main PDS of the automotive spoiler component. Five main elements in the spoiler PDS, which are structural strength, lightweight, aerodynamic, recyclability, and low cost, were recognized based on market research. However, the morphological chart developed in this case study also involved the combination of design features such as core material and sandwich configuration, and nondesign features like fiber treatment, fiber impregnation, painting, core material, sandwich configuration, and resin with curing agent have been considered in this case study, which further extends the conventional design that drives the function of the morphological chart itself. The integration of TRIZ and the morphological chart is shown in Fig. 2.6 and the conceptual design is illustrated in Fig. 2.7.

Mastura, Sapuan, Mansor, and Nuraini (2017) utilized a hybrid approach of Quality Function Deployment for Environment (QFDE), TRIZ, Free-Body

TRIZ solution principles and specific strategy	Feature	Solution 1	Solution 2	Solution 3
#10. Preliminary action i) perform fiber treatment method such as alkali treatment using NaOH ii) perform fiber impregnation process iii) the finished composite spoiler is coated (painted)	Fiber treatment	Yes	No	
	Fiber impregnation	Yes	No	
	Painting	Yes	No	
#40. Composite materials i) sandwich construction for the spoiler	Core material	Wood	Foam	Honeycomb
	Sandwich configuration	single core layer	double core layer	
#28. Mechanics substitution i) using resin with self-curing agents such as MEKP	Resin with curing agent	Yes	No	

FIGURE 2.6 Integration of TRIZ and the morphological chart for kenaf polymer composites automotive spoiler (Mansor et al., 2015).

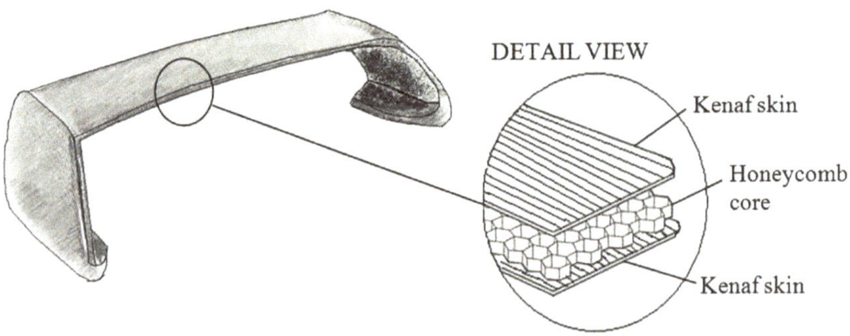

FIGURE 2.7 New conceptual design of kenaf polymer composites automotive spoiler (Mansor et al., 2014).

Diagram, Blue Ocean Strategy (BOS), and a morphological chart in the development of the conceptual design for a sugar palm reinforced composite automotive antiroll bar. Concept generation began with consideration of five mechanical design elements: materials, function, forces, failure mode, and geometry. Based on the Voice of Customers and Voice of Environment, PDS is developed and suitable natural fibers for the design of automotive anti roll bar (ARB) are selected using three-phase QFDE. The assembly of ARB is identified and the function of each assembled components is defined using a function analysis diagram from TRIZ and the main problems in the design of the antiroll bar are identified. The solutions to the problems are generated using engineering contradiction and 40 inventive principles from TRIZ using the related technical formulation. The suggested solution strategies are enhanced by the four-action framework from BOS.

BOS TRIZ	Eliminate	BOS TRIZ	Raise
#2 Taking out	Critical point that gained the highest stress concentration should be eliminated	#40 Composite materials	Fiber orientation could be varied for better mechanical properties of the composite materials
BOS TRIZ	**Reduce**	**BOS TRIZ**	**Create**
#2 Taking out	Cross section of the ARB could be a hollow shape to reduce the weight while maintaining the performance of the component	#3 Local quality	The bar could be designed with reinforced features to strengthen the structure

FIGURE 2.8 BOS and TRIZ four-action framework (Mastura et al., 2017).

Fig. 2.8 shows the BOS and TRIZ four-action framework. Then, the morphological chart is used to present all the ideas and all the potential solutions from the morphological chart are combined and presented in a list of conceptual designs. The combination of all the ideas from all the assembly parts of the component generated 42 potential conceptual designs for the automotive antiroll bar. Fig. 2.9 shows the morphological chart and Fig. 2.10 illustrates the conceptual designs of the antiroll bar.

Azammi, Sapuan, Ishak, and Sultan (2018) developed the conceptual design of an automobile engine rubber mounting composite using TRIZ and a morphological chart. The automobile engine rubber mounting contains kenaf fiber polymer composites. The conceptual design of an engine rubber mounting is developed to replace the existing rubber-based engine rubber mount in order to reduce the component's weight while maintaining the required structural strength for safety and functionality performances. In this case study, the TRIZ solution principle is used to suit the conceptual design of an engine rubber composite mount by integrating the problem using 39 engineering parameters and identify the appropriate TRIZ solution principle using the contradiction matrix while the morphological chart is used to blend the solution principle together to achieve the conceptual designs for the product. Fig. 2.11 shows the morphological chart of the TRIZ solution principle and their related design of engine composite rubber mounting design features and Fig. 2.12 illustrates the proposed conceptual design of the engine rubber composite mounting.

Sapuan, Purushothman, Sanyang, and Mansor (2018) designed a portable laptop table using kenaf fiber reinforced polymer composites as the material. Through PDS, the important features and requirements to design the portable laptop table are identified. Eight elements have been chosen to be implemented in the design of the composite laptop

	Actions	Solution			
		1	2	3	4
A	Eliminate- Critical point that gained the highest stress concentration should be eliminated.		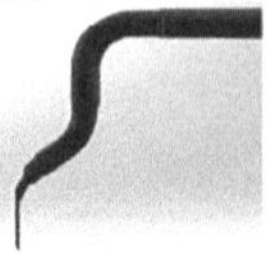		rib
B	Raise- Fiber orientation could be varied to improve the mechanical properties of the composite materials.	Woven	Uni-directional	Random	Multi directional
C	Reduce- Cross section of the ARB could be a hollow shape to reduce the weight while maintaining the performance of the component [20].	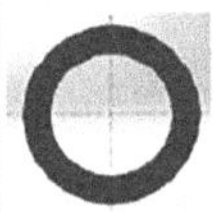	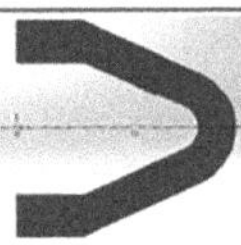	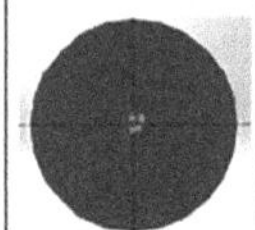	
D	Create- The bar could be designed with reinforced features to strengthen the structure [7].	I-rib	V-rib	X-rib	

FIGURE 2.9 Morphological chart for the antiroll bar (Mastura et al., 2017).

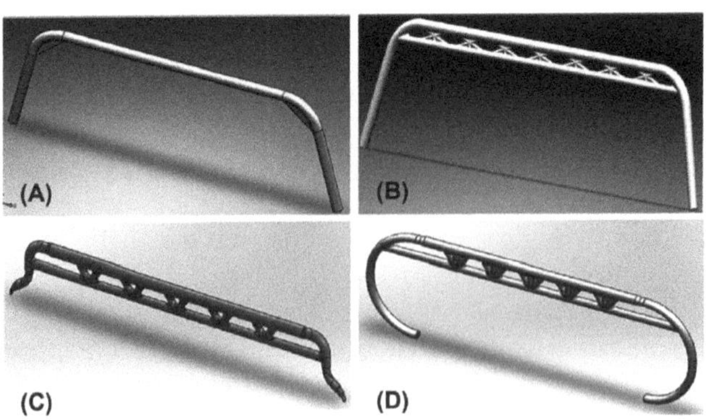

FIGURE 2.10 Conceptual designs for the antiroll bar (Mastura et al., 2017).

TRIZ solution principle and design strategy	Design features	Solution 1	2	3	4	5	6
#36 parameter changes	Type of engine mount absorbent	Elesthomers		Hydraulic		Spring	
-Degree flexiblity is determined by the composition of material used, type of absorbant and position of engine mount	Type of material mount	Natural rubber		Natural rubber+PU		Teflon	
	Support position	Engine		Engine+transmission		Transmission	
#6 universality	Axis support type	Vertical	Horizontal	Vertical+ horizontal	torsion	Vertical +torsion	Horizontal +torsion
-The capability of adapting various force direction ,the design of support type and the position of inner tube to cater the multidirection force	Design of rubber mount support	2 support vertical	2 support horizontal	3 support (Y)	4 support (X)	5 support	Full
	Inner tube location	Center			Offset		

FIGURE 2.11 Morphological chart of the TRIZ solution principle of the engine composite rubber mounting design features (Azammi et al., 2018).

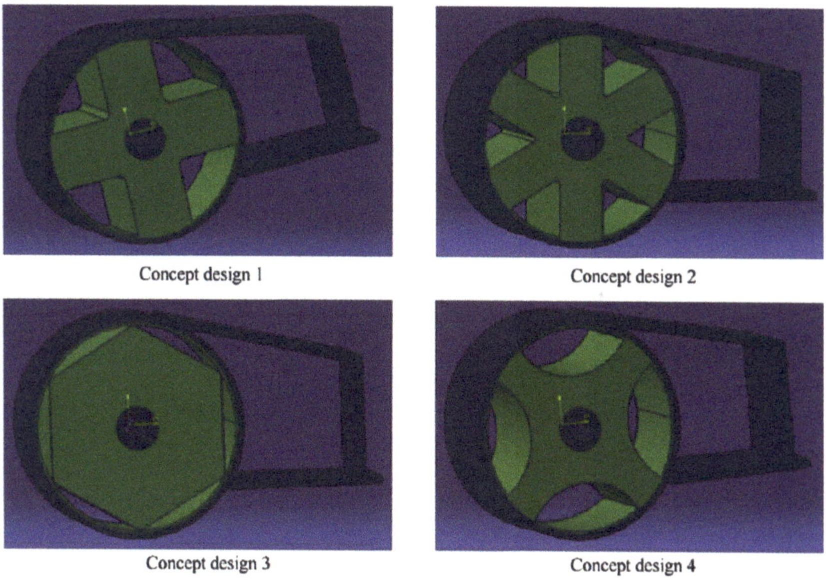

FIGURE 2.12 Conceptual design of the engine rubber composite mounting (Azammi et al., 2018).

table, which are performance, weight, product cost, materials, safety, aesthetics, size, and manufacturing process. Seven design concepts are sketched as alternative solutions that are expected to meet the PDS. Fig. 2.13 shows the conceptual designs of the portable laptop table.

Ishak, Sivakumar, and Mansor (2018) designed the car front hood using natural fiber metal laminate to reduce its weight. In this case study, the

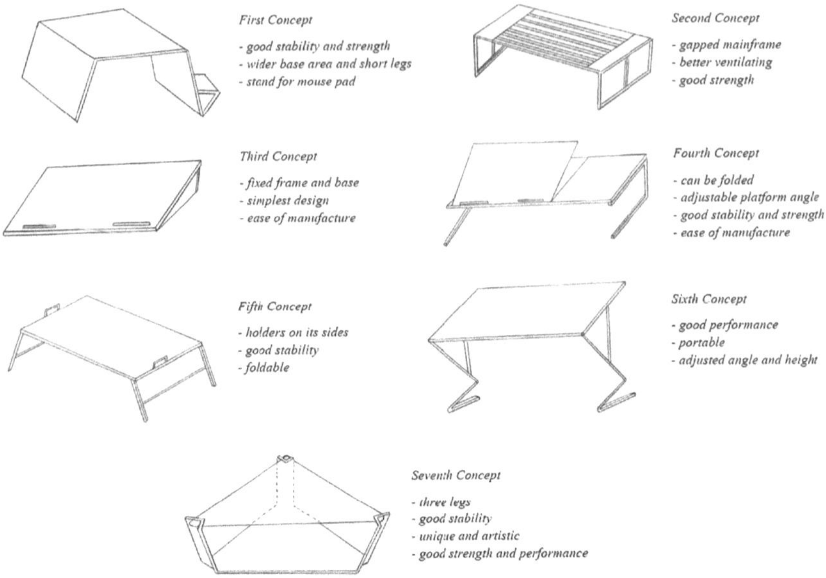

FIGURE 2.13 Conceptual designs of the portable laptop table (Sapuan et al., 2018).

TRIZ method has been applied to generate a suitable design for the car front hood. This method was applied as the solution tool for idea generation and the idea was not limited to design only but involved the nondesign features such as the stacking configuration of the fiber reinforced composite in the FML layering, fiber weight fraction, and fiber volume fraction of the natural fiber reinforced composite. The contradictions associated with using natural fiber metal laminate as the car front hood are identified using *"If," "Then,"* and *"But"* keywords to determine the parameters used for the selection of the 40 Inventive Principles via the TRIZ contradiction matrix. Function analysis of the car front hood was developed to visualize the information about the way the car front hood technical system functioned. The proposed design of the car front hood is shown in Fig. 2.14.

The aforementioned case study showed that the conceptual design development process of the green product is similar to the development of other products. However, in some cases, there are some features which are unique to green product conceptual design, where the manufacturing and nondesign features such as the green materials' composition, core material, configuration, fiber treatment, and fiber impregnation must be considered.

2.4 Conceptual design selection process

The conceptual design selection process is a systematic approach to evaluate a set of concepts to satisfy customer needs and engineering

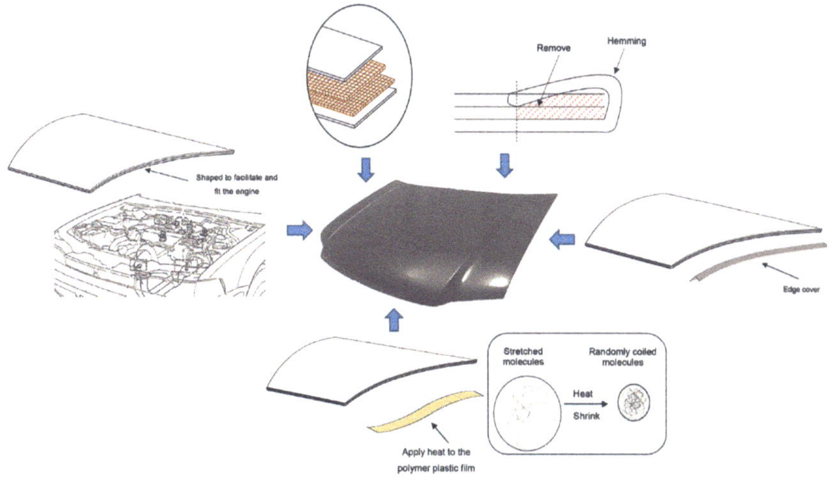

FIGURE 2.14 Proposed design of the car front hood (Ishak et al., 2018).

specifications (Davoodi et al., 2011). It is vital that the best initial concepts are selected, as they determine the direction of the design embodiment stage (King & Sivaloganathan, 1999). Concept screening and scoring are the methods that are most commonly used in the conceptual design selection process, as well as in the development of green products. Sapuan and Maleque (2005) developed conceptual designs for a household telephone stand using green material—banana woven fabric reinforcement epoxy composite. A concept scoring matrix, also known as an evaluation matrix, was implemented in the selection of the suitable conceptual design. Using this method, the designs were evaluated with weightage factors and evaluation of each concept on a scale of 0–5. The weight factor is based on how important the criteria are and the concept with the highest score is chosen as the final design. In this case study, concept 1 was chosen as the best design to be developed for the household telephone stand using a banana woven fabric reinforcement epoxy composite. Fig. 2.15 shows the concept scoring matrix for the conceptual design selection process of the household telephone stand.

The application of the concept scoring matrix was also explored in the conceptual design selection process for a multipurpose table (Sapuan, Harun, & Abbas, 2007). The same natural composite was utilized in this case study, however the green material was only focused on the banana trunk. Eight concepts were developed and all criteria for the designs were evaluated based on the stability of the product, cost of production, ease of manufacture, reliability of service, ergonomic aspects, low setup time, and ease of transport. Concept 8 had the highest score and was finalized as the

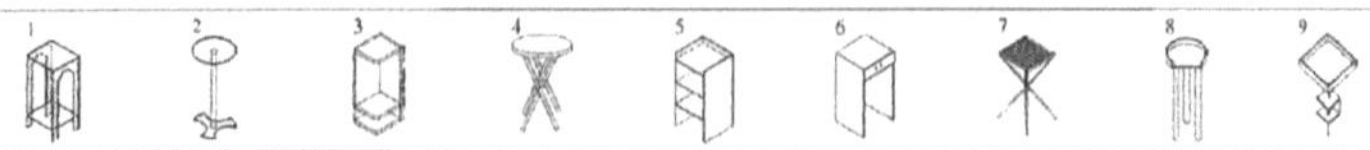

Concept		1		2		3		4		5		6		7		8		9	
Characteristic	C	W	S	W	S	W	S	W	S	W	S	W	S	W	S	W	S	W	S
Stability	5	5	25	4	20	4	20	5	25	3	15	5	25	5	25	4	20	3	15
Cost effective and cheap	5	4	20	4	20	4	20	5	25	5	25	3	15	5	25	3	15	4	20
Ergonomic design	5	5	25	5	25	3	15	5	25	4	20	5	25	3	15	4	20	3	15
Low setup time	4	4	16	4	16	3	12	4	16	4	16	3	12	4	16	3	12	5	20
Less space needed	4	4	16	4	16	5	20	4	16	5	20	4	16	3	12	3	12	4	16
Less complex	3	5	15	5	15	3	9	4	12	3	9	4	12	2	6	2	6	5	15
Portability	3	5	15	4	12	4	12	3	9	4	12	4	12	3	9	3	9	2	6
Total score		132		124		108		128		117		117		108		94		107	

Note. C means the weightage for the characteristic. W means the score of the concept. S is the product of weightage and score of the concept.

FIGURE 2.15 Concept scoring matrix for the household telephone stand (Sapuan & Maleque, 2005).

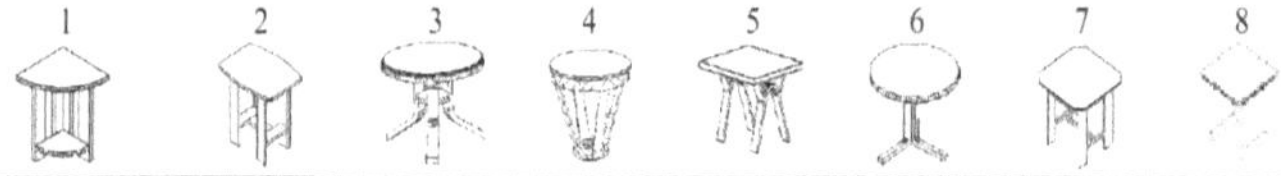

Concepts		1		2		3		4		5		6		7		8	
Characteristic	W	R	S	R	S	R	S	R	S	R	S	R	S	R	S	R	S
Stability of product	5	4	20	5	25	5	25	3	15	5	25	2	10	5	25	4	20
Cost of manufacture	5	3	15	2	10	2	10	2	10	2	10	3	15	2	10	5	25
Ease of manufacture	4	3	12	3	12	2	8	2	8	3	12	5	20	3	12	5	20
Reliability of service	5	4	20	4	20	4	20	4	20	4	20	3	15	4	20	3	15
Ergonomic to user	3	2	6	3	9	4	12	4	12	2	6	3	9	4	12	4	12
Low setup time	3	3	9	2	6	1	3	1	3	3	9	4	12	2	6	4	12
Ease to carry and light	3	3	9	3	9	3	9	4	12	3	9	4	12	2	6	4	12
Total score		91		91		87		80		91		93		91		116	

Note: W is the weight factor for a characteristic; the higher weight factor for a characteristic means the said characteristic is more important to the design. R is the rating for a characteristic. S is the product of W and R. The total score for a concept is obtained by the summation of all values of S.

FIGURE 2.16 Concept scoring matrix for the multipurpose table (Sapuan et al., 2007).

most suitable design for multipurpose table banana trunk fiber-woven fabric-reinforced composites. Fig. 2.16 shows the concept scoring matrix for the multipurpose table.

MCDM is a method that has been widely studied by researchers in the conceptual design selection process. Researchers conducting the conceptual design development process of green product mentioned in the previous section have furthered their studies on the conceptual design selection process to determine the most suitable design. The development of a bumper beam by Davoodi et al. (2011) applied the technique for order of preference by similarity to ideal solution (TOPSIS), which is an MCDM technique. This method selects the best geometrical bumper beam concept that meets the safety parameters of the defined PDS. Six criteria—deflection, strain energy, mass, cost, easy manufacturing, and the rib possibility—were evaluated for eight alternative concepts and weighted values were appointed for each criterion to form an evaluation matrix. The eight alternative concepts were named as reverse C profile, closed oblique profile, curved C profile,

No.	Concepts	Name	Material cost 0.15	Easy manufacturing 0.1	Product weight 0.2	Strain energy 0.3	Rib possibility 0.1	Minimum deflection 0.15
1		RCP	24.40	2	2.44	2482.82	2	16.92
2		COP	29.00	1	2.9	43419.9	1	29.86
3		CCP	18.60	4	1.86	38825.1	5	21.34
4		DHP	25.50	3	2.55	76106.5	5	18.34
5		DCC	29.40	2	2.94	63671.6	4	25.72
6	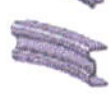	DCP	25.60	4	2.56	44910.3	5	21.15
7		SHP	21.90	3	2.19	47231.5	4	22.92
8		SCP	22.50	5	2.25	2137.62	5	16.73

FIGURE 2.17 Evaluation matrix for selecting the best profile concept for the bumper beam (Davoodi et al., 2011).

double hat profile (DHP), double C closed, double C profile, simple hat profile, and simple C profile. The evaluation matrix was then normalized, the positive ideal and negative ideal solution calculated, the separation measurements were determined, and the relative closeness to the ideal solution rank in the preference order in decreasing order to determine the most suitable design of the hybrid biocomposite bumper beam was calculated. Fig. 2.17 shows the evaluation matrix for selecting the best profile concept for the bumper beam. Through the TOPSIS process, concept design four, the DHP was selected as the most suitable design for the hybrid biocomposite bumper beam that fulfilled the PDS.

Conceptual design of the kenaf fiber polymer composite for automotive parking brake lever by Mansor et al. (2014) in which the selection process was carried out using the Analytic Hierarchy Process (AHP) which is also an MCDM technique. This method provides a systematic process through the present pair-wise comparison technique as well as a consistency index indicator during the judging and analysis process. In this study, four levels of AHP hierarchy were developed, as shown in Fig. 2.18. The goal of the study was defined in level 1 and the design selection main criteria—performance, weight, and cost—were outlined at level 2 followed by the level 3 subcriteria—stress, deformation, mass, volume raw material cost, and shape complexity—which were defined for the design selection. Design concepts for the polymer composites parking brake lever were listed in level 4. Based on the highest priority vector value, concept design 2 was selected as the final design concept which scored the highest priority value—20.8% compared to the other concept designs, as shown in Fig. 2.19.

Mastura et al. (2017) also implemented AHP for the sugar palm reinforced composite automotive antiroll bar conceptual design selection process. Selection of the final conceptual design is based on the PDS as per

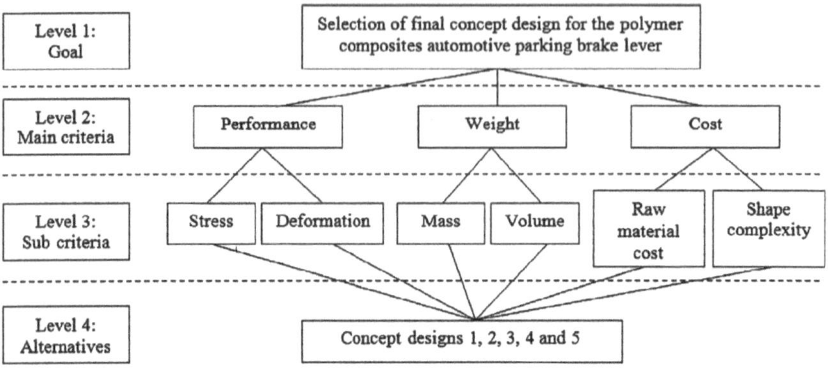

FIGURE 2.18 Analytic Hierarchy Process (AHP) hierarchy framework concept design selection for the kenaf fiber polymer composites parking brake lever (Mansor et al., 2014).

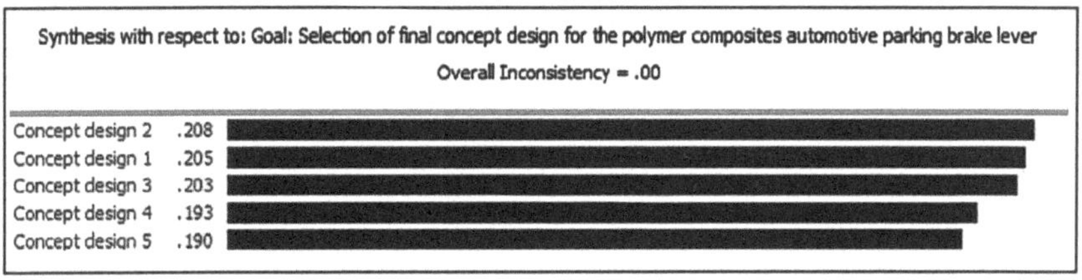

FIGURE 2.19 Overall AHP results of the concept design selection (Mansor et al., 2014).

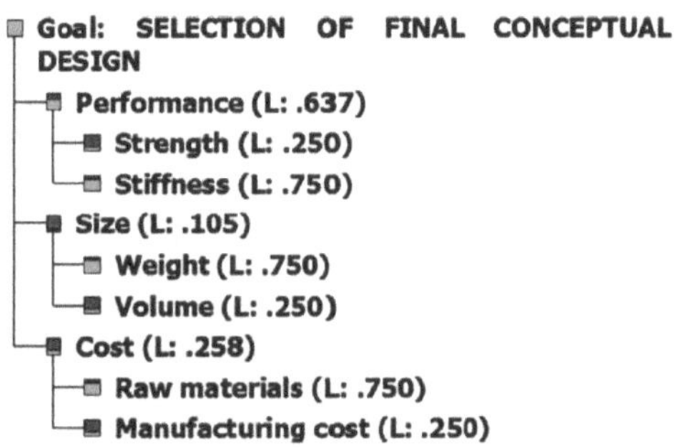

FIGURE 2.20 Hierarchy framework concept design selection of the sugar palm reinforced composite automotive antiroll bar (Mastura et al., 2017).

the AHP hierarchy framework, as shown in Fig. 2.20. The AHP hierarchy started with the goal of the study, followed by the main criteria design selection where performance, size, and cost, and subcriteria—strength, stiffness, weight, volume, raw materials, and manufacturing cost—were outlined for the concept design selection process. Each criterion at every level was evaluated based on the relative importance on a pairwise

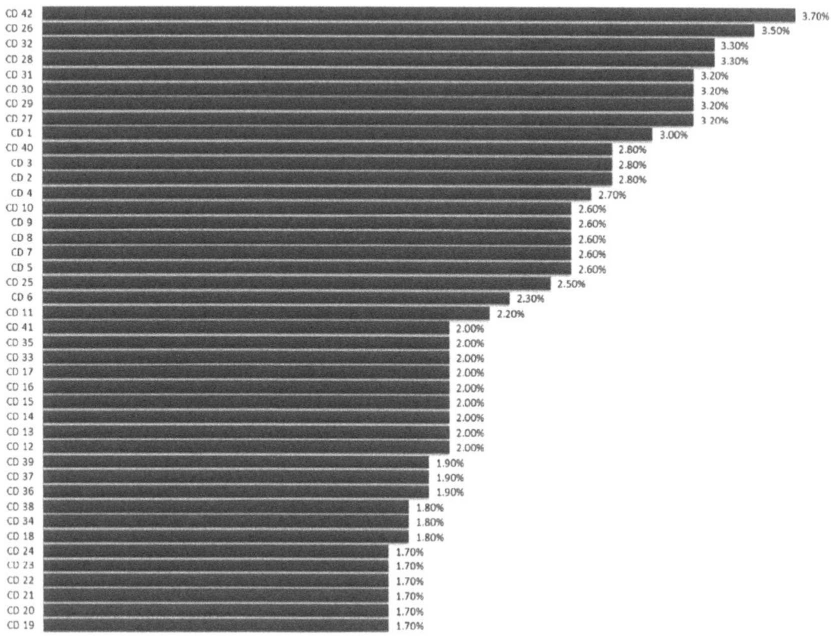

FIGURE 2.21 Ranking of the antiroll bar conceptual designs (Mastura et al., 2017).

comparison basis. The design that gained the highest score (3.7%), and was thus chosen as the final conceptual design, was the one with tapered arms reinforced with a corner rib to stiffen the structure of the design, as shown in Fig. 2.21.

The conceptual design of the automobile engine rubber mounting composite by Azammi et al. (2018) applied the analytic network process (ANP) technique to performed the conceptual design selection process. A four-level AHP hierarchy was formulated using the available information, as shown in Fig. 2.22. At level 1, the goal of the project was defined and followed by the design selection main criteria and subcriteria at levels 2 and 3, respectively. All design concepts of the kenaf fiber polymer composite engine rubber mounting are listed as alternatives at level 4 of the hierarchy. A pair-wise comparison technique was implemented for the judgment process between all concepts designs with respect to each selection main criterion and subcriterion. Based on the ANP conceptual design selection process, concept design no. 3 had an overall good result compared to the other concept designs, as shown in Fig. 2.23.

The conceptual design selection process for the kenaf fiber reinforced polymer portable laptop table studied by Sapuan et al. (2018) was done using the Pugh evaluation method. Through this method, each design

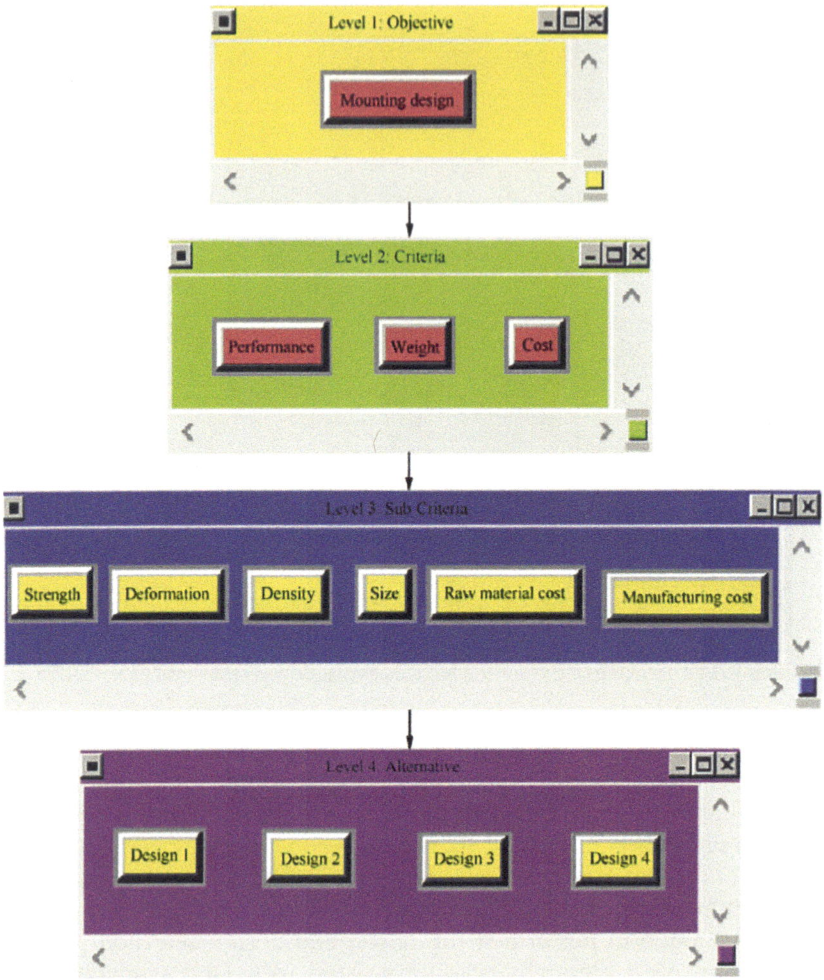

FIGURE 2.22 Analytic network process (ANP) hierarchy framework for the kenaf fiber polymer composite engine rubber mounting (Azammi et al., 2018).

was evaluated based on the baseline design as the datum using three indicators to evaluate the data which were (+) for better improvement, (−) for less improvement, and (S) for equal improvement. The concept which had the largest positive difference between plus and minus was chosen as the best concept. In this case study, concept design no. 4 had the highest total number of positive differences and was selected as the most suitable design to develop the kenaf fiber reinforced polymer portable laptop table. Fig. 2.24 shows the conceptual design selection process using the Pugh evaluation method for the kenaf fiber reinforced polymer portable laptop table.

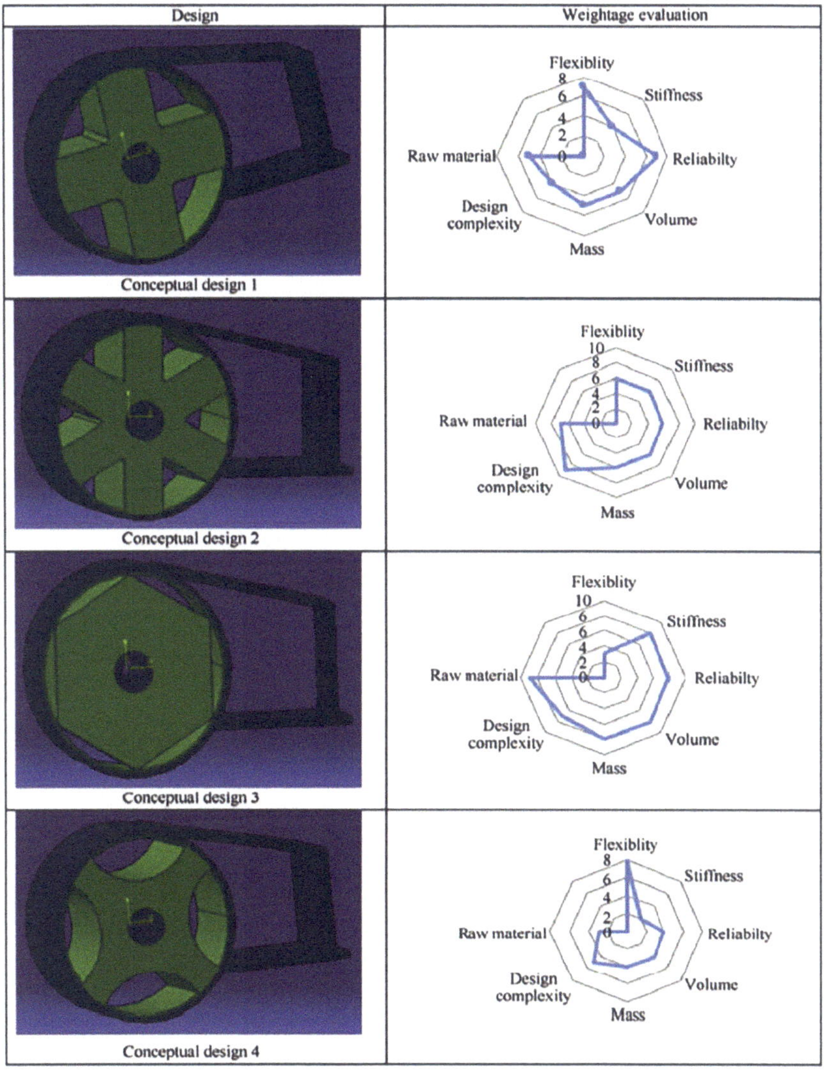

FIGURE 2.23 Conceptual design evaluation data (Azammi et al., 2018).

In the conceptual design selection process, concept screening and scoring are the methods most commonly used to determine the most suitable concept design. However, based on the aforementioned case study in the conceptual design development process of green product, most researchers implement the MCDM method as this method is concerned with structuring, considers ideal and nonideal solutions, evaluates ranking, and selects the best solution involving multiple criteria. In addition, the concept design selection process was based on the PDS and this document

Concept	Baseline 	First	Second	Third	Fourth	Fifth	Sixth	Seventh
Performance	D	–	+	–	+	S	+	+
Stability	A	+	+	+	+	+	–	+
Strength	T U	–	+	+	S	+	–	+
Easy to manufacture	M	+	–	+	+	–	–	–
Portability		–	–	–	+	+	+	S
Weight		–	–	S	S	S	+	–
Cost of manufacture		S	–	S	+	–	–	–
Total score	$\sum +$	2	3	3	5	3	3	3
	$\sum -$	4	4	2	0	2	4	3
	$\sum S$	1	0	2	2	2	0	1
Positive difference	$\sum + (-) \sum -$	−2	−1	1	5	1	−1	0

Note Total score is obtained by obtaining the difference between '$\sum +$' and '$\sum -$'

FIGURE 2.24 Conceptual design selection process using the Pugh evaluation method (Sapuan et al., 2018).

was essential for the project progress as it has the necessary information regarding the product result.

2.5 Case study on the conceptual design development of a green product

To demonstrate the conceptual design development of a green product, a simplified product development on a natural fiber reinforced composite water bottle was conducted. The product development consisted of PDS, conceptual design development using the TRIZ method, and the conceptual design selection process using the MCDM method. The MCDM method included criteria weighting by entropy and a ranking process by the VIseKriterijumska Optimizacija I Kompromisno Resenje (VIKOR) technique. Both techniques were integrated with fuzzy set because the selection process involved a human evaluation task.

2.5.1 Product design specification

PDS is important in all fields of design activity to ensure that the design process goes smoothly as is lays out the groundwork for all design activities and ensures that all relevant factors are taken into account. The new PDS was developed by considering the requirements for the product functional performance and satisfying the customer requirements. Fig. 2.25 shows the PDS of the natural fiber reinforced composite water bottle.

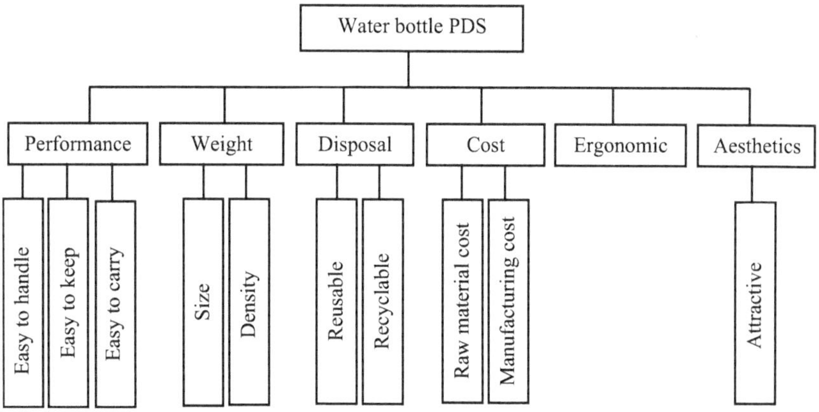

FIGURE 2.25 PDS of the natural fiber reinforced composite water bottle.

The conceptual design selection process was based on the PDS of the natural fiber reinforced composite water bottle. From the overall PDS document, three main elements (performance, ergonomic, aesthetics) and their subsequent subelements (easy to handle, easy to keep, easy to carry) are selected for the concept design selection purpose. These elements are evaluated using the fuzzy set technique which involves a logical human judgment process for identifying and choosing alternatives based on the values and preferences of the decision makers as the data are not presented in numbers.

2.5.2 Conceptual design development using TRIZ

The contradictions of a natural fiber reinforced composite water bottle is identified using *If, Then,* and *But* keywords that are compatible with the constraint that can be manipulated to acquire the solution where 39 systems parameters will be used. Then, the appropriate parameter is translated for the selection of 40 inventive principles through the TRIZ contradiction matrix. The application of TRIZ for this case study is simplified in Table 2.1.

Through the solution principles suggested by TRIZ, the potential solutions to solve the contradictions could assist in generating ideas. In this case study, solution principle #7 has been chosen to generate the specific solution ideas. Fig. 2.26 illustrates the conceptual designs of the natural fiber reinforced composite water bottle.

Based on the inventive principles suggested by the TRIZ method, three conceptual designs for the natural fiber reinforced composite water bottle are generated that are inspired by the nested doll design. Each design concept was designed to fulfill the problem statement where there is a large size of water bottle, which can hold a lot of water but the water bottle does not take up much space.

TABLE 2.1 Application of TRIZ in natural fiber reinforced composite water bottle development.

Problem statement	*If* the water bottle is large, *Then* can fill a lot of water, *But* take space to keep the water bottle.	
The chosen engineering parameter describing the condition	*Improving features* # 12 Shape	*Worsening features* #8 Volume of stationary object
TRIZ Inventive principles	# 7 nested doll # 2 taking out # 35 parameter changes	

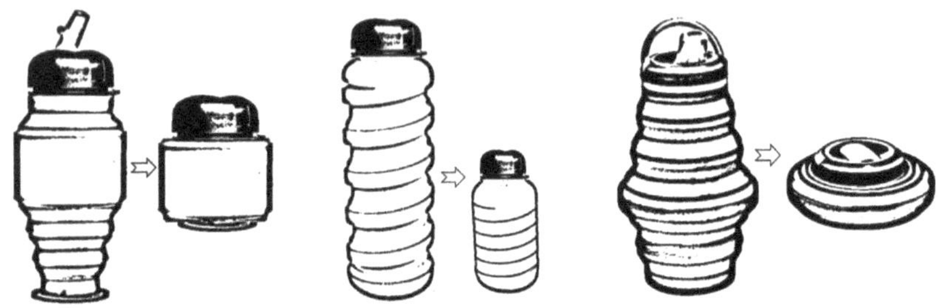

Concept design 1 Concept design 2 Concept design 3

FIGURE 2.26 Conceptual designs of the natural fiber reinforced composite water bottle.

2.5.3 Weighting of criteria using the entropy method

The conceptual design selection process is one of the major challenges in product development. Improper selection may lead to a failure result with customer dissatisfaction and cost inefficiency. Weighting of criteria plays a very significant role in the ranking results of the conceptual design selection process, where every criterion of the design concept is evaluated to ensure the reliability and accuracy of the decision. In this case study, weighting criteria were based on attribute preference information by expert or decision maker's evaluation. A fuzzy set was integrated with the entropy method since this technique involves a logical human judgment process for identifying and choosing alternatives based on the values and preferences of the decision makers. Table 2.2 shows the linguistic variables and corresponding fuzzy number to evaluate the weight of criteria of the conceptual designs of the natural fiber reinforced composite water bottle.

To determine the weighting criteria of the natural fiber reinforced composite water bottle concept design the fuzzy data were aggregated. The aggregation of fuzzy weight W_{ij} for each criterion was calculated

TABLE 2.2 Linguistic variables and corresponding fuzzy number to evaluate the weight of criteria of the conceptual designs (Ishak, Sivakumar, & Mansor, 2017).

Linguistic variable	Fuzzy number
Unimportant (UI)	(0.0, 0.0, 0.1, 0.2)
Low importance (LI)	(0.1, 0.2, 0.2, 0.3)
Slightly important (SI)	(0.2, 0.3, 0.4, 0.5)
Moderate importance (MI)	(0.4, 0.5, 0.5, 0.6)
Important (I)	(0.5, 0.6, 0.7, 0.8)
Very important (VI)	(0.7, 0.8, 0.8, 0.9)
Extremely important (EI)	(0.8, 0.9, 1.0, 1.0)

using Eqs. 2.1 and 2.2. The aggregate weight for each criterion W_j is calculated as follow:

$$W = [w_1, w_2, \ldots w_n] \tag{2.1}$$

$$X_{ij} = \frac{1}{n}\left(\sum_{e=1}^{n} X_{ij}^{e}\right), i = 1, 2, \ldots m \tag{2.2}$$

The decision matrix needs to be normalized to gain the projection value of each criterion $C_j (j = 1, 2, \ldots n): P_{ij}$

$$P_{ij} = \frac{x_{ij}}{\sum_{j=1}^{m} x_{ij}} \tag{2.3}$$

After normalization, the entropy value e_j is calculated as

$$e_j = -k \sum_{j=1}^{n} P_{ij} \ln P_{ij} \tag{2.4}$$

k is constant, let $k = (\ln(m))^{-1}$. The degree of divergence d_i of the basic information of each criterion is calculated as

$$d_{ij} = 1 - e_j \tag{2.5}$$

The higher d_j is, the more important the criterion C_j is for the problem. The objective weight for each criterion can be obtained

$$W_j = \frac{d_j}{\sum_{k--=1}^{n} d_k} \tag{2.6}$$

To determine the weighting criteria of the natural fiber reinforced composite water bottle concept design, data were obtained via a questionnaire which was administered to a group of decision makers consisting of four experienced designers, designated as DM1, DM2, DM3, and DM4. Significant criteria evaluated by decision makers were designated as easy to keep (C1), easy to carry (C2), easy to handle (C3), ergonomic (C4), and attractive (C5). Table 2.3 shows the linguistic variables of the importance weight (W) criteria of the natural fiber reinforced composite water bottle concept design assessed by the decision makers and Table 2.4 shows the respective fuzzy values. Table 2.5 summarizes the overall scores based on

TABLE 2.3 Linguistic variables of importance weight criteria assessed by decision makers.

DM	C1	C2	C3	C4	C5
1	EI	VI	VI	SI	MI
2	EI	VI	VI	SI	SI
3	VI	VI	I	I	I
4	VI	EI	I	MI	MI

TABLE 2.4 Fuzzy value of importance weight criteria assessed by decision makers.

DM	C1	C2	C3	C4	C5
1	(0.8, 0.9, 1, 1)	(0.7, 0.8, 0.8, 0.9)	(0.7, 0.8, 0.8, 0.9)	(0.2, 0.3, 0.4, 0.5)	(0.4, 0.5, 0.5, 0.6)
2	(0.8, 0.9, 1,1)	(0.7, 0.8, 0.8, 0.9)	(0.7, 0.8, 0.8, 0.9)	(0.2, 0.3, 0.4, 0.5)	(0.2, 0.3, 0.4, 0.5)
3	(0.7, 0.8, 0.8, 0.9)	(0.7, 0.8, 0.8, 0.9)	(0.5, 0.6, 0.7, 0.8)	(0.5, 0.6, 0.7, 0.8)	(0.5, 0.6, 0.7, 0.8)
4	(0.7, 0.8, 0.8, 0.9)	(0.8, 0.9, 1, 1)	(0.5, 0.6, 0.7, 0.8)	(0.4, 0.5, 0.5, 0.6)	(0.4, 0.5, 0.5, 0.6)

TABLE 2.5 Fuzzy value of importance weight criteria assessed by decision makers.

DM	C1	C2	C3	C4	C5
X_{ij}	(0.75, 0.85, 0.9, 0.95)	(0.73, 0.83, 0.85, 0.93)	(0.6, 0.7, 0.75, 0.85)	(0.33, 0.43, 0.5, 0.58)	(0.38, 0.48, 0.53, 0.63)
P_{ij}	0.86	0.83	0.73	0.46	0.5
e_j	0.09	0.11	0.17	0.26	0.25
d_j	0.91	0.89	0.83	0.74	0.75
W_j	0.22	0.22	0.20	0.18	0.18

the entropy method for the weighting criteria of the natural fiber reinforced composite water bottle concept design.

2.5.4 Conceptual design selection process using VIKOR

VIKOR is the suitable tool in MCDM when decision makers are unable to interpret their preferences at the early stage of product development as it provides the maximum group utility (S) of the majority, and the minimum of the individual regret (R) of the opponent to obtain the compromise solution. The maximum group utility of the majority represents the concordance of the compromise solution, while the minimum of the individual regret of the opponent represents the discordance of the compromise solution.

Fuzzy set was integrated with the VIKOR method to evaluate the conceptual designs. Table 2.6 was used by the decision makers to evaluate the three conceptual designs of the natural fiber reinforced composite water bottle with respect to the weight of criteria.

To determine the suitable design of the natural fiber reinforced composite water bottle concept design, the fuzzy data were then aggregated. The aggregation of fuzzy weight W_{ij} for each criterion was calculated using Eq. 2.7.

$$W_j = \left\{ W_{j1}, W_{j2}, W_{j3}, W_{j4} \right\} \tag{2.7}$$

where,

$$W_{j1} = \min\left\{ W_{jk1} \right\}$$

$$W_{j2} = \frac{1}{k}\sum W_{jk2}$$

$$W_{j3} = \frac{1}{k}\sum W_{jk3}$$

$$W_{j4} = \max\left\{ W_{jk4} \right\}$$

TABLE 2.6 Linguistic variables and corresponding fuzzy number to evaluate the water bottle conceptual designs with respect to weight of criteria (Ishak et al., 2017).

Linguistic variable	Fuzzy number
Very poor (VP)	(0.0, 0.0, 0.1, 0.2)
Poor (P)	(0.1, 0.2, 0.2, 0.3)
Medium poor (MP)	(0.2, 0.3, 0.4, 0.5)
Fair (F)	(0.4, 0.5, 0.5, 0.6)
Medium good (MG)	(0.5, 0.6, 0.7, 0.8)
Good (G)	(0.7, 0.8, 0.8, 0.9)
Very good (VG)	(0.8, 0.9, 1.0, 1.0)

Thus, the aggregate matrix for the weight of criteria and concept design ratings lead to the formation of the weight of the criteria matrix (Eq. 2.8).

$$W = \begin{bmatrix} W_1 & W_2 & W_3 & W_4 \end{bmatrix} \tag{2.8}$$

The aggregated matrix for the weight of criteria and concept design ratings were then defuzzified to derive their crisp value. The defuzzification was based on Eq. 2.9.

$$\begin{aligned} Defuzz(X_{ij}) &= \frac{\int \mu(x) \cdot xdx}{\int \mu(x)dx} \\ &= \frac{\int_{x_{ij1}}^{x_{ij2}} \left(\frac{x - x_{ij1}}{x_{ij2} - x_{ij1}}\right) \cdot xdx + \int_{x_{ij2}}^{x_{ij3}} xdx + \int_{x_{ij3}}^{x_{ij4}} \left(\frac{x_{ij4} - x}{x_{ij4} - x_{ij3}}\right) \cdot xdx}{\int_{x_{ij1}}^{x_{ij2}} \left(\frac{x - x_{ij1}}{x_{ij2} - x_{ij1}}\right)dx + \int_{x_{ij2}}^{x_{ij3}} dx + \int_{x_{ij3}}^{x_{ij4}} \left(\frac{x - x_{ij1}}{x_{ij2} - x_{ij1}}\right)dx} \\ &= \frac{-x_{ij1}x_{ij2} + x_{ij3}x_{ij4} + \frac{1}{3}\left(x_{ij4} - x_{ij3}\right)^2 - \frac{1}{3}\left(x_{ij2} - x_{ij1}\right)^2}{-x_{ij1} - x_{ij2} + x_{ij3} + x_{ij4}} \end{aligned} \tag{2.9}$$

The integration VIKOR with entropy performance $(u_{ij} \otimes w_s^j)$ is

$$\begin{aligned} F &= \left[f_{ij}\right]_{m \times n} \\ f_{ij} &= defuzz(u_{ij} \otimes w_j^s) \end{aligned} \tag{2.10}$$

$$f_{ij} = \frac{-(u_{ij1}u_{ij2})(w_{j1}^s w_{j2}^s) + (u_{ij3}u_{ij4})(w_{j3}^s w_{j4}^s) + \frac{1}{3}(u_{ij4}w_{j4}^s - u_{ij3}w_{j3}^s)^2 - \frac{1}{3}(u_{ij2}w_{j2}^s - u_{ij1}w_{j1}^s)^2}{-u_{ij1}w_{j1}^s - u_{ij2}w_{j2}^s + u_{ij3}w_{j3}^s + u_{ij4}w_{j4}^s} \tag{2.11}$$

Through the crisp value, the best value $\left(f_i^*\right)$ and worst value $\left(f_i^-\right)$ were next identified as this value was used in the measurement of the utility index (S_i). The utility index referred to the separation measure of i-th alternative with fuzzy best value. Eq. 2.10 was used to determine the utility index.

$$S_i = \sum_{j=1}^{n} \frac{W_j\left(f_i^*\right)}{\left(f_i^* - f_i^-\right)} \tag{2.12}$$

Then, the regret index (R_i), which was the separation measure of i-th alternative with fuzzy worst value was determined using Eq. 2.11.

$$R_i = \max\left(\frac{w_j\left(f_i^* - f_i^-\right)}{\left(f_i^* - f_i^-\right)}\right) \tag{2.13}$$

Next, the comprehensive utility value, or the VIKOR value (Q_i) was calculated using Eq. 2.12.

$$Q_i = \left(\frac{v(S_i - S^*)}{(S^- - S^*)}\right) + \frac{(1 - v)(R_i - R^*)}{R^- - R^*} \tag{2.14}$$

The utility index (S_i), regret index (R_i), and VIKOR value (Q_i) were then ranked in ascending order as the smallest alternative Q_i value was determined to be the most suitable of the natural fiber reinforced composite water bottle concept designs. Then, the next step dealt with proposing the compromise solution. The compromise solution was considered, if only Conditions 1 and 2 were satisfied. The two conditions were:

Condition 1:

Acceptable advantage: $(A^{(2)}) - Q(A^{(1)}) \geq \frac{1}{(m-1)}$, where $A^{(1)}$ and $A^{(2)}$ are the first and second positions in the alternatives ranked by Q, and m is the numbers of alternatives.

Condition 2:

Acceptable stability in decision-making: alternative $A^{(1)}$ must also be best ranked by S or/and R. When one of the conditions is not satisfied, a set of compromise solutions is detected. The set of compromised solutions is composed of:

1. Alternatives A^1 and A^2 if only condition C2 is satisfied.
2. Alternatives $A^{(1)}, A^{(2)}, \ldots A^{(m)}$ if condition C1 is not satisfied. $A^{(M)}$ is calculated using the relation $Q(A^{(M)}) - Q(A^{(1)}) < \frac{1}{(m-1)}$ for maximum M.

Through linguistic variables, decision makers evaluate the three conceptual designs for the natural fiber reinforced composite water bottle with respect to the weight of criteria. Table 2.7 shows the linguistic

TABLE 2.7 Evaluation of water bottle concept designs by decision makers (linguistic variable).

	C1	C2	C3	C4	C5
DM1					
Design 1	VI	VI	EI	MI	EI
Design 2	VI	I	I	VI	EI
Design 3	EI	EI	EI	MI	EI
DM2					
Design 1	EI	EI	EI	EI	EI
Design 2	VI	EI	VI	EI	MI
Design 3	EI	EI	EI	EI	EI
DM3					
Design 1	VI	VI	VI	EI	EI
Design 2	I	VI	VI	VI	EI
Design 3	EI	EI	VI	EI	EI

(Continued)

TABLE 2.7 (Continued)

	C1	C2	C3	C4	C5
DM4					
Design 1	VI	MI	VI	VI	MI
Design 2	MI	SI	VI	VI	I
Design 3	EI	EI	VI	EI	VI

variables of the decision maker's evaluation results and Table 2.8 shows the respective fuzzy value. Tables 2.9, 2.10, 2.11, 2.12, and 2.13 summarize the overall scores based on the VIKOR method for the natural fiber reinforced composite water bottle conceptual design selection process.

As per the VIKOR method, the smallest alternative value is determined to be the best solution, therefore the utility index, regret index, and VIKOR value are rank in ascending order of the water bottle concept designs shown in Table 2.14.

In this solution, Design 3 of the water bottle design satisfied both conditions where the value of Condition 1 for the first position is more than that of the second position, and for Condition 2, Design 3 is the best rank for both utility index (S_i) and regret index (R_i).

Through the development of the natural fiber reinforced composite water bottle, application of the TRIZ method to develop the conceptual design has provided an opportunity to develop ideas and solve problems in creative and innovative approaches. Integration of VIKOR and entropy gives a higher degree of confidence to decision makers since the selection process involving the MCDM method provides a systematic and scientific selection method. In addition, combination of these two approaches can increase the success probability of the product since, through determination of the weighting criteria, the decision results are more reliable and accurate when considering the product performance and satisfaction of the customer requirements.

2.6 Conclusion and future directions in the conceptual design of green products

In conclusion, the conceptual design stage of green products involved several major stages, starting with identification of customer requirements, generating PDS, developing solutions for the product by conceptual design, selecting the best solutions among the design alternatives, as well as improvising the features of the final design in terms of manufacturability, recyclability, and ergonomics. The conceptual design development process of green products can be performed using various methods, such

TABLE 2.8 Evaluation of water bottle concept designs by decision makers (fuzzy set).

	C1	C2	C3	C4	C5
DM1					
Design 1	(0.7, 0.8, 0.8, 0.9)	(0.7, 0.8, 0.8, 0.9)	(0.8, 0.9, 1, 1)	(0.4, 0.5, 0.5, 0.6)	(0.8, 0.9, 1, 1)
Design 2	(0.7, 0.8, 0.8, 0.9)	(0.5, 0.6, 0.7, 0.8)	(0.5, 0.6, 0.7, 0.8)	(0.7, 0.8, 0.8, 0.9)	(0.8, 0.9, 1, 1)
Design 3	(0.8, 0.9, 1, 1)	(0.8, 0.9, 1, 1)	(0.8, 0.9, 1, 1)	(0.4, 0.5, 0.5, 0.6)	(0.8, 0.9, 1, 1)
DM2					
Design 1	(0.8, 0.9, 1, 1)	(0.8, 0.9, 1, 1)	(0.8, 0.9, 1, 1)	(0.8, 0.9, 1, 1)	(0.8, 0.9, 1, 1)
Design 2	(0.7, 0.8, 0.8, 0.9)	(0.8, 0.9, 1, 1)	(0.7, 0.8, 0.8, 0.9)	(0.8, 0.9, 1, 1)	(0.4, 0.5, 0.5, 0.6)
Design 3	(0.8, 0.9, 1, 1)	(0.8, 0.9, 1, 1)	(0.8, 0.9, 1, 1)	(0.8, 0.9, 1, 1)	(0.8, 0.9, 1, 1)
DM3					
Design 1	(0.7, 0.8, 0.8, 0.9)	(0.7, 0.8, 0.8, 0.9)	(0.7, 0.8, 0.8, 0.9)	(0.8, 0.9, 1, 1)	(0.8, 0.9, 1, 1)
Design 2	(0.5, 0.6, 0.7, 0.8)	(0.7, 0.8, 0.8, 0.9)	(0.7, 0.8, 0.8, 0.9)	(0.7, 0.8, 0.8, 0.9)	(0.8, 0.9, 1, 1)
Design 3	(0.8, 0.9, 1, 1)	(0.8, 0.9, 1, 1)	(0.7, 0.8, 0.8, 0.9)	(0.8, 0.9, 1, 1)	(0.8, 0.9, 1, 1)
DM4					
Design 1	(0.7, 0.8, 0.8, 0.9)	(0.4, 0.5, 0.5, 0.6)	(0.7, 0.8, 0.8, 0.9)	(0.7, 0.8, 0.8, 0.9)	(0.4, 0.5, 0.5, 0.6)
Design 2	(0.4, 0.5, 0.5, 0.6)	(0.2, 0.3, .4, 0.5)	(0.7, 0.8, 0.8, 0.9)	(0.7, 0.8, 0.8, 0.9)	(0.5, 0.6, 0.7, 0.8)
Design 3	(0.8, 0.9, 1, 1)	(0.8, 0.9, 1, 1)	(0.7, 0.8, 0.8, 0.9)	(0.8, 0.9, 1, 1)	(0.7, 0.8, 0.8, 0.9)

TABLE 2.9 Aggregated fuzzy value of water bottle concept designs.

	C1	C2	C3	C4	C5
W	(0.7, 0.85, 0.9, 1)	(0.7, 0.83, 0.85, 1)	(0.5, 0.7, 0.75, 0.9)	(0.2, 0.43, 0.5, 0.8)	(0.2, 0.48, 0.53, 0.8)
Design 1	(0.7, 0.83, 0.85, 1)	(0.4, 0.75, 0.78, 1)	(0.7, 0.85, 0.9, 1)	(0.4, 0.78, 0.83, 1)	(0.4, 0.8, 0.88, 1)
Design 2	(0.4, 0.68, 0.7, 0.9)	(0.2, 0.65, 0.73, 1)	(0.5, 0.75, 0.78, 0.9)	(0.7, 0.83, 0.85, 1)	(0.4, 0.73, 0.8, 1)
Design 3	(0.7, 0.9, 1, 1)	(0.8, 0.9, 1, 1)	(0.7, 0.85, 0.9, 1)	(0.4, 0.80, 0.88, 1)	(0.7, 0.88, 0.95, 1)

TABLE 2.10 Integration VIseKriterijumska Optimizacija I Kompromisno Resenje with entropy $\left(u_{ij} \otimes w_s^j\right)$.

	C1	C2	C3	C4	C5
Design 1	(0.49, 0.7, 0.77, 1)	(0.28, 0.62, 0.66, 1)	(0.35, 0.6, 0.68, 0.9)	(0.08, 0.33, 0.41, 0.8)	(0.08, 0.38, 0.46, 0.8)
Design 2	(0.28, 0.57, 0.63, 0.9)	(0.14, 0.54, 0.62, 1)	(0.25, 0.53, 0.58, 0.81)	(0.14, 0.35, 0.43, 0.8)	(0.08, 0.34, 0.42, 0.8)
Design 3	(0.49, 0.77, 0.9, 1)	(0.32, 0.74, 0.85, 1)	(0.35, 0.6, 0.68, 0.9)	(0.08, 0.34, 0.44, 0.08)	(0.14, 0.42, 0.5, 0.8)

TABLE 2.11 Crisp value of water bottle concept designs.

	C1	C2	C3	C4	C5
Design 1	0.66	0.58	0.48	0.38	0.29
Design 2	0.54	0.53	0.42	0.28	0.28
Design 3	0.69	0.63	0.48	0.27	0.30

TABLE 2.12 Best and worst values of water bottle concept designs.

	C1	C2	C3	C4	C5
(f_i^*)	0.69	0.63	0.48	0.38	0.30
(f_i^-)	0.54	0.58	0.42	0.27	0.28

TABLE 2.13 The utility index, regret index, and VIKOR value of water bottle concept designs.

	(S_i)	(R_i)	(Q_i)
Design 1	1.25	0.58	0.39
Design 2	2.90	0.85	1.05
Design 3	0.44	0.44	0.06

TRIZ and a morphological chart. Meanwhile, the conceptual design selection process is often performed using MCDM methods such as AHP and VIKOR. The case study included in this chapter for the redesigning of a new eco-friendly water bottle which was able to reduce storage space while using green composite materials also showed in practice how conceptual design development and selection process can be applied.

TABLE 2.14 Ranking of ascending order of the utility index, regret index, and VIKOR value.

Ranking	(S_i)	(R_i)	(Q_i)
1	Design 3	Design 3	Design 3
2	Design 1	Design 1	Design 1
3	Design 2	Design 2	Design 2

Integration of the TRIZ and VIKOR methods in the case study successfully created a new solution for a new eco-friendly water bottle, which has a unique feature of being able to store a large amount of water without requiring a large storage space (and is easy to carry by the user).

Furthermore, new methods for conducting conceptual design development and selection process in green products continue to be explored, to provide more options for product designers in getting better and innovative solutions to meet customer requirements. Among these is the use of a biomimetics (or biomimicry) approach in generating new and innovative conceptual design ideas, which is based on adopting how nature works. Living creatures (plants and animals) have demonstrated many ways of obtaining ideal features for various conditions, which can be adopted, directly or indirectly, into green product design features.

Acknowledgment

The authors thank Universiti Teknikal Malaysia Melaka for providing support throughout the conduct of this study.

References

Azammi, A. M. N., Sapuan, S. M., Ishak, M. R., & Sultan, M. T. H. (2018). Conceptual design of automobile engine rubber mounting composite using TRIZ-morphological chart-analytic network process technique. *Defence Technology, 14*(4), 268–277.

Davoodi, M. M., Sapuan, S. M., Ahmad, D., Aidy, A., Khalina, A., & Jonoobi, M. (2011). Concept selection of car bumper beam with developed hybrid bio-composite material. *Materials and Design, 32*(10), 4857–4865.

Hambali, A., Sapuan, S. M., Ismail, N., & Nukman, Y. (2009). Application of analytical hierarchy process in the design concept selection of automotive composite bumper beam during the conceptual design stage. *Scientific Research and Essay, 4*(4), 198–211.

Ishak, N. M., Sivakumar, D., & Mansor, M. R. (2017). Thermoplastic matrix selection for fibre metal laminate using fuzzy VIKOR and entropy measure for objective weighting. *Journal of Engineering Science and Technology, 12*(10), 2792–2804.

Ishak, N. M., Sivakumar, D., & Mansor, M. R. (2018). The application of TRIZ on natural fibre metal laminate to reduce the weight of the car front hood. *Journal of the Brazilian Society of Mechanical Sciences and Engineering, 40*(105), 105.

King, A. M., & Sivaloganathan, S. (1999). Development of a methodology for concept selection in flexible design strategies. *Journal of Engineering Design, 10*(4), 329–349.
Mansor, M. R., Sapuan, S. M., Hambali, A., Zainudin, E. S., & Nuraini, A. A. (2015). Conceptual design of kenaf polymer composites automotive spoiler using TRIZ and morphology chart methods. *Applied Mechanics and Materials, 761*, 63–67.
Mansor, M.R., Sapuan, S. M., Salim, M. A., Akop, M.Z., Musthafah, M. T., & Shaharuzaman, M. A. (2016). Concurrent design of green composites. Green approaches to biocomposite materials science and engineering. *IGI Global*. (pp. 48–75).
Mansor, M. R., Sapuan, S. M., Zainudin, E. S., Nuraini, A. A., & Hambali, A. (2014). Conceptual design of kenaf fiber polymer composite automotive parking brake lever using integrated TRIZ-morphological chart-analytic hierarchy process method. *Materials and Design, 54*, 473–482.
Mastura, M. T., Sapuan, S. M., Mansor, M. R., & Nuraini, A. A. (2017). Conceptual design of a natural fibre-reinforced composite automotive anti-roll bar using a hybrid approach. *International Journal of Advanced Manufacturing Technology, 91*(5–8), 2031–2048.
Maxwell, D., & Van der Vorst, R. (2003). Developing sustainable products and services. *Journal of Cleaner Production, 11*(8), 883–895.
Pahl G., Beitz, W., Feldhusen, J., & Grote, K. H. (2007). Conceptual design. Engineering design. London: Springer.
Sapuan, S. (2010). Concurrent engineering for composites. Universiti Putra Malaysia Press.
Sapuan, S. M., & Maleque, M. A. (2005). Design and fabrication of natural woven fabric reinforced epoxy composite for household telephone stand. *Materials and Design, 26*(1), 65–71.
Sapuan, S. M., Harun, N., & Abbas, K. A. (2007). Design and fabrication of a multipurpose table using a composite of epoxy and banrna pseudostem fibres. *Journal of Tropical Agriculture, 45*(1–2), 66–68.
Sapuan, S. M., & Mansor, M. R. (2014). Concurrent engineering approach in the development of composite products: A review. *Materials & Design, 58*, 161–167.
Sapuan, S. M., Purushothman, K. R., Sanyang, M. L., & Mansor, M. R. (2018). Design and fabrication of Kenaf fibre reinforced polymer composites for portable laptop table. In S. Kalia (Ed.), *Lignocellulosic composite materials* (pp. 323–356). Cham: Springer, Springer Series on Polymer and Composite Materials. Available from: https://doi.org/10.1007/978-3-319-68696-7_8.
Sun, X., Houssin, R., Renaud, J., & Gardoni, M. (2019). A review of methodologies for integrating human factors and ergonomics in engineering design. *International Journal of Production Research, 57*(15-16), 4961–4976.
United Nations, (2020). Sustainable development goals, accessed from <https://www.un.org/sustainabledevelopment/>.
United States General Service Administration (GSA), (2020). Sustainable design, accessed from <https://www.gsa.gov/real-estate/design-construction/design-excellence/sustainability/sustainable-design>.

CHAPTER

3

Sustainable materials selection: principles and applications

M.A. Shaharuzaman[1], *S.M. Sapuan*[2,3] *and Muhd Ridzuan Mansor*[1]

[1]Faculty of Mechanical Engineering, Universiti Teknikal Malaysia Melaka, Melaka, Malaysia, [2]Laboratory of Biocomposite Technology, Institute of Tropical Forestry and Forest Products (INTROP), Universiti Putra Malaysia, Selangor, Malaysia, [3]Advanced Engineering Materials and Composites Research Centre (AEMC), Department of Mechanical and Manufacturing Engineering, Faculty of Engineering, Universiti Putra Malaysia, Selangor, Malaysia

3.1 Introduction to materials selection in design for sustainability

Design for sustainability (DfS) is a holistic design approach to the activities that emphasize the well-being of people and the environment as the outcome. It focuses on resource efficiency and the use of environmentally friendly materials to develop products and processes (Manzini, 2020). The DfS approach is also considered as a lifecycle design approach, whereby the approach incorporates information about the total lifecycle phases of products and their effect both on the environment and living things. Currently, the drive toward sustainability by consumers and industries has increased the use of the DfS approach, especially for new product and process development purposes. One of the areas in which the approach is widely applied is product innovation. Compared to traditional product innovation strategies which focused on profitability alone, current new product innovation also focuses on the effects on people and the environment and is aligned with the DfS objectives (Crul & Diehl, 2005).

Design for Sustainability
DOI: https://doi.org/10.1016/B978-0-12-819482-9.00001-0

Among the strategies in the DfS of products and processes are the use of eco-friendly or green materials. The use of green materials in product development reduces the environmental impact by enabling them to be reused, recycled, and naturally degradable at their end of life. Furthermore, green materials are often derived from plants, which provide the advantage of unlimited resources and renewability compared to conventional engineering materials such as minerals which are nonrenewable and depleting resources. These characteristics of green materials when applied to products and processes have greatly helped to reduce the effects of the carbon footprint and high energy consumption throughout the entire product lifecycle (Clark, Kosoris, Hong, & Crul, 2009).

Based on the above advantages, many types of green materials have been developed by researchers and put into the commercial market to be used by various industries. Among them are natural fiber reinforced polymer composites as substitute materials, especially for synthetic polymer composites. However, due to the vast number of material options available to be applied for new products, an issue has also been created of selecting the best material which is able to complement successfully the design intent of the product. The complexity in choosing the best type of material to perform the desired function is also created by the various attributes available within the same class of material (Mansor et al., 2016).

Therefore, materials selection is one of the most important factors to resolve the aforementioned issues for successful implement of DfS in developing products and processes. Various methods of decision-making have been developed and applied for the materials selection process by researchers to achieve the desired goal of selecting the best candidate material in green product development. These modern and scientific methods have the ability to aid decision makers in performing the judgment process of selecting the best candidate material by simultaneously analyzing problems involving multiple selection criteria and multiple candidate materials with varying attributes between them. Hence, the use of these multicriteria decision-making (MCDM) methods has enabled more confidence and better accuracy in the final decision made as compared to traditional intuitive decision-making methods (which are often not reliable). Furthermore, modern MCDM methods applied in materials selection, especially involving green materials, can also incorporate and later consolidate judgments from various stakeholders such as the product designer, end user, people involved in the production process, etc. This advantage helps to reduce the conflict which often arises due to a mismatch in expectations between the product stakeholders during the materials selection process (Sapuan & Mansor, 2014).

In this chapter, the principles and applications of the materials selection process toward DfS of green products and processes are discussed, with a focus on green composite materials. The discussions are divided into two categories for materials selection purpose—screening and ranking. In addition, an explanation of the weighing process of the selection criteria is also included. Furthermore, this chapter also discusses recent literature reviews on the applications of MCDM methods in materials selection involving green composite materials. The reviews are further divided into three categories, namely natural fiber selection, polymer matrix selection, and final natural fiber composite selection. To aid readers in the practical applications of the green composite materials selection process using the MCDM method, a case study on natural fiber composite materials selection is also added. Finally, the chapter concludes with several remarks on the future directions of materials selection involving green materials.

3.2 Overview of multiple criteria decision-making for materials selection

Materials selection is known as the strategies to perform choosing of the best materials that have different attributes in DfS for product design and development. Materials selection should be taken into consideration in the early stage of the conceptual design process in the product development process. Fig. 3.1 shows the product development process developed by Pugh (1991) which consists of (1) market investigation, (2) product design specification, (3) conceptual design, (4) detail design,

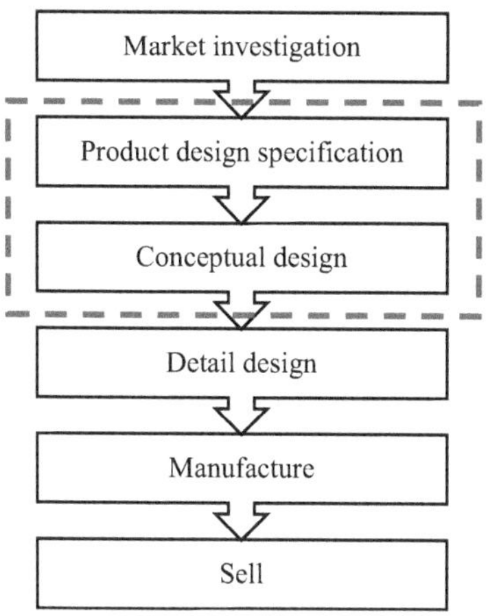

FIGURE 3.1 New product development process (Pugh, 1991).

(5) manufacture, and (6) selling/marketing. This chapter focuses on two stages of the product development process, as highlighted in the Fig. 3.1, which are determination of the weightage criteria for product design specification and materials selection in the conceptual design stage.

In designing a new product, the product design specification is one of the most important stages that can helps engineers to fulfill all the criteria throughout the process of design and development. In this stage, engineers will try to consider all of the criteria for the product specification to ensure the final product gives a good result to consumers. In Total Design, Pugh (1991) listed 32 product design specifications that can be consider in developing a new product. The idea is to generate the specification before moving to the next stage. It can be added or removed depending on the suitability of the product. In addition, the product design specification can have subcriteria and subsubcriteria to perform. This groups the specification under the same criteria. As examples, Singh, Pant, Godiyal, and Kumar Sharma (2020) listed three main criteria, namely chemical properties, morphological properties, and physical properties for selecting fibers in papermaking industries, with 12 subcriteria under chemical properties, five subcriteria under morphological properties, and four subcriteria under physical properties. Mansor et al. (2013a) and Mansor, Sapuan, Zainudin, Nuraini, & Hambali (2013b) listed three main criteria and five subcriteria in designing the hybrid composites of an automotive parking brake lever. On the other hand, Singh, Patnaik, Chauhan, and Chauhan (2018) listed eight subcriteria without grouping them under the physical, chemical, and mechanical properties for the product design. It was shown that the product design specification is not rigid, but rather total consideration needs to be given by engineers before starting the product design and development process. After the product design specification has been finalized, the conceptual design process can be started.

The modern tools approach in conceptual design gives a systematic and collective method to the product design and development process compared to trial-and-error methods from the past (Sapuan & Mansor, 2014). In this chapter, the materials selection process of green composite design is discussed to show that the systematic approach to help designers and engineers in making a decision at the early stage of product design and development can be done. The materials selection process can be divided into two methods, which are the screening and ranking methods, as shown in Fig. 3.2. The screening method uses screening tools such as Cambridge Engineering Selector (CES) software, knowledge-based system (KBS), and expert system, while in the ranking method, there are two steps involved, namely weight determination and ranking of the materials using multiple criteria decision-making tools. If there is more than one material from the screening method, it

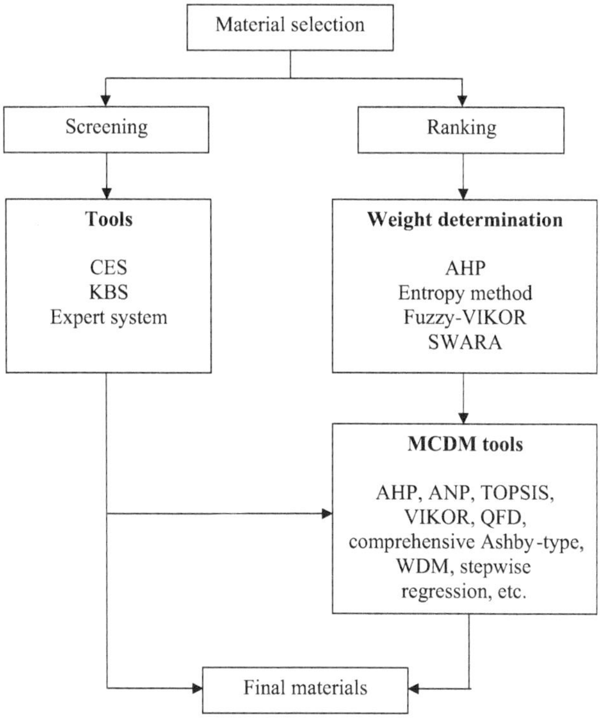

FIGURE 3.2 Material selection process in conceptual design for green composites.

can be rank in the ranking method to perform the analysis and select the best materials for product design and development. Finally, the highest ranking material is chosen before proceeding to the next step of product design and development. A detailed explanation of both steps is given in the next section.

3.2.1 Screening

The materials selection process has been used by designers and engineers for many years in the product development process. The materials selection process requires a lot of materials data in order to make a wise choice in selecting the most suitable and best materials for the product designed. Materials data sheets or data in a book are the conventional methods which have become outdated because the data are rarely updated and published only occasionally. As a consequence, a computerized database of materials is now important to engineers in selecting the best materials that suit their products. Therefore Sapuan (2001) developed a KBS for polymeric-based composite materials using rule-based reasoning for automotive components. The architecture of the KBS system uses a Knowledge Engineering Environment toolkit (Sapuan & Abdalla, 1998). The materials selection process is based on the product design specification with the predefined criteria and constraints by successfully reducing the weight of a metal pedal box by 31% by replacing it with a composite pedal box.

The KBS is also known as an expert system where the selection of materials is done through the input in a computerized database. Using the programming language of IF-THEN relations, Sapuan et al. (2011a, 2011b) developed an expert system for the polymeric composite materials selection via PHPMyAdmin and MySQL over the world wide web. The mechanical properties data required from the product design specification were then analyzed by the system to feature the best materials from the database. In addition, Ahmed Ali, Sapuan, Zainudin, and Othman (2015) developed another expert system for polymeric composite materials selection using Java programming language integrated with the weighted range method to enhance the performance of the search engine. The case study to find the most suitable natural fiber composites for an automotive interior door panel shows that sisal 30%/polypropylene (PP) was the best material. A computerized database of the polymeric composites really helps engineers and designers in making a decision in the conceptual design stage, especially in materials selection.

Despite using a data sheet and data from a book for plant fiber reinforced polymer composites, Shah (2013) developed a comprehensive Ashby chart to help engineers in making a selection of materials based on their tensile properties. The considerations of the chart include the fiber type, fiber extraction process, surface modification technique, fiber volume fraction, reinforcement geometry, interfacial properties, reinforcement arrangement and orientation, matrix type, and manufacturing technique. Besides the plant fiber composites, synthetic fibers such as glass fiber, Kevlar, and carbon fiber are also plotted in the chart for a comparison between plant fibers and synthetic fibers. The extended version of the Ashby charts includes optimal strength and stiffness, weight, cost, and eco-impact (Shah, 2014). The Ashby charts comprise the data for the plant fiber reinforced polymer composites for engineers to screen the materials that satisfy the product design specification and proposed the materials in the product development process.

CES Edupack developed by Granta Design is the Ashby materials selection available commercially to assist engineers for selecting materials (Selector, 2013). Mustafa, Abdollah, Ismail, Amiruddin, and Umehara (2014) screened alternative friction materials using CES software that can replace currently used asbestos materials. The materials should satisfy the product design specification with six criteria and 11 subcriteria. Comparing with the Pugh method, both methods result in kenaf being identified as the most promising material that can reduce the impact on the environment, reduce weight, and reduce the cost. On the other hand, Mastura, Sapuan, Mansor, and Nuraini (2018) used the CES software to produce five thermoplastics materials in a data table with complete mechanical, physical, and chemical properties to be selected using the Analytic Hierarchy Process (AHP) method in product design development. Besides using CES software to screen the materials, Morini, Ribeiro, and Hotza (2019) proposed CES EcoAudit 2017 to pre-assess materials for the environmental impact including currently available data on embodied energy and carbon footprint to develop a product that is ecologically benign. The screening method using a huge computerized database really helps engineers and designers in performing the materials selection process at the conceptual design stage. Therefore the data collection for the natural fiber composites database is important for the screening method to update the data in the system so that engineers and designers are enabled to be best informed in the materials selection process.

3.2.2 Ranking process using the multiple criteria decision-making method

The early stage of product design usually requires materials selection, geometrical design selection, and manufacturing process selection (Mansor et al., 2013a, 2013b). The six stages in product design and development developed by Pugh (1991) in Total Design are market investigation, product design specification, conceptual design, detail design, manufacture, and sale. These stages can be used by engineers to produce a quality product that will satisfy customer needs. In addition, Sapuan (2017) added a valuable step in the conceptual design stage which is preconcept generation that includes the concept generation and concept evaluation process. The conceptual design stage consumed below five percent of the total budget and around fifteen percent of total time in the whole process of product design and development (Ulrich & Eppinger, 2012). Owing to that, it can be noted that materials selection was consider as a crucial step in the conceptual design stage to ensure that the end product developed was lower in cost, shorter in time, and high in quality.

Measurement of qualitative to quantitative value in the conceptual design stage is difficult. Therefore, the suggested approach of using the MCDM method on the specification of product design as the criteria, and the availability of materials as the alternatives could provide systematic numerical data to solve the qualitative to quantitative problems face by engineers. There has been research that shows significant findings in the conceptual design generation process, especially when using the MCDM method. Davoodi et al. (2011) used the Technique for Order Preferences by Similarity to an Ideal Solution (TOPSIS) method to select the best design for a biocomposite car bumper beam with eight concepts and six criteria. (Mansor et al., 2014b) and Mastura, Sapuan, Mansor, and Nuraini (2017) utilized the AHP method to generate the best conceptual design for an automotive parking brake lever and anti-roll bar, respectively.

In MCDM, determination of the weight criteria needs to be defined as early as the process is started to ensure the smooth numerical process of ranking using the MCDM method in the conceptual design stage. There are many MCDM methods that have been used by researchers and engineers where the weight of criteria is needed which is TOPSIS, AHP, VIseKriterijumska Optimizacija I Kompromisno Resenje (VIKOR), Simple Additive Weighting (SAW) method, Multiobjective Optimization model based on the Ratio Analysis (MOORA), Complex Proportional Assessment (COPRAS), and Preference Ranking Organizations METHods for Enrichment Evaluations (PROMETHEE), which are shown in Eqs. 3.1–Eq. 3.8.

TOPSIS (Hwang & Yoon, 1981):

$$v_{ij} = w_{ij} r_{ij} \tag{3.1}$$

where v_{ij} is a weighted normalized value, w_{ij} is the weight of the i-th, and r_{ij} is the normalized value.

AHP (Saaty, 1980):

$$w = \frac{1}{n} \sum_{j=1}^{n} \frac{a_{ij}}{\sum_{i=1}^{a} a_{ij}} \tag{3.2}$$

where w is priority vector, a_{ij} is importance scale, and n is the number of criteria.

VIKOR (Opricovic & Tzeng, 2002):

$$S_i = \sum_{j=1}^{n} w_j \frac{\left(f_j^* - f_{ij}\right)}{\left(f_j^* - f_j^-\right)} \tag{3.3}$$

$$R_i = \max w_j \frac{\left(f_j^* - f_{ij}\right)}{\left(f_j^* - f_j^-\right)} \tag{3.4}$$

where w_j are the weights of criteria, S_i represents the distance rate of the i-th alternative to the positive ideal solution (best combination), R_i represents the distance rate of the i-th alternative to the negative ideal solution (worst combination), and f_{ij} is the normalized value of the matrix, where f_j^* and f_j^- are the positive ideal solution and negative ideal solution values, respectively.

SAW (Tzeng & Huang, 2011):

$$u_i(x) = \sum_{j=1}^{n} w_j r_{ij}(x) \tag{3.5}$$

where $u_i(x)$ is the utility of i-th alternative, w_j is weights of the j-th criterion, and $r_{ij}(x)$ is the normalized ratings of the i-th alternative.

MOORA (Brauers & Zavadskas, 2006):

$$y_i = \sum_{j=1}^{g} w_j r_{ij} - \sum_{j=g+1}^{n} w_j r_{ij} \tag{3.6}$$

where w_j is the compromised weight, g is the number of beneficial criteria, and n is the number of nonbeneficial criteria.

COPRAS (Zavadskas, Kaklauskas, & Sarka, 1994):

$$y_{ij} = w_j \times r_{ij} \tag{3.7}$$

where y_{ij} is the weighted normalized value, w_j is the weight of the j-th, and r_{ij} is the normalized value.

PROMETHEE (Tzeng & Huang, 2011):

$$\pi(a, b) = \sum_{j=1}^{n} w_j f_j(a, b) \tag{3.8}$$

where w_j is the criterion weight in criterion j.

The weight of criteria as shown in the MCDM method above is needed to determine either from the point of view of designers and engineers or from the systematic calculation method that can prevent the results of the materials selection being biased and based on the opinion of only one person. Therefore the systematic numerical calculation of the weight determination is suggested in this process to ensure the best final product at the end of the product design and development. The next section discusses the available research and techniques that can be used to determine the weightage of criteria in the MCDM method in the materials selection process.

3.3 Weight determination on criteria for the multiple criteria decision-making method

Eqs. 3.1–3.8 show that there are weight criteria for each criterion in the equation and there are no clear procedures for how to obtain the values. As an example, Davoodi et al. (2011) used the TOPSIS method to select the best concept design for a car bumper beam. However, they did not explain the analysis required to determine the applied weight of their criteria. In another report, Mustafa et al. (2015) applied the Weighted Decision Matrix (WDM) method to select the best reinforcement material for an automotive friction component. They also did not include any information on how the weightage value for the criteria was chosen. There are several methods that can be implemented to develop and calculate the value of the weightage criteria, and eliminate bias in the decision due to individual preferences. Ishak, Malingam, and Mansor (2017) showed the usage of an entropy method with five decision makers to calculate the weightage criteria of a fiber metal laminate for a car's front hood.

Eight types of magnesium alloys for automotive applications were analyzed by Sasanka and Ravindra (2015) where their weight of criteria were compared using three techniques, namely equal weight, nonequal weight, and entropy method. It can be seen from the results that the weightage of criteria influences the results of the ranking process analysis, and indicates the relative important of the attributes in the selection process. In other research, a compromised method has been used to determine the weight of criteria for thermoplastics materials for an automotive brake booster valve body where a combination of the entropy method and AHP method was implemented. The compromised weight of criteria then were used in the MCDM method to rank the thermoplastics materials (Moradian, Modanloo, & Aghaiee, 2019). The compromised method was used to take advantage of both methods in defining the weight of criteria. Mansor et al. (2014) used the AHP method to determine the weight of criteria, while Ishak, Malingam, and Mansor (2016) performed the fuzzy-VIKOR method to analyze the weightage criteria based on a comparison of five decision makers. In the fuzzy-VIKOR method, the linguistic variables comparison from decision makers was converted to a fuzzy set and the steps included aggregation and defuzzification to obtain the crisp value of the weightage criteria.

In another report, the principal component analysis and preference selection index were used to avoid biases in selecting the weight of materials for different design attributes which can balance the constraints (technological, ecological, economic, and societal) for sustainable automotive bodies (Roy, Tadele, Defersha, Misra, & Mohanty, 2019). Elsewhere, the Stepwise Weight Assessment Ratio Analysis (SWARA) method was also applied for weighting of the selection criteria (Yazdani, Zavadskas, Ignatius, & Abad, 2016). The SWARA method was used to determine the weightage of the selection criteria involving two case studies on material selection, involving the application for hard magnetic material and microelectromechanical systems. In can be concluded that it is important to analytically determine the weight of importance of the selection criteria prior to the ranking or prioritizing process, so that the values are more accurate in representing the decision makers' judgment and do not respond to the perception of designers as traditionally practiced.

3.4 Materials selection for green composites product development

Materials selection of natural fiber reinforced polymer composites in conceptual design can be divided into three categories which are materials selection of natural fibers, materials selection of the polymer matrix, and materials selection of the polymer composites. Each of these selection categories depends on the engineers and designers in fulfilling the requirements in the early stage of the product development process.

3.4.1 Natural fiber materials selection

In biocomposites, natural fibers act as the filler in the composite materials and will affect the mechanical and chemical properties of the composites. Natural fiber can be classified into two categories depending on the fiber length, where natural fiber with long fibers is classified as continuous fiber, while short fiber is discontinuous fiber (Rajak, Pagar, Menezes, & Linul, 2019). Selection of the natural fiber types will also affect the performance of the natural fiber composites. Therefore it is important to select the most suitable natural fiber materials to perform as needed by the design. There are various methods available in selecting the natural fiber materials for the composites, mostly using MCDM methods such as AHP, TOPSIS, and VIKOR.

Mansor et al. (2013a, 2013b) performed the selection of natural fibers using the AHP method for an automotive parking brake lever. Three criteria and four subcriteria were analyzed in this method, with 13 natural fiber alternatives showing that kenaf was the best natural fiber to be hybridized with the glass fiber for the automotive parking brake lever. The same method was used by Naveen, Jawaid, Zainudin, Sultan, and Yahaya (2018) to determine the best natural fiber among the 14 alternatives listed. Seven criteria were chosen and analyzed in the study showing that *Cocos nucifera* sheath was the most promising natural fiber that can be hybridized with Kevlar29 in designing personal body armor. In other research, Mastura, Sapuan, Mansor, and Nuraini (2016) used the integrated method of AHP and Quality Function Deployment for Environment (QFDE) in performing the natural fiber materials selection analysis. Fourteen criteria were listed in making a decision for nine alternatives of natural fibers and the results showed that sugar palm was the best natural fiber for the application of an automotive antiroll bar. Shaharuzaman, Sapuan, Mansor, and Zuhri (2019) performed the integrated method of AHP-VIKOR to select the best natural fiber for an automotive side-door impact beam. The study used the AHP method in determining the weight of criteria, while VIKOR was used as the ranking method for the natural fibers. Three criteria with seven subcriteria of natural fibers were analyzed and the results showed that kenaf was the most promising material for the natural fiber composites of an automotive side-door impact beam.

Another method used by researchers in selecting natural fiber materials is by using the fuzzy-VIKOR method. In this method, Ishak et al. (2016) converted the linguistic variables made by five decision makers for the weight of criteria and material ratings which were then analyzed by the VIKOR method for their ranking for the best natural fiber materials in the fiber metal laminate for a car front hood. Singh et al. (2020) identified the best natural fiber for pulp and paper industries using the TOPSIS method based on 21 criteria and found that acacia was the best for wood fiber materials, while it was jute for nonwood materials. Another MCDM method used by researchers is the WDM which verified that kenaf fiber is the best material as an alternative friction material to replace asbestos (Mustafa et al., 2015).

Furthermore, the natural fiber material selection process was also applied using a comprehensive Ashby-type chart (Shah, 2014). The selection made was based on a performance graph created using the Ashby-type, based on the decision makers' criteria. Elsewhere, another MCDM method used for natural fiber materials selection was by using the stepwise regression method (Noryani, Sapuan, Mastura, Zuhri, & Zainudin, 2019). This a unique work in materials selection, whereby it utilized the statistical approach to perform the prioritizing of the candidate materials. Table 3.1 summarizes several recent works in natural fiber material selection using various MCDM methods.

3.4.2 Matrix materials selection

The selection process for green composite materials also involved another important element forming the composite structure, which is the matrix material. For green polymer composites, the matrix materials can be selected based on the thermoset or thermoplastic type. Considering the vast variety of thermoset and thermoplastic matrices available in the market and developed by researchers, the use of the MCDM method truly helped to expedite the selection process in determining the best type of matrix for the intended green composites application.

TABLE 3.1 Summary of natural fiber material selection using various multicriteria decision-making (MCDM) methods.

MCDM method	Applications	References
AHP	Hand brake lever	Mansor et al. (2013a, 2013b)
AHP	Body armor	Naveen et al. (2018)
AHP and QFDE	Antiroll bar	Mastura et al. (2016)
AHP-VIKOR	Side-door impact beam	Shaharuzaman et al. (2019)
Fuzzy-VIKOR	Car front hood	Ishak et al. (2016)
TOPSIS	Pulp papermaking	Singh et al. (2020)
WDM	Friction material	Mustafa et al. (2015)
Comprehensive Ashby-type	-NIL-	Shah (2014)
Stepwise regression	Hand brake lever	Noryani et al. (2019)

AHP, Analytic hierarchy process; *QFDE*, quality function deployment for environment; *TOPSIS*, technique for order preferences by similarity to an ideal solution; *VIKOR*, VIseKriterijumska Optimizacija I Kompromisno Resenje; *WDM*, weighted decision matrix.

Thermoplastic matrix application for green composites has been steadily increasing, due to the environmental advantage of its recyclability and its low raw material cost. Thermoplastic matrix materials selection was conducted using the AHP method to produce a new sustainable parking brake lever component for a passenger car using hybrid kenaf/glass fiber reinforced polymer composites (Mansor et al., 2013a, 2013b). The results showed that the polypropylene matrix was the best candidate material for the hybrid kenaf composites. The authors also stated that PP is also one of the main thermoplastic matrix commodities used in the automotive industries currently, and has further strengthened the decision to use a similar material for achieving more sustainable design of the parking brake lever component. Furthermore, thermoplastic matrix materials section was also performed using the integration of the QFDE-AHP method for automotive hybrid composite antiroll bar application (Mastura et al., 2018). The novel selection method integrates the environmental criteria gathered through customer requirements and later applied it to the product development process. The results from the selection process revealed that thermoplastic polyurethane (PU) is the best candidate material to be applied in the hybrid sugar palm/glass fiber composites for an antiroll bar component.

Meanwhile, several MCDM methods were applied and compared for the purposes of thermoplastic materials selection in producing an automotive brake booster valve body component (Moradian et al., 2019).

The author first applied three MCDM methods, which are AHP, entrophy, and a combination AHP-entrophy method, to determine the selection criteria weightage, and later applied VIKOR, MOORA, and TOPSIS methods for ranking the process criteria. Four selection criteria (tensile strength, heat deflection temperature, density, and cost) and thirteen candidate thermoplastic materials, including their composites, were taken into consideration in the selection process. The results from the ranking stage using all three MCDM methods showed consistent findings whereby the best material for the valve body was found to be polyethylene terephthalate reinforced with 35 weight percent (wt%) glass fiber. The selected material offered advantages in terms of extremely high strength, very good dimensional stability, and very good process stability compared to the other candidate materials. In another report, the hybrid PROMETHEE and TOPSIS MCDM methods were also applied for thermoplastic matrix materials selection to produce a composite automotive bumper component (Ilangkumaran, Avenash, Balakrishnan, Kumar, & Raja, 2013). In the study, fuzzy AHP (FAHP) was integrated with PROMETHEE. FAHP was used to compute the criteria weight, while PROMETHEE II was used to find the total ranking of the candidate material. A total of five candidate materials for the automotive bumper beam were involved, which are polyethylene (PE), PP, acrylonitrile butadiene styrene (ABS), polyamide (PA), and polystyrene (PS). Furthermore, the selection process also included consideration of six evaluation criteria, which are compressive yield strength, flexural modulus, hardness, Charpy impact strength, elongation, and cost, in selecting the appropriate materials for the bumper beam application. On the other hand, matrix materials selection for a similar bumper beam application was also performed using the TOPSIS method, but focusing on the thermoset matrix (Mansor et al., 2014). Despite having less environmental advantages compared with the thermoplastic matrix, the thermoset matrix is still widely applied to produce polymer composites for high load-bearing applications. In their study, three thermoset candidate materials (namely polyester, vinyl ester, and epoxy resins) were evaluated to be used as the matrix in producing more sustainable hybrid natural fiber/synthetic fiber composites. Eight selection criteria related to bumper beam requirements were involved in the selection process, which are tensile strength, Young's modulus, elongation, compressive strength, impact strength, density, water absorption, and cost. The results from the selection process revealed that vinyl ester is the best thermoset matrix to be chosen for the intended application.

Elsewhere, the use of the MCDM method for thermoplastic materials selection was also demonstrated for producing 3D printed bipolar plates (BP) in a direct methanol fuel cells (DMFCs) application (Santiago, Raso, Navarro, & Leo, 2019). The project objective was to determine the best candidate thermoplastic materials as the BP for three different DMFC applications, which are stationary, portable (smartphones), and on-board unmanned aerial vehicles (UAVs). Six candidate materials for the BP were analyzed, these being ABS, copolyester (CPE and CPE +), nylon (NY), polycarbonate (PC), and thermoplastic polyurethane (TPU 95 A), involving seven selection criteria: cost, density, solution absorption, degradation rate, flexural strength, thickness change, and surface change. Owing to the situation of the problem which involved complex decision-making involving many criteria and candidate materials with varying attributes between them, four MCDM methods were selected to perform the judgment task: TOPSIS, SAW, COPRAS, and AHP. The AHP method was also applied in all three DMFC applications to determine the weight of the selection criteria. At the end of the ranking process, it was found that ABS was the best candidate material to produce the BP for stationary and portable applications, while PC was the best candidate material for the on-board UAV application.

Furthermore, MCDM methods were also applied for thermoplastic matrix materials selection, to produce hybrid natural fiber/glass fiber composites (Mansor et al., 2014a) (Mansor, Sapuan, Zainudin, Nuraini, & Hambali, 2014c). The hybridization between synthetic and natural fibers enabled final composites with higher load-bearing capability to be produced, while at the same time reducing the synthetic composite material environmental burden using a portion of renewable, recyclable, and biodegradable fiber. In the report, the weighted sum method (WSM) and signal-to-noise (S/N) approach were implemented to determine the best candidate material between six of the most common synthetic thermoplastic matrix materials, which are polypropylene (PP), low-density polyethylene (LDPE), high-density polyethylene, polystyrene (PS), and two polyamide variants, namely nylon 6 and nylon 6,6. The results showed that LDPE was the best thermoplastic material for the intended application based on the design requirements.

The use of the MCDM method to select the best matrix material by combining thermoset and thermoplastic candidates for new natural fiber composites for the hand brake lever design was also conducted (Noryani, Sapuan, Mastura, Zuhri, & Zainudin, 2020). In their study, a statistical approach using the stepwise regression method was used to perform the judgment process. Three mechanical properties were selected to represent the performance of the ten candidate materials, which are impact strength, elongation at break, and tensile strength. The results obtained showed that polyamide (PA6), polyurethanes (PUR), and polypropylene (PP) are the potential candidates for the manufacture of hand-brake levers. Table 3.2 summarizes several recent works in green composites matrix selection using various MCDM methods.

TABLE 3.2 Summary of green composites matrix selection using various multicriteria decision-making (MCDM) methods.

Type of matrix	MCDM method	Applications	References
Thermoplastic	AHP	Parking brake lever	Mansor et al. (2013a, 2013b)
Thermoplastic	Integrated QFDE and AHP	Automotive antiroll bar	Mastura et al. (2018)
Thermoplastic	MOORA, TOPSIS, and VIKOR	Brake booster valve body	Moradian et al. (2019)
Thermoplastic	Integrated FAHP and PROMETHEE	Automobile bumper	Ilangkumaran et al. (2013)
Thermoset	TOPSIS	Automotive bumper beam	Mansor et al. (2014)
Thermoplastic	TOPSIS, COPRAS, SAW, and AHP	Bipolar plate for DMCF	Santiago et al. (2019)
Thermoplastic	WSM and S/N	Hybrid natural fiber composites	Mansor et al. (2014c)
Thermoplastic and thermoset	Stepwise regression	Hand brake lever	Noryani et al. (2020)

AHP, Analytic hierarchy process; *COPRAS*, complex proportional assessment; *DMCF*, direct methanol fuel cell; *FAHP*, fuzzy AHP; *MOORA*, multiobjective optimization model based on the ratio analysis; *PROMETHEE*, Preference Ranking Organizations METHods for Enrichment Evaluations; *QFDE*, quality function deployment for environment; *SAW*, simple additive weighting; *S/N*, signal-to-noise; *TOPSIS*, technique for order preferences by similarity to an ideal solution; *VIKOR*, VIseKriterijumska Optimizacija I Kompromisno Resenje; *WSM*, weighted sum method.

3.4.3 Natural fiber composite materials selection

Apart from the individual element of the composites (reinforcement and matrix material), several researchers have also demonstrated the application of MCDM for final natural fiber composites selection. One of the reported examples was for automotive interior parts using the AHP method (Al-Oqla, Sapuan, Ishak, & Nuraini, 2016). In their study, the AHP method was used in both the weighting and ranking processes during the evaluation. The selection was performed involving 15 different natural fiber/polypropylene composites, and six evaluation criteria, namely tensile strength, tensile modulus, flexural strength, flexural modulus, impact strength, and the maximum water absorption of the composites. The results from the ranking process revealed that treated flax/polypropylene composite with 30 wt% fiber loading alternative was found to be the best candidate with regards to the overall evaluation criteria. Similarly, selection of the best composites for an automotive bumper beam using the AHP method was also performed

(Hambali, Sapuan, Ismail, & Nukman, 2010). In their study, six different types of composite materials were considered involving eight main selection factors and 12 subcriteria. The final result from the selection process showed that glass fiber reinforced epoxy composites at 25.7%, mass fraction is the most appropriate material compared with other materials. The author also performed sensitivity analysis to validate the consistency of the ranking produced. In another report, the AHP method was also used to select the optimum environmentally friendly natural fiber composite materials for automotive components (Ahmed Ali et al., 2015). In their study, four main criteria (namely physical, mechanical, environmental, and cost) and seven subcriteria (namely density, tensile strength, Young's modulus, biodegradability, toxic level, raw material cost, and manufacturing cost) were involved in the selection process. The candidate materials were kenaf, sisal, flax, hemp, jute, coir, and palm empty fruit bunch natural fibers, whereby all of them were reinforced with 60% PP matrix to form the final composites. The selection results obtained showed that hemp-reinforced PP composites were the best candidate materials based on the design specifications.

Apart from AHP, the use of computer-based programming solutions such as Expert System was also reported for biocomposite materials selection (Ali, Sapuan, Jawaid, & Sanyang, 2017). The Expert System is a computer application that emulates the decision-making ability of a human expert for a specific task. Similarly, a KBS computer-based approach was also employed to perform composite materials selection for an automotive pedal box (Sapuan & Abdalla, 1998). The KBS system is implemented using the rule-based reasoning method, based on the rule-based technique "If–Then." Several rules were later constructed to perform the selection process based on all the criteria identified and the succession was described as a chaining process.

Meanwhile, a statistical approach was also employed to perform selection of the best natural fiber composites (Noryani, Sapuan, Mastura, Zuhri, & Zainudin, 2018). The performance of potential materials was investigated using a statistical analysis such as the coefficient of correlation, coefficient of determination, and analysis of variance. The statistical approach was used to determine the best natural fiber composites for an automotive hand brake lever component. Another approach applied for green composite materials selection was with the integration of MCDM methods (Shaharuzaman, Sapuan, Mansor, & Zuhri, 2018). In their research works, the AHP and VIKOR methods were combined to perform natural fiber reinforced PP composite materials selection for an automotive side-door impact beam. Thirteen types of natural fiber reinforced PP composites were evaluated, based on three selection criteria, namely density, stiffness, and strength, to represent the operational requirements for the structural component. The AHP method was used to determine the

TABLE 3.3 Summary of final composite materials selection using various multicriteria decision-making (MCDM) methods.

MCDM method	Applications	References
AHP	Automotive interior parts	Al-Oqla et al. (2016)
AHP	Automotive bumper beam	Hambali et al. (2010)
AHP	Automotive components	Ahmed Ali et al. (2015)
Expert system	Automotive interior components	Ali et al. (2017)
Knowledge-based system	Automotive components	Sapuan & Abdalla (1998)
Regression model	Hand brake lever	Noryani et al. (2018)
AHP-VIKOR	Side-door impact beam	Shaharuzaman et al. (2018)

AHP, Analytic hierarchy process; *VIKOR*, VIseKriterijumska Optimizacija I Kompromisno Resenje.

weight of the criteria, while VIKOR was used to rank the performance of the candidate composites. The selection results obtained showed that kenaf-reinforced PP composites at 50 wt.% are the optimum materials to construct a side-door impact beam compared to other candidates. Table 3.3 summarizes examples of research works regarding final composite materials selection using various MCDM methods.

3.5 Case study: natural fiber composite materials selection using a TOPSIS method

To demonstrate the application of the MCDM method in performing green materials selection toward DfS, a case study on material selection involving natural fiber composites for a coffee table was conducted. Currently, the coffee table is generally made either from solid wood or steel due to its reliable strength and aesthetic appearance. Despite the advantages, they are relatively heavy and more expensive compared to plastic-based coffee tables. In this case study, a new substitute material for a coffee table which is more lightweight and cheaper was investigated, using natural fiber composites as the candidate material. Natural fiber composites have been acknowledged as promising green materials for creating more sustainable products, whereby the materials offer advantages such as high specific strength and stiffness, low density making them lighter, low cost due to the cheap and renewable plant fibers used as the reinforcement agent, as well as being able to be recycled when combined with a thermoplastic matrix to produce the final composites.

TABLE 3.4 Summary of the physical and mechanical properties of selected natural fiber reinforced polypropylene (PP) composites (Sapuan et al., 2011a).

No.	Natural fiber + polypropylene (PP)	Density (kg m^{-3})	Tensile modulus (GPa)	Tensile strength (MPa)
1	Pineapple leaf 10.8% + PP	920.52	0.687	37.28
2	Coir 40% + PP	1023	1.300	10
3	Flax 20% + PP	991	1.502	17.9
4	Hemp 40% + PP	1076	6.800	52
5	Jute 40% + PP	1036	3.700	28
6	Kenaf 30% + PP	1027	5.800	27
7	Kenaf 40% + PP	1072	6.800	29
8	Kenaf 50% + PP	1120	7.500	35
9	Kenaf 60% + PP	1174	10.100	110
10	Kenaf 85% + PP	1332	7.430	75
11	Sisal 40% + PP	1044	5.500	34
12	Vetiver grass 20% + PP	991	1.500	28
13	Vetiver grass (powder) 20% + PP	991	1.250	22

Based on the above design intent, 13 different types of natural fiber composites were selected as the potential substitute materials to produce the coffee table. The list of materials was focused on the use of a polypropylene (PP) thermoplastic matrix, due to its recyclable property, high availability as a commodity material in current biocomposites market, and also its competitive price. A list of the natural fiber reinforced PP composite candidate materials and their physical and mechanical properties is summarized in Table 3.4.

In this case study, three main performance requirements were selected to be included in the materials section process, these being the final composite density, tensile modulus, and tensile strength. These three main material properties were selected due to their direct influence on the coffee table performance and improvements, such as lighter furniture while at the same time being able to maintain the structural strength and rigidity of the product shape under static loading condition. Since all candidate materials are made from natural fiber composites, they also contribute to the design intent of having environmentally friendly characteristics in terms of recyclability and using renewable substances. These three points were later used as the selection criteria in this case study. The best candidate material for the coffee table application should have the lowest density, and highest tensile modulus and tensile strength properties.

Since the materials selection process involved multiple selection criteria as well as varying types of materials with different attributes between them, an MCDM method was the most appropriate tool to be used in executing the task of finding the best type of natural fiber reinforced PP composites to make the new coffee table. Hence, the TOPSIS method was selected to perform the task. In general, the process of materials selection using the TOPSIS method involved six major steps, as described in detail below.

Step 1: To formulate the overall TOPSIS decision matrix using Eq. 3.9

$$D = \begin{matrix} & C_1 & C_2 & \cdots & C_n \\ A_1 & X_{11} & X_{12} & \cdots & X_{1n} \\ A_2 & X_{21} & X_{22} & \cdots & X_{2n} \\ \vdots & \vdots & \vdots & \vdots & \vdots \\ A_m & X_{m1} & X_{m2} & \cdots & X_{mn} \end{matrix} \tag{3.9}$$

where $A_1, A_2, \ldots, A_n$ are potential alternatives that decision makers need to select and $C_1, C_2, \ldots, C_n$ are criteria, which evaluated the alternative performance and were calculated, X_{ij} is the rating of alternative A_i with respect to criterion C_j when w_j is the weight of criterion C_j (Davoodi et al., 2011).

Step 2: To calculate the normalized decision matrix using Eq. 3.10

$$n_{ij} = \frac{X_{ij}}{\sqrt{\sum_{j=1}^{m} X^2{}_{ij}}} \text{ where } i = 1, \ldots, m, \text{ and } j = 1, \ldots, \tag{3.10}$$

Step 3: To construct the weighted normalized decision matrix using Eq. 3.11

$$V = N_D \cdot W_{n \times n} = \begin{vmatrix} V_{1i} & \cdots & V_{1j} & \cdots & V_{1n} \\ \vdots & & \vdots & & \vdots \\ V_{m1} & \cdots & V_{mj} & \cdots & V_{mn} \end{vmatrix} \tag{3.11}$$

where w_j is the weight of the i-th attribute or criterion, and $\sum_{j=1}^{n} w_j = 1$.

Step 4: To determine the positive ideal and negative ideal solutions using Eqs. 3.12 and 3.13:

$$A^+ = \left\{ \left({}^{max}_{j} v_{ij}; i \in I \right) \left({}^{min}_{j} v_{ij}; i \in J \right); i = 1, 2, \ldots, n \right\} \tag{3.12}$$

$$A^- = \left\{ \left({}^{min}_{j} v_{ij}; i \in I \right) \left({}^{max}_{j} v_{ij}; i \in J \right); i = 1, 2, \ldots, n \right\} \tag{3.13}$$

where I is associated with a benefit criterion, and J is associated with cost criterion.

Step 5: To calculate the separation measures using the n-dimensional Euclidean distance. The separation of each alternative from the ideal solution is calculated using Eq. 3.14:

$$d_{i^+} = \left\{ \sum_{j=1}^{n} \left(v_{ij} - v_{j^+}\right)^{1/2}; \ i = 1, 2, \ldots, m \right\} \tag{3.14}$$

Similarly, the separation from the negative ideal solution is given as Eq. 3.15

$$d_{i^-} = \left\{ \sum_{j=1}^{n} \left(v_{ij} - v_{j^-}\right)^{1/2}; \ i = 1, 2, \ldots, m \right\} \tag{3.15}$$

Step 6: To determine the relative closeness to the ideal solution values for every alternative, whereby the relative closeness of the alternative A_i with respect to A^+ is determined using Eq. 3.16. The rank between the candidate materials under investigation is made by ranking the preference in decreasing order based on the indices. The candidate material having the highest relative closeness values is the recommended best candidate material for the project.

$$cl_{i^+} = \frac{d_{i^-}}{(d_{i^+} - d_{i^-})}, \ 0 \leq cl_{i^+} \leq 1; \ i = 1, 2, \ldots, m \tag{3.16}$$

3.5.1 Final materials selection results and discussion

The results of the analyzing process to determine the best natural fiber reinforced PP composites using the TOPSIS method are shown in Tables 3.5–3.8. The selection process used the material attributes as listed in Table 3.4. Furthermore, the selection criteria were given a specific weightage value to represent the relative importance between the criteria with respect to the objective of the case study.

The overall relative closeness value of the candidate materials obtained using the TOPSIS method is shown in Fig. 3.3. It can be observed that the natural fiber reinforced PP composites formulated using 60 wt% of kenaf fiber (kenaf 60% + PP) scored the highest relative closeness value of 0.90 compared to the other candidate materials, followed by kenaf 85% + PP (0.64) and hemp 40% + PP (0.51). The lowest relative closeness value was found to be for coir 40% + PP composites. The results obtained finally indicated that kenaf 60% + PP is the best substitute material to produce the new coffee table, based on the design requirements. The exercise also showed the simplicity of the TOPSIS method in providing systematic and justified judgment in a complex decision-making process to any decision makers.

TABLE 3.5 Normalized TOPSIS decision matrix for selecting the best natural fiber reinforced PP composites for the coffee table.

No.	Candidate materials	Selection criteria and weightage		
		Density 0.4	Tensile modulus 0.3	Tensile strength 0.3
1	Pineapple leaf 10.8% + PP	0.240	0.035	0.222
2	Coir 40% + PP	0.266	0.066	0.060
3	Flax 20% + PP	0.258	0.076	0.107
4	Hemp 40% + PP	0.280	0.344	0.309
5	Jute 40% + PP	0.270	0.187	0.167
6	Kenaf 30% + PP	0.267	0.293	0.161
7	Kenaf 40% + PP	0.279	0.344	0.173
8	Kenaf 50% + PP	0.291	0.379	0.208
9	Kenaf 60% + PP	0.305	0.510	0.655
10	Kenaf 85% + PP	0.347	0.375	0.446
11	Sisal 40% + PP	0.272	0.278	0.202
12	Vetiver grass 20% + PP	0.258	0.076	0.167
13	Vetiver grass(powder) 20% + PP	0.258	0.063	0.131

TABLE 3.6 Weighted normalized decision matrix.

No.	Candidate materials	Selection criteria		
		Density	Tensile modulus	Tensile strength
1	Pineapple leaf 10.8% + PP	0.096	0.010	0.067
2	Coir 40% + PP	0.106	0.020	0.018
3	Flax 20% + PP	0.103	0.023	0.032
4	Hemp 40% + PP	0.112	0.103	0.093
5	Jute 40% + PP	0.108	0.056	0.050
6	Kenaf 30% + PP	0.107	0.088	0.048
7	Kenaf 40% + PP	0.112	0.103	0.052
8	Kenaf 50% + PP	0.117	0.114	0.062
9	Kenaf 60% + PP	0.122	0.153	0.196
10	Kenaf 85% + PP	0.139	0.113	0.134
11	Sisal 40% + PP	0.109	0.083	0.061
12	Vetiver grass 20% + PP	0.103	0.023	0.050
13	Vetiver grass(powder) 20% + PP	0.103	0.019	0.039

TABLE 3.7 The positive and negative ideal solution matrix.

	Selection criteria		
Attributes	**Density**	**Tensile modulus**	**Tensile strength**
Positive ideal solution	0.096	0.153	0.196
Negative ideal solution	0.139	0.010	0.018

TABLE 3.8 Summary of separation value for each candidate materials and its relative closeness to the ideal solution.

No.	Candidate materials	Separation from positive ideal solution, d_{i^+}	Separation from negative ideal solution, d_{i^-}	Relative closeness from ideal solution, cl_{i^+}
1	Pineapple leaf 10.8% + PP	0.193	0.065	0.25
2	Coir 40% + PP	0.223	0.034	0.13
3	Flax 20% + PP	0.210	0.041	0.16
4	Hemp 40% + PP	0.116	0.122	0.51
5	Jute 40% + PP	0.176	0.064	0.27
6	Kenaf 30% + PP	0.162	0.090	0.36
7	Kenaf 40% + PP	0.153	0.103	0.40
8	Kenaf 50% + PP	0.141	0.115	0.45
9	Kenaf 60% + PP	0.026	0.229	0.90
10	Kenaf 85% + PP	0.085	0.155	0.64
11	Sisal 40% + PP	0.153	0.090	0.37
12	Vetiver grass 20% + PP	0.196	0.050	0.20
13	Vetiver grass(powder) 20% + PP	0.206	0.043	0.17

3.6 Conclusion

As discussed in this chapter, materials selection is an essential part of the DfS effort to develop products and processes which are sustainable, both in terms of environmental aspects as well as product cost and fulfilling end-user expectations. Materials selection is a complex decision-making process due to the involvement of various selection criteria

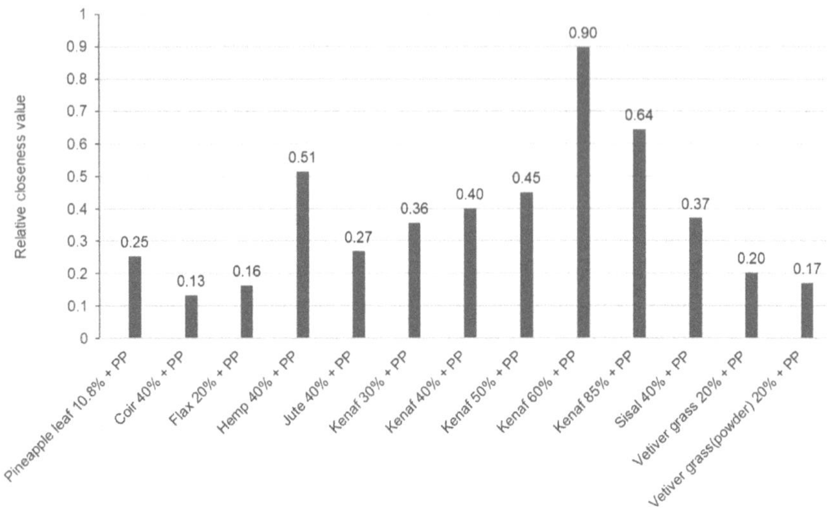

FIGURE 3.3 Overall relative closeness value of the candidate materials obtained using the Technique for Order Preferences by Similarity to an Ideal Solution method.

from different stakeholders and multiple candidate materials with varying attributes. Hence, materials selection in DfS needs to be performed based on scientific and systematic methods so that accurate judgments and decisions can be made. The use of different MCDM methods in sustainable material selection was also discussed in this chapter. Finally, a case study on green composite materials selection for sustainable design of a coffee table was included showing that the systematic approach of the MCDM method really helps engineers and designers in making justified decisions when multiple attributes with multiple criteria problems are involved.

Acknowledgment

The authors wish to express their sincere gratitude to Universiti Teknikal Malaysia Melaka, Universiti Putra Malaysia, and Ministry of Higher Education Malaysia for their kind support provided throughout the preparation of this chapter.

References

Ahmed Ali, B. A., Sapuan, S. M., Zainudin, E. S., & Othman, M. (2015). Implementation of the expert decision system for environmental assessment in composite materials selection for automotive components. *Journal of Cleaner Production, 107*, 557–567. Available from https://doi.org/10.1016/j.jclepro.2015.05.084.

Ali, B. A. A., Sapuan, S. M., Jawaid, M., & Sanyang, M. L. (2017). Expert material selection for manufacturing of green bio composites. In M. Jawaid, S. M. Sapuan, & Othman Y. Alothman (Eds.), *Green energy and technology green biocomposites: manufacturing and properties* (pp. 1–12). Switzerland: Springer International Publishing AG. Available from https://doi.org/10.1007/978-3-319-46610-1.

Al-Oqla, F. M., Sapuan, S. M., Ishak, M. R., & Nuraini, A. A. (2016). A decision-making model for selecting the most appropriate natural fiber-Polypropylene-based composites for automotive applications. *Journal of Composite Materials, 50*(4), 543–556. Available from https://doi.org/10.1177/0021998315577233.

Brauers, W. K. M., & Zavadskas, E. K. (2006). The MOORA method and its application to privatization in a transition economy. *Control and Cybernetics, 35*(2), 445–469.

CES Selector, (2013). Granta design. UK, version 2013.

Clark, G., Kosoris, J., Hong, L. N., & Crul, M. (2009). Design for sustainability: Current trends in sustainable product design and development. *Sustainability, 1*(3), 409–424. Available from https://doi.org/10.3390/su1030409.

Crul, M. R. M., & Diehl, J. C. (2005). *Design for sustainability: a practical approach for developing economies.* Retrieved from <http://www.d4s-de.org/manual/d4stotalmanual.pdf>.

Davoodi, M. M., Sapuan, S. M., Ahmad, D., Aidy, A., Khalina, A., & Jonoobi, M. (2011). Concept selection of car bumper beam with developed hybrid bio-composite material. *Materials and Design, 32*(10), 4857–4865. Available from https://doi.org/10.1016/j.matdes.2011.06.011.

Hambali, A., Sapuan, S. M., Ismail, N., & Nukman, Y. (2010). Material selection of polymeric composite automotive bumper beam using analytical hierarchy process. *Journal of Central South University of Technology, 17*(2), 244–256. Available from https://doi.org/10.1007/s11771-010-0038-y.

Hwang, C.-L., & Yoon, K. (1981). In M. Beckmann, & H. P. Kunzi (Eds.), *Multiple attribute decision making: methods and application.* New York, Berlin Heidelberg: Springer-Verlag.

Ilangkumaran, M., Avenash, A., Balakrishnan, V., Kumar, S. B., & Raja, M. B. (2013). Material selection using hybrid MCDM approach for automobile bumper. *International Journal of Industrial and Systems Engineering, 14*(1), 20–39. Available from https://doi.org/10.1504/IJISE.2013.052919.

Ishak, N. M., Malingam, S. D., & Mansor, M. R. (2016). Selection of natural fibre reinforced composites using fuzzy VIKOR for car front hood. *International Journal of Materials and Product Technology, 53*(3/4), 267–285. Available from https://doi.org/10.1504/IJMPT.2016.079205.

Ishak, N. M., Malingam, S. D., & Mansor, M. R. (2017). Weighting of natural fibre criteria for fibre metal laminate using entropy method for car front hood utilization. *Key Engineering Materials, 740*, 75–80. Available from https://doi.org/10.4028/www.scientific.net/KEM.740.75.

Mansor, M. R., Hambali, A., Azaman, M. D., Sapuan, S. M., Zainudin, E. S., & Nuraini, A. A. (2013a). Material selection of thermoplastic matrix for hybrid natural fiber/glass fiber polymer composites using analytical hierarchy process method. International Symposium on the Analytic Hierarchy Process 2013, 1–8.

Mansor, M. R., Sapuan, S. M., Zainudin, E. S., Nuraini, A. A., & Hambali, A. (2013b). Hybrid natural and glass fibers reinforced polymer composites material selection using analytical hierarchy process for automotive brake lever design. *Materials and Design, 51*, 484–492. Available from https://doi.org/10.1016/j.matdes.2013.04.072.

Mansor, M. R., Sapuan, S. M., Hambali, A., Zainudin, E. S., & Nuraini, A. A. (2014). Materials selection of hybrid bio-composites thermoset matrix for automotive bumper beam application using TOPSIS method. *Advances in Environmental Biology, 8*(Spec. Issue 3), 3138–3142.

Mansor, M. R., Sapuan, S. M., Zainudin, E. S., Nuraini, A. A., & Hambali, A. (2014a). Application of integrated AHP-TOPSIS method in hybrid natural fiber composites materials selection for automotive parking brake lever component. *Australian Journal of Basic and Applied Sciences, 8*(5), 431–439.

Mansor, M. R., Sapuan, S. M., Zainudin, E. S., Nuraini, A. A., & Hambali, A. (2014b). Conceptual design of kenaf fiber polymer composite automotive parking brake lever using integrated TRIZ-morphological chart-analytic hierarchy process method. *Materials and Design, 54*, 473–482. Available from https://doi.org/10.1016/j.matdes.2013.08.064.

Mansor, M. R., Sapuan, S. M., Salim, M. A., Akop, M. Z., Musthafah, M. T., & Shaharuzaman, M. A. (2016). Concurrent design of green composites. In D. Verma, S. Jain, X. Zhang, & P. C. Gope (Eds.), *Green Approaches to Biocomposite Materials Science and Engineering* (pp. 48–75). Hershey, PA: IGI Global. Available from https://doi.org/10.4018/978-1-5225-0424-5.ch003.

Mansor, M. R., Sapuan, S. M., Zainudin, E. S., Nuraini, A. A., & Hambali, A. A. (2014c). Thermoplastic matrix material selection using multi criteria decision making method for hybrid polymer composites. *Applied Mechanics and Materials, 564*(October), 439–443. Available from https://doi.org/10.4028/www.scientific.net/AMM.564.439.

Manzini, E. (2020). Design for sustainability. Available from https://doi.org/10.1201/b19041-14.

Mastura, M. T., Sapuan, S. M., Mansor, M. R., & Nuraini, A. A. (2016). Environmentally conscious hybrid bio-composite material selection for automotive anti-roll bar. *The International Journal of Advanced Manufacturing Technology, 89*(5–8), 2203–2219. Available from https://doi.org/10.1007/s00170-016-9217-9.

Mastura, M. T., Sapuan, S. M., Mansor, M. R., & Nuraini, A. A. (2017). Conceptual design of a natural fibre-reinforced composite automotive anti-roll bar using a hybrid approach. *International Journal of Advanced Manufacturing Technology, 91*(5–8), 2031–2048. Available from https://doi.org/10.1007/s00170-016-9882-8.

Mastura, M. T., Sapuan, S. M., Mansor, M. R., & Nuraini, A. A. (2018). Materials selection of thermoplastic matrices for 'green' natural fibre composites for automotive anti-roll bar with particular emphasis on the environment. *International Journal of Precision Engineering and Manufacturing-Green Technology, 5*(1), 111–119.

Moradian, M., Modanloo, V., & Aghaiee, S. (2019). Comparative analysis of multi criteria decision making techniques for material selection of brake booster valve body. *Journal of Traffic and Transportation Engineering (English Edition), 6*(5), 526–534. Available from https://doi.org/10.1016/j.jtte.2018.02.001.

Morini, A. A., Ribeiro, M. J., & Hotza, D. (2019). Early-stage materials selection based on embodied energy and carbon footprint. *Materials and Design, 178*, 107861. Available from https://doi.org/10.1016/j.matdes.2019.107861.

Mustafa, A., Abdollah, M. F., Shuhimi, F. F., Ismail, N., Amiruddin, H., & Umehara, N. (2015). Selection and verification of kenaf fibres as an alternative friction material using Weighted Decision Matrix method. *Materials and Design, 67*, 577–582. Available from https://doi.org/10.1016/j.matdes.2014.10.091.

Mustafa, A., Abdollah, M. F. B., Ismail, N., Amiruddin, H., & Umehara, N. (2014). Materials selection for eco-aware lightweight friction material. *Mechanics & Industry, 15*(4), 279–285. Available from https://doi.org/10.1051/meca/2014039.

Naveen, J., Jawaid, M., Zainudin, E. S., Sultan, M. T. H., & Yahaya, R. B. (2018). Selection of natural fiber for hybrid kevlar/natural fiber reinforced polymer composites for personal body armor by using analytical hierarchy process. *Frontiers in Materials, 5*(August), 1–13. Available from https://doi.org/10.3389/fmats.2018.00052.

Noryani, M., Sapuan, S. M., Mastura, M. T., Zuhri, M. Y. M., & Zainudin, E. S. (2018). A Statistical framework for selecting natural fibre reinforced polymer composites based on regression model. *Fibers and Polymers, 19*(5), 1039–1049. Available from https://doi.org/10.1007/s12221-018-8113-3.

Noryani, M., Sapuan, S. M., Mastura, M. T., Zuhri, M. Y. M., & Zainudin, E. S. (2019). Material selection of natural fibre using a stepwise regression model with error analysis. *Journal of Materials Research and Technology, 8*(3), 2865–2879. Available from https://doi.org/10.1016/j.jmrt.2019.02.019.

Noryani, M., Sapuan, S. M., Mastura, M. T., Zuhri, M. Y. M., & Zainudin, E. S. (2020). Statistical inferences in material selection of a polymer matrix for natural fiber composites. *Polimery, 65*(02), 105–114. Available from https://doi.org/10.14314/polimery.2020.2.4.

Opricovic, S., & Tzeng, G. H. (2002). Multicriteria planning of post-earthquake sustainable reconstruction. *Computer-Aided Civil and Infrastructure Engineering, 17*(3), 211–220. Available from https://doi.org/10.1111/1467-8667.00269.

Pugh, S. (1991). *Total design: integrated method for successful product engineering* (1st ed.). Essex, England: Addison-Wesley Publishers Ltd.

Rajak, D. K., Pagar, D. D., Menezes, P. L., & Linul, E. (2019). Fiber-reinforced polymer composites: Manufacturing, properties, and applications. *Polymers, 11*(10). Available from https://doi.org/10.3390/polym11101667.

Roy, P., Tadele, D., Defersha, F., Misra, M., & Mohanty, A. K. (2019). Environmental and economic prospects of biomaterials in the automotive industry. *Clean Technologies and Environmental Policy, 21*(8), 1535–1548. Available from https://doi.org/10.1007/s10098-019-01735-8.

Saaty, T. L. (1980). *The analytic hierarchy process*. New York: McGraw-Hill.

Santiago, Ó., Raso, M. A., Navarro, E., & Leo, T. J. (2019). Selection of thermoplastic polymers for use as bipolar plates in direct methanol fuel cell applications. *Materials & Design, 183*, 108148. Available from https://doi.org/10.1016/j.matdes.2019.108148.

Sapuan, S. (2001). A knowledge-based system for materials selection in mechanical engineering design. *Materials & Design, 22*(8), 687–695. Available from https://doi.org/10.1016/S0261-3069(00)00108-4.

Sapuan, S. M. (2017). *Conceptual design in concurrent engineering for composites. Composite materials: concurrent engineering approach* (1st ed., pp. 141–207). Oxford: Butterworth-Heinemann. Available from https://doi.org/10.1016/B978-0-12-802507-9/00005-2.

Sapuan, S. M., & Abdalla, H. S. (1998). A prototype knowledge-based system for the material selection of polymeric-based composites for automotive components. *Composites Part A: Applied Science and Manufacturing, 29*(7), 731–742. Available from https://doi.org/10.1016/S1359-835X(98)00049-9.

Sapuan, S. M., & Mansor, M. R. (2014). Concurrent engineering approach in the development of composite products: A review. *Materials & Design, 58*, 161–167. Available from https://doi.org/10.1016/j.matdes.2014.01.059.

Sapuan, S. M., Kho, J. Y., Zainudin, E. S., Leman, Z., Ahmed Ali, B. A., & Hambali, A. (2011a). Materials selection for natural fiber reinforced polymer composites using analytical hierarchy process. *Indian Journal of Engineering & Materials Sciences, 18*(August), 255–267. Available from http://nopr.niscair.res.in/handle/123456789/12892.

Sapuan, S. M., Mun, N. K., Hambali, A., Lok, H. Y., Fairuz, A. M., & Ishak, M. R. (2011b). Prototype expert system for material selection of polymeric composite automotive dashboard. *International Journal of the Physical Sciences, 6*(25), 5988–5995. Available from https://doi.org/10.5897/IJPS11.939.

Sasanka, C. T., & Ravindra, K. (2015). Implementation of VIKOR method for selection of magnesium alloy to suit automotive applications. *International Journal of Advanced Science and Technology, 83*, 49–58.

Shah, D. U. (2013). Developing plant fibre composites for structural applications by optimising composite parameters: A critical review. *Journal of Materials Science, 48*(18), 6083–6107. Available from https://doi.org/10.1007/s10853-013-7458-7.

Shah, D. U. (2014). Natural fibre composites: Comprehensive Ashby-type materials selection charts. *Materials and Design, 62*, 21–31. Available from https://doi.org/10.1016/j.matdes.2014.05.002.

Shaharuzaman, M.A., Sapuan, S.M., Mansor, M.R., & Zuhri, M.Y.M. (2018). Materials selection of natural fiber/polypropylene composites for automotive side door impact beam using integrated AHP-VIKOR method. In *6th Postgraduate Seminar on Natural*

Fiber Reinforced Polymer Composites 2018 (pp. 8–11). Selangor: Institute of Tropical Forestry and Forest Product (INTROP).

Shaharuzaman, M. A., Sapuan, S. M., Mansor, M. R., & Zuhri, M. Y. M. (2019). Decision support strategy in selecting natural fiber materials for automotive side-door impact beam composites. *Journal of Renewable Materials, 7*(10).

Singh, M., Pant, M., Godiyal, R. D., & Kumar Sharma, A. (2020). MCDM approach for selection of raw material in pulp and papermaking industry. *Materials and Manufacturing Processes, 35*(3), 241–249. Available from https://doi.org/10.1080/10426914.2020.1711917.

Singh, T., Patnaik, A., Chauhan, R., & Chauhan, P. (2018). Selection of brake friction materials using hybrid analytical hierarchy process and vise Kriterijumska Optimizacija Kompromisno Resenje approach. *Polymer Composites, 39*(5), 1655–1662. Available from https://doi.org/10.1002/pc.24113.

Tzeng, G.-H., & Huang, J. J. (2011). *Multiple attribute decision making: methods and application* (1st ed.). FL: CRC Press.

Ulrich, K. T., & Eppinger, S. D. (2012). *Product design and development* (5th ed.). New York: McGraw-Hill/Irwin.

Yazdani, M., Zavadskas, E. K., Ignatius, J., & Abad, M. D. (2016). Sensitivity analysis in MADM methods: Application of material selection. *Engineering Economics, 27*(4), 382–391.

Zavadskas, E. K., Kaklauskas, A., & Sarka, V. (1994). The new method of multicriteria complex proportional assessment of projects. *Technological and Economic Development of Economy, 1*(3), 131–139.

CHAPTER

4

Implementation of design for sustainability in developing trophy plaque using green kenaf polymer composites

S.M. Sapuan[1,2], *G. Hemapriya*[1], *R.A. Ilyas*[3,4], *M.S.N. Atikah*[5], *M.R.M. Asyraf*[6] *and Muhd Ridzuan Mansor*[7]

[1]Advanced Engineering Materials and Composites Research Centre (AEMC), Department of Mechanical and Manufacturing Engineering, Faculty of Engineering, Universiti Putra Malaysia, Selangor, Malaysia, [2]Laboratory of Biocomposite Technology, Institute of Tropical Forestry and Forest Products (INTROP), Universiti Putra Malaysia, Selangor, Malaysia, [3]School of Chemical and Energy Engineering, Faculty of Engineering, Universiti Teknologi Malaysia, Johor, Malaysia, [4]Centre for Advanced Composite Materials, Universiti Teknologi Malaysia, Johor, Malaysia, [5]Department of Chemical and Environmental Engineering, Universiti Putra Malaysia, Serdang, Selangor, Malaysia, [6]Department of Aerospace Engineering, Universiti Putra Malaysia, Selangor, Malaysia, [7]Faculty of Mechanical Engineering, Universiti Teknikal Malaysia Melaka, Melaka, Malaysia

4.1 Introduction

There is a need to maintain a healthy and environmentally sustainable raw material as experts predict renewable energy sources, primarily petroleum supplies, will run out as soon as 2050 (Atikah et al., 2019;

Design for Sustainability
DOI: https://doi.org/10.1016/B978-0-12-819482-9.00013-7

Ilyas, Sapuan, Ibrahim, Abral, et al., 2019; Nishino, Hirao, Kotera, Nakamae, & Inagaki, 2003). Finding alternatives to petroleum is crucial for global energy use to meet the current demand of increasing technology (Norizan, Abdan, Ilyas, & Biofibers, 2020; Nurazzi, Khalina, Sapuan, & Ilyas, 2019a; Nurazzi et al., 2019b). In addition, depletion of other natural resources, for instance coal, wood, and iron, is also rapidly accelerating. The need for renewable resources has awakened the consciences of governments, nongovernmental organizations and also the public following warnings about the impact of global warming (Jumaidin, Ilyas, Saiful, Hussin, & Mastura, 2019a; Jumaidin et al., 2019b; Jumaidin, Khiruddin, Asyul Sutan Saidi, Salit, & Ilyas, 2020). The situation has led both graduate engineers and university researchers to search for a solution in order to substitute petroleum-based materials, and this has succeeded in many areas. Researchers have found that natural fibers are a suitable substitute for petroleum-based plastic materials (Bachtiar, Sapuan, & Hamdan, 2008; Halimatul, Sapuan, Jawaid, Ishak, & Ilyas, 2019; Ilyas, Sapuan, Atiqah, et al., 2019; Ilyas, Sapuan, & Ishak, 2018; Ilyas, Sapuan, Ishak, & Zainudin, 2019; Ilyas, Sapuan, Ishak, & Zainudin, 2018a; Ilyas, Sapuan, Ishak, & Zainudin, 2018b; Ilyas, Sapuan, Ishak, & Zainudin, 2018c; Jaafar, Rizal, & Zainol, 2018a; Jaafar, Zainol, & Rizal, 2018b).

Design for sustainability (DfS) is an approach toward achieving better environmentally friendly performances of products and processes. Furthermore, DfS also aims for the well-being of living things on the planet. As products and processes are developed for the consumer to gain profitability, the DfS approach also caters to the need for achieving a balance between profit, human wellness, and the environment. Among the elements recommended in DfS for products and processes development are optimized use of resources, use of renewable resources, and opting for solutions which are can reduce energy consumption (Crul, 2006). In terms of renewable resource usage for developing new products and processes, natural fiber composites are one of the promising alternative which strongly support that aim.

Natural fibers originate from many parts of plants or animals, and they have been traditionally used to make textiles, paper, cloth, or rope (Atiqah, Jawaid, Sapuan, Ishak, & Alothman, 2018; Ilyas, Sapuan, Ishak, Zainudin, & Atikah, 2018; Ilyas, Sapuan, Ibrahim, Atikah, et al., 2019; Ilyas et al., 2021; Sapuan et al., 2018; Shahroze et al., 2019). The main advantage of natural fibers is that they, together with other synthetic polymers/biopolymers, can produce lightweight composites and inexpensive products compared to glass fibers (Asyraf, Ishak, Sapuan, Yidris, & Ilyas, 2020; Asyraf, Ishak, Sapuan, Yidris, Shahroze, et al., 2020; Asyraf, Ishak, Razman, & Chandrasekar, 2019a; Ashraf et al., 2019b; Jaafar et al., 2018a, 2018b; Johari et al., 2019). Therefore natural fibers can be better than some synthetic fibers in terms of cost, and lightweight with the help of research and development. Moreover, the biodegradability of natural fibers solves the issues of overfilling landfills because when burned they do not produce toxic chemicals; this is particularly true for natural fiber reinforced biopolymer composites (Abral et al., 2020a, 2020b; Nazrin et al., 2020; Syafri et al., 2019). Natural fibers also are much cheaper than E-glass and synthetic fibers (Ilyas & Sapuan, 2020). Natural fiber composites have a respectable acoustic damping value compared to E-glass or carbon fiber composites and therefore can be used for noise attenuation, which is an increasingly important design necessity for automotive interior components. Moreover, natural fibers have high mechanical properties, especially in resisting creep rupture (Asyraf, Ishak, Sapuan, Yidris, Ilyas, et al., 2020; Jaafar et al., 2018a, 2018b) and creep loading (Asyraf et al., 2019a, 2019b) after being treated with specific chemicals and the inclusion of additives such as nanosilica and nanoclay.

Kenaf has existed for over 4000 years, with its roots traced back to Africa. The kenaf plant is part of the family of hibiscus (*Hibiscus cannabinus L.*) (Mazani, Sapuan, Sanyang, Atiqah, & Ilyas, 2019), and it is closely related to the jute, cotton, and okra family. It is grown in India, Bangladesh, Malaysia, the United States, Indonesia, Vietnam, Thailand, South Africa and other parts of Africa, and some parts of south-eastern Europe. Kenaf grows rapidly in the United States to a height of about 3.5–4.5 m in a short time, while in Malaysia it grows to a height of 1.5–3.5 m in a short time (Jaafar et al., 2018a). Kenaf has recently been recognized in Malaysia as an important natural raw fiber in the manufacture of many goods in the construction, automotive, and textile industries (Akil et al., 2011). Many researchers have also reviewed the potential for kenaf fiber to be used as reinforcements in polymer composites. In terms of its mechanical properties, kenaf reinforced composites were found to have inferior strength and modulus, which hinders their use in applications that require good strength. Kenaf fiber also exhibits good water resistance.

Unsaturated polyester (UP) resin is a resin that is very simple and the most inexpensive system of resins that is the fastest to process and produce products. Each year thousands of gallons of this material are used all over the world. They are formed by transposing through polymerizing various idols (alcohols) and dibasic acids (e.g., malefic anhydride or numeric acid) to produce esters, a liquid that is very high in viscosity, which is then melted into a reactive monomer, styrene. Styrene decreases a liquid's viscosity to a specific level for impregnation or plastic molding. The uses of UPs include transport for heavy industries (large body parts for cars, buses, trailers, trucks), marine industries (minor and major casings of boats and additional marine machinery), construction (panels, baths and shower shells), and other equipment (Hashmi, 2017).

Prior to the manufacturing process, a product design specification (PDS) is crucial to assisting a researcher or developer in ensuring that the product's requirements match the project maturity. A design is usually required to start with a basic framework that contains the basic design information and it is helpful in leading the entire development process (Asyraf, Ishak, Sapuan, Yidris, Ilyas, et al., 2020). The PDS is essential for the success of product, as it has the necessary information about the final product. Therefore all areas relevant to the product should be included in this document which can only be modified if all members of the design team agree.

Many new emerging environmental regulations have instigated the use of natural fibers. Nevertheless, most plaques available in the market are made from plastics, metal, a variety of woods, and other materials, with basic and simple designs. These materials are generally quite heavy and not environmentally friendly. In addition, consumers demand the plaque design to be smarter, more fashionable, and flexible. Hence, natural fiber reinforced composite materials might be promising candidates for the fabrication of a plaque that is stronger, lighter, and environmentally friendly. Therefore this chapter focuses on designing a composite plaque and its fabrication using a kenaf fiber reinforced UP composite using the DfS approach.

4.2 Product development of kenaf composite trophy plaque using DfS approach

The main objective of this project was to produce a new trophy plaque product using kenaf composites using the DfS approach. The project outcome is intended to be an alternative to current products in the market, as well as demonstrating the application of green composite materials to users and manufacturers in the merchandise market. The scope of the design is to produce a kenaf composites trophy plaque for use by the authors' institution, which is Universiti Putra Malaysia. As described in the previous section, the use of kenaf composites as an alternative material to produce a trophy plaque can help to improve the product's environmentally friendly and lightweight properties, while retaining the required structural strength so that it can function as desired. The overall kenaf composites plaque development process for this project involved market investigation, creating PDSs, developing a plaque conceptual design, detailed design including plaque mold, manufacturing of the plaque mold and plaque component, and finally product assembly. Details for each process are given in the subsequent sections.

4.2.1 Product background and market investigation

The word "plaque" originated from a word meaning ornamental plate or tablet, thus attaching importance to its appearance. It is a product which represents an award given in commemoration of a person or event. The award is also a means of evidence to giving credit to the person or group who have accomplished a certain degree of excellence. A plaque is usually a plate made from metal, ceramic, stone, wood, etc. that bears text. Newer designs of plaque design also integrate a trophy feature to add more aesthetic appeal to the product.

Based on the Pugh total design method, the new product development process starts with market analysis (Pugh, 1996). In order to develop and provide a high-quality product, all relevant information about the planned product design is gathered during market analysis, such as customer requirements, existing product features, and product cost (Sapuan, 2014). In this project, gathering of information was conducted using web browsing on plaque manufacturers and informal interviews with users that had experience dealing with the product. One notable finding from the market analysis performed was that, up to date, a similar trophy plaque product has not yet been produced using NFC material, in particular using kenaf composites. Furthermore, there are very few trophy plaque designs available in the market with an engineering element. For this project in particular, the authors' institution (Universiti Putra Malaysia) did not have any trophy plaque products using similar materials.

Hence, there is a potential new market segment to be explored using kenaf composites for this product, especially using the DfS approach. Apart from being environmentally friendly and lightweight, kenaf composites can also be produced with low production cost and consume less energy during manufacturing. Moreover, adding an engineering element into the product feature may also attract customer acceptance of the new kenaf composites trophy plaque when sold in the market.

4.2.2 Product design specification

In general, the PDS is a document created based on the information obtained from the market analysis conducted during the product development process. It details the requirements that must be met for the product to be well accepted by the customer. There are, in general, 32 elements listed in a PDS (Mazani et al., 2019; Pugh, 1991). Among them, nine elements were identified for implementation in the design of the kenaf composites trophy plaque as shown in Fig. 4.1.

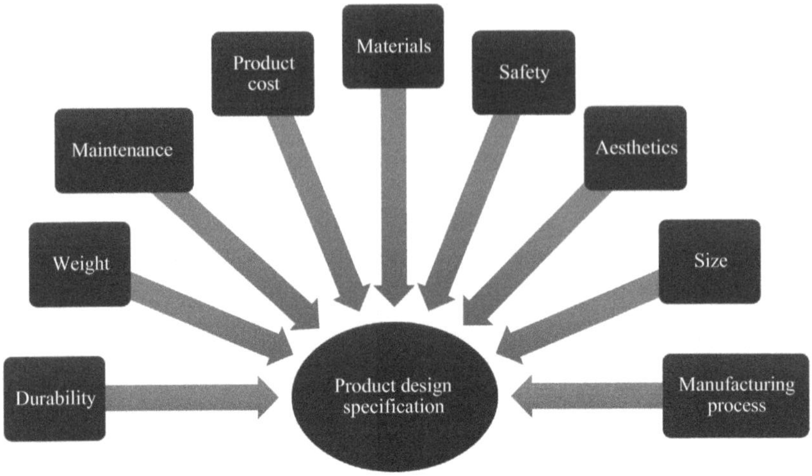

FIGURE 4.1 Elements of product design specification considered for the kenaf composites trophy plaque.

Among the identified PDS is durability performance. Durability of kenaf composites trophy plaque performance must be comparable to conventional plaques made of wood or plastics. Quality can be seen when plaques are proved to have an equal longevity with other plaques. The weight and size of the plaques are other factors that must be taken into account. Such elements constitute the key factors when consumers consider buying a plaque. The material costs and manufacturing cost must also be kept to a minimum to ensure the product cost remains competitive.

Safety is another important aspect for the PDS for any merchandise. The kenaf composites trophy plaque must be free from sharp edges to prevent harming the user while handling and should not hurt the user if it were to fall. The product must have high aesthetic value, such as being highly attractive and with unique design as well as having a smooth surface finish in order to be marketable. The fabrication process must be considered in order to produce and assemble all the components. This dimension influences the design outcome and the manufacturing capacity. Details of the kenaf composites trophy plaque PDS are summarized in Table 4.1.

4.2.3 Conceptual design development of kenaf composites trophy plaque

Based on the generated PDS, the next stage in kenaf composites trophy plaque production is developing a new product conceptual design

TABLE 4.1 Description of kenaf composites trophy plaque product design specification.

PDS elements	Description
1. Durability	The plaque must be long lasting because it is a memorable item that recognizes a person for his/her achievement (at least 25 years)
2. Weight	The plaque weight must not exceed 500 g
3. Maintenance	The plaque must be maintenance free
4. Product cost	The total cost of the plaque must not exceed RM 50
5. Material	The plaque shall be made from kenaf fiber reinforced unsaturated polyester composites
6. Safety	The plaque must have a smooth surface. To avoid unwanted accidents, any sharp edges must be trimmed. The joints must be closely hidden to prevent it from protruding
7. Aesthetics	The plaque must be presentable, as well as having high aesthetic value (highly attractive and unique design) and quality end finishing
8. Size	The maximum height of the plaque must be 254 mm and width must not exceed 208 mm
9. Manufacturing process	Easy to manufacture. Able to be fabricated using existing fabrication techniques available either manually or using an automated molding process

RM, Malaysian Ringgit (currency used in Malaysia).

and later selecting the best conceptual design from the ideas generated. Details for each activity are explained below.

4.2.3.1 Generating a conceptual design using the brainstorming method

There are many idea generation methods for developing the conceptual design of products. One of the most simple and widely applied methods is the brainstorming method. The design brainstorming process was carried out through a design focus group, with members from the Advanced Engineering Materials and Composites Research Centre, Department Mechanical and Manufacturing Engineering, Universiti Putra Malaysia. All conceptual design ideas generated were based on the PDS outlined in the previous design stage. At the end of the brainstorming process, seven new conceptual designs for a kenaf composites trophy plaque were developed. Descriptions for each conceptual design are given in Table 4.2. In general, all conceptual designs have the name of the institution displayed on the trophy plaque exterior surface. Creative and innovative variations were made to add value to the ideas

TABLE 4.2 New conceptual designs of the kenaf composites trophy plaque and its description.

Conceptual design		Description
1.	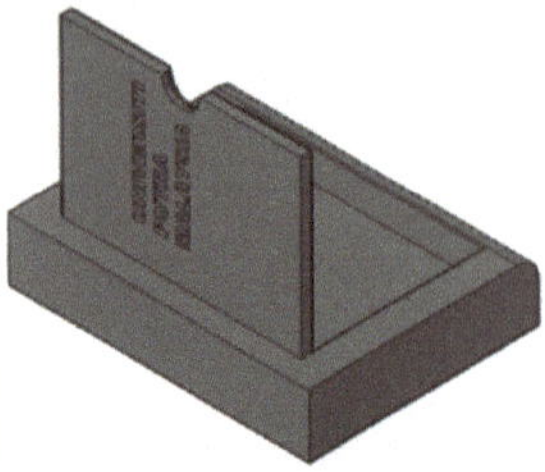	High strength. High stability due to wider baseline. Concept inspired by book design. Easy to manufacture. Has a cover feature to protect plaque content
2.	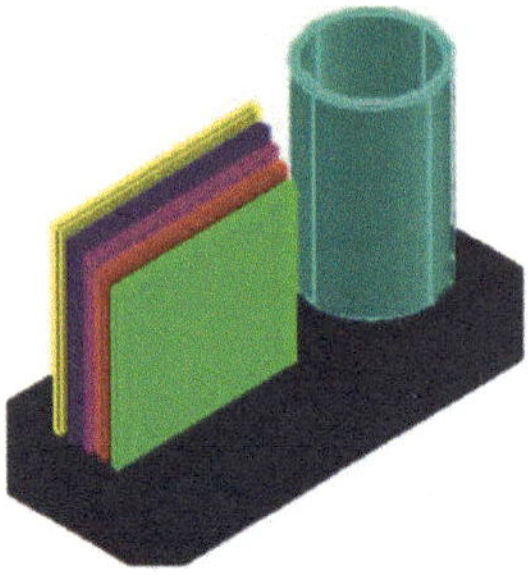	Good aesthetic structure. High stability and strength. Cylinder holder for placing stationery. Able to hold cards or papers in stacks. Multifunctional
3.		Concept inspired by a landmark built in Engineering Faculty of Universiti Putra Malaysia. The "AE" wording indicates "Agricultural Engineering"
4.		Concept inspired by a spring (a spring has dynamic features and is a mechanical energy-saving solution)
5.		Concept inspired by a spring (a spring has dynamic features and is a mechanical energy-saving solution). Two cylindrical holders to place stationery (i.e., pens, pencils, etc.)

(Continued)

TABLE 4.2 (Continued)

Conceptual design		Description
6	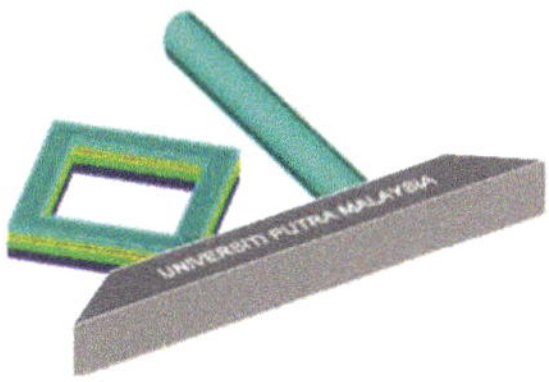	A rectangular photo frame feature (user can display small pictures). Long rod for aesthetics
7.		Concept based on several engineering nuts joined together (joining component) to represent unity and strength. Isometric design, hence easy to fabricate

in addition to being manufactured using kenaf composites. All solutions were also created using DfS recommendations which are to optimize the use of material to gain the desired functionality.

4.2.3.2 Conceptual design selection using the Pugh selection method

In this stage, the best concept design among the seven alternative ideas for the kenaf composites trophy plaque was performed using the Pugh concept selection method. In this method, all conceptual designs are evaluated based on all the selection criteria gathered from the PDS. It should noted that the PDS for material is excluded from the conceptual design selection criteria since all ideas used similar types of material. Performance scoring for each design was made by comparing the performance to the datum design (existing product with similar function selected as a reference for comparison). In this project, the datum selected was a trophy plaque made from plastic material. Furthermore, performance scoring are given based on three criteria, which are " + " for performance better than datum, "−" for worse performance than datum, and "S" for equal/similar performance as datum. The best concept design was finally selected based on the highest net score (total of " + " ratings minus total of "−" ratings). The Pugh selection method was selected due to its ease of use, and ability to perform simultaneous and systematic decision-making processes involving problems with multiple selection criteria and multiple alternative solutions. Table 4.3

TABLE 4.3 Summary of conceptual design selection using the Pugh selection method.

Selection criteria	Conceptual design of kenaf composites trophy plaque							
	Existing	1	2	3	4	5	6	7
			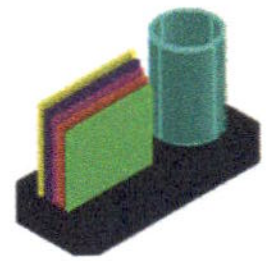			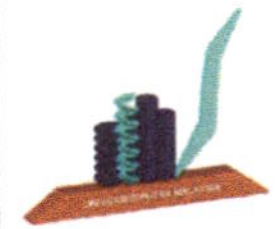	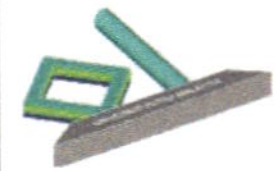	
Durability	DATUM	+	S	S	S	S	S	S
Weight		–	–	S	S	S	+	+
Maintenance		–	S	S	S	S	S	S
Product cost		–	–	+	–	–	+	+
Safety		+	S	S	–	–	–	+
Aesthetics		–	+	+	+	+	+	+
Size		–	–	S	S	–	–	S
Manufacturing process (easy to manufacture)		–	–	S	–	–	–	S
Total score	$\sum +$	2	1	2	1	1	3	4
	$\sum -$	6	4	0	3	4	3	0
	$\sum S$	0	3	6	4	3	2	2
Nett Score	$\sum + (-) \sum -$	0	–3	2	–2	–3	0	4

FIGURE 4.2 Best conceptual design selected for the kenaf composites trophy plaque.

shows the selection of the best design concepts using the Pugh selection method.

Based on Table 4.3, it can be observed that conceptual design no. 7 (engineering nut concept) scored the highest net score of +4 compared with the remainder of the design candidates. Concept design no. 7 had better performance in four aspects of the selection criteria, which were weight, cost, safety, and aesthetics, compared with the existing trophy plaque product. Meanwhile, minimum ratings were obtained for conceptual design nos. 3 and 5. Hence, conceptual design no. 7 was used as the final design for the kenaf composites trophy plaque, as shown in Fig. 4.2.

There are several features included in conceptual design no. 7 which make it preferable compared to the rest of the concept design candidates evaluated, especially with regards to weight, cost, safety, and aesthetics. The design was developed in a symmetrical manner, which makes the design much simpler compared to a nonsymmetrical shape. The symmetrical feature also helps to reduce the complexity of the mold required to produce the kenaf composites trophy plaque, which in return reduced the manufacturing cost of the product. Furthermore, the upper part design of the product (which consists of multiple engineering nuts) also has through-holes features. These through-holes helps to reduce the weight of the product by removing areas which did not contribute to supporting the load without compromising on its strength and stiffness. The through-holes also function as a gripping point for the user when handling the product, making it easier to carry. The upper part of the trophy plaque also has similar all-round fillet edges as in conventional engineering nut design, hence eliminating sharp edges and improving the product safety when use. All these features aligned with another design approach, called Design for Excellence (DfX). In DfX, designers are encouraged to incorporated elements such as design for ease of assembly, design for safety, design for manufacturing, and ergonomics in their conceptual design solution. Implementation of DfX in the conceptual design stage of green product development helps to improve the end-product quality and customer acceptance as well as eliminate design errors as early as possible (Jack, 2013).

4.2.4 Materials selection

The materials selected for producing the trophy plaque are kenaf reinforced UP composites. Kenaf is among the class of plant type fibers with good mechanical properties, low density, and low raw material cost. Kenaf resources are also abundant. The use of kenaf fiber is highly desirable to produce green products as in the case of a trophy plaque. Its environmentally friendly properties, such as being recyclable and renewable, aligned successfully with the recommended element in DfS, which prefers to reduce the use of nonrenewable resources. In terms of matrix selection, UP resin is used due to matching structural properties with the low load-bearing requirement of the product, as well as the lower raw material cost compared to other types of thermoset material.

Furthermore, environmentally friendly poly(lactic) acid (PLA) resin is also used to produce the kenaf composites mold for manufacturing the end-product. PLA is a biodegradable type of resin, which safely decomposes in the environment. The use of environmentally friendly PLA material to fabricate the product mold also successfully aligned with the DfS approach selected for this project.

Kenaf fiber in powder form was used to reinforce the thermoset matrix. Kenaf was initially purchased in roving form, then cut and ground into powder. Meanwhile, the liquid UP resin used in this project was purchased from Multifilla (M) Sdn. Bhd, Balakong, Selangor, Malaysia, and the PLA-based 3D-print filament (commercial name PolyLite™ PLA) was purchased from RS Components Sdn. Bhd., Kuala Lumpur, Malaysia. Table 4.4 summarizes the selected material properties of kenaf fiber, UP resin, and PLA resin used in this project.

TABLE 4.4 Material properties of kenaf fiber, unsaturated polyester resin, and poly(lactic) acid resin used to fabricate the kenaf composites trophy plaque.

Material properties	Kenaf fiber	Unsaturated polyester	Polylactic acid
Density (g cm^{-3})	1.4	1.12	1.24
Young's modulus (MPa)	53,000	3100	2636
Tensile strength (MPa)	930	62	46.6
Elongation at break (%)	1.6	2.2	11
Melting temperature (°C)	–	280	150

4.2.5 Detailed design of kenaf composites trophy plaque

The next activity performed in DfS of the kenaf composites is detailed design. In this stage, the necessary information to fabricate the final conceptual design is defined. Among the information added are the product dimensions, tolerance, and bill of materials. The information is conveyed using engineering drawings, modeled in either 3D or 2D views. In this project, a commercial computer-aided design software (AutoCAD) was used to create the engineering drawings for the kenaf composites trophy plaque final design. Similar software was also used to develop the mold design to fabricate the product.

4.2.6 Kenaf composites trophy plaque fabrication

After the detailed design stage, the final activity performed is manufacture of the kenaf composites trophy plaque. The fabrication process is divided into two procedures, which are mold fabrication and end-product fabrication.

4.2.6.1 Mold fabrication

Fig. 4.3 shows the kenaf composites trophy plaque manufacturing process sequence, starting from the mold design, followed by the pre-molding process and the final manual pressing process. A compression molding process was used to produce the final product. The mould was designed in two halves for ease of molding operation. Part 1 in Fig. 4.3A shows the lower fixed mold half which was fixed in place during molding and part 2 in Fig. 4.3B shows the upper removable mold half. Fig. 4.3C shows the condition when pressing is completed. A mold release agent was applied to the mold surface prior to the molding process so that the end-product could be easily removed.

A 3D printing manufacturing method was used to produce the mold, since the size of the mold was small it was compatible to be printed using a 3D printer (Ultimaker 2 + , Ultimaker, Netherlands). PLA was used as

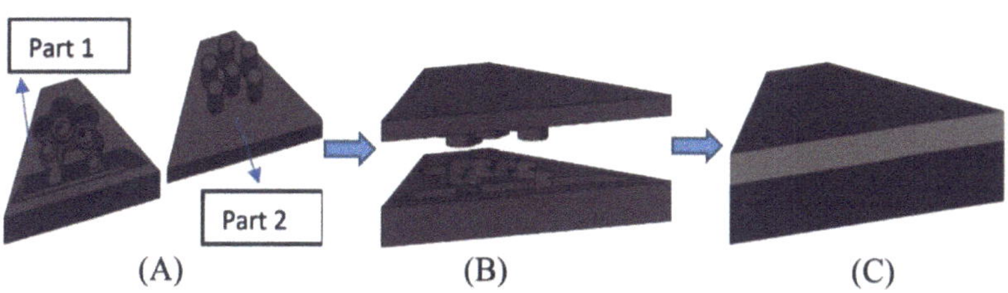

FIGURE 4.3 (A) Mold of composite plaque; (B) premolding process; and (C) manual pressing.

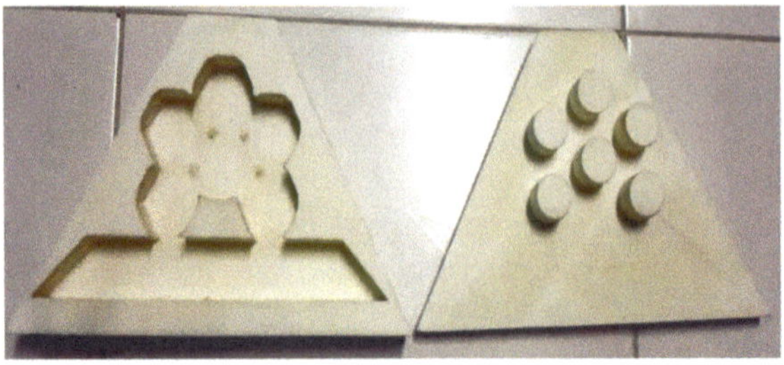

FIGURE 4.4 Poly(lactic) acid-based trophy plaque product.

TABLE 4.5 Required amounts of unsaturated polyester, methyl ethyl ketone peroxide catalyst, and kenaf fiber.

Item	Weight
Volume of composite mold	$= 21 \times 13 \times 1.2$ cm
	$= 327.6\ \text{cm}^3$
60% neat unsaturated polyester	$= 0.6 \times 327.6\ \text{cm}^3 \times 1.12\ (\text{g cm}^{-3})$
	$= 220.15$ g
2% MEKP catalyst	$= 0.02 \times 327.6\ \text{cm}^3 \times 1.55\ (\text{g cm}^{-3})$
	$= 10.16$ g
40% kenaf powder	$= 0.40 \times 327.6\ \text{cm}^3 \times 1.4\ (\text{g cm}^{-3})$
	$= 183.46$ g
Total weight	$= 220.15 + 10.16 + 183.46$ g
	$= 413.77$ g

the filament for the 3D printer. The use of a 3D printing process together with PLA-based filament was selected due to its suitability for a low-quantity production process which requires less energy consumption and a low amount of waste material generated. Furthermore, PLA material can be easily disposed through a natural decomposition process at the end of its life cycle. Both features strongly adhere to the DfS approach of developing green products. Fig. 4.4 shows the fabricated mold for the trophy plaque.

4.2.6.2 Composites trophy plaque fabrication

As for the composites trophy plaque fabrication, neat UP resin with methyl ethyl ketone peroxide (MEKP) as a catalyst were used to bind kenaf fiber. Unsaturated UP, kenaf fiber, and MEKP catalyst were manually mixed based on the composition shown in Table 4.5. The MEKP

FIGURE 4.5 (A) Location of the initial trophy plaque section and the added platform, (B) final assembly of the kenaf composites trophy plaque.

catalyst used was 2% of the resin quantity. Both chemicals were measured to a suitable quantity and poured into a well-cleaned container. The solution was gently agitated to prevent air bubbles from forming. This agitation was carried out so that there was no porosity in the final product.

UP resin and MEKP catalyst were mixed in a separate container. Then, around 40% powdered kenaf fiber was also prepared in a small container. The resin mix and powdered kenaf fiber were premixed before the mixture was inserted into the mold. The mixture was then inserted carefully into part 1 of the mold as shown in Fig. 4.4 while monitoring for the formation of bubbles in the mixture. Then, the upper removable mold half was placed on top of the lower fixed mold half and manually pressed. Following that, the composite was put under room temperature to be cured for 24 hours. Subsequently, the composite was removed from the mold. Sandpaper was later used to remove uneven and excess edges for suitable surface finishing for the composite end-product.

Furthermore, to provide higher stability for the trophy plaque when in use, an additional base component (the platform) was fabricated using a similar type of kenaf reinforced UP composites. Once the platform was fabricated, it was joined together with the initial composites trophy plaque using a simple adhesive. Finally, the whole assembly was coated using spray paint to give the desired aesthetic appeal for the end-product as shown in Fig. 4.5.

4.3 Conclusions

The main objective of this project was to produce a new trophy plaque product using kenaf composites using the DfS approach, as outlined in this chapter. The kenaf composites trophy plaque was aimed at being an alternative green solution to current products in the market. Among the activities performed throughout the product development process based on the DfS approach were market investigation, creating the PDS, developing the plaque conceptual design, detailed design including plaque mold, manufacturing of the plaque mold and plaque component, and finally product assembly. The success of the DfS approach implemented in this project is hoped to spur greater interest from product manufacturers in developing more sustainable products in the future.

Acknowledgments

The authors would like to thank the Ministry of Higher Education, Malaysia, for financial support through HICoE Grant Scheme (vote number 6369107) and Fundamental Research Grant Scheme project number FRGS/1/ 2017/TK05/UPM/01/1 (vote number 5540048).

References

Abral, H., Ariksa, J., Mahardika, M., Handayani, D., Aminah, I., Sandrawati, N., ... Ilyas, R. A. (2020a). Transparent and antimicrobial cellulose film from ginger nanofiber. *Food Hydrocolloids, 98*, 105266. Available from: https://doi.org/10.1016/j.foodhyd.2019.105266.

Abral, H., Ariksa, J., Mahardika, M., Handayani, D., Aminah, I., Sandrawati, N., ... Ilyas, R. A. (2020b). Highly transparent and antimicrobial PVA based bionanocomposites reinforced by ginger nanofiber. *Polymer Testing, 81*, 106186, *October*. Available from: https://doi.org/10.1016/j.polymertesting.2019.106186.

Akil, H. M., Omar, M. F., Mazuki, A. A. M., Safiee, S., Ishak, Z. A. M., & Abu Bakar, A. (2011). Kenaf fiber reinforced composites: A review. *Materials and Design, 32*(8–9), 4107–4121. Available from: https://doi.org/10.1016/j.matdes.2011.04.008.

Ashraf, W., Ishak, M.R., Zuhri, M.Y.M., Yidris, N., Yaacob, A.M.B., & Asyraf, M.R.M. (2019b). Investigation of different facesheet materials on compression properties of honeycomb sandwich composite. In *Proceedings of Seminar Enau Kebangsaan 2019b*, Bahau, Negeri Sembilan, Malaysia, 1–2 April, 129–132.

Asyraf, M. R. M., Ishak, M. R., Razman, M. R., & Chandrasekar, M. (2019a). Fundamentals of creep, testing methods and development of test rig for the full-scale crossarm: A review. *Jurnal Teknologi, 81*(4), 155–164.

Asyraf, M. R. M., Ishak, M. R., Sapuan, S. M., Yidris, N., Ilyas, R. A., Rafidah, M., & Razman, M. R. (2020). Evaluation of design and simulation of creep test rig for full-scale cross arm structure. *Advances in Civil Engineering, 2020*. Available from: https://doi.org/10.1155/2019/6980918.

Asyraf, M.R.M., Ishak, M.R., Sapuan, S.M., Yidris, N., Johari, A.N., Ashraf, W., ... Mazlan, R. (2019b). Creep test rig for full-scale composite crossarm: simulation modelling and analysis. In Proceedings of Seminar Enau Kebangsaan 2019, Bahau, Negeri Sembilan, Malaysia, 1–2 April, 34–38.

Asyraf, M. R. M., Ishak, M. R., Sapuan, S. M., Yidris, N., Shahroze, R. M., Johari, A. N., ... Ilyas, R. A. (2020). Creep test rig for cantilever beam: Fundamentals, prospects and present views. *Journal of Mechanical Engineering and Sciences, 14*(2), 6869–6887.

Asyraf, M. R. M. R. M., Ishak, M. R. R., Sapuan, S. M. M., Yidris, N., & Ilyas, R. A. (2020). Woods and composites cantilever beam: A comprehensive review of experimental and numerical creep methodologies. *Journal of Materials Research and Technology, January*. Available from: https://doi.org/10.1016/j.jmrt.2020.01.013.

Atikah, M. S. N., Ilyas, R. A., Sapuan, S. M., Ishak, M. R., Zainudin, E. S., Ibrahim, R., ... Jumaidin, R. (2019). Degradation and physical properties of sugar palm starch/sugar palm nanofibrillated cellulose bionanocomposite. *Polimery, 64*(10), 27–36. Available from: https://doi.org/10.14314/polimery.2019.10.5.

Atiqah, A., Jawaid, M., Sapuan, S. M., Ishak, M. R., & Alothman, O. Y. (2018). Thermal properties of sugar palm/glass fiber reinforced thermoplastic polyurethane hybrid composites. *Composite Structures, 202*. Available from: https://doi.org/10.1016/j.compstruct.2018.05.009.

Bachtiar, D., Sapuan, S. M. M., & Hamdan, M. M. M. (2008). The effect of alkaline treatment on tensile properties of sugar palm fibre reinforced epoxy composites. *Materials & Design, 29*(7), 1285–1290. Available from: https://doi.org/10.1016/j.matdes.2007.09.006.

Crul, M. R. M. (2006). *Design for sustainability: A practical approach for developing economies*. UNEP/Earthprint.

Halimatul, M. J., Sapuan, S. M., Jawaid, M., Ishak, M. R., & Ilyas, R. A. (2019). Water absorption and water solubility properties of sago starch biopolymer composite films filled with sugar palm particles. *Polimery, 64*(9), 27–35. Available from: https://doi.org/10.14314/polimery.2019.9.4.

Hashmi, S. (2017). In S. Hashmi (Ed.), *Comprehensive materials finishing* (1st ed.). Elsevier.

Ilyas, R. A., & Sapuan, S. M. (2020). The preparation methods and processing of natural fibre bio-polymer composites. *Current Organic Synthesis, 16*(8), 1068–1070. Available from: https://doi.org/10.2174/157017941608200120105616.

Ilyas, R. A., Sapuan, S. M., Atiqah, A., Ibrahim, R., Abral, H., Ishak, M. R., … Ya, H. (2019). Sugar palm (*Arenga pinnata [Wurmb.] Merr*) starch films containing sugar palm nanofibrillated cellulose as reinforcement: Water barrier properties. *Polymer Composites*, 1–9. Available from: https://doi.org/10.1002/pc.25379.

Ilyas, R. A., Sapuan, S. M., Ibrahim, R., Abral, H., Ishak, M. R., Zainudin, E. S., … Jumaidin, R. (2019). Effect of sugar palm nanofibrillated cellulose concentrations on morphological, mechanical and physical properties of biodegradable films based on agro-waste sugar palm (*Arenga pinnata* (Wurmb.) Merr) starch. *Journal of Materials Research and Technology, 8*(5), 4819–4830. Available from: https://doi.org/10.1016/j.jmrt.2019.08.028.

Ilyas, R. A., Sapuan, S. M., Ibrahim, R., Atikah, M. S. N., Atiqah, A., Ansari, M. N. M., & Norrrahim, M. N. F. (2019). *Production, processes and modification of nanocrystalline cellulose from agro-waste: A review. Nanocrystalline Materials* (pp. 3–32). IntechOpen. Available from: https://doi.org/10.5772/intechopen.87001.

Ilyas, R. A., Sapuan, S. M., & Ishak, M. R. (2018). Isolation and characterization of nanocrystalline cellulose from sugar palm fibres (*Arenga pinnata*). *Carbohydrate Polymers, 181*, 1038–1051. Available from: https://doi.org/10.1016/j.carbpol.2017.11.045.

Ilyas, R. A., Sapuan, S. M., Ishak, M. R., & Zainudin, E. S. (2018a). Water transport properties of bio-nanocomposites reinforced by sugar palm (*Arenga pinnata*) nanofibrillated cellulose. *Journal of Advanced Research in Fluid Mechanics and Thermal Sciences Journal, 51* (2), 234–246.

Ilyas, R. A., Sapuan, S. M., Ishak, M. R., & Zainudin, E. S. (2018b). Development and characterization of sugar palm nanocrystalline cellulose reinforced sugar palm starch bionanocomposites. *Carbohydrate Polymers, 202*, 186–202. Available from: https://doi.org/10.1016/j.carbpol.2018.09.002.

Ilyas, R. A., Sapuan, S. M., Ishak, M. R., & Zainudin, E. S. (2018c). Sugar palm nanocrystalline cellulose reinforced sugar palm starch composite: Degradation and water-barrier properties. *IOP Conference Series: Materials Science and Engineering, 368*(1). Available from: https://doi.org/10.1088/1757-899X/368/1/012006.

Ilyas, R. A., Sapuan, S. M., Ishak, M. R., & Zainudin, E. S. (2019). Sugar palm nanofibrillated cellulose (*Arenga pinnata* (Wurmb.) Merr): Effect of cycles on their yield, physic-chemical, morphological and thermal behavior. *International Journal of Biological Macromolecules, 123*, 379–388. Available from: https://doi.org/10.1016/j.ijbiomac.2018.11.124.

Ilyas, R. A., Sapuan, S. M., Ishak, M. R., Zainudin, E. S., & Atikah, M. S. N. (2018). *Characterization of sugar palm nanocellulose and its potential for reinforcement with a starch-based composite. Sugar Palm Biofibers, Biopolymers, and Biocomposites* (pp. 189–220). CRC Press. Available from: https://doi.org/10.1201/9780429443923-10.

Ilyas, R. A., Sapuan, S. M., Norrrahim, M. N. F., Yasim-Anuar, T. A. T., Kadier, A., Kalil, M. S., … Asyraf, M. R. M. (2021). Nanocellulose/starch biopolymer nanocomposites: Processing, manufacturing, and applications. In F. M. Al-Oqla (Ed.), *Advanced processing, properties, and application of strach and other bio-based polymer* (1st ed.). Elsevier Inc. Available from: https://doi.org/10.1016/B978-0-12-189661-8.00006-8.

Jaafar, C. N. A., Rizal, M. A. M., & Zainol, I. (2018a). Effect of kenaf alkalization treatment on morphological and mechanical properties of epoxy/silica/kenaf composite. *International Journal of Engineering and Technology*, *7*, 258–263. Available from: https://doi.org/10.14419/ijet.v7i4.35.22743.

Jaafar, C.N.A., Zainol, I., & Rizal, M.A.M. (2018b). Preparation and characterisation of epoxy / silica / kenaf composite using hand lay-up method. In *27th Scientific Conference of the Microscopy Society Malaysia (27th SCMSM 2018)*, 2–6.

Jack, H. (2013). *Universal design topics*. In *Engineering design, planning, and management* (1st ed., pp. 290–323). Academic Press.

Johari, A. N., Ishak, M. R., Leman, Z., Yusoff, M. Z. M., Asyraf, M. R. M., Ashraf, W., & Sharaf, H. K. (2019). Fabrication and cut-in speed enhancement of savonius vertical axis wind turbine (SVAWT) with hinged blade using fiberglass composites. *Seminar Enau Kebangsaan*, *1997*, 978–983.

Jumaidin, R., Ilyas, R. A., Saiful, M., Hussin, F., & Mastura, M. T. (2019a). Water transport and physical properties of sugarcane bagasse fibre reinforced thermoplastic potato starch biocomposite. *Journal of Advanced Research in Fluid Mechanics and Thermal Sciences*, *61*(2), 273–281.

Jumaidin, R., Khiruddin, M. A. A., Asyul Sutan Saidi, Z., Salit, M. S., & Ilyas, R. A. (2020). Effect of cogon grass fibre on the thermal, mechanical and biodegradation properties of thermoplastic cassava starch biocomposite. *International Journal of Biological Macromolecules*, *146*, 746–755. Available from: https://doi.org/10.1016/j.ijbiomac.2019.11.011.

Jumaidin, R., Saidi, Z. A. S., Ilyas, R. A., Ahmad, M. N., Wahid, M. K., Yaakob, M. Y., … Osman, M. H. (2019b). Characteristics of cogon grass fibre reinforced thermoplastic cassava starch biocomposite: Water absorption and physical properties. *Journal of Advanced Research in Fluid Mechanics and Thermal Sciences*, *62*(1), 43–52.

Mazani, N., Sapuan, S. M., Sanyang, M. L., Atiqah, A., & Ilyas, R. A. (2019). *Design and fabrication of a shoe shelf from kenaf fiber reinforced unsaturated polyester composites. Lignocellulose for future bioeconomy* (pp. 315–332). Elsevier Issue 2000. Available from: https://doi.org/10.1016/B978-0-12-816354-2.00017-7.

Nazrin, A., Sapuan, S. M., Zuhri, M. Y. M., Ilyas, R. A., Syafiq, R., & Sherwani, S. F. K. (2020). Nanocellulose reinforced thermoplastic starch (TPS), polylactic acid (PLA), and polybutylene succinate (PBS) for food packaging applications. *Frontiers in Chemistry*, *8* (213), 1–12. Available from: https://doi.org/10.3389/fchem.2020.00213.

Nishino, T., Hirao, K., Kotera, M., Nakamae, K., & Inagaki, H. (2003). Kenaf reinforced biodegradable composite. *Composites Science and Technology*, *63*(9), 1281–1286. Available from: https://doi.org/10.1016/S0266-3538(03)00099-X.

Norizan, M. N., Abdan, K., Ilyas, R. A., & Biofibers, S. P. (2020). Effect of fiber orientation and fiber loading on the mechanical and thermal properties of sugar palm yarn fiber reinforced unsaturated polyester resin composites. *Polimery*, *65*(2), 34–43. Available from: https://doi.org/10.14314/polimery.2020.2.5.

Nurazzi, N. M., Khalina, A., Sapuan, S. M., & Ilyas, R. A. (2019a). Mechanical properties of sugar palm yarn / woven glass fiber reinforced unsaturated polyester composites:

Effect of fiber loadings and alkaline treatment. *Polimery, 64*(10), 12–22. Available from: https://doi.org/10.14314/polimery.2019.10.3.

Nurazzi, N. M., Khalina, A., Sapuan, S. M., Ilyas, R. A., Rafiqah, S. A., & Hanafee, Z. M. (2019b). Thermal properties of treated sugar palm yarn/glass fiber reinforced unsaturated polyester hybrid composites. *Journal of Materials Research and Technology*. Available from: https://doi.org/10.1016/j.jmrt.2019.11.086.

Pugh, S. (1991). Total design: Integrated methods for successful product engineering. *Quality and Reliability Engineering International, 7*(2), 119–119. Available from: https://doi.org/10.1002/qre.4680070210.

Pugh, S. (1996). Concept selection–A method that works. In D. Clausing, & R. Andrade (Eds.), *Creating innovative products using total design*. Addison-Wesley Publishing Company.

Sapuan, S. M. (2014). *Tropical natural fibre composites*. Singapore: Springer. Available from: https://doi.org/10.1007/978-981-287-155-8.

Sapuan, S. M., Ilyas, R. A., Ishak, M. R., Leman, Z., Huzaifah, M. R. M., Ammar, I. M., & Atikah, M. S. N. (2018). *Development of sugar palm-based products: A community project. Sugar palm biofibers, biopolymers, and biocomposites* (1st ed., pp. 245–266). CRC Press. Available from: https://doi.org/10.1201/9780429443923-12.

Shahroze, R. M., Chandrasekar, M., Senthilkumar, K., Senthilmuthukumar, T., Ishak, M. R., & Asyraf, M. R. M. (2019). A review on the various fibre treatment techniques used for the fibre surface modification of the sugar palm fibres. *Seminar Enau Kebangsaan, 1*, 48–52.

Syafri, E., Sudirman., Mashadi., Yulianti, E., Deswita., Asrofi, M., ... Fudholi, A. (2019). Effect of sonication time on the thermal stability, moisture absorption, and biodegradation of water hyacinth (*Eichhornia crassipes*) nanocellulose-filled bengkuang (*Pachyrhizus erosus*) starch biocomposites. *Journal of Materials Research and Technology, 8*(6), 6223–6231. Available from: https://doi.org/10.1016/j.jmrt.2019.10.016.

CHAPTER

5

Dynamic analysis of a laminated rubber–metal spring vibration isolator for sustainable design

Mohd Azli Salim[1,2], *Azma Putra*[2], *Mohammed Hussin Ahmed Al-Mola*[3], *Norbazlan Mohd Yusof*[4] *and Muhd Ridzuan Mansor*[2]

[1]Advanced Manufacturing Centre, Universiti Teknikal Malaysia Melaka, Melaka, Malaysia, [2]Faculty of Mechanical Engineering, Universiti Teknikal Malaysia Melaka, Melaka, Malaysia, [3]Faculty of Petroleum and Mining Engineering, University of Mosul, Mosul, Iraq, [4]Centre of Excellence, Projek Lebuhraya Usahasama Berhad, Menara Korporat, Selangor, Malaysia

5.1 Introduction

The concept of good product development from green materials and concept of eco-design were introduced in the early 1990s. Many researchers and scientists stated that both concepts are highly viable strategies to reduce environmental impacts while supporting sustainable product manufacturing processes for many applications, such as building and construction, automotives, and consumer goods. Both concepts were later integrated into a more holistic approach toward sustainable design called design for sustainability (DfS) (Caïd, 2006; Clark, Kosoris, Hong, & Crul, 2009).

There are many strategies that have been developed to implement the DfS approach in sustainable product development, especially for building and construction applications. Among these are the use of nonhazardous and environmentally friendly (renewable, recyclable,

Design for Sustainability
DOI: https://doi.org/10.1016/B978-0-12-819482-9.00018-6

biodegradable) building materials (Claramunt, Fernández-Carrasco, Ventura, & Ardanuy, 2016; Taiwo, Yahya, & Haron, 2019), optimizing the building material contents and reducing waste materials (Ahmad, Chen, Haque, & Shah, 2020; Hossain & Ng, 2019), as well as reducing energy usage during building construction and operation processes (Moran, O'Connell, & Goggins, 2020). Current reports show that these strategies are also being widely adapted as early as during the building design process through the aid of efficient and user-friendly applications such as life cycle assessment (LCA) and building information modeling (BIM) methods (Liu, Zhang, Guo, Osmani, & Demian, 2019; Röck, Hollberg, Habert, & Passer, 2018; Santos, Costa, Silvestre, & Pyl, 2019). Their applications during the building design stage help to support the decision-making process in assessing optimum construction options and the environmental impact, such as materials selection for different building elements. In return, the ecological advantages gained through implementation of DfS strategies also help to reduce the building construction, operation, and demolition costs throughout the life cycle phases (Carvalho, Bragança, & Mateus, 2019; Fregonara, Giordano, Rolando, & Tulliani, 2016).

One area of interest in building and construction applications from a DfS perspective is the development of a sustainable vibration isolator, particularly made from a rubber-based material. In general, vibration isolators have been applied for vibration isolation to reduce vibration transmission and are widely used in high-rise buildings, heavy bridges, engine vehicle mountings, aerospace, aeronautical, as well as for sensitive instruments including for earthquake protection (Chang, 2002; Dickens, 2000; Ibrahim, 2008). Rubber material has the advantage of being used as a vibration isolator because it has very high damping to ensure sufficient dissipation of vibration energy from the seismic waves. Mordini and Strauss (2008) showed that the strength of a high damping isolator can be further improved by adding some glass fiber fabrics into the rubber blocks.

Further advancement on laminated rubber bearings has also been made to dissipate the vibration energy coming from earthquakes. The embedded metal plates are intended to protect the rubber blocks of the conventional isolator from an excessive bulging effect which otherwise reduces the performance of the isolator as well as its durability (Chang, 2002; Xi et al., 2003). Kelly and Konstantinidis (2011) reported that the idea to develop multilayer reinforcing rubber blocks borne by thin steel plates first came from a French engineer called Eugene Freyssinet. This invention was continued in the 1960s at the Malaysian Rubber Producer's Research Association. This type of isolator is known as a laminated rubber spring.

Fuller, Gough, Pond, and Ahmadi (1997) further reported that the use of rubber compounds in the laminated rubber bearing showed reliable long-term stiffness stability after 40 years of service as a seismic isolator.

As it is designed to work well for earthquake protection, the rubber bearing is predominantly intended to provide isolation when subjected to a shear force at very low frequency (Dickens, 2000). For other applications such as for engine mountings, isolation of buildings from ground-borne vibration near railway lines or train vehicle suspensions, where the input force comes in a vertical direction and can extend up to higher frequencies, there appears to be a lack of discussion for this situation. Investigation into the possibility of using the laminated rubber spring for these applications is therefore of interest.

A laminated rubber spring vibrator isolator consists of several features. Metal plate stacks are added between the rubber rod using a vulcanization process. The purpose of the metal plates is to minimize the bulging effect of the rubber rod during a large excitation force applied at the top of the isolator as well as to divide the isolator to increase the number of degrees-of-freedom (DOF). In addition, it is also used to increase the vertical stiffness value of the rubber isolator (Salim et al., 2018).

In this chapter, a dynamic analysis of a laminated rubber–metal spring (LR-MS) vibrator isolator using analytical method is presented. The aim of this chapter is to evaluate the characteristics of internal resonance in a finite rod (representing the LR-MS model) for one and multi-DOF systems in longitudinal, rotational, torsional, and bending moment conditions. In this method, the internal resonance of the rubber rod was identified using a wave propagation method. It is an important contribution from this study as the influence of internal resonance inside the isolator has not yet been implemented by other researchers. Furthermore, the mathematical model developed using a wave propagation method in this study also included the application of dynamic forces on the top surface of the vibration isolator. Later, the dynamic stiffness matrix for the rubber isolator was derived by adopting the equation from internal resonance. Next, the model of rubber isolator measuring from one- to multi-DOF was developed. The results obtained were finally plotted into a transmissibility graph, followed by producing a validity model.

The construction of the LR-MS mathematical modeling conducted in this study is aimed at speeding up the change from a conservative design and development of vibration isolator toward a more sustainable development process based on the DfS approach. A conservative method to develop LR-MS usually involved an intense experimentation effort to determine the material optimum composition using a highly pressurized process up to 500 MPa, which also required careful handling to ensure safety during sample production. The highly pressurized process also required high energy usage. The destructive type of experimentation process used often resulted in many vibration isolator samples being wasted before the optimum LR-MS composition was obtained. Unfortunately, these samples cannot be recycled for reuse purposes because of the sulfur, paraffin, and Santoflex added to strengthen the LR-MS composition. Hence, the use of an innovative mathematical modeling method to develop the LR-MS vibration isolator as demonstrated in this study helps to predict the optimum composition of LR-MS based on the customer-specific design intent. Subsequently, a similar mathematical model can also be used to study the behavior of the LR-MS vibration isolator at many operating conditions without the need for intense experimentation. The advantages provided by the analytical method resulted in a more sustainable design and development process of the LR-MS vibration isolator in terms of optimum use of material, lowering the material waste, as well saving valuable energy during sample production. The analytical method also helps to speed up the LR-MS development process by reducing the number of experiments and experimentation time required, which in return significantly lowered the overall product cost.

5.2 Mathematical modeling method of a laminated rubber–metal spring

5.2.1 Internal resonance in the longitudinal direction

In recent years, there has been an increasing interest in longitudinal vibration. To date, there has been little agreement on the longitudinal wave motion in an elastic body whose length is much greater than its cross-sectional dimension. A finite rod is an example of this kind of elastic body. Most studies in longitudinal waves have only been carried out in the longitudinal direction and neglected any lateral contraction on expansion of the rod. According to this theory, it is an approximation based only on the strength of materials theory. Several studies have produced an approximation theory and the results can be concluded as very good if the wavelengths of the motion are long compared with the rod's cross-sectional dimension. In addition, plane cross-sections of the rod remain as a plane and parallel during deformation by the wave propagation method.

In the construction sector, the purpose of the green standards is to speed up the transition from conventional design and construction of buildings and structures toward sustainable development that adheres to the green principles:

- Safety and wholesome favorable conditions of human life;
- Curbing negative impacts on the environment;
- Taking the interests of future generations into account.

Basically, for the first step in developing a mathematical expression of internal resonance, a nondispersive finite rod is considered as a uniform finite rod having length L subjected to a propagating longitudinal force as shown in Fig. 5.1.

$$u(x,t) = U(x)e^{j\omega t} = \left(Ae^{jk_l x} + Be^{-jk_l x}\right)e^{j\omega t} \tag{5.1}$$

where A_l and B_l are the complex wave amplitudes and $k_l = \omega/c_l$ is the longitudinal wavenumber where $c_l = \sqrt{E(1+j\eta)/\rho}$ is the longitudinal wave speed with E the Young's modulus and ρ the density. Note that damping has been introduced in the Young's modulus E where η is the damping loss factor. The impedance at a particular location can be obtained by unrestraining the motion at which force is applied but restraining the motion elsewhere.

The point localized impedances at each end of the isolator are

$$Z_{11} = \left.\frac{F_e}{\dot{u}(0)}\right|_{\dot{u}(L)=0} = \left.\frac{F_e}{j\omega u(0)}\right|_{\dot{u}(L)=0} \tag{5.2}$$

and

$$Z_{22} = \left.\frac{F_t}{\dot{u}(L)}\right|_{\dot{u}(0)=0} = \left.\frac{F_t}{j\omega u(L)}\right|_{\dot{u}(0)=0} \tag{5.3}$$

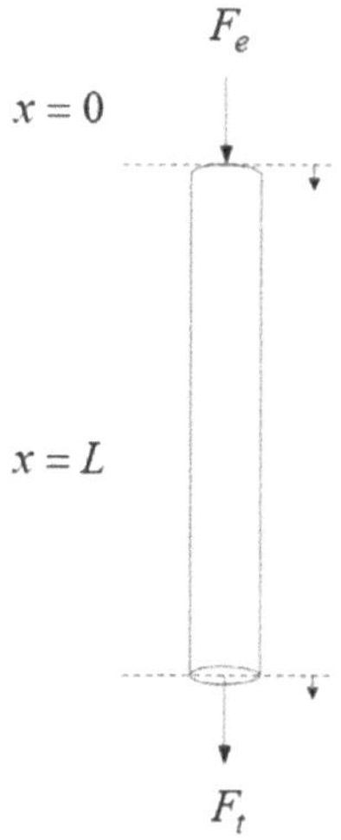

FIGURE 5.1 Uniform finite rod undergoing longitudinal force.

where $\dot{u} = du/dt$. From Hooke's law at $x = 0$ where $\partial u(0)/\partial x = -F_e/\kappa_l$, Eq. (5.1) gives

$$A - B = \frac{jF_e}{\kappa_l k_l} \tag{5.4}$$

where $\kappa_l = E(1 + j\eta)S$ is the longitudinal rigidity and S is the cross-sectional area of the finite rod.

From the boundary condition in Eq. (5.2) at $x = L$ when $u(L) = 0$ yields

$$B = -Ae^{2jk_lL} \tag{5.5}$$

By substituting Eq. (5.5) into Eq. (5.4) yields

$$A = \frac{jF_e}{k_l\kappa_l}\left(\frac{1}{e^{2jk_lL} + 1}\right) \tag{5.6}$$

and

$$B = \frac{-jF_e}{k_l\kappa_l}\left(\frac{e^{2jk_lL}}{e^{2jk_lL} + 1}\right) \tag{5.7}$$

From the boundary condition stated in Eqs. (5.2) and (5.3), the displacement at $x = 0$ is

$$u(0) = A + B = \frac{F_e}{jk_l\kappa_l}\left(\frac{e^{2jk_lL} - 1}{e^{2jk_lL} + 1}\right) = \frac{F_e}{k_l\kappa_l}\tan(k_lL) \tag{5.8}$$

From the principle of reciprocity, where $Z_{11} = Z_{22}$, the point impedance in Eqs. (5.2) and (5.3) is thus given by

$$Z_{11} = Z_{22} = \frac{F_e}{j\omega u(0)} = \frac{k_l\kappa_l}{j\omega\tan(k_lL_l)} = \frac{S\sqrt{E\rho}}{j\tan[k_lL]} \tag{5.9}$$

The transfer impedance is designed by restraining the location where meanwhile the force is applied and unrestrained at the other end of the rod. It is written as

$$Z_{12} = \left.\frac{F_e}{\dot{u}(L)}\right|_{\dot{u}(0)=0} \tag{5.10}$$

and

$$Z_{21} = \left.\frac{F_t}{\dot{u}(0)}\right|_{\dot{u}(L)=0} \tag{5.11}$$

From Eqs. (5.2) and (5.3), where $\dot{u}(0) = 0$, this yields $A = -B$. Substituting this into Eq. (5.4) gives

$$A = \frac{F_e}{2jk_l\kappa_l} \tag{5.12}$$

and

$$B = \frac{-F_e}{2jk_l\kappa_l} \tag{5.13}$$

Substituting Eqs. (5.12) and (5.13) into Eq. (5.1), the displacement at the end of the isolator, that is, at $x = L$, can be given by

$$u(L) = \frac{F_e}{2jk\kappa}\left(e^{-jk_lL} - e^{jk_lL}\right) = \frac{-F_e}{k_l\kappa_l}\sin(k_lL) \tag{5.14}$$

From the principle of reciprocity, where $Z_{21} = Z_{12}$, the transfer impedance in Eqs. (5.10) and (5.11) is

$$Z_{21} = Z_{12} = \frac{F_e}{j\omega u(L)} = \frac{-k_l\kappa_l}{j\omega\sin(k_lL)} = \frac{-k_lS\sqrt{E\rho}}{j\sin(k_lL)} \tag{5.15}$$

Note that here, the damping is presumed to be very small, $\eta << 1$.

The impedance matrix for a nondispersive finite rod due to a longitudinal wave can therefore be written as

$$Z = \begin{bmatrix} Z_{11} & Z_{12} \\ Z_{21} & Z_{22} \end{bmatrix} = \frac{S\sqrt{E\rho}}{j\omega\sin(k_lL)}\begin{bmatrix} \cos(k_lL) & -1 \\ -1 & \cos(k_lL) \end{bmatrix} \tag{5.16}$$

By substituting appropriate k_l and κ_l into Eq. (5.16), the new equation can be expressed as

$$Z = \begin{bmatrix} Z_{11} & Z_{12} \\ Z_{21} & Z_{22} \end{bmatrix} = \frac{S\sqrt{E\rho}}{j\sin(k_lL)} \cdot \begin{bmatrix} \cos(k_lL) & -1 \\ -1 & \cos(k_lL) \end{bmatrix} \tag{5.17}$$

By expending each element in Eq. (5.17) into a standalone equation, it becomes

$$Z_{11} = Z_{22} = \frac{S\sqrt{E\rho}}{j\sin(k_lL)} \cdot \cos(k_lL) = \frac{S\sqrt{E\rho}\cos(k_lL)}{j\sin(k_lL)} = jS\sqrt{E\rho}\cot(k_lL) \tag{5.18}$$

and

$$Z_{12} = Z_{21} = \frac{S\sqrt{E\rho}}{j\sin(k_lL)} \cdot -1 = -\frac{S\sqrt{E\rho}}{j\sin(k_lL)} \tag{5.19}$$

The impedance matrix of a nondispersive finite rod also can be changed into a stiffness matrix. It only can be changed at low frequencies, when the wavelength is much greater than the length of the finite rod. At this moment, the finite rod reduces to a simple spring so that the impedance matrix for the first and fourth elements can be expressed by

$$Z_{11} = Z_{22} = \frac{K_{11}}{j\omega} = \frac{K_{22}}{j\omega} \tag{5.20}$$

and the second and third elements can be written as

$$Z_{11} = Z_{22} = \frac{-K_{12}}{j\omega} = \frac{-K_{21}}{j\omega} \tag{5.21}$$

where $K = SA/L$ is the longitudinal stiffness of the finite rod.

Based on the statement given, the stiffness matrix for a nondispersive finite rod can be expanded as

$$K_{11} = K_{22} = \frac{ES}{L}\left(\frac{k_l L}{\tan(k_l L)}\right) \tag{5.22}$$

and

$$-K_{12} = -K_{21} = \frac{ES}{L}\left(\frac{k_l L}{\sin(k_l L)}\right) \tag{5.23}$$

In matrix form, the stiffness matrix for a nondispersive finite rod is

$$K = \begin{bmatrix} K_{11} & -K_{12} \\ -K_{21} & K_{22} \end{bmatrix} = \begin{bmatrix} \frac{ES}{L}\left(\frac{k_l L}{\tan(k_l L)}\right) & \frac{ES}{L}\left(\frac{k_l L}{\sin(k_l L)}\right) \\ \frac{ES}{L}\left(\frac{k_l L}{\sin(k_l L)}\right) & \frac{ES}{L}\left(\frac{k_l L}{\tan(k_l L)}\right) \end{bmatrix} \tag{5.24}$$

The longitudinal wave of a nondispersive finite rod associated with the impedance and stiffness matrix is shown in Fig. 5.2. According to this figure, the internal resonance waves for both results started at 108 Hz and fluctuated until reaching high frequency. Only two patterns of internal resonance wave occurred in a longitudinal system, one from positive and the other one in the opposite position. For impedance matrix, the wave starts at point 10^5 which represents the impedance value of the system itself. However, the value of impedance rapidly decreases associated with the increasing quantum of frequency. The fluctuated wave is huge at a range of 108–120 Hz and it is slowly stable by increasing the frequency. According to this figure, the internal resonance for the impedance matrix happens in at a greater amount at a certain frequency only (Du, Burdisso, Nikolaidis, & Tiwari, 2003).

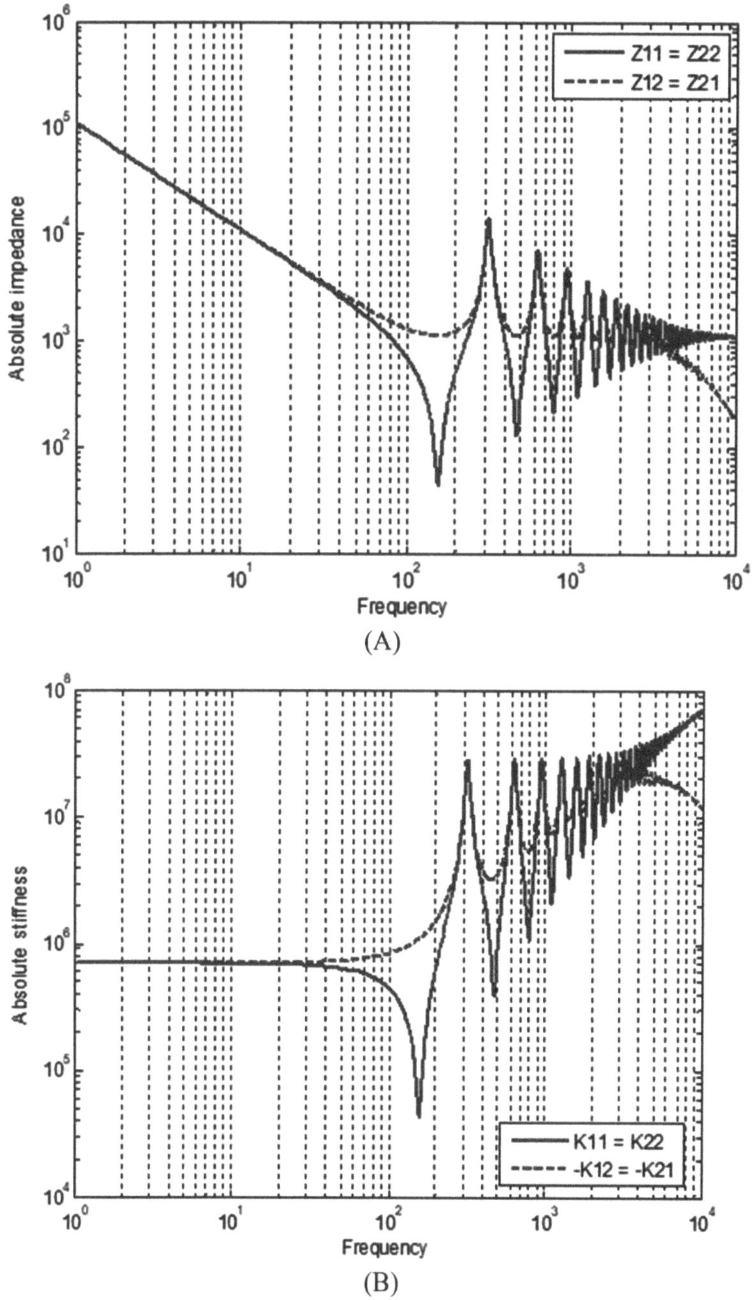

FIGURE 5.2 Longitudinal wave (A) impedance and (B) stiffness matrix.

Fig. 5.2(B) shows the internal resonance in a stiffness matrix. The starting point of internal resonance happens at 7×10^5 where the amount represents the total amount of the stiffness of nondispersive finite rod in a longitudinal condition. The fluctuated wave remains starting at 108 Hz and is constantly in a sinusoidal pattern, however, the amplitude rapidly decreases with a frequency increase. It has a reasonably close reaction for the internal resonance pattern between these two matrices, but using an analytical approach it is totally different.

5.2.2 Lumped parameter system

In this section, the baseline LR-MS model is modeled using a lumped parameter system consisting of mass, damper, and spring components. To assess the vibration isolation performance, the spring is loaded with a lumped mass M excited with a harmonic force F_e. The rubber is ideally modeled as a massless component having a constant stiffness k and damping coefficient c. The embedded plate is treated as a rigid solid mass m without damping. The LR-MS model is attached to a structure and the transmissibility is derived. Only vertical motion is taken into account, with the rotational motion neglected. Fig. 5.3 shows a schematic diagram of the model of the LR-MS model with one layer of a metal plate which creates a two-DOF system.

Two equations of motion can be derived as follows

$$M\ddot{u}_1 + c_1(\dot{u}_1 - \dot{u}_2) + k_1(u_1 - u_2) = F_e \tag{5.25}$$

and

$$m\ddot{u}_2 + c_1(\dot{u}_2 - \dot{u}_1) + c_2\dot{u}_2 + k_1(u_2 - u_1) + k_2u_2 = 0 \tag{5.26}$$

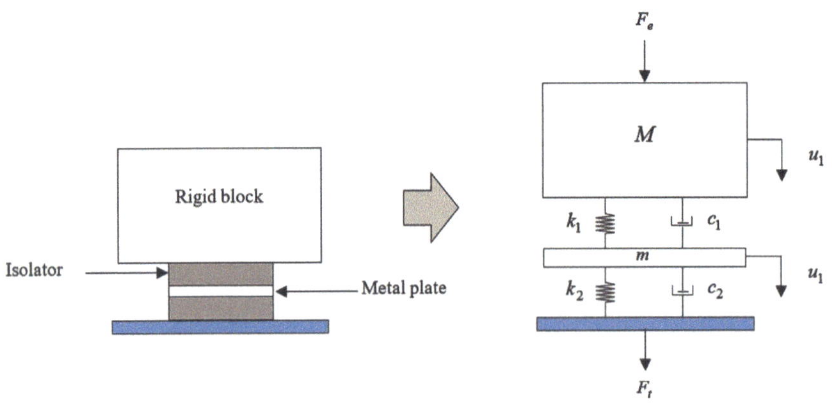

FIGURE 5.3 Mass-damper-spring model of a laminated rubber metal spring.

Due to a harmonic force, the resulting motion of the system is also harmonic. Substituting $u = Ue^{j\omega t}$ in Eqs. (5.25) and (5.26) with U the complex amplitude and ω the frequency, the equations of motion can be expressed in matrix form as

$$\left(-\omega^2\begin{bmatrix} M & 0 \\ 0 & m \end{bmatrix} + j\omega\begin{bmatrix} c_1 & -c_1 \\ -c_1 & c_1+c_2 \end{bmatrix} + \begin{bmatrix} k_1 & -k_1 \\ -k_1 & k_1+k_2 \end{bmatrix}\right)\begin{Bmatrix} U_1 \\ U_2 \end{Bmatrix} = \begin{Bmatrix} F_e \\ 0 \end{Bmatrix} \tag{5.27}$$

or in general form

$$\left[-\omega^2\mathbf{M} + j\omega\mathbf{C} + \mathbf{K}\right] \sim\mathbf{U} = \sim\mathbf{F} \tag{5.28}$$

where $\mathbf{M}$ is the mass matrix, $\mathbf{C}$ is the damping matrix, $\mathbf{K}$ is the stiffness matrix, and $\sim\mathbf{U}$ and $\sim\mathbf{F}$ are the vectors of complex displacement of amplitude and force, respectively.

The displacements U_1 and U_2 at each frequency ω can therefore be obtained by

$$\sim\mathbf{U} = [-\omega^2\mathbf{M} + j\omega\mathbf{C} + K]^{-1} \sim\mathbf{F} \tag{5.29}$$

where A^{-1} indicates the inverse of matrix A.

For the case of the model in Fig. 5.3, the force transmitted to the receiver structure can be written as a function of the displacement of the bottom metal layer. For an excitation force of unit amplitude $F_e = 1$, the transmissibility, that is, the amplitude ratio of transmitted force to excitation force, is given by

$$T = \left|\frac{F_t}{F_e}\right| = |F_t| = Y_2\sqrt{(k_2^2 + \omega^2 c_2^2)} \tag{5.30}$$

For two layers of metal plates inside the rubber blocks (three-DOF system), the mass, damping, and stiffness matrices in Eq. (5.28) are given by

$$\mathbf{M} = \begin{bmatrix} M & 0 & 0 \\ 0 & m_1 & 0 \\ 0 & 0 & m_2 \end{bmatrix}, \ \mathbf{C} = \begin{bmatrix} c_1 & -c_1 & 0 \\ -c_1 & c_1+c_2 & -c_2 \\ 0 & -c_2 & c_2+c_3 \end{bmatrix} \text{ and}$$
$$\mathbf{K} = \begin{bmatrix} k_1 & -k_1 & 0 \\ -k_1 & k_1+k_2 & -k_2 \\ 0 & -k_2 & k_2+k_3 \end{bmatrix} \tag{5.31}$$

By observing the structural pattern of the matrices in Eqs. (5.27) and (5.31), for N layers of metal plates, the matrices thus have dimensions of $(N+1)\times(N+1)$ and are expressed as

$$\mathbf{M} = \begin{bmatrix} M & 0 & 0 & 0 \\ 0 & m & 0 & 0 \\ \vdots & \cdots & \ddots & \vdots \\ 0 & \cdots & 0 & m_N \end{bmatrix},$$
$$\mathbf{C} = \begin{bmatrix} c_1 & -c_1 & 0 & 0 & \cdots & 0 \\ -c_1 & c_1+c_2 & -c_2 & 0 & \cdots & 0 \\ 0 & -c_2 & c_2+c_3 & -c_3 & \cdots & \vdots \\ \vdots & \vdots & \vdots & \vdots & \ddots & c_N \\ 0 & 0 & \cdots & -c_{N-1} & & c_{N-1}+c_N \end{bmatrix} \text{ and} \tag{5.32}$$
$$\mathbf{K} = \begin{bmatrix} k_1 & -k_1 & 0 & 0 & \cdots & 0 \\ -k_1 & k_1+k_2 & -k_2 & 0 & \cdots & 0 \\ 0 & -k_2 & k_2+k_3 & -k_3 & & \vdots \\ \vdots & \vdots & & & \ddots & k_N \\ 0 & \cdots & \cdots & 0 & -k_{N-1} & k_{N-1}+k_N \end{bmatrix}$$

Fig. 5.4 plots the transmissibility for different numbers of layers N of embedded metal plates. The calculation is made assuming the mass of each plate is the same as the loaded mass. The damping is also assumed to be very small. Peaks indicating amplification of the injected force to the receiver structure can be seen at low frequencies, for example at 9 Hz in Fig. 5.4(A), which appear at the natural frequencies of the system. The more plates are embedded in the rubber, the higher the number of DOF and the more natural frequencies the system has. The results also show that the resonant peaks shift toward lower frequencies as the number of intermediate plates increases.

According to this analysis, the isolation region (where the transmissibility is less than unity) can be seen to be roughly above 25 Hz. It is interesting to observe that the transmissibility curve rapidly decreases with the increasing number of embedded plates in the rubber. This indicates that the LR-MS model improves the vibration isolation at high frequency. However, this lumped parameter model ignores the mass of the rubber which can contribute to more internal resonances. The width of the rubber layer is neglected as, at high frequencies when the wavelength is much smaller than the rubber layer thickness, wave effects from various directions affected the isolation performance. This wave effect has been highlighted in the classical rubber rod isolator (Dickens, 2000; Du et al., 2003).

5.2.3 Wave and mass effect in a finite rod

The performance of the rod to isolate the vibration energy transmitted to the mass is assessed. Fig. 5.5 shows a schematic diagram of a rod loaded with rigid mass M and excited by an excitation force F_e and

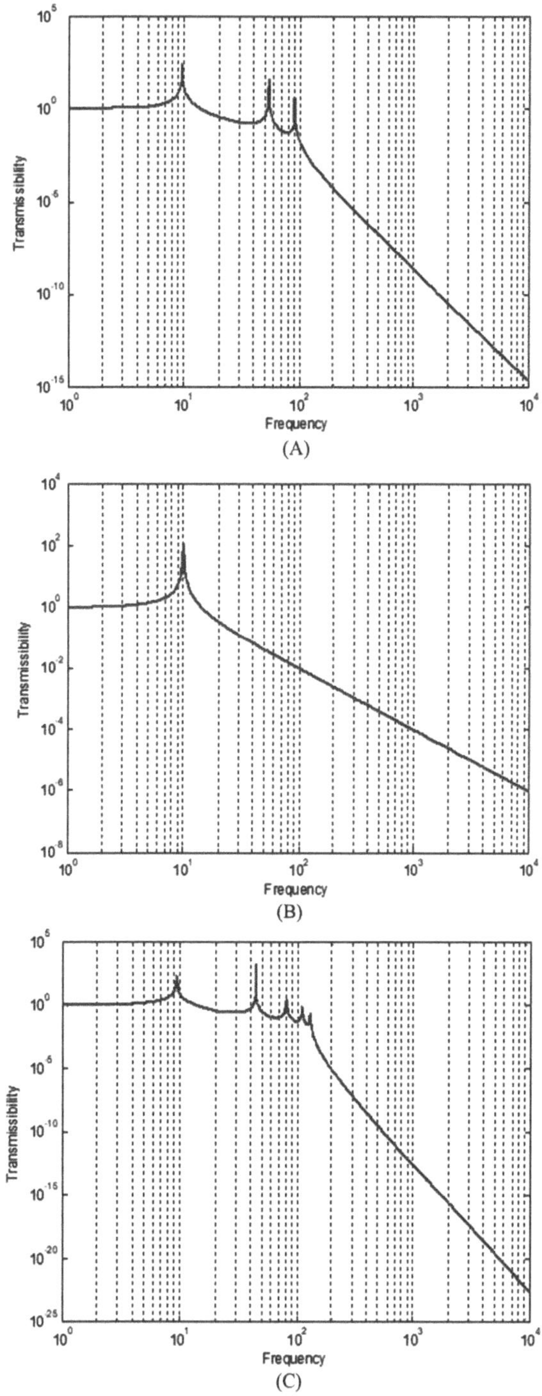

FIGURE 5.4 Lumped parameter model with numbers of layers of plate (A) $N = 0$, (B) $N = 2$, and (C) $N = 4$.

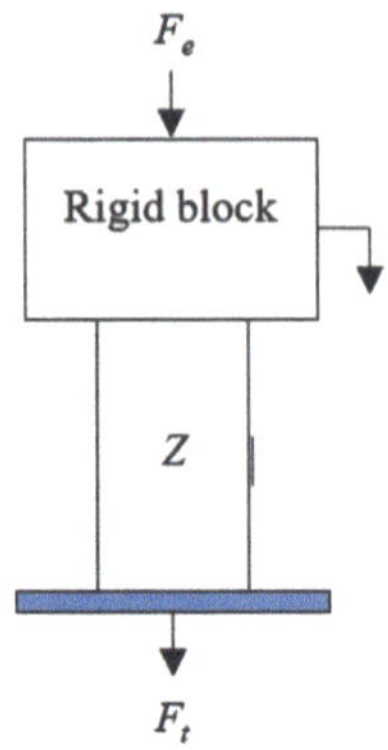

FIGURE 5.5 Schematic diagram of a rigid mass with a finite rod.

transmitted force F_t to the base structure. The equation of motions can be expressed by

$$-\omega^2 M U_1 + i\omega Z_{11} U_1 = F_e \tag{5.33}$$

and

$$i\omega Z_{21} U_1 = F_t \tag{5.34}$$

where Z_{11} and Z_{21} are impedance equations for a nondispersive finite rod.

The transmissibility T_r for the finite rod can be written as

$$T_r = \left|\frac{F_t}{F_e}\right| = \left|\frac{Z_{21}}{i\omega M + Z_{11}}\right| \tag{5.35}$$

Fig. 5.6 shows the transmissibility from Eq. (5.35) and is plotted together with that from the lumped parameter system (massless isolator) of a single-DOF. Both systems can be seen to have the same peak at around 4.5 Hz, which corresponds to the natural frequency proportional to the stiffness of the system. For the rod spring, this is where the vibration wavelength is still much larger than the length of the rod. As the frequency increases, the wavelength becomes much smaller and the incoming and reflected waves in the rod can therefore superimpose each other, developing the internal resonances. These are indicated in the figure by the peaks appearing at higher frequency above 100 Hz. The effect of waves in the rod can also be seen with the presence of terms $\cos(kL) \approx 1$ and $\sin(kL) \approx 1$ in the impedance as in Eq. (5.17).

Consider now that the nondispersive finite rod isolator is inserted with a rigid metal plate having mass M creating a multi-DOF system as seen in Fig. 5.7. For simplification, here it is assumed that the wavelength is much greater than the dimension of the rod, that is, $kL \rightarrow 0$

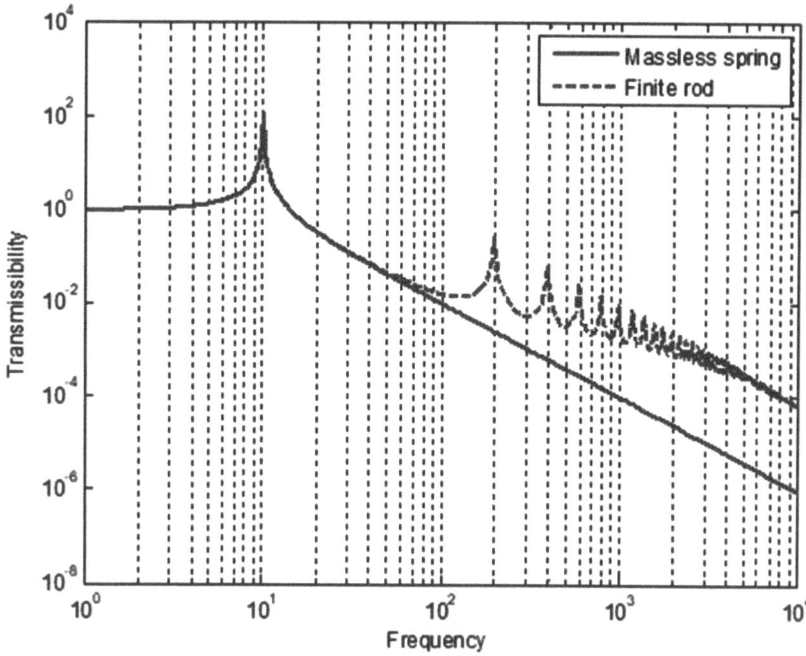

FIGURE 5.6 Comparison between a massless spring and finite rod.

where in consequence $\cos(kL) \approx 1$ and $\sin(kL) \approx 1$. The impedance matrix in Eq. (5.17) can now be written as

$$Z = \frac{\kappa}{j\omega L}\begin{bmatrix} 1 & -1 \\ -1 & 1 \end{bmatrix} \tag{5.36}$$

As the impedance is proportional to the stiffness constant K, that is, $Z = -jK/\omega$, Eq. (5.36) therefore implies that at low frequency the finite rod system has stiffness constant

$$K = \frac{\kappa}{L} = \frac{E(1 + j\eta)S}{L} \tag{5.37}$$

which is the stiffness of the rod due to the longitudinal wave with hysteretic damping.

For the rod with N layers of rigid mass, the equation of motion is given by

$$\left[-\omega^2\mathbf{M} + \mathbf{K}\right] \sim\mathbf{U} = \sim\mathbf{F} \tag{5.38}$$

where $\mathbf{M}$ is the mass matrix, $\mathbf{K}$ is the stiffness matrix, and $\sim\mathbf{U}$ and $\sim\mathbf{F}$ are the vectors of complex displacement amplitude and force, respectively, given by

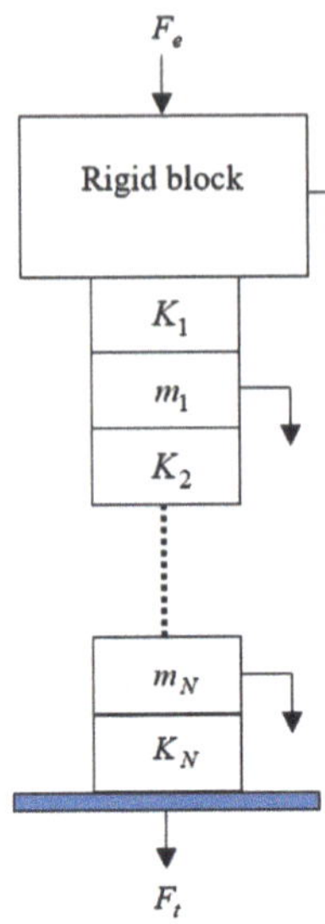

FIGURE 5.7 Schematic diagram of a rod isolator inserted with rigid masses.

$$\mathbf{M} = \begin{bmatrix} M & 0 & 0 & 0 \\ 0 & m_1 & 0 & 0 \\ \vdots & \cdots & \ddots & \vdots \\ 0 & \cdots & 0 & m_N \end{bmatrix},$$
$$\mathbf{K} = \begin{bmatrix} K_1 & -K_1 & 0 & 0 & \cdots & 0 \\ -K_1 & K_1+K_2 & -K_2 & 0 & \cdots & 0 \\ 0 & -K_2 & K_2+K_3 & -K_3 & & \vdots \\ \vdots & \vdots & & & \ddots & K_N \\ 0 & \cdots & \cdots & 0 & -K_{N-1} & K_{N-1}+K_N \end{bmatrix} \text{ and} \tag{5.39}$$

$$\sim U = \{u_1 \quad u_2 \quad \cdots \quad \cdots \quad u_{N-1}\}^T, \ \sim F = \{F_e \quad 0 \quad \cdots \quad \cdots \quad 0\}^T$$

Note that the matrix has dimensions of $(N+1)\times(N+1)$. The solution for the displacements at each frequency ω can therefore be obtained by

$$\sim\mathbf{U} = \left[-\omega^2\mathbf{M}+\mathbf{K}\right]^{-1}\sim\mathbf{F} \tag{5.40}$$

where A^{-1} indicates the inverse of matrix A.

For an excitation force of unit amplitude $F_e = 1$, the transmissibility is given by

$$T = \left|\frac{F_t}{F_e}\right| = |F_t| = |K_N U_{N-1}| \tag{5.41}$$

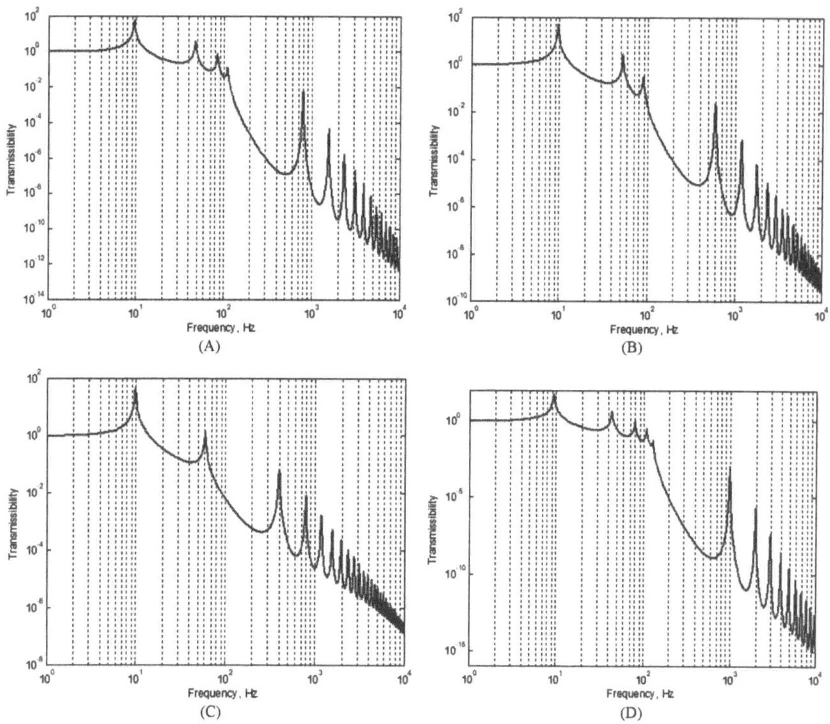

FIGURE 5.8 Transmissibility of multilayers rigid masses in the isolator: (A) $N = 1$, (B) $N = 2$, (C) $N = 3$, and (D) $N = 4$.

The transmissibility is shown in Fig. 5.8 with two to four layers of rigid masses in the isolator. Note that the total length of the isolator is kept constant. Therefore, if N number of masses are added, the length of each rod becomes $L/(N+1)$, which then reduces the stiffness of each element of the rod. This, in consequence, increases the internal natural frequencies of the system apart from the fundamental frequency. The peaks appear in the isolation frequency area ($T < 1$) which degrade the isolation performance of the laminated spring system. The situation might become worse if the wave effects are considered in the model. This phenomenon is investigated in the next section.

5.3 Results and discussion

5.3.1 Mathematical modeling results of a laminated rubber–metal spring model

According to the previous section, a lumped parameter model is developed to represent the basic relationship mass-damper-spring model for an LR-MS model. Two mathematical formulations are derived, which are wave and mass effect in the finite rod model. Regarding these two formulations, both derivations give an effect to finite rod model as shown in a graphical plot. This section discusses more about the complete mathematical formulation for the wave and mass effect for axial direction in the LR-MS model as shown in Fig. 5.9. This consists of alternating layers of natural rubber (NR) vulcanized to reinforce a metal plate. The basic element in the LR-MS model is one NR layer laminated on a metal plate in the middle.

It is shown in Fig. 5.10 where F_e is the force generated by applying mass while F_t is the transmitted force. The applied mass, M, represents a total mass effect from preload to the laminated layer and plate. The impedance equation for a one-dimensional element for a nondispersive finite rod model in longitudinal direction has been derived in the previous section and the equation is represented in matrix form. Basically,

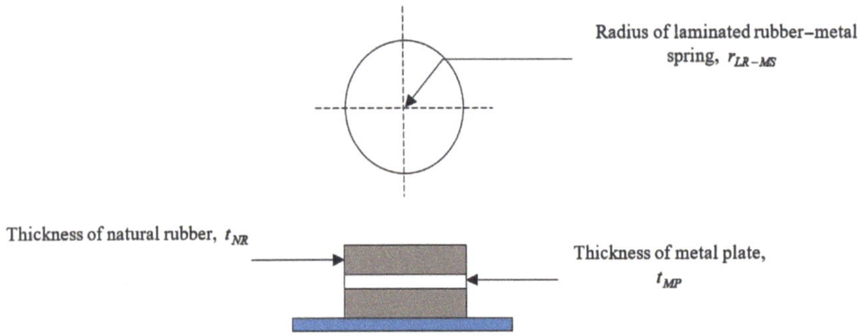

FIGURE 5.9 Specification of a laminated nondispersive finite rod.

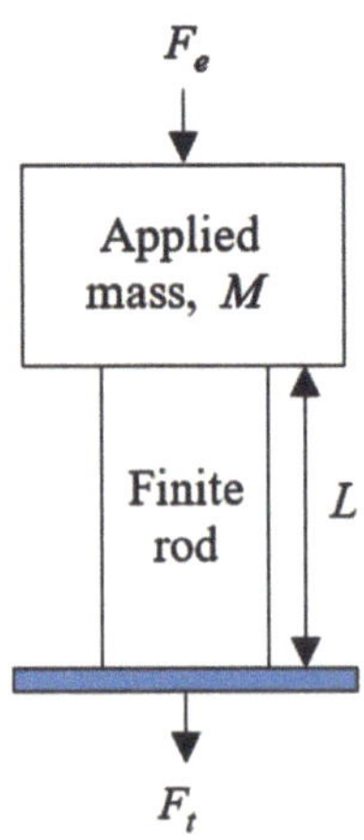

FIGURE 5.10 Applied mass–finite rod system.

the impedance method has been used for more 10 years and the purpose of the study is to identify the relationship between electrical-mechanical or to simplify the mechanical systems formulation.

In this study, the derivation of one-dimensional elements of a nondispersive finite rod has developed using a wave propagation method. There are advantages to using this technique where it is possible to relate the dynamic behavior of a finite structure and allowed exact impedance equation derived. On the other hand, this technique is particularly appropriate for a high damped system and it can operate well in the high-frequency range. The frequency domain ratio of the force transmitted F_t at the applied mass–finite rod system to a rigid foundation to the force excitation F_e and the ratio of the equipment displacement to the displacement of the host structure turns out to be the same and it is called force transmissibility T_f, this term is also known as a displacement transmissibility T_x. These two variables can be expressed in mathematical form as in Eq. (5.42). The mass system is shown in Fig. 5.11.

$$T = T_f = T_x = \frac{1 + j\eta}{1 - \Omega^2 + j\eta} \tag{5.42}$$

where $\Omega = \omega/\omega_n$ is the forcing frequency and it is normalized by the undamped natural frequency $\omega_n = \sqrt{k/m}$ of the equipment on the isolator and η is the loss factor of the isolator.

The equation of motion for a mass system in one-DOF can be expressed by

$$m\ddot{x} + k(1 + j\eta)u = F_e \tag{5.43}$$

where k is stiffness and u is displacement.

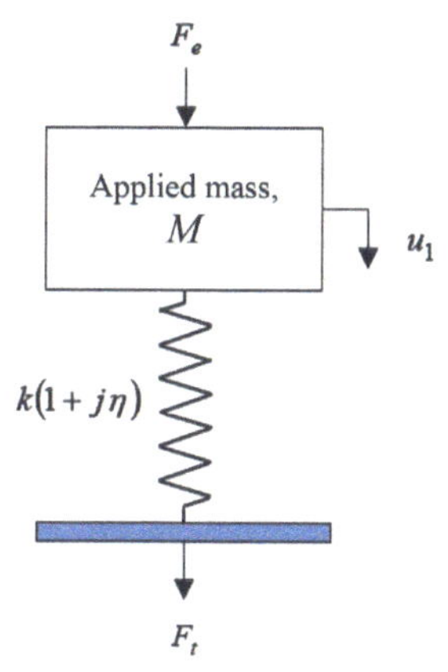

FIGURE 5.11 Mass system in one degree of freedom.

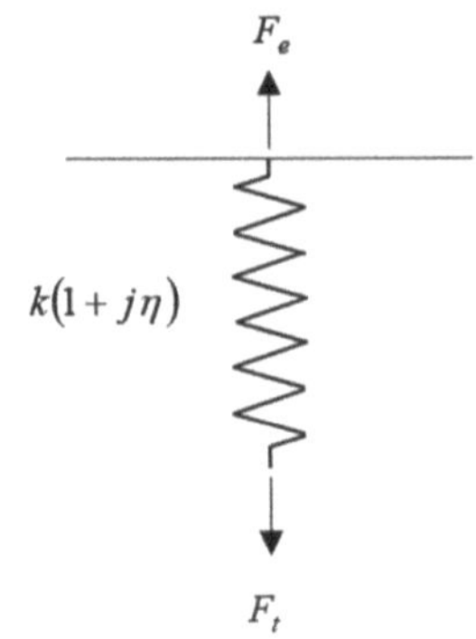

FIGURE 5.12 Transmitted force in one degree of freedom.

In matrix form, it can be represented as

$$[m]\{\ddot{u}\} + [k(1 + j\eta)]\{u\} = \{F_e\} \tag{5.44}$$

For harmonic motion at circular frequency, ω, the equation becomes

$$(-\omega^2 m + k(1 + j\eta))u = F_e \tag{5.45}$$

For transmitted force F_t, the schematic diagram is shown in Fig. 5.12.

The transmitted force for a one-DOF mass system can be represented as

$$F_t = k(1 + j\eta)u \tag{5.46}$$

In harmonic motion, the equation is still the same as the equation of transmitted force because the mass is not involved at the system. The force transmissibility, T_f, can be represented as

$$T_f = \left|\frac{F_t}{F_e}\right| = \left|\frac{k(1 + j\eta)}{k(1 + j\eta) - \omega^2 M}\right| \tag{5.47}$$

A force $F_e(t)$ excites the rigid mass and is transferred to the base structure as a transmitted force $F_t(t)$. Consider now the nondispersive finite rod attached to a structure that is loaded with an applied mass M as seen is Fig. 5.10. In this section, there are two methods that can be used to represent the transmissibility characteristics which are impedance and stiffness equations of motion. Let us consider the schematic diagram for the LR-MS model in one-DOF (adapted from Fig. 5.10) in terms of displacement as shown in Fig. 5.13. The equations of motion of the system are given by

$$F_e(t) = M\ddot{u}_1 + Z_{12}u_2(t) + Z_{11}u_1(t) \tag{5.48}$$

and

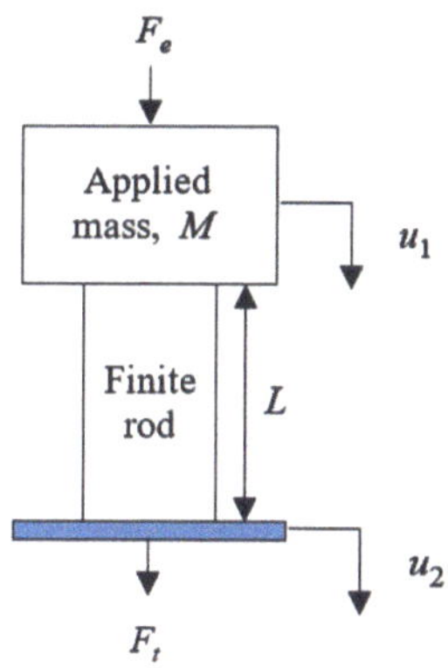

FIGURE 5.13 Schematic diagram for a laminated rubber–metal spring model in terms of displacement.

$$F_t(t) = Z_{22}u_2(t) + Z_{21}u_1(t) \tag{5.49}$$

where $\ddot{u} = d^2u/dt$, $u(t) = Ue^{j\omega t}$ and U is the complex amplitude.

By substituting $f(t) = Fe^{j\omega t}$ and $u(t) = Ue^{j\omega t}$ into Eqs. (5.48) and (5.49), the equations of motion can be expressed in matrix form by

$$\begin{Bmatrix} F_e \\ F_t \end{Bmatrix} = -\omega^2 \begin{bmatrix} M & 0 \\ 0 & 0 \end{bmatrix} \begin{Bmatrix} u_1 \\ u_2 \end{Bmatrix} + j\omega \begin{bmatrix} Z_{11} & Z_{12} \\ Z_{21} & Z_{22} \end{bmatrix} \begin{Bmatrix} u_1 \\ u_2 \end{Bmatrix} \tag{5.50}$$

The transmissibility for the LR-MS model in the impedance equation can be written as

$$T_r = \left|\frac{F_t}{F_e}\right| = \left|\frac{Z_{22}u_2 + Z_{21}u_1}{-\omega^2 M u_1 + Z_{12}u_2 + Z_{11}u_1}\right| \tag{5.51}$$

For transmissibility subjected in a stiffness equation, the derivation is similar to the impedance. The final equation can be represented as

$$T_r = \left|\frac{F_t}{F_e}\right| = \left|\frac{K_{22}u_2 + K_{21}u_1}{-\omega^2 M u_1 + K_{12}u_2 + K_{11}u_1}\right| \tag{5.52}$$

The results for the LR-MS model for both impedance and stiffness equation are shown in Fig. 5.14. According to the results, there is no difference. This shows that the impedance and stiffness equation shows the same result in terms of transmissibility. The peak remains at the same location and at the frequencies of 9 Hz. The graph shifts down directly until 200 Hz and then slightly fluctuates up to 10 kHz. The fluctuated graph remains and slowly goes down until the end of the graph. This range represents the internal resonances of the LR-MS model. Based on previous discussion, the impedance and stiffness equation

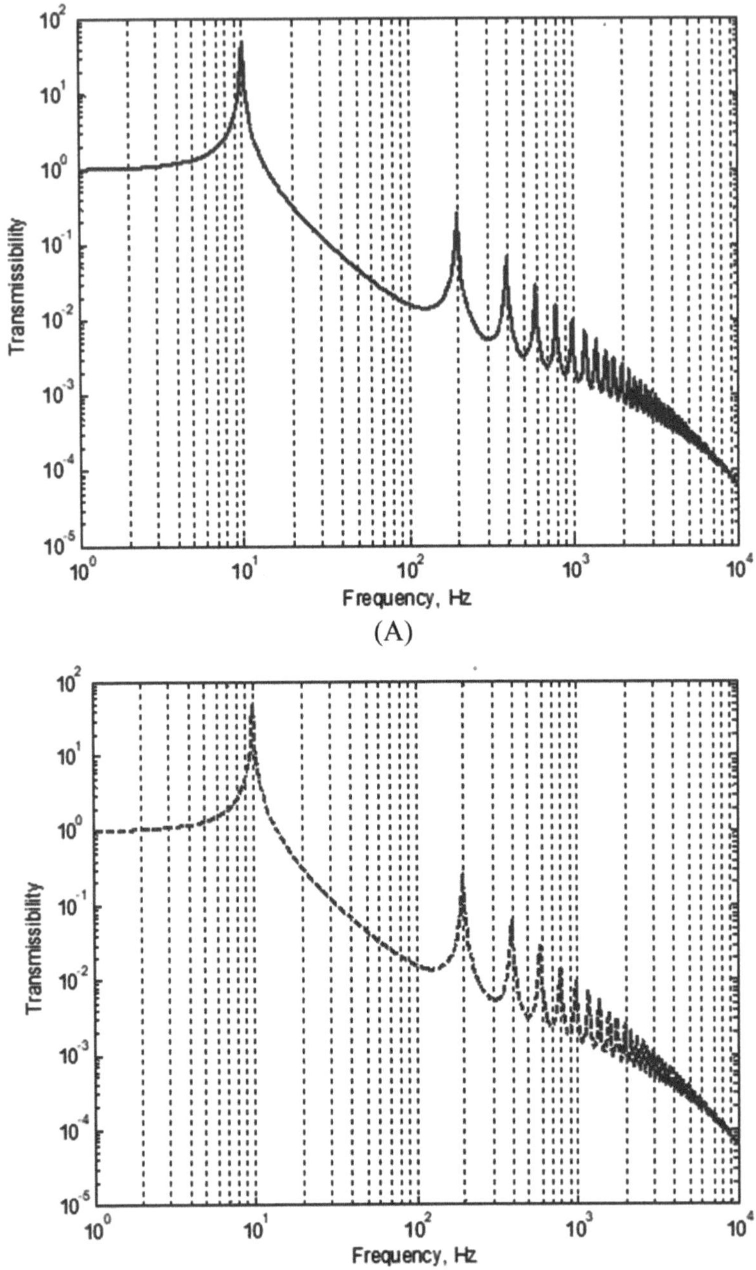

FIGURE 5.14 Transmissibility for a one-degree-of-freedom laminated rubber–metal spring model: (A) impedance and (B) stiffness equation.

gives the same results in terms of transmissibility. Therefore, in the axial direction for multi-DOF for the LR-MS model, only the stiffness equation is involved.

Consider the schematic diagram for two-DOF of the LR-MS model in Fig. 5.15.

The stiffness matrix for each element in a nondispersive finite rod can be represented as $\begin{bmatrix} \overset{(1)}{K}_{11} & -\overset{(1)}{K}_{12} \\ -\overset{(1)}{K}_{21} & \overset{(1)}{K}_{22} \end{bmatrix}$ for element one and $[\overset{(2)}{K}_{11}]$ for element two, respectively. The question is, why at element two is the stiffness matrix only $[\overset{(2)}{K}_{11}]$? This can be answered by referring to Fig. 5.16. According to this figure, $u_3 = 0$, which means there is no displacement occurring at this point by applying force F_2. In order to solve the equation of motion at that element, refer back to Fig. 5.14, in which the equation can be written as

$$\overset{(2)}{K}_{11} u_2 - \overset{(2)}{K}_{11} u_3 = F_2 \tag{5.53}$$

and

$$-\overset{(2)}{K}_{21} u_2 + \overset{(2)}{K}_{22} u_3 = F_3 \tag{5.54}$$

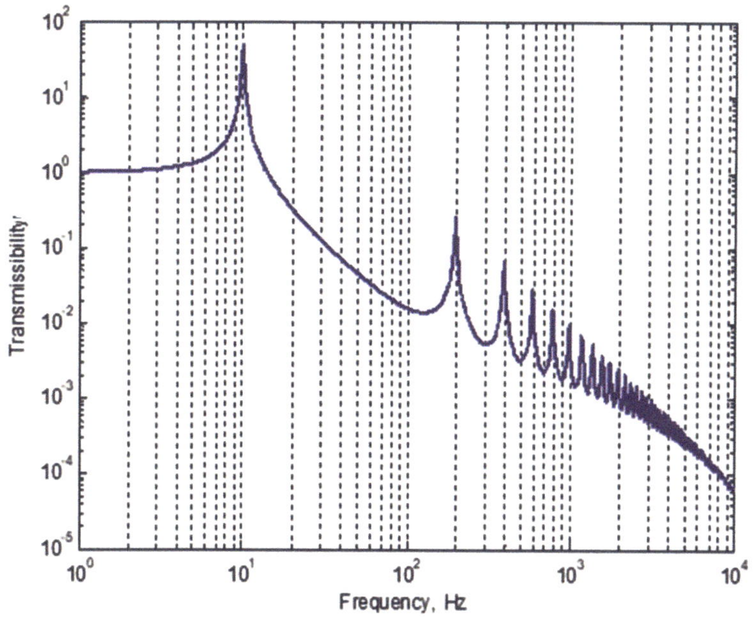

FIGURE 5.15 Schematic diagram for two degrees-of-freedom of a laminated rubber–metal spring.

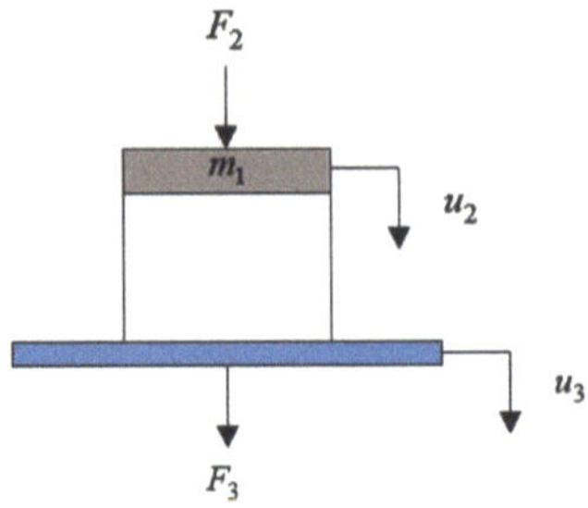

FIGURE 5.16 One element in a nondispersive finite rod.

Previous discussion mentioned that $u_3 = 0$, so the equations of motion in Eqs. (5.53) and (5.54) become

$$\overset{(2)}{K_{11}} u_2 = F_2 \tag{5.55}$$

and

$$-\overset{(2)}{K_{21}} u_2 = F_3 \tag{5.56}$$

In matrix form, this can be represented as

$$\begin{bmatrix} \overset{(2)}{K_{11}} & -\overset{(2)}{K_{12}} \\ -\overset{(2)}{K_{21}} & \overset{(2)}{K_{22}} \end{bmatrix} \begin{Bmatrix} u_2 \\ u_3 \end{Bmatrix} = \begin{Bmatrix} F_2 \\ F_3 \end{Bmatrix} \tag{5.57}$$

where $u_2 \neq 0$ and $u_3 = 0$.

The displacement only occurred at u_2. From Eq. (5.57) the stiffness matrix for element two becomes $[\overset{(2)}{K_{11}}]$.

The global stiffness matrix for a two-DOF LR-MS model in harmonic motion equation is

$$\left(-\omega^2 \begin{bmatrix} M & 0 \\ 0 & m_1 \end{bmatrix} + \begin{bmatrix} \overset{(1)}{K_{11}} & -\overset{(1)}{K_{12}} \\ -\overset{(1)}{K_{21}} & K_{22}^{(1)} + \overset{(2)}{K_{11}} \end{bmatrix} \right) \begin{Bmatrix} u_1 \\ u_2 \end{Bmatrix} = \begin{Bmatrix} F_e \\ 0 \end{Bmatrix} \tag{5.58}$$

By referring to Figs 5.13 and Fig. 5.14 and Eq. (5.58), the transmitted force for the system is

$$F_t = -\overset{(2)}{K_{21}} u_2 \tag{5.59}$$

The transmissibility of the system is

$$T_r = \left| \frac{F_t}{F_e} \right| \tag{5.60}$$

However, to solve the unknown u_1 and u_2 need to be discovered first. By using Eq. (5.58), two new equations can be derived, and they are

$$-\omega^2 + (M)u_1 + \left(\overset{(1)}{K_{11}}\, u_1\right) - \left(\overset{(1)}{K_{12}}\, u_2\right) = F_e \tag{5.61}$$

and

$$-\omega^2 + (m_1)u_1 - \left(\overset{(1)}{K_{21}}\, u_1\right) + \left(\overset{(1)}{K_{22}} + \overset{(2)}{K_{11}}\right)u_2 = 0 \tag{5.62}$$

Then, Eqs. (5.59) and (5.61) are substituted into Eq. (5.60) to observe the transmissibility characteristics of two-DOF for the LR-MS model. The transmissibility equation can have written as

$$T_r = \left|\frac{F_t}{F_e}\right| = \left|\frac{-\overset{(2)}{K_{21}}u_2}{-\omega^2 + (M)u_1 + \left(\overset{(1)}{K_{11}}u_1\right) - \left(\overset{(1)}{K_{12}}u_2\right)}\right| \tag{5.63}$$

Fig. 5.17 shows the result of two-DOF for LR-MS model. Based on this figure, the first peak remains at 9 Hz. A comparison between this figure and

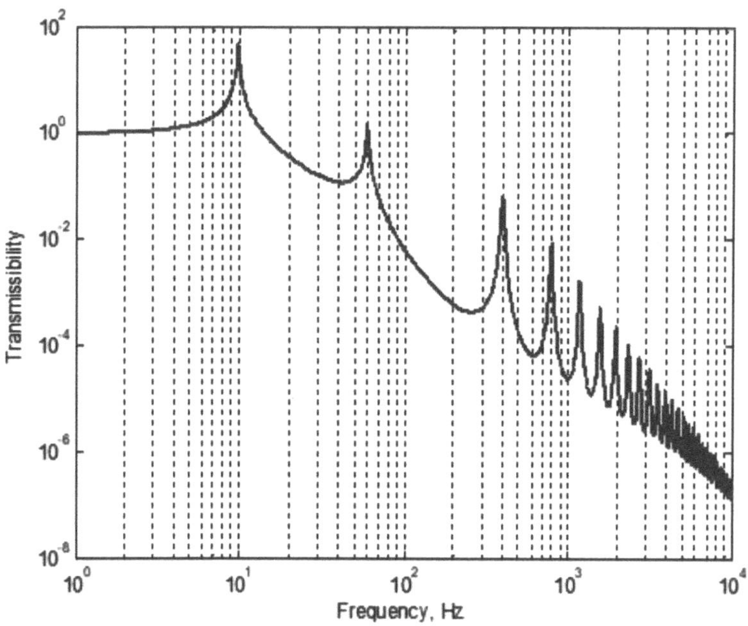

FIGURE 5.17 Transmissibility result for a two-degrees-of-freedom laminated rubber–metal spring model.

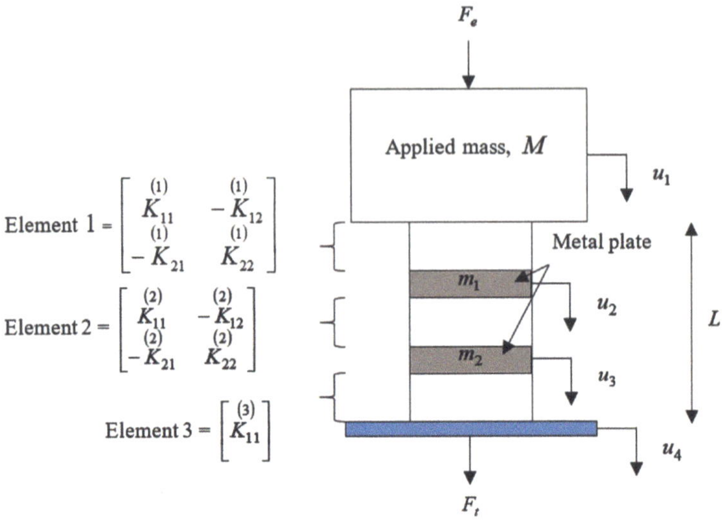

FIGURE 5.18 Schematic diagram for three degrees-of-freedom of a laminated rubber–metal spring.

one-DOF of LR-MS model, the second peak appeared at 60 Hz. These two peaks represent the number of degrees-of-freedom inside the LR-MS model. The fluctuated wave occurs at 400 Hz and remains, but the amplitude slowly decreases to 10 kHz. By comparing one DOF, by increasing the DOF by adding a metal plate, the range of internal resonance slightly increases. In addition, the internal resonance wave goes far below the unity comparing graph with less DOF.

For three-DOF of the LR-MS model, Fig. 5.18 is considered as a schematic diagram of the system.

According to Fig. 5.18, the stiffness matrix for element one is $\begin{bmatrix} \overset{(1)}{K}_{11} & -\overset{(1)}{K}_{12} \\ -\overset{(1)}{K}_{21} & \overset{(1)}{K}_{22} \end{bmatrix}$, element two is $\begin{bmatrix} \overset{(2)}{K}_{11} & -\overset{(2)}{K}_{12} \\ -\overset{(2)}{K}_{21} & \overset{(2)}{K}_{22} \end{bmatrix}$, and element three is $[\overset{(2)}{K}_{11}]$, respectively. The problem faced in this system is the same as in two-DOF. By referring to Fig. 5.19, presuming the value of u_4 is equal to zero, it locates at the receiver structure. The equation of motion for this system can be derived as

$$\overset{(3)}{K}_{11} u_3 - \overset{(3)}{K}_{12} u_4 = F_3 \tag{5.64}$$

and

$$-\overset{(3)}{K}_{21} u_3 + \overset{(3)}{K}_{22} u_4 = F_4 \tag{5.65}$$

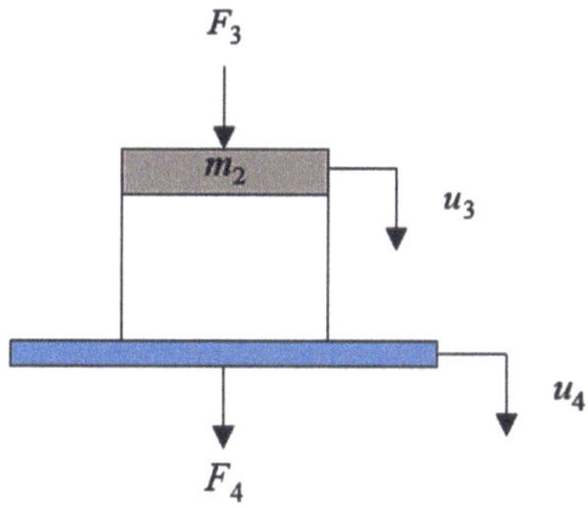

FIGURE 5.19 Two elements in a nondispersive finite rod.

The value of u_4 has been declared, so the new equations of motion become

$$\overset{(3)}{K}_{11}\,u_3 = F_3 \tag{5.66}$$

and

$$-\overset{(3)}{K}_{21}\,u_3 = F_4 \tag{5.67}$$

Eqs. (5.66) and (5.67) can be expressed in matrix form and are shown in Eq. (5.68).

$$\begin{bmatrix} \overset{(3)}{K}_{11} & -\overset{(3)}{K}_{12} \\ -\overset{(3)}{K}_{21} & \overset{(3)}{K}_{22} \end{bmatrix} \begin{Bmatrix} u_3 \\ u_4 \end{Bmatrix} = \begin{Bmatrix} F_3 \\ F_4 \end{Bmatrix} \tag{5.68}$$

By referring again to Figs. 5.17 and Fig. 5.18 and Eq. (5.68), the transmitted force can be simplified as

$$F_t = -\overset{(3)}{K}_{12}\,u_3 \tag{5.69}$$

The global stiffness matrix for the three-DOF of LR-MS model can be represented as

$$\left(-\omega^2 \begin{bmatrix} M & 0 & 0 \\ 0 & m_1 & 0 \\ 0 & 0 & m_2 \end{bmatrix} + \begin{bmatrix} \overset{(1)}{K}_{11} & -\overset{(1)}{K}_{12} & 0 \\ -\overset{(1)}{K}_{21} & \overset{(1)}{K}_{22} + \overset{(2)}{K}_{11} & -\overset{(2)}{K}_{12} \\ 0 & -\overset{(2)}{K}_{21} & \overset{(2)}{K}_{22} + \overset{(3)}{K}_{11} \end{bmatrix}\right) \begin{Bmatrix} u_1 \\ u_2 \\ u_3 \end{Bmatrix} = \begin{Bmatrix} F_e \\ 0 \\ 0 \end{Bmatrix} \tag{5.70}$$

Eq. (5.70) can be expanded into three equations as shown here

$$-\omega^2(M)u_1 + (\overset{(1)}{K}_{11}\,u_1) - (\overset{(1)}{K}_{12}\,u_2) = F_e \tag{5.71}$$

$$-\omega^2(m_1)u_2 - \left(\overset{(1)}{K_{21}}\,u_1\right) + \left(\left(\overset{(1)}{K_{22}} + \overset{(2)}{K_{11}}\right)u_2\right) - \left(\overset{(2)}{K_{12}}\,u_3\right) = 0 \tag{5.72}$$

and

$$-\omega^2(m_2)u_3 - \left(\overset{(2)}{K_{21}}\,u_2\right) + \left(\left(\overset{(2)}{K_{22}} + \overset{(3)}{K_{11}}\right)u_3\right) = 0 \tag{5.73}$$

Then, Eqs. (5.69) and (5.71) are substituted into Eq. (5.60) to observe the transmissibility characteristic of three-DOF for the LR-MS model. The transmissibility equation can be written as

$$T_r = \left|\frac{F_t}{F_e}\right| = \left|\frac{-\overset{(3)}{K_{21}}u_3}{-\omega^2 + (M)u_1 + \left(\overset{(1)}{K_{11}}u_1\right) - \left(\overset{(1)}{K_{12}}u_2\right)}\right| \tag{5.74}$$

Transmissibility characteristics for three-DOF of the LR-MS model are shown in Fig. 5.20. The first peak remained at 9 Hz which is located at the same point as in one- and two-DOF, respectively. The second peak is located at 52 Hz, which it is less than 8 Hz for the two-DOF systems. This happens because, in three-DOF, the length between two metal plates is shorter than

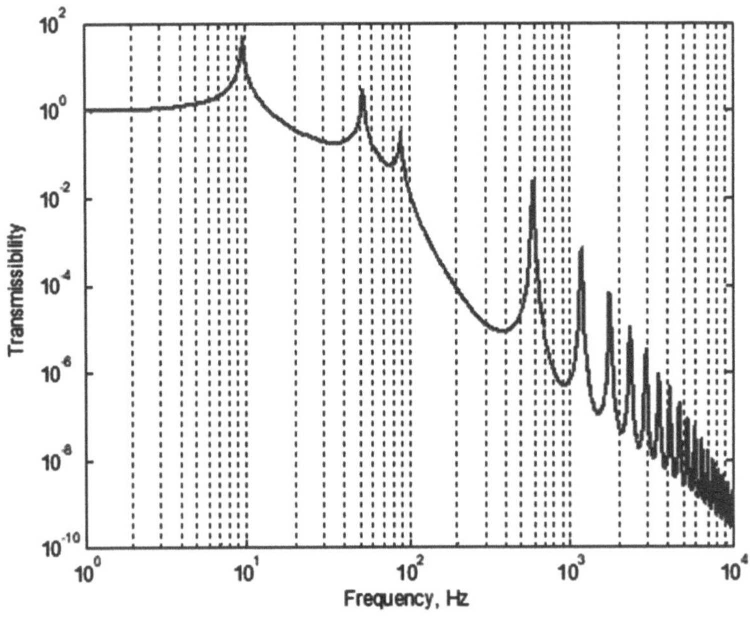

FIGURE 5.20 Transmissibility result for three-degrees-of-freedom laminated rubber–metal spring model.

in two-DOF. According to this, the internal resonance fluctuates easily and gives a response in the low-frequency range. However, the fluctuated wave remains constant, but the amplitude decreases by increasing the quantum of frequency. In three-DOF, an internal resonance wave is going far below the unity compares with the one- and two-DOF, respectively. The value of final transmissibility in the order of 10 kHz is located at 1×10^{-9}.

Based on the previous discussion on the two- and three-DOF of the LR-MS model, the transmissibility characteristics of this system only involved the reaction between transmitted and excitation force. The middle effect is not affected by the transmissibility characteristics. This happens because transmissibility is evaluated in terms of a ratio. Based on this statement, the equation for four- and more-DOF of the LR-MS model can be adopted from Eq. (5.74) and, regarding this equation, the value of stiffness equation and displacement change by the number of DOF. Then, Eqs. (5.75) and (5.76) represent the transmissibility equation for four- and five-DOF, respectively, and Figs. 5.21(A) and (B) show the transmissibility results for both equations.

$$T_r = \left|\frac{F_t}{F_e}\right| = \left|\frac{-\overset{(4)}{K_{21}}u_4}{-\omega^2 + (M)u_1 + \left(\overset{(1)}{K_{11}}u_1\right) - \left(\overset{(1)}{K_{12}}u_2\right)}\right| \tag{5.75}$$

and

$$T_r = \left|\frac{F_t}{F_e}\right| = \left|\frac{-\overset{(5)}{K_{21}}u_5}{-\omega^2 + (M)u_1 + \left(\overset{(1)}{K_{11}}u_1\right) - \left(\overset{(1)}{K_{12}}u_2\right)}\right| \tag{5.76}$$

5.3.2 Results and discussion on the laminated rubber–metal spring model

The basic formulation of mathematical modeling for the LR-MS model in this study was developed using the classical theory of a lumped parameter system. This system consists of mass, spring, and damper components. To investigate the isolation performance, a lumped mass was added with some harmonic force. The rubber material was chosen as a vibration isolator, and ideally, to model the mathematical modeling, it was modeled as a massless component with a constant stiffness and damping coefficient.

An embedded metal plate was used in the LR-MS, and it was treated as a rigid solid mass without damping. This embedded plate was used to create two-DOF for a lumped parameter system. In this development, only vertical motion was considered. For multi-DOF, more embedded

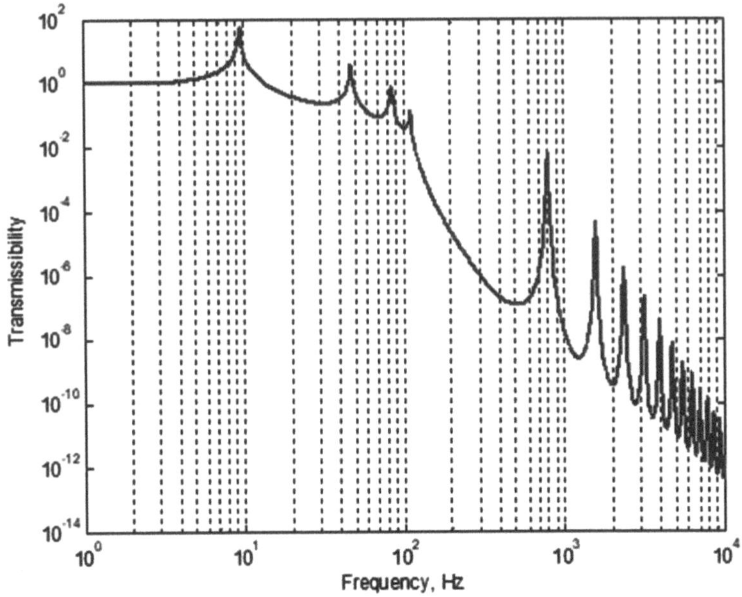

(A)

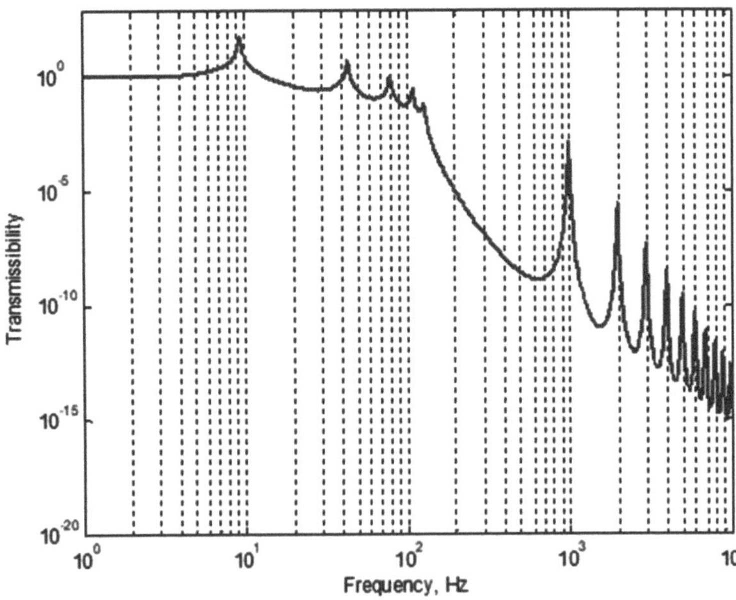

(B)

FIGURE 5.21 Transmissibility result for a laminated rubber–metal spring model: (A) four- and (B) five-degrees-of-freedom.

plate was used and finally, the matrices for multi-DOF were derived. The transmissibility analysis was done by taking the amplitude ratio of transmitted and excitation force. All results were obtained assuming the mass of each embedded plate was the same as the loaded mass and the damping value was also presumed to be very small.

The wave effect in the finite rod was investigated using the equation of motion for internal resonance for a nondispersive finite rod and the basic formulation of mathematical modeling for the LR-MS model. The transmissibility results were verified with a lumped parameter system and it was shown that the natural frequencies for both models have good agreement. However, the internal resonance behavior appeared for the finite rod and was close to unity and went far below by increasing the frequency. The internal resonance only exists in higher frequency because the wavelength in the finite rod becomes much smaller, and finally the incoming and reflected waves in the finite rod occurred.

The mass effect in the finite rod was also investigated in this study. The nondispersive finite rod was inserted with a rigid metal plate which might contribute a larger role than expected, and because of that this could be further investigated. First, some simplifications were made to make it easier to develop the mathematical expressions. Here, the wavelength was presumed to be much greater than the dimension of the finite rod, but the total length was kept constant. Finally, the multi-DOF of the matrices was derived and the transmissibility graphs plotted. The results obtained showed that by increasing the number of metal plates, the stiffness value for each element of the finite rod was reduced. Therefore it increased the internal natural frequencies of the finite rod apart from the fundamental frequency.

Overall, the lumped parameter system represented the basic relationship between the mass-spring-damper model and the LR-MS model. Two effects were later studied which were wave and mass in the finite rod. Based on three equations of motion and mathematical modeling, the complete model of the LR-MS vibration isolator was finally developed. However, the limitation of the developed LR-MS model is that it only investigates on an axial direction. The modeling results obtained also showed that the internal resonance wave was located at unity and was going far below as the frequency increased.

5.4 Conclusions

The mathematical model for the LR-MS vibration isolator using dynamic analysis is presented in this chapter. The equation of motion for internal resonance subjected to longitudinal wave in a finite rod was derived based on the wave propagation method. This equation of motion was used to investigate the longitudinal wave of a nondispersive finite rod. To investigate the internal resonance, free-free boundary conditions were chosen. Impedance and stiffness matrices were developed and shown to be capable of detecting and predicting the internal resonance behavior of a nondispersive finite rod.

The LR-MS mathematical model created in this study is expected to rapidly improve current vibration isolator development by adding the concept of internal resonance to the analytical analysis which has not yet been investigated by other researchers. Using a similar method, the vibration isolator samples development process will also be more effective and faster based on accurate prediction of optimum materials composition and performance parameters. An equally important advantage obtained is that it helps to create a sustainable design and development process for the LR-MS vibration isolator in terms of optimum use of material, lowering the material waste, as well as saving valuable energy during sample production and in return resulting in significantly lower overall product cost. The solution developed can be applied for many applications such as building and construction, automotives, and others.

Acknowledgment

This work is ostensibly supported by the Advanced Manufacturing Centre and Fakulti Kejuruteraan Mekanikal, Universiti Teknikal Malaysia Melaka.

References

Ahmad, M. R., Chen, B., Haque, M. A., & Shah, S. F. A. (2020). Utilization of industrial and hazardous waste materials to formulate energy-efficient hygrothermal biocomposites. *Journal of Cleaner Production, 250*, 119469.

Caïd, N. (2006). *Decoupling the environmental impacts of transport from economic growth*. Paris: OECD Publishing.

Carvalho, J. P., Bragança, L., & Mateus, R. (2019). Optimising building sustainability assessment using BIM. *Automation in Construction, 102*, 170–182.

Chang, C. H. (2002). Modeling of laminated rubber bearings using an analytical stiffness matrix. *International Journal of Solids and Structures, 39*(24), 6055–6078.

Claramunt, J., Fernández-Carrasco, L. J., Ventura, H., & Ardanuy, M. (2016). Natural fiber nonwoven reinforced cement composites as sustainable materials for building envelopes. *Construction and Building Materials, 115*, 230–239.

Clark, G., Kosoris, J., Hong, L. N., & Crul, M. (2009). Design for sustainability: Current trends in sustainable product design and development. *Sustainability, 1*(3), 409–424.

Dickens, J. D. (2000). Dynamic model of vibration isolator under static load. *Journal of Sound and Vibration, 236*(2), 323–337.

Du, Y., Burdisso, R. A., Nikolaidis, E., & Tiwari, D. (2003). Effects of isolators internal resonances on force transmissibility and radiated noise. *Journal of Sound and Vibration, 268*(4), 751–778.

Fregonara, E., Giordano, R., Rolando, D., & Tulliani, J. M. (2016). Integrating environmental and economic sustainability in new building construction and retrofits. *Journal of Urban Technology, 23*(4), 3–28.

Fuller, K. N. G., Gough, J., Pond, T. J., & Ahmadi, H. R. (1997). High damping natural rubber seismic isolators. *Journal of Structural Control, 4*(2), 19–40.

Hossain, M. U., & Ng, S. T. (2019). Influence of waste materials on buildings' life cycle environmental impacts: Adopting resource recovery principle. *Resources, Conservation and Recycling, 142*, 10–23.

Ibrahim, R. A. (2008). Recent advances in nonlinear passive vibration isolators. *Journal of Sound and Vibration, 314*(3-5), 371–452.

Kelly, J. M., & Konstantinidis, D. (2011). *Mechanics of Rubber Bearings for Seismic and Vibration Isolation*. United Kingdom: John Wiley & Sons.

Liu, Z., Zhang, C., Guo, Y., Osmani, M., & Demian, P. (2019). A building information modelling (BIM) based water efficiency (BWe) framework for sustainable building design and construction management. *Electronics, 8*(6), 599.

Moran, P., O'Connell, J., & Goggins, J. (2020). Sustainable energy efficiency retrofits as residenial buildings move towards nearly zero energy building (NZEB) standards. *Energy and Buildings, 211*, 109816.

Mordini, A., & Strauss, A. (2008). An innovative earthquake isolation system using fibre reinforced rubber bearings. *Engineering Structures, 30*(10), 2739–2751.

Röck, M., Hollberg, A., Habert, G., & Passer, A. (2018). LCA and BIM: Visualization of environmental potentials in building construction at early design stages. *Building and Environment, 140*, 153–161.

Salim, M. A., Sheikh Md Fadzullah, S. H., Omar, G., Akop, M. Z., Nurfaizey, A. H., & Tamaldin, N. (2018). *Challenges and developments of rubber materials as vibration isolator. Saleem Hashmi (editor-in-chief), Reference Module in Materials Science and Materials Engineering* (pp. 1–17). Oxford: Elsevier.

Santos, R., Costa, A. A., Silvestre, J. D., & Pyl, L. (2019). Integration of LCA and LCC analysis within a BIM-based environment. *Automation in Construction, 103*, 127–149.

Taiwo, E. M., Yahya, K., & Haron, Z. (2019). Potential of using natural fiber for building acoustic absorber: A review. *Journal of Physics: Conference Series, 1262*(1), 012017.

CHAPTER

6

Design for sustainability integration in education

Ainur Zaireen Zainudin, Norhidayah Md Yunus, Siti Radiaton Adawiyah Zakaria and Aminah Mohsin

Faculty of Built Environment and Surveying, Universiti Teknologi Malaysia, Johor Bahru, Malaysia

6.1 Introduction

The world is facing many interrelated issues such as the impact of climate change, hunger and malnourishment, income inequality, rapid urbanization, and recurrence of financial crises. These issues have led to an understanding of the need for sustainable development (SD). The concerns over understanding SD suggest a good understanding on how to balance the linkage between social, economic, and environmental objectives or needs when making decisions today. Social, economic, and environmental are the three key pillars of the SD concept which need to be supported by proper institutional arrangements to make the SD agenda achievable. Education is among other important institutions to make the arrangement for enhancing SD understanding and applicability among communities. As a matter of fact, it becomes phenomena since it is believed that SD education has the capacity to share concerns for the future of the planet and local communities all over the world (Jucker & Mathar, 2015). This can be seen clearly with the launch of the Global Action Program by UNESCO in 2014 to promote education for sustainable development (ESD) in policy, educational institutions, local communities, and among educators and youth.

Design for Sustainability
DOI: https://doi.org/10.1016/B978-0-12-819482-9.00002-2

Education for sustainability forms approaches that aim to develop students and communities with the values and motivation to take action for sustainability. It thus requires the commitment of academics, especially in higher education (HE) institutions, where theories and practices are highly integrated. Academics can make a significant impact to the advancement of SD as they take on various roles and responsibilities to pioneer designs for sustainability. Embedding SD in HE is a strategy to bring about changes in society that are underpinned by an ethic of solidarity, equality, and mutual respect among people, countries, cultures, and generations; it is development in harmony with nature, meeting the needs of the present generation without compromising the ability of future generations to meet their own needs (United Nations Economic Commission for Europe, UNECE, 2012).

In addition, according to UNESCO, 2019, ESD empowers learners to take informed decisions and responsible actions for environmental integrity, economic viability, and a just society for present and future generations, while respecting cultural diversity. It is about lifelong learning and is an integral part of quality education that allows students to think critically and creatively, and work collaboratively, to appreciate multiple perspectives and act constructively in a way to design for sustainability (DfS).

Nevertheless, how to make sustainability operational has become the starting point (Dawe, Jucker, & Martin, 2005). In education, it can be achieved from a variety of different orientations in teaching. UNESCO aims to improve access to quality education on SD at all levels and in all social contexts, to transform society by reorienting education, and help people develop the knowledge, skills, values, and behaviors needed for SD. It is about including SD issues, such as climate change and biodiversity, into teaching and learning.

In the digital era, education has been enhanced with the integration of technology along with raising awareness on sustainable development goals (SDGs). The delivery of a blended curriculum has been incorporating different strategies taking advantage of the recently available technology to give students and lecturers a greater variety of experiences related to teaching and learning to design for SD.

In this chapter, not only the details on the ESD concepts and challenges are presented, but a number of cases are also reported to show how teaching and learning in HE institutions has been designed and delivered using a blended-mode curriculum to execute the ESD. The first exemplary design offers forms of process to engage students and communities in community action toward SD practices or ways of life. Meanwhile, the second exemplary teaching and learning approaches allow students to develop a product to create and increase communities' awareness of environmental issues. At the end of the teaching and learning activities, both examples have enabled students to develop products that link the balance between the social–environment–economy objectives and needs, thus helping achieve the SDGs.

6.2 Concept of sustainable design and sustainable development in education

Many researchers have using generic terminology in describing efforts made toward introducing the concept of sustainability in education, which is "sustainable development." This may be due to the broader definition of "sustainable development," which caters to three aspects of sustainability, which are environment, social, and economy. The broad definition therefore can be integrated into various fields in education, such as in the areas of social sciences, environmental sciences, and technology. On the other hand, the terminology of "design for sustainability" (DfS) or sustainable design has often been related to technical applications of sustainability toward developing products and processes. DfS implementation involves three aspects in product development, which are environment, people, and profit (Crul & Jan Carel Diehl, 2006). Nevertheless, both terminologies complement each other closely in the context of sustainability, whereby DfS is an approach toward achieving SD. Both terminologies are based on the whole development life cycle, starting from the birth (or design) stage to the end-of-life stage of any activities. The holistic approach is a massive improvement compared to the traditional development approach, which addresses the issue of sustainability only at certain stages in the development process. Furthermore, both sustainable development and DfS include the objective of simultaneously care of the environment as well as the well-being of living things in the same ecosystem in terms of their social and economic needs. Therefore both terminologies can be applied hand-in-hand to describe the current works into how sustainability is integrated into the education system.

Specifically, the Brundtland Report defines SD as "development that meets the needs of the present without compromising the needs of the future generation" (World Commission on Environment and Development, WCED, 1987). The key principle of SD underlying all others is the integration of environmental, social, and economic concerns into all aspects of decision-making. This definition remains valid in the sense that the international debate on the Agenda 2030 and the SDGs does not provide a new definition of sustainability or sustainable development. It reflects the UNCED (Rio) definition of the three dimensions of SD, stated as "We are committed to achieving SDin its three dimensions—economic, social and environmental—in a balanced and integrated manner" (United Nations, 2015).

Fundamental sustainability principles include (Huge, Wass, Eggermont, & Verbruggen, 2011):

- Global responsibility (in tackling global socioecological issues), which refers to international cooperation in a spirit of "shared but differentiated responsibility."
- Integration (of ecological, social, and economic impacts and their interactions), which refers to the reconciliation and integration of environmental and developmental objectives.
- Inter and intragenerational equity, which refers to the needs and preferences of current and future generations (intergenerational equity), and also includes geographical (global North–global South) and social intragenerational equity.
- Precaution (in the face of uncertainty): the precautionary principle states that the lack of full scientific certainty shall not be used as a reason for postponing measures to prevent environmental degradation.
- Participation, which refers to the involvement of all concerned stakeholders in decision-making for SD.

Realizing the 2030 Agenda under SDG 4, to ensure inclusive and equitable quality education and promote lifelong learning opportunities for all, it is essential to refer to the detailed prescription of the targets and indicators, especially of SDG 4.7 which reads: "by 2030 ensure all learners acquire knowledge and skills needed to promote SD, including among others through ESD and sustainable lifestyles, human rights, gender equality, promotion of a culture of peace and non-violence, global citizenship, and appreciation of cultural diversity and of culture's contribution to sustainable development" (United Nations, 2015).

To emphasize, education should play an important role in enabling people to live together in ways that contribute to SD. Especially in recent years, HE institutions have assumed a privileged position as key drivers of education for the SD of the new generation (Rowe, 2007). Education is an instrument to support the SD process and it can generally be stated that ESD covers all kinds of educational concepts, steps, and processes, which are suitable to fostering the individual and/or collective contributions toward SD.

Assessing the ESD first requires clarifying what is meant by SD in HE. The concept of SD has been around for many years. According to Lipscombe (2008), ESD used to mean learning that promotes SD. United Nations Economic Commission for Europe, UNECE (2012) says that ESD promotes sustainable thinking and acting. It enables children and adults to make decisions and at the same time understand how those decisions affect future generations and the lives of others. The interpretation of the Department for Environment Food and Rural Affairs (1999) also reported that ESD is about the learning needed to maintain and improve our quality of life and the quality of life of generations to come without damaging the planet for the future. One Subject Center interpreted Brundtland as follows: "In terms of SD in HE we understand this to mean 'development of curricula and pedagogy to equip students with the skills and knowledge to live and work sustainably. This recognizes the importance of sustainability literacy among students and the growing demand for sustainability skills among employers'" (Dawe et al., 2005).

According to UNESCO (2005), ESD:

- is based on the principles and values that underlie SD;
- deals with the well-being of all four dimensions of sustainability—environment, society, culture, and economy;
- uses a variety of pedagogical techniques that promote participatory learning and higher order thinking skills;
- promotes lifelong learning;
- is locally relevant and culturally appropriate;
- is based on local needs, perceptions, and conditions, but acknowledges that fulfilling local needs often has international effects and consequences;
- engages formal, nonformal, and informal education;
- accommodates the evolving nature of the concept of sustainability;
- addresses content, taking into account context, global issues, and local priorities;
- builds civil capacity for community-based decision-making, social tolerance, environmental stewardship, an adaptable workforce, and a good quality of life;
- is interdisciplinary. No single discipline can claim ESD for itself; all disciplines can contribute to ESD.

This means that ESD allows every human to acquire the knowledge, skills, attitudes, and values necessary to shape a sustainable future. It involves rethinking the curriculum, campus operations, organizational culture, student participation, leadership and management, community relationships, and research (UNESCO, 2014). In this way, the institution itself functions as a role model for the learners.

The implementation process of the ESD depends on a country's needs and the needs of the local people (UNESCO, 2017). In this context, Malaysia also gives priority toward SD through, for example, the Local Agenda 21 in 2001, which was entrusted to the Ministry of Housing and Local Government (Hanifah & Shaharudin, 2016). The steps taken by Malaysia are in tandem with the opinion of researchers (see Gough, 2005; Henderson & Tilbury, 2004; Hopkins, 2013), which stressed that the theory and practice of ESD require the participation of the school community in terms of its practice and implementation.

In fact, according to Aini and Laily (2010), teachers at preschools should start by introducing the first steps to understanding the concept of sustainability to children.

Based on Peter, Libunao, and Abdul Latif (2016), there are several ways by which universities and other HEIs (community colleges and polytechnics included) can contribute to SD:

- By giving SD a place in all curricular, educational, and research programs;
- By playing an important role as local knowledge centers for SD in order to help society meet the challenges of SD at the local level;
- By making SD a leading principle in their own logistics and managerial processes (UNESCO, 2002).

For example, in Malaysia, as part of HEIs, the college community contributes to help promote the awareness of SD using their communications, services, contracts, and partnerships to their clients and stakeholders (Peter et al., 2016). To support ESD, the Treaty promotes eight principles to emphasize a holistic and transformative perspective for sustainable HE (www.sustainabilitytreaties.org):

1. To be transformative, HE must transform itself;
2. Efforts across the HE system must be aligned;
3. Partnership underpins progress;
4. SD is an institutional and sector-wide learning process;
5. Facilitating access to the underprivileged;
6. Inter- and transdisciplinary learning and action;
7. Redefining the notion of quality HE;
8. SD as a whole-of-institution commitment.

In conclusion, such definitions were used creatively. However, according to Burns (2009), although the terms are used alternately, they reflect the same goal. Here students were encouraged to think critically, not so much about the underlying concept of sustainability, but about how it was being used in practice.

6.3 Sustainable design and development: teaching methods and approaches in higher education

Different kinds of models for teaching and learning are required to meet UNESCO's challenge; the very nature of SD is very complex and is multidisciplinary at many levels (UNESCO, 2014). ESD is not easy to teach in a traditional sense, but there is a growing number of examples of new teaching orientations or approaches, which support the development of such skills as interdisciplinary thinking, problem solving, and team working.

The implementation of ESD in HE institutions would require different approaches to curriculum integration. The simplest form of integration consists of an insertion of sustainability-related topics in conventional learning settings such as a lecture series. In improvising, new learning approaches like self-directed and problem-oriented learning that support competence development should be introduced.

Research by Dawe et al. (2005) concluded that teaching methods and approaches are as follows:

- *Conventional methods:* ESD was delivered by conventional methods; lectures, seminars and tutorials, and Internet resources as being essential for ESD's delivery.
- *Personal:* mentioned issues around personal perspectives (e.g., tutors as role models). This orientation places an emphasis on how the tutor can act as a role model for students in order to offer a credible and authoritative perspective on the realities of putting sustainability principles into practice.
- *Reconnecting to reality*: This orientation focuses on real and practical life issues and actual experiences as learning situations. Included within this category were such items as: real-life and real-time interpretation; field trips; environmental audits; the use of role-play; a year abroad; and development of case studies. Some also mentioned assessing the performance of the HEI, in terms of its environment, or social connections, or the HEI's overall performance in relation to ESD, as having value, although they doubted that a formal connection could be made between this and their disciplinary area.
- *Holistic thinking*: this is an approach to systems thinking, stating that it explored the "big" questions about globalization, political systems, and the nature of science. Many of the skills and knowledge for SD are associated with complex, multilayered, and interconnected systems. This approach encompasses a more open-ended exploration of interdependency and transdisciplinary connections between subjects as well as including approaches to developing and honing critical thinking.

Pedagogy is defined as "the art or science of teaching" (OED, 2007). The variation in pedagogical approaches offered is also important: given the diversity of students (e.g., gender or cultural background) within a program, it is desirable and necessary that various approaches be used (UNESCO, 2012). Rodrigo, Michelle, Kaisu, Kim, and Francisco (2017) listed pedagogical approaches selected from those that have well-cited references in ESD literature or are known to be broadly used. The pedagogical approaches have been separated into:

- *Universal*: broadly applicable pedagogies that have been used in many disciplines and contexts (case studies, interdisciplinary team

teaching, lecturing, mind and concept maps, and project and/or problem-based learning);
- *Community and social justice*: pedagogies developed specifically for use in addressing social justice and community-building (community service learning, jigsaw/interlinked teams, participatory action research); and
- *Environmental education*: pedagogies emerging from environmental sciences and environmental education practices (eco-justice and community, place-based environmental education, supply chain/life cycle analysis, and traditional ecological knowledge).

Seatter and Ceulemans (2017) states that three of the approaches to sustainability are viewed through the lenses of the three teaching styles of Roberts and Silva (1968) as in the following table (Table 6.1).

The above discussion shows many ways to embed sustainability in education. However, based on Cottona, Ian, Martyn, and Susie (2009), learning and teaching in the context of sustainability should be based more on investigative learning rather than reproductive learning; as well as on exploring reality rather than reading books; which means it should be based on active learning rather than passive reception of information and on gaining experience rather than acquiring knowledge.

TABLE 6.1 Sustainability curriculum and pedagogy table for higher education (Seatter & Ceulemans, 2017).

Three approaches to sustainability	Three basic teaching styles
1. Emphasis on sustainability as a discourse—A contested discourse that is spoken and claimed by competing groups and cultures, rather than a concept that can be pinned down and identified in the real world	*Trialogue teaching style* – Constructivist learning – Inquiry methodology – Authentic "field" excursions – Case study – Community service learning – Active learning Critical thinking, creativity, and transformative learning are fostered
2. Emphasis on *sustainability implementations*—What is unsustainable, how to make practices more sustainable, and how to evaluate sustainable outcomes	*Imposition teaching style* – Lecture – "As if" situation problem solving. Little critical thinking, creativity, or transformative learning are fostered
3. Emphasis on *definitions of sustainability*—Where they have emerged from, what they attempt to achieve, and how they can be compared	*Imposition teaching style* – Lecture – Traditional lab exercises – Question and answer – Assessment: exam and/or essay. Neither critical thinking, creativity, nor transformative learning are fostered

Teaching methods and approaches including pedagogical teaching styles can be utilized as a starting point in HE. However, the improvements in the future HE for Sustainable Development (HESD) program should be aware of the emphases and teaching styles regarding sustainability.

6.4 Embedding the design for sustainability concept into education: example of practices in a higher education institution

"Sustainability" has become a global focus, particularly after the reconciliation of the social, economic, and ecological dimensions of change in the Brundtland Report 1987. Since then, various initiatives have been put forward by local as well as international communities committing themselves to the movement for a better quality of life. To support a SD process, social actors are engaged in many programs associated with sustainability agendas and education is seen to be among the most important policy instruments that could build up better engagement or community action of the social actors. Consequently, the SD field of study has been included into the curriculum at schools and HE institutions. In the remaining sections of this chapter, examples of teaching and learning designs embedding both DfS and the SD concept are presented based on the activities undertaken by the lecturers and students in Universiti Teknologi Malaysia (UTM).

6.4.1 Experiential learning of design for sustainability: design of a floating wetland product for the Riverdale Project

Experiential learning through a practical design and development project is a very beneficial method of giving hands-on experience to students about DfS and SD concepts in general. Furthermore, the approach also created a better understanding to students about the real practical challenges in implementing DfS solutions to solve daily problems, especially related to environmental issues. Transferring students' knowledge from a classroom setting to a practical setting which involved community members within the same project could also enhance their interpersonal skills, so that they can encourage better acceptance by communities in adopting their solution toward gaining the intended environmental improvement.

HE institutions are also increasingly required to introduce elements of sustainability into their curricular activities. Therefore the search for an approach capable of assisting educators to embed the sustainability context into their syllabus becomes highly relevant. In teaching sustainability to HE learners, the delivery of the sustainability context requires the "reconnecting to reality" teaching approach for effective learning. In order to create a "reconnecting to reality" teaching approach in sustainability-related academic courses, educators should undertake a real-life project. Through this kind of project, the teaching and learning methods not only emphasize students' achievements in cognitive aspects but also in developing their soft skills. In this context, soft skills are seen to be useful to integrate and thus enable students to live together in ways that contribute to SD.

Basically, soft skills refer to the personalities, attributes, qualities, and personal behaviors of individuals (Majid, Liming, Tong, & Raihana, 2012). In 2006, the Ministry of Higher Education, Malaysia launched Modules of Soft Skills Development for Malaysian HE Institutions which aims at producing quality human capital in term of skills and competitiveness at an international level (Mohd Shafie & Ahmad, 2015). The Module has been implemented by many HE institutions, including by UTM. It has been embedded in various in-class learning activities, out-of-class activities, as well as curriculum activities (Rozlin et al., 2018).

As an example, at UTM, SD is one of the elective academic courses offered to undergraduate students of the Land Administration and Development Program. This academic course is offered in the final semester of the undergraduate study, in which, besides knowledge, there is a focus on developing students' practical, critical thinking and problem solving as well as life-long learning skills. In the context of knowledge, students are not only provided with the principles of SD, but most importantly they are taught to better understand the importance of this model to be adopted in developing every nation in the world. As previously mentioned, sustainability needs community action to be successfully achieved (Diesendorf, 2000). It is thus very important for this community action context to be included into the syllabus of an SD course.

It is believed that community action is best delivered to students not only through conventional lectures but also by engaging students in experiential learning activities. Experiential learning is "the process whereby knowledge is created through the transformation of experience" (Kolb & Kolb, 2005). For this SD course, the transformation of experience has been executed through an innovative experiential learning methodology that integrates two teaching and learning methods, namely the role-playing and service-learning methods. In addition, the service learning was set up through a synergy made with the campus community and industry in order to minimize the financial implications to the students of this course. It is believed that, by executing these strategies, experiential education could be enriched and better achieve the targeted learning outcomes of the course.

The main objective of these experiential learning activities is to achieve the first learning outcome of this SD course, which is to enhance

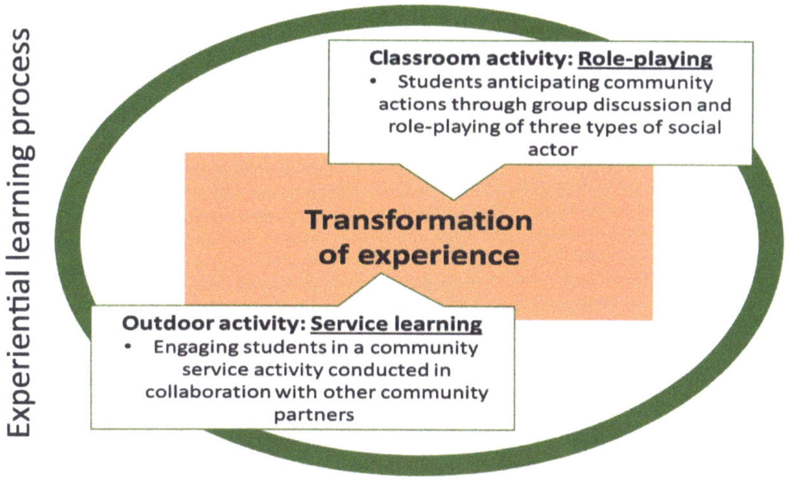

FIGURE 6.1 Experiential learning process framework.

students' understanding on the importance of SD. Considering the broad area of sustainability literature, "community action" was opted to be the focus area for achieving this learning outcome, thus being the theme for the entire experiential learning process. This process has been designed to have two main phases in order to better execute the transformation of experience during the learning process. The first phase was conducted as a classroom learning activity and the latter phase was then carried out by students outside the classroom by engaging them in an actual community service program. The framework of this experiential learning process is demonstrated in Fig. 6.1.

As a classroom learning activity, the community action is intended to be learnt by students using the role-playing method. Role-playing was utilized to allow students to select, organize, and integrate information (Mayer, 1989) by themselves from acting activities. For diverse experiences on community action, students were divided into three social actor groups, namely the government, the service provider agency, and the society. Before the acting began, students were instructed to conduct a short brainstorming session to anticipate the (community) actions of the social actor that they would represent toward the Solid Waste Separation Program as enforced by the government. For this purpose, students were guided with the given issues to keep students' focus on the chosen theme of the learning content. Upon the brainstorming session, the role-playing activity was conducted in a sequence starting with the society group, followed by the service provider agency group, and finally the government group. Fig. 6.2 shows one of the groups acting like having a meeting session for their role-playing. In this role-playing activity, each group was required to act

FIGURE 6.2 Role-playing by a group of students.

on community action based on the role of the social actor they represented. They were also required to follow closely the community actions performed by other groups before them as they need to show their response to those actions through their actions when it was their turn.

The outcomes of the role-playing activity were then discussed by focusing on the different community actions shown by each social actor. In the discussion, the lecturer led the students to compare the anticipated actions of students gathered in the brainstorming session with the actual outcomes of their acting when the students played the roles themselves. Eventually, it can be seen that the activity has allowed students to relate the community actions with the importance of SD initiative according to different social actors not only through critical thinking skills they acquired during the brainstorming session, but they were also able to experience them through their own acting activities.

In the second phase of the experiential learning process, students were intended to further experience the community action context using a service-learning method. Service learning is one of the many experiential education approaches that allows students to learn from real experiences (Furco, 1996; Sigmon, 1979). At UTM, service learning has been identified as one of the New Academia Learning Initiatives that is highly recommended as part of student-centric learning modes. By using service learning, students' experience which was once simulated in the classroom through the role-playing activity was then transformed to a real project. Nevertheless, using the service-learning method is not without challenges. The common main barriers of service learning are time constraints, other commitments, inadequate funding, and lack of institutional support (Kolenko, Porter, Wheatley, & Colby, 1996). To avoid service-learning failure, the synergy concept has been adopted. Literally, synergy is the combined power of a group of things when they are working together that is greater than the total power achieved by each working separately. Synergy can be achieved by determining activities with minimal fund and resources and exploring existing resources that can be used effectively for greater project efficiency. In order to minimize the financial implications to students of this sustainability course, synergized service learning has been designed in collaboration with the organizer of the Riverdale Project.

The Riverdale Project was a community service project organized by the Rahman Putra College Student Committee (KRP) in collaboration with UTM Centre of Sustainability (UTMCS), Center for Environmental Sustainability and Water Security (IPASA), and Iskandar Puteri City Council (MBIP). It was aimed to raise the awareness and concern of the UTM campus community to river cleanliness and water resources, besides fostering a volunteerism spirit among the campus community. This project involved four phases of activities, namely the Plastic Water Bottle Recycling Campaign, Floating Wetland Design Competition (Fig. 6.3), Floating Wetland Production (Fig. 6.4), and Launching of Riverdale Project (Fig. 6.5).

The Plastic Water Bottle Recycling Campaign aimed to get 2000 plastic water bottles to be recycled as one of the materials for the floating wetland production. This was where the synergy between the Riverdale Project and SD Course was formed. The students of the SD Course were engaged with the Riverdale Project as volunteers to assist the Riverdale's organizer to collect the targeted number of plastic water bottles (Fig. 6.6).

Fig. 6.7 shows that after the bottles were collected, the bottles will be cut in the middle and at its bottom so that a tree could easily be put into the bottle. In addition, each student was required to interview two participants of the Riverdale Project, which had also contributed water bottles. The interview was aimed at gathering participants' feedback on their participation in this community service project and the importance of the Riverdale Project in the pursuit of SD. The results of the interview were then submitted as an individual assignment report for SD Course assessment purposes.

The Riverdale project also received national media coverage such as Astro Awani, Sinar Harian, Kampus Uols, and News UTM. Iskandar Puteri City Council also included the Riverdale Project in the Sungai Skudai Rejuvenation Plan, which involves behavior change through community awareness and a participation campaign. The Riverdale Project was also exhibited in the Sustainable Campus Convention at Universiti Kebangsaan Malaysia (23 April, 2019 to 25 April, 2019), Coast Exhibition at Desaru Johor (5 July, 2019 to 7 July, 2019) and Good Earth Exhibition 2019 at Sutera Mall, Johor Bahru (13 September, 2019 to 29 September, 2019).

FIGURE 6.3 Floating Wetland Design Competition.

FIGURE 6.4 Floating Wetland Production.

FIGURE 6.5 Launching of Riverdale Project.

FIGURE 6.6 Used plastic water bottles collected by students.

FIGURE 6.7 Used plastic water bottles cutting and production phase.

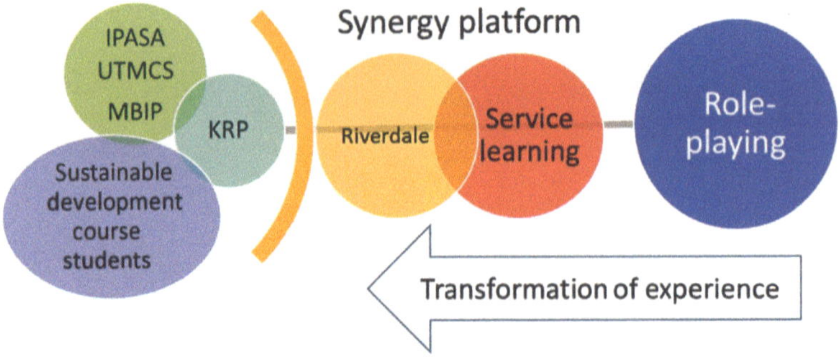

FIGURE 6.8 Strategies in executing the designed teaching and learning modes.

Fig. 6.8 shows clearly the whole idea of innovation made throughout this experiential learning process. This innovation was designed to enrich the experiential education by combining classroom learning activities through role-playing, and the learning process was then strengthened by a real service-learning project on a volunteerism basis. The novelty of this whole idea was on the use of a synergy concept to overcome service-learning barriers. This approach is important to ensure the students' experience in the simulated learning outcome could be well transformed into a real experience through a real community service project with minimal financial implications.

Although the contribution made by the students in the Riverdale Project seems small, the benefits of the overall learning process were tremendous in the following ways:

1. Role-playing activity enables students to gain knowledge not through conventional lectures but through a student-centric learning process which includes critical thinking and response-based activities.
2. The synergy concept adopted in this experiential education teaches students the importance of social networks as one of the success factors for SD initiatives.
3. The transformation of experience occurs in the learning process from a classroom environment to the real-world environment and has enriched students' learning on civic responsibility that may foster their civic engagement to further contribute to the community in which they live.
4. The instruction for the students to work together with the Riverdale's organizer has benefited all partners of the project and its targeted beneficiaries.
5. The students have the opportunity to learn to design and produce products for environmental sustainability.
6. The instruction for students to gather other participants' views on the Riverdale Project enables students not only learn the importance of SD initiative from their own experience after participating in the program but also from the experience of others.

6.4.2 Creating awareness of a sustainable design solution through social media video development: plastic recycling and solid waste segregation

In-line with a 21st-century teaching and learning approach, the sustainable design concept in education can be promoted and delivered using the internet medium, in particular, through social media platforms such as YouTube, Facebook, and Instagram. As many people have better access to the internet through various gadgets, especially cell phones, the use of these social media platforms is one of the most effective methods to spread awareness about the sustainable design concept to educate students as well as communities. The awareness can be communicated in the form of a creative poster and an interactive video presentation (combining visual and audio). Through this method, information about sustainable design can be delivered, for example, in terms of reducing waste material, reuse and recycling of waste materials, and using biodegradable alternative materials made from renewable resources.

Learning based on video development plays a role in promoting student-centered learning strategies (Sidek & Hashim, 2016). Using video can facilitate learning activities while making it fun for students to learn. Student-centered learning enables students to be more active in gaining knowledge. Also, video can be a very powerful educational tool for this purpose. However, much of its power lies not in the video itself but in the way it is used. The video itself is not a goal or an end in itself, but a way of achieving learning objectives and goals (Embi, 2011). Effective educational videos are not television-based teaching for students but lecturers teaching students using video as a tool for disclosure (Duffy, 2008).

YouTube is increasingly being used by educators as a source of pedagogy. Its use includes educational videos to online spaces and student-written content sharing. Some general guidelines by Clark and Mayer (2002) in considering appropriate use of any media to improve teaching suggest that the media should be in line with the desired learning outcomes or performance, reduce cognitive load, contain meaningful text or graphics, according to the literacy level of student learning and students find that video facilitates the learning process. The methodology used in the project is the video recording method during the project and needs to be uploaded on YouTube. There are two exemplary projects of video development to create awareness for SD during teaching and learning activity which are the UTM Skudai toward Sustainable Campus and the Save the Earth program.

The UTM Sustainable Campus project aimed to achieve awareness to reduce the use of plastic bags among UTM students. In order to support

FIGURE 6.9 Distribution of questionnaire forms among UTM students.

the project, the undergraduate students of the Land Administration and Development Program who enrolled in the SD Course have to develop a video about UTM Skudai toward Sustainable Campus using the "Reduce Plastic Bag" theme and upload it on YouTube. Fig. 6.9 shows, before the video development, the students distributing questionnaire forms in several areas within UTM such as at the Meranti Arcade, Cengal Arcade, Space Arcade, Azah Grocery Store, and Mak Ngah Café. Each student played a role in distributing questionnaires at the targeted areas.

Before video development, students were required to search articles using various keywords through Google Scholar, news sources, and others. Fig. 6.10 shows one of the groups acting as a student buying things without plastic bags but using recycled bags. They were also needed to act in ways such as giving advice to other students to reduce the use of plastic bags. After that, the video was uploaded on YouTube (as Fig. 6.11).

The second example is the Save The Earth project. In this project, students were required to develop a video through the Save The Earth project using the "We Care We Clean" slogan and to subsequently upload it on YouTube. The Save The Earth project is designed to provide students with the opportunity to better understand SD. The program also exposes students to development related to environmental preservation and conservation in line with the concept of SD. This program is one of the steps for students to better understand and practice what has being taught in the classroom.

This project is a community service program conducting solid waste segregation activities in Kampung Pulau Semut, Malacca, Malaysia. The program is a collaboration between lecturers and final year students of Bachelor of Science (Land Administration and Development) UTM with Solid Waste Corporation (SWCorp) Alor Gajah Branch, Alor Gajah

FIGURE 6.10 Reduce the use of plastic bags.

FIGURE 6.11 Video uploaded to YouTube about the "UTM Skudai toward Sustainable Campus."

Municipal Council, and residents of Kampung Pulau Semut, Malacca. In this project, students were divided into four groups to collect solid wastes in the village. During the collection, each group explained to the villagers about the 3R practice (reduce, reuse, recycle) in managing their solid wastes. Fig. 6.12 shows the video about the Save The Earth project. During the project, the villagers gave good cooperation in carrying out solid waste segregation activities and appreciated how the community service program benefited them.

FIGURE 6.12 Video in YouTube for "Save The Earth."

YouTube invites users to enjoy the video and experience engaging in video content as a reviewer and creator, an activity that improves student visual literacy, as an important skill in the current electronic culture. This application encourages experimentation using new media among students. Most educators believe that the effort of content building is a useful learning exercise, which helps to improve understanding of study materials and equipment used to build the video content. In addition, it has the potential to expose students to new skills at the same time as connecting them to different communities online. As a social software application, YouTube has become part of the trend among Net Generation students to replace passive learning with active participation, where all students can voice their views, anyone can make a contribution, and the value is not in the content itself, but rather to a network of students working together to form content and support each other in achieving their learning goals (Embi, 2011).

6.4.3 Student-community-based activities on the product end-of-life stage: solid waste separation and recycling campaign

As another example of current practices in HE institution in embedding DfS and sustainable design approach in their curriculum design, this section presents a real-life project executed by an academic course in the Faculty of Geoinformation and Real Estate, UTM, in 2015.

Basically, the real-life project was conducted to achieve one of the academic course learning outcomes which is to create awareness on practicing sustainable design and developments, especially to cater for the end-of-life stages of product use. In general, sustainable design caters for the whole product and process life cycle stages, starting from raw material extraction, production, produce use, transportation, until disposal. There are many methods of product disposal, such as landfilling, recycling, and composting. Hence, knowledge and experience in choosing and applying a sustainable disposal method is critical to educating students in handling product end-of-life stages. Furthermore, applying the knowledge about an appropriate strategy in product disposal and sharing the knowledge to product users especially in the local community is also important toward successful implementation of DfS and SD. The project also included the participation of other stakeholders involved in product end-of-life processing such as the local authority and waste management industries.

For this purpose, the students chose to conduct a community service program called "Save the Earth" with the theme "We Care We Clean" to help the government to promote the solid waste separation program in the country. The solid waste separation program is also one of the agendas promoted by the Ministry of Urban Wellbeing, Housing, and Local Government in Malaysia. Starting from 1 September, 2015 all premises in selected states in the country are required to separate solid waste at the source. This implementation is pursuant to regulations under Solid Waste and Public Cleansing Management Act 2007 (Act 672) enforced in several states and territories in Malaysia, that is, in Johor, Melaka, Negeri Sembilan, Pahang, Kedah, Perlis, Kuala Lumpur, and Putrajaya. According to Mohammad Arshad, Zainudin, Zakaria, Mansor, and Md Yunus (2017), solid waste separation at source is the practice of separating or setting aside goods and post-consumer materials produced by households to prevent them from entering mixed-waste streams. These wastes are stored in designated containers or bins in order to facilitate recycling and disposal. This particular program was chosen by the students due to the result of the Solid Waste Separation at Source Program that was disappointing ever since its implementation in the country due to the low participation and involvement of Malaysian society. Therefore the students saw that their involvement was vital in order to help the government.

The program that organized by the students was conducted in collaboration with Alor Gajah City Council and the villagers of Kampung Pulau Semut, Melaka, Malaysia. Basically, this program was designed to enable students to create awareness on the importance of solid waste separation and recycling among the villagers. Fig. 6.13 shows the villagers of Kampung Pulau Semut that were involved in the Save the Earth program.

FIGURE 6.13 Villagers of Kampung Pulau Semut who participated in the Save the Earth program.

FIGURE 6.14 Activities conducted during the Save the Earth program.

In addition, through this program, the villagers were taught and assisted by the students on how to properly manage their solid waste, particularly for separation and recycling purposes (Fig. 6.14). Therefore, SWCorp, the waste collection service provider, was also invited to be involved in the program to support the waste separation activities, particularly to buy the recyclable wastes from the villagers. This collaboration was indirectly able to show to the villagers how solid waste segregation would benefit them economically.

Through this program, students and villagers were not only trained to contribute to environmental sustainability, but the students especially were able to enhance their soft skills in practicing good values in society. Here, the volunteerism spirit was indirectly developed. The spirit of volunteerism is a core value that must exist in modern society, especially among the young generation (Wilson, 2000). Through a volunteerism program like the Save the Earth, good values can be applied to produce quality graduates. The learning outcome from this activity was also designed in such a way as to enhance the students' soft skills. The targeted soft skill was the ability to practice good ethics and positive values in society which are crucial in the sustainability movement.

6.5 Challenges of integrating the sustainable design concept in higher education

In the face of the 21st millennium and with the advent of Industrial Revolution 4.0 technology, the field of education should be able to move along as education becomes more relevant to today's generation. The traditional education design may no longer be effective in this century. With the need to provide sustainable education there are several challenges to be faced, as shown in Fig. 6.15.

Fig. 6.15 shows that there are five main challenges to be considered for designing sustainable education, which are policy, educator, environment, economy, and classroom activity. The details are described in the following subsections.

6.5.1 Policy

Antonio, Teresa, and Carlos (2006) explained that an education policy comprises the standards and government arrangements in the instructive circle just as the gathering of laws and principles that administer the activity of education frameworks. Education occurs in numerous structures for various reasons through numerous institutions. This includes early youth education, kindergarten, multiyear schools or colleges, graduate, and professional education. In this manner, education policy should be straightforwardly influencing education at all ages.

The education policy has to cover all aspects, including explicitly from the field of schools, including school size, class size, safe and healthy environment in school, educator's requirement, educator's payment, teaching techniques, curricular substance, graduation necessities, school foundation venture, and the qualities that schools are relied upon to maintain and demonstrate (Bush et al., 2019). In fact, the implementation of the policy indirectly may change the structure of an institution and administration in

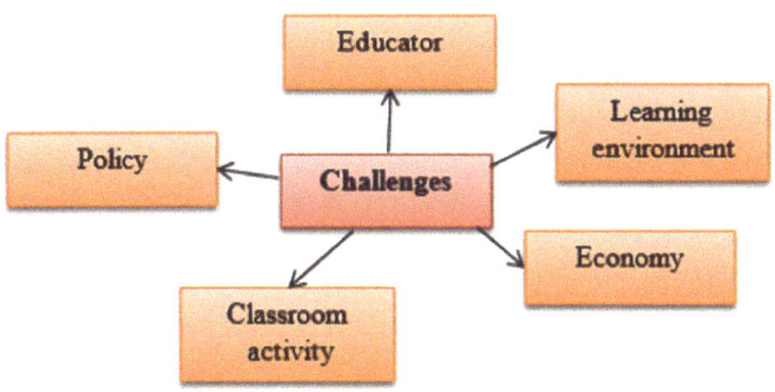

FIGURE 6.15 The challenges for designing sustainability in education.

order to align with the standards of supportable improvement. These must be changed to ensure a coordinated way to deal with sustainable education development. Sustainable education will enable educators and students to take educational challenges with a comprehensive methodology, going past the conventional perspectives on learning and connections to reach an agreeable structure, arranging, and generation of information. Therefore education policy needs to be revised to achieve quality education that is universally accessible.

6.5.2 Educator

The country's largest challenge in the education system is to provide quality and knowledgeable educators to produce outstanding students in line with the vision of 2020 that seeks to enhance fully developed industries by 2020. The role of educators is significant in nation building. The major role of the educator is to engage teachers and leaders to decipher and adapt policy rather than simply enforce it. Some countries have educator affirmation necessities and others have educator training program prerequisites. In Malaysia, for instance at UTM, new educators are obliged to attend a program called "Sijil Pengajaran Pengajian Tinggi" and now recognized as "Baseline Competencies for Differentiated Career Pathway" (BC4DCP). This program is aimed at exposing various teaching skills to lecturers to convey more holistic approaches and stresses the significance of comprehension of the underlying causes of the unsustainability of current trends.

The characteristics of the younger generation are rich in information (including science and technology), mind power (creative and proactive), high and pure values (including caring), self-esteem, entrepreneurship, and skills. In order to determine the requirements of these human qualities to be produced, educators must also be equipped with the following features such as mastering the subject (curriculum content), skilled in pedagogy (teaching and learning), understanding the students' progress and caring about them, understanding cognitive psychology, have counseling skills, use the latest technology, be able to refine the findings of recent research and research, can work with colleagues and others, and have confidence in the role and contribution of educators (Baharom et al., 2013).

The staff development program must be implemented by institutions of higher learning that want to grow progressively and dynamically. For example, institutions of higher learning should inculcate and strengthen the value of integrity among academics, administrative officers, support staff, and students because institutions of higher learning are centers of knowledge development that will produce a generation of smart people. In addition, the institutions of higher learning should provide relevant technology to the educators, so that they are able to choose the appropriate use of technology to create interactive and collaborative learning without time and place constraints. This can be achieved with the Internet which allows online and blended learning. The supportive learning environment provided by an institution of higher learning can support educators in their journey toward transformational teaching capability and learning engagement (Leal Filho et al., 2018).

In fact, the times have changed. Consequently, the old style of lectures may not be applicable to the current generation. Subsequently, the great educator cannot simply pass on knowledge and content, but needs to always expound them critically, creating enthusiasm, so as to create ever-new information and culture that can cope with the new millennium in the 21st century.

6.5.3 Learning environment

The learning environment alludes to the physical area, context, and culture in which students learn. It makes sense that students will improve when they are learning in a positive environment. In fact, students are highly influenced by the learning environment, and educators play a significant role in controlling the setting. To make a positive environment, an important aspect is the set up of the classroom. This is an important component of the learning environment as it is an integral part of classroom management to support teaching and learning.

For example, Fig. 6.16 shows an unsupportive classroom setting which shows the students' seats and desks fixed to the floor in rows. This setting makes discussion among students in the classroom difficult. In addition, small classrooms also cause limited activity.

The physical atmosphere of the classroom should help prevent behavior issues as well as promote and improve learning among students. Furthermore, a supportive environment encourages and enhances the social, scholarly, physical, and passionate development of students. Therefore the classroom setting should support a learning environment that encourages educators and students to share the focus in the classroom. A supportive

FIGURE 6.16 Unsupportive learning environment.

learning environment is more than the physical classroom and assets, but also about the qualities and connections among the students.

The learning environment is different from class to class. Generally, there are two main types of learning environment, which are teaching-centered learning and student-centered learning. Teaching-centered learning is a strategy whereby the educators are effectively in class conveying all the topics and once in a while allow the students to communicate in the classroom. In this scenario, the students are in a latent, receptive mode listening to the educator's instructions. Meanwhile, the student-centered learning approach is where educator and student discourses are exceptionally empowered in the class. Table 6.2 demonstrates the contrasts between teaching-centered and student-centered learning.

As indicated in Table 6.2, there are differences between teaching-centered and student-centered learning. The teaching-centered learning method is a traditional method used in teaching, in other words traditional ways where teachers play a more active role in the classroom.

TABLE 6.2 The differences between teaching-centered and student-centered learning (Essays, 2018; Hamdi, 2018).

Differences	Teaching-centered learning	Student-centered learning
Concept	Students put the majority of their emphasis on the educator. The educator talks, while the students only listen in. During exercises, students work alone, collaboration is discouraged	Educator and students share the core interest. Rather than listening to the educator solely, students and educator collaborate similarly. Group work is encouraged and students figure out how to team up and speak with each other
Advantages	The classroom remains deliberate. Students are quiet, and the educator holds full control of the classroom and its exercise. Additionally, since students learn alone, they figure out how to be autonomous and settle on their choices	Students learn significant communicative and community-oriented abilities through group work. Indeed, students are increasingly keen on learning activities when they can interface with each other and participate actively
Disadvantages	When students work alone, they do not figure out how to team up with other students and communication skills may suffer. Additionally, this approach can be exhausting for students. Their brains may wander, and they may miss significant facts	Since students are talking, classrooms are regularly busy, noisy, and disordered. Educators must endeavor to deal with each students' activities at once, which can be troublesome when students are at various phases of a similar project. Additionally, a few students like to work alone, so group work can end up being problematic

However, in order to realize sustainable education, student-centered learning methods are more suitable because they not only educate students in academic matters, but also develop students' skills such as collaboration and communication. However, to create a student-centered learning environment, it is a major challenge for teachers in particular, as they need to change their teaching methods to fit the 21st-century student. In addition, it presents a challenge for HE providers to provide a proactive learning environment for today's learning.

6.5.4 Economy

The economy is a great challenge to realizing sustainable education. This is because every change that is made has a certain cost to keep it running smoothly, for example, the cost of providing a proactive learning environment and the cost of sending teachers to 21st-century teaching and learning courses. This is not to mention, today's student-centered learning approach may increase the cost of providing interactive activities in the classroom. Therefore the economy of an institution must be stable in order to implement sustainable education.

6.5.5 Classroom activities

Activities in the classroom are important for getting students to interact and converse with each other. The purpose of interactive classroom activities is to prevent students from becoming exhausted/bored and losing focus on their learning. Therefore, the role of teachers is significant in giving intelligent activities such as learning experiences, learning questions, solving community problems, and playing games. Fig. 6.17 shows examples of interactive activities in the classroom, so that students enjoy learning and can engage until the end of the lecture.

To sum up, there are five challenges that need to be faced in designing sustainable education including challenges to policies, educators,

FIGURE 6.17 Interactive classroom activities.

learning environments, economics, and classroom activities. All of these challenges need to be addressed so that changes in the training paradigm, which are institutional changes, educational programs, and training processes, enable them to better respond to the requirements of current and future generations.

6.6 Conclusions

Education has a tremendous impact on the various opportunities in obtaining and maintaining quality of life. Quality education is very effective in combating poverty, building democracy, and building prosperous communities. Through education individuals are able to build their confidence in the face of a challenging world. Quality education is a concept that is dynamic, changing, and evolving over time and changing in social, economic, and local contexts. Therefore, to produce quality education in schools, colleges, and universities, sustainable education is required. Sustainable education is an emphasis on both the knowledge and the soft skills to shape a sustainable future. In realizing sustainable education, educators play a very important role. This is because educators are the people closest to educating students. To date, various methods have been used in teaching and learning to ensure sustainable education such as experience learning, community engagement, and problem solving. These examples show that current teaching is more about student-centered learning, which means that students and educators are actively discussing specific topics. This approach not only means that students gain knowledge but also develop many soft skills such as communication skills, collaboration, and leadership as well as enabling them to develop designs of processes and products that will benefit the wider communities toward achieving SDGs. As a result, the educational institution has become a prominent producer of pioneering designs for sustainability. Nevertheless, in order to implement sustainable education, there are some challenges that need to be addressed such as policy transformation, the availability of educators to improve or change their teaching approaches, budgets, and the extent to which the current learning environment supports sustainable education. However, all these challenges need to be overcome to provide quality teaching and learning and directly produce highly competitive students, who are self-reliant and able to adapt to this challenging world.

Acknowledgment

The authors thank the Faculty of Built Environment and Surveying, Universiti Teknologi Malaysia, for providing valuable support throughout the writing of this chapter.

References

Aini, M., & Laily, P. (2010). Preparedness of Malaysian pre-school educators for environmental education. *Pertanika Journal of Social Sciences & Humanities, 18*(2), 271–283.

Antonio, A. M., Teresa, M. M., & Carlos, A. V. C. (2006). Education for sustainable: Challenges and trends. *Clean Technology and Environment Policy Journal, 8*(1), 31–37.

Baharom, M., Roslee, T., Muhammad, S. T., Abdul, S. A., Shukri, Z., & Mohd Yusof, A. (2013). Cabaran Transformasi Agenda Pendidikan Negara [Challenges of Transformation of National Education Agenda]. Utusan Borneo, 4.

Burns, H. (2009). *Education as sustainability: An action research study of the Burns model of sustainability pedagogy*. Portland State University.

Bush, T., Ng, Y. M., Wei, K. T., Chay, J., Glover, D., & Lei, M. T. (2019). *Educational policy in Malaysia: Implementation challenges and policy proposal*. Singapore: The HEAD Foundation.

Clark, R. C., & Mayer, R. E. (2002). *E-learning and the science of instruction: Proven guidelines for consumers and designers of multimedia learning*. San Francisco: Wiley.

Cottona, D., Ian, B., Martyn, W., & Susie, B. (2009). Revolutions and second-best solutions: education for sustainable development in higher education. *Studies in Higher Education, 34*(7), 719–733.

Crul, M., & Jan Carel Diehl, J. C. (2006). *Design for sustainability: A practical approach for developing economies*. UNEP/Earthprint accessed online from. Available from http://www.d4s-de.org/manual/d4stotalmanual.pdf.

Dawe, G., Jucker, R. & Martin, S. (2005). *Sustainable development in higher education: Current practice and future developments*. A report for The Higher Education Academy, York.

Department for Environment Food and Rural Affairs (1999). *Sustainable development education panel: First annual report 1998*. Retrieved from <https://webarchive.nationalarchives.gov.uk/20080306221400/> <http://www.defra.gov.uk/environment/sustainable/educpanel/1998ar/ann4.htm> Accessed 14.10. 2019.

Diesendorf, M. (2000). Sustainability and sustainable development. In D. Dunphy, J. Benveniste, A. Griffiths, & P. Sutton (Eds.), *Sustainability: The corporate challenge of the 21st century* (pp. 19–37). Sydney: Allen & Unwin, chap. 2.

Duffy, P. (2008). Engaging the YouTube Google-eyed generation: Strategies for using Web 2.0 in teaching and learning. *The Electronic Journal of e-Learning, 6*(2), 119–130.

Embi, M.A. (2011). e-Learning in Malaysian institutions of higher learning: status, trends and challenges. Keynote address presented at the International Lifelong Learning Conference (ICLLL 2011), November 14–15, Kuala Lumpur: Seri Pacific Hotel.

Essays, U. K. (2018). Difference between teacher and learner centered approach. Retrieved from <https://www.ukessays.com/essays/education/teacher-centered-and-student-centered-learning-education-essay.php?vref = 1>.

Furco, A. (1996). Service-learning: A balanced approach to experiential education. Retrieved from <digitalcommons.unomaha.edu>. Accessed 27.06. 2019.

Gough, A. (2005). Education for sustainable development: Challenges for schools, curriculum and instruction. In *JPPG Education Conference*, 28–30 August, The Shangri-La Hotel, Penang.

Hamdi, S. (2018). A comparison of teacher-centered and student-centered approaches in educational settings. *International Journal of Social Sciences and Educational Studies, 5*(1), 164–167.

Hanifah, M., & Shaharudin, I. (2016). Education for sustainable development in Malaysia: A study of teacher and student awareness. *Malaysian Journal of Society and Space, 12*(6), 77–88, ISSN 2180-2491.

Henderson, K. & Tilbury, D. (2004) Whole-school approaches to sustainability: An international review of sustainable school programs. *Report Prepared by the Australian Research Institute in Education for Sustainability (ARIES) for The Department of the Environment and Heritage*, Australian Government.

Hopkins, C. (2013). Educating for sustainability an emerging purpose of education. *Kappa Delta Pi Record, 49*(3), 122–125.

Huge, J., Wass, T., Eggermont, G., & Verbruggen, A. (2011). Impact assessment for sustainable energy future—Reflections and practical experiences. *Energy Policy, 39*(10), 6243–6253.

Jucker, R., & Mathar, R. (2015). Introduction: from a single project to a systemic approach to sustainability—An overview of developments in Europe. In R. Jucker, & R. Mathar (Eds.), *Schooling for sustainable development in Europe* (pp. 3–14). Springer International Publishing.

Kolb, A. Y., & Kolb, D. A. (2005). Learning styles and learning spaces: Enhancing experiential learning in higher education. *Academy of Management Learning & Education, 4*(2), 193–212.

Kolenko, T. A., Porter, G., Wheatley, W., & Colby, M. (1996). A critique of service learning projects in management education: Pedagogical foundations, barriers, and guidelines. *Journal of Business Ethics, 15*(133), 142.

Leal Filho, W., Raath, S., Lazzarini, B., Vargas, V. R., De Souza, L., Anholon, R., ... Orlovic, V. L. (2018). The role of transformation in learning and education for sustainability. *Journal of Cleaner Production, 199*, 286–295.

Lipscombe, B. P. (2008). Exploring the role of the extra-curricular sphere in higher education for sustainable development in the United Kingdom. *Environmental Education Research, 14*(4), 455–468.

Majid, S., Liming, Z., Tong, S., & Raihana, S. (2012). Importance of soft skills for education and career success. *International Journal for Cross-Disciplinary Subjects in Education, 2*(2), 1036–1042.

Mayer, R. E. (1989). Models for understanding. *Review of Educational Research, 59*(1), 43–64.

Mohammad Arshad, N. A., Zainudin A. Z., Zakaria, S. R. A., Mansor, M. R. & Md Yunus, N. (2017) A review of barriers that hinder households to separate solid waste at source. In *Proceedings of Social Sciences Postgraduate International Seminar*, 29 November, Penang, Malaysia.

Mohd Shafie, M. S. A., & Ahmad, N. (2015). In S. Che Lah, & N. Esa (Eds.), *Ilmu, Tradisi dan Kelestarian Dalam Kearifan Tempatan*. Pulau Pinang: Penerbit Universiti Sains Malaysia.

OED. (2007). *Shorter Oxford English Dictionary* (6th ed.). New York: Oxford University Press Inc.

Peter, C. J., Libunao, W. H., & Abdul Latif, A. (2016). Extent of education for sustainable development (ESD) integration in Malaysian community colleges. *Journal of Technical Education and Training (JTET), 8*(1).

Roberts, D., & Silva, D. (1968). Curriculum design, teaching styles and consequences for students. *Samplings, 1*(4), 16–28.

Rodrigo, L., Michelle, Y. M., Kaisu, S., Kim, C., & Francisco, J. L. (2017). Connecting competences and pedagogical approaches for sustainable development in higher education: A literature review and framework proposal. *Journal Sustainability, 9*(10).

Rowe, D. (2007). Education for a sustainable future. *Science, 317*, 323–324.

Rozlin, R., Ismail, F., Idris, N., Mustaffa, N., Mohamed Saat, M., Mohd. Jamal, N., & Ramakrishnan, S. (2018). Generic skills of the undergraduates: A case study of Faculty of Built Environment in Universiti Teknologi Malaysia. *International JPournal of Engineering & Technology, 7*(2.29), 297–302.

Seatter, C. S., & Ceulemans, K. (2017). Teaching sustainability in higher education: Pedagogical styles that make a difference. *Canadian Journal of Higher Education, 47*(2), 47–70.

Sidek, S., & dan Hashim, M. (2016). Pengajaran Berasaskan Video dalam Pembelajaran Berpusatkan Pelajar: Analisis dan Kajian Kritikal. *Journal of ICT Education (JICTIE), 3*, 24–33.

Sigmon, R. L. (1979). Service-learning: Three principles. *ACTION*, *8*(1), 1–9.

UNESCO (2002). Higher education for sustainable development. Retrieved from <http://portal.unesco.org/pv_obj_cache/pv_obj_id_654B8DCFEE8A0875B86F293C20CA405DD9C70200/filename/21a5450c515bab552176c98215fffaf8brief + Higher + Education.pdf> Accessed 14.10.2019.

UNESCO (2005). Education for sustainable development. Retrieved from <https://en.unesco.org/themes/education-sustainable-development> Accessed 14.10.2019.

UNESCO (2012). *Education for sustainable development sourcebook. Education for sustainable development in action. Learning training tools*. Paris, France: United Nations Educational, Scientific and Cultural Organization.

UNESCO (2014). Shaping the future we want. UN decade of education for sustainable development (2005-2014). Final report. <http://unesdoc.unesco.org/images/0023/002301/230171e.pdf> Accessed 18.05.2020.

UNESCO (2017). *Education for sustainable development goals learning objectives*. Paris: United Nations Educational, Scientific and Cultural Organization. Available from https://unesdoc.unesco.org/ark:/48223/pf0000247444. (Accessed 14.10.2019).

UNESCO (2019). What is education for sustainable development? Retrieved from < https://en.unesco.org/themes/education-sustainable-development/what-is-esd > Accessed 14.10.2019.

United Nations (2015). 2030 Agenda for sustainable development. <https://sustainabledevelopment.un.org/post2015/transformingourworld> Accessed 14.10.2019.

United Nations Economic Commission for Europe, UNECE. (2012). *Learning for the future: Competences in education for sustainable development*. Switzerland: United Nations Economic Commission for Europe.

Wilson, J. (2000). Volunteering. *Annual Review of Sociology*, *26*, 215–240.

World Commission on Environment and Development, WCED. (1987). *Report of the world commission on environment and development: Our common future*. England: Oxford University Press.

PART II

Green materials

CHAPTER

7

A review of natural fiber reinforced recycled thermoplastic polymer composites

M.J. Taufiq[1], *Muhd Ridzuan Mansor*[1] *and Zaleha Mustafa*[2]

[1]Faculty of Mechanical Engineering, Universiti Teknikal Malaysia Melaka, Melaka, Malaysia, [2]Faculty of Manufacturing Engineering, Universiti Teknikal Malaysia Melaka, Melaka, Malaysia

7.1 Introduction

The philosophy of composites combines two or more elements to produce a new material. These two elements play significant roles, one is the matrix, and the other is a reinforcing material that is embedded in the matrix. As one of composite structures, wood plastic composite is a part of a natural fiber reinforced polymer (NFRP) composite, which is most widely used, including for building construction, automotive, marine, and even aeronautics applications (Ashori, 2008). The typical reinforcing material in NFRP is natural fiber, while the matrix can be either thermoplastic or thermoset polymers. The matrix acts as a load transfer medium from the outer source onto the fibers as well as to protect the fiber, while the role of the fiber is to carry the load that is applied along the dispersed fiber (Elanchezhian et al., 2018; Pandey & Sreekumar, 2015; Radzi, Sapuan, & Jawaid, 2018).

In the automotive industry, metal matrix composite (MMC) is another type of composite that commonly can be found in engine components and braking systems (Stojanović & Ivanović, 2013). Ceramic is

DOI: https://doi.org/10.1016/B978-0-12-819482-9.00016-2

used as the reinforcing agent in MMC for high-pressure and high-temperature components such as engine pistons. In addition, the use of ceramic significantly increases the piston resistivity toward wear and improves the thermal conductivity properties (Miracle et al., 2001). On the other hand, glass reinforced polymer (GRP) composite has been used in marine applications. Starting in 1947, the United Stated Navy made an attempt to develop a boat hull using GRP. In addition, glass reinforced polyester and aramid reinforced vinyl ester are among the FRP composites used in maritime craft (Selvaraju & Ilaiyavel, 2011).

This chapter carries out a review of natural fiber reinforced recycled polymer composite and discusses the terminology of natural fiber, matrix, and recycled plastic. Moreover, the literature on natural fiber composite (NFC) made using kenaf fiber and recycled plastic is also discussed in this chapter.

7.2 Natural fibers

There two main sources of natural fiber, namely animal-based and plant-based natural fibers. Animal-based natural fibers may be derived from animal hair and extruded protein fibers, which contain amino acids in their chemical structures for both protein and animal hair fibers (Zhang, 2014). Animal-based natural fibers have been used in several applications including in both biomedical and engineering fields. A study done by Fraunhofer and Sichinat (1992) revealed that silk fiber retained a high degree of stiffness over a long time under dynamic mechanical analysis compared to other surgical suture materials. Animal-based natural fiber have also been used for drug-delivery production such as in drug capsules. Jao, Xue, Medina, and Hu (2017) discussed the importance of using water-soluble, biocompatible, biodegradable, and nontoxic polymeric materials for drug consumers, especially for chronically ill patients. Silk, keratin, collagen, and elastin are protein fibers that are suitable for drug-delivery applications as they have the characteristics of release behavior, degradation profile, and mucoadhesive nature. In addition, the silk sericin can be used as a biocomposite material for drug delivery as it can be cross-linked, copolymerized, and blended with other macromolecular materials. Also, materials modified with sericin are useful as degradable biomedical materials (Zhang, 2002). Fig. 7.1 shows part of the classification of natural fibers.

Unlike animal-based natural fibers, the major part of plant-based natural fiber is cellulose. Plant natural fiber is a natural polymeric chain containing cellulose, hemicellulose, lignin, and waxes, as presented in Fig. 7.2, and a schematic can be seen in Fig. 7.3. The characteristics of

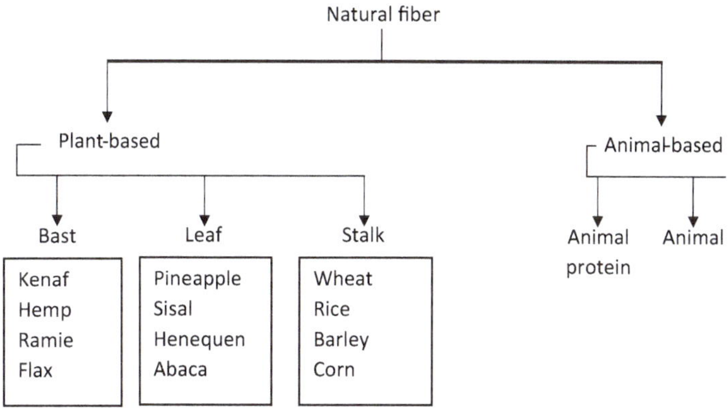

FIGURE 7.1 Classification of natural fibers (Djafari Petroudy, 2017).

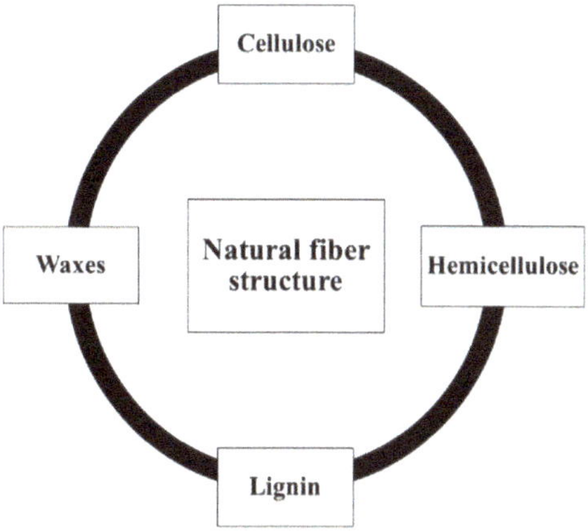

FIGURE 7.2 The structure of natural fiber.

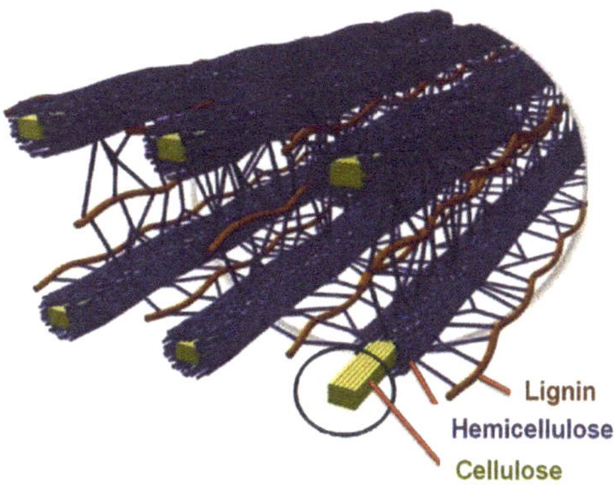

FIGURE 7.3 Lignin, hemicellulose, and cellulose positions in natural fiber (Doherty, Mousavioun, & Fellows, 2011).

plant natural fiber depend on the source of the fiber, cultivation condition, plant age, and digestion process (Fuqua, Huo, & Ulven, 2012). There are two different situations on the bulk and surface. In bulk fiber, the structure is closely packed with a large amount of covalent and hydrogen bonds, making them stiff, insoluble, and chemically resistant. However, for the surface fiber, the hydroxyl group on the cellulose and hemicellulose are exposed toward chemical reactions. This provides the potential surface modification through a chemical reaction (Chan et al., 2018).

7.2.1 Celluloses

Cellulose is the major part of plant natural fiber containing pyranose ring and OH groups, and is linear macromolecular consisting of β-D glucose units. The β-D glucose units are linked by a β-1,4-glycosidic linkage at the C_1 and C_4 positions through a hydrogen bond and van der Waals forces, to form linear polymer chains (shown in Fig. 7.4) (Bledzki & Gassan, 1999). Each repeating cellulose unit contains six hydroxyl groups, which form intermolecular and intramolecular hydrogen bonds, thus creating all types of lignocellulosic fibers that are hydrophilic in nature (Satyanarayana, Arizaga, & Wypych, 2009). The existence of OH groups in cellulose also leads to a large number of hydrogen bonds. Due to these hydrogen bonds and van der Waals forces, some of the cellulose molecules align together, forming a highly ordered region (crystalline) and a less ordered region (amorphous) (Fuqua et al., 2012). There are four crystalline forms of cellulose, specifically types I, II, III, and IV. Type I can be differentiated into two, cellulose 1_α, the rarest form which exists only in some green algae, and cellulose I_β which can be found in almost all types of plant fiber (Hori & Wada, 2005). On the other hand, cellulose type II is obtained from the derivation of the cellulose type I_β through aqueous treatment of an alkali solution such as sodium hydroxide (NaOH). This process is irreversible since the hemicellulose is removed during treatment. Treating the fiber with ammonium hydroxide (NH_4OH) will produce cellulose type III (Wada, Nishiyama, & Langan, 2006). Interestingly, cellulose type III can be reversed to type II using hydrothermal (160°C) or

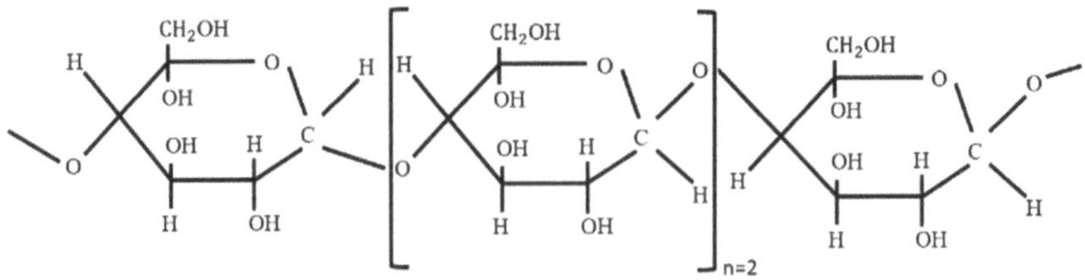

FIGURE 7.4 The chemical structure of cellulose (Bledzki & Gassan, 1999).

thermal treatment (200°C) (Wada, 2001). Finally, cellulose type IV can be obtained from heat treatment in glycerol of cellulose type III at 260°C (Newman, 2008).

7.2.2 Hemicelluloses

Hemicellulose is the second major part of natural fiber. It has lower molecular weight polysaccharides, a hydrophilic nature, and can be easily hydrolyzed by dilute acids and bases (Azwa, Yousif, Manalo, & Karunasena, 2013). In addition, hemicellulose is also more complex than cellulose, containing only a 1,4-β-D glucose linkage (Gurunathan, Mohanty, & Nayak, 2015). Hemicellulose is made up of branches of amorphous polysaccharides, such as xylose, mannose, glucose, galactose, and arabinose, as shown in Fig. 7.5 (Fuqua et al., 2012). The hemicellulose extraction method can be performed using various techniques such as autohydrolysis and prehydrolysis hydrothermal treatments. These methods contribute to the production of oligomeric and monomeric sugars, and high-purity dissolving pulp (Santucci, Maziero, Rabelo, Curvelo, & Pimenta, 2015; Testova et al., 2014; Tian, Fu, Zhuang, Wang, & Li, 2017). However, these methods are performed in high temperatures (170°C–190°C), causing significant degradation of carbohydrates into furanic compounds. Thus, treatment using organic acids has been developed as a feasible method in order to promote the extraction efficiency of the hemicellulose from lignin (Xu et al., 2018).

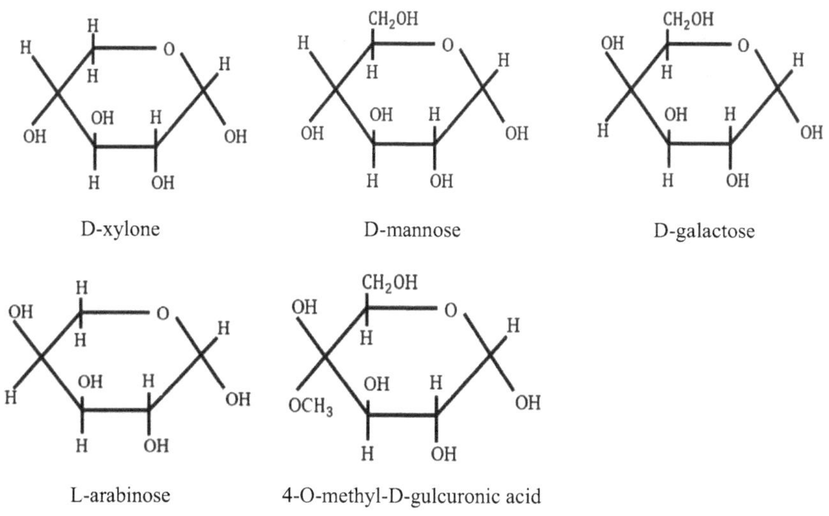

FIGURE 7.5 Chemical structure of hemicellulose (Fuqua et al., 2012).

7.2.3 Lignin

Lignin is one of the major components of natural fiber, consisting of extremely complex three-dimensional polymers formed by radical coupling polymerization of p-hydroxycinnamyl, coniferyl, and sinapyl alcohols. Lignin functions as the structural material which adds strength and rigidity to the cell wall, and is able to resist most forms of biological attack better than cellulose is able to (Fan, Ghaffar, & Fan, 2013; Xu, Sun, Sun, Fowler, & Baird, 2006). The structure and characteristics of lignin are significantly dependent on the chemical functional groups such as hydroxyl, methoxyl, carbonyl, and carboxylic groups. Phenolic hydroxyl has significant effects on the physical and chemical properties of lignin polymer and can be determined using UV spectroscopy (Aulin-Erdtman, 1958). On the other hand, methoxyl groups ($-OCH_3$) can be found in lignin from all plants, and depend on the plant species and method of isolation. The amount of the methoxyl groups is often used as a measure of the purity of a lignin preparation, since isolated lignin can be contaminated with hydrocarbons. There are two methods to determine methoxyl, one is using hydroiodic acid which combines demethylation and gas chromatography, and the other is using proton nuclear magnetic resonance (^1H NMR) (Aberu & Freire, 1995).

7.3 Recycled plastic as an alternative to virgin plastic

The approach to using recycled plastic in several areas of research has been published by various researchers. Bertin and Robin (2002) performed a comparison study of the mechanical, thermal, and morphological properties between virgin low-density polyethylene/polypropylene (LDPE/PP) and post-consumer LDPE/PP blends. In their results, the yield tensile strength of post-consumer LDPE/PP was 7% higher compared with virgin LDPE/PP. This section briefly explains the reported studies and findings on the use of recycled plastic in NFC production.

The approach to understanding the mechanical behavior of virgin and recycled high-density polyethylene (HDPE) has been described by Adhikary, Pang, and Staiger (2008). Recycled HDPE is obtained from a local plastic recycling plant. In their study, the tensile strength of recycled HDPE was 23.2 MPa, which is 8% higher than virgin HDPE.

Meanwhile, Turku, Kärki, Rinne, and Puurtinen (2016) reported the results of producing a polymer blend from recycled materials which was later used as a composite matrix. There are two dissimilar sources, one is construction plastic waste and the other is municipal plastic waste. Both types of plastics were separated from other materials, leaving only PP and polyethylene. The results indicated that the polymer blend from construction waste has better mechanical properties than municipal waste because of the multiphase coarse surface morphology and poor adhesion which was clearly seen in the blend using SEM.

In addition, Taufiq, Mustafa, and Mansor (2017) discovered potential recycled thermoplastics that can be used as an alternative to virgin thermoplastics. The recycled thermoplastics were derived from rejected and unused disposable diapers, which contain PP and polyethylene (PE). Due to difficulties in segregating these two materials, they were blended together and produced a recycled PP/PE blend (rPP/PE). In order to identify the performance of the recycled blend, the tensile properties of rPP/PE were compared with virgin PP/PE (vPP/PE). The tensile strength of the recycled blend was found to be 57% lower than the virgin blend, however the tensile modulus for both blends was similar. The poor tensile strength of rPP/PE may be caused by the impurities found in the blend which induced phase separation and reduced stress transfer efficiency between PP and PE elements. SEM examination showed the phase separations as well as the poor miscibility and homogeneity in the recycled blend (Fig. 7.6).

7.4 Recycled plastic applications in natural fiber composites

Natural fiber is an alternative synthetic fiber that can be obtained from either plant, wood, or animals. The aim of sustainable product development is a key point in the tremendous utilization of natural fiber for composite production, including for automotive, building construction, marine, and aeronautics uses (Sapuan, Leenie, Harimi, & Beng, 2006; Singha & Thakur, 2008). In addition, the development of natural fiber-based products also has disadvantages of synthetic fibers which are nonrenewable, nonrecyclable, high energy consuming during the manufacturing process, with a health risk when inhaled, and nonbiodegradable (Cheung, Ho, Lau, Cardona, & Hui, 2009). Previous researchers have revealed the applicability of recycled plastic as a virgin plastic alternative through experimental approaches. The increase of using recycled plastics for NFC development may reduce the dependency on nonrenewable resources. The following subsection describes earlier studies conducted by several researchers.

7.4.1 Recycled polystyrene-based composites

Polystyrene (PS) is a commodity plastic used in many applications. The suggestion for using recycled PS for several applications has been reported by previous researchers. In the effort to reduce solid waste originating from plastic waste, recycled PS was reported as an alternative for natural aggregates in concrete manufacturing. The potential applications of this type of concrete include use in partition walls, floorings, ceiling concretes, briquettes or bricks, and plaster (Kaya & Kar, 2016). Recycled PS can also

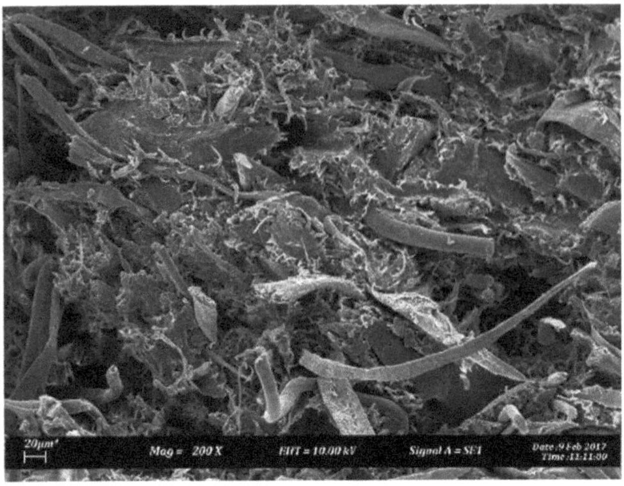

(A)

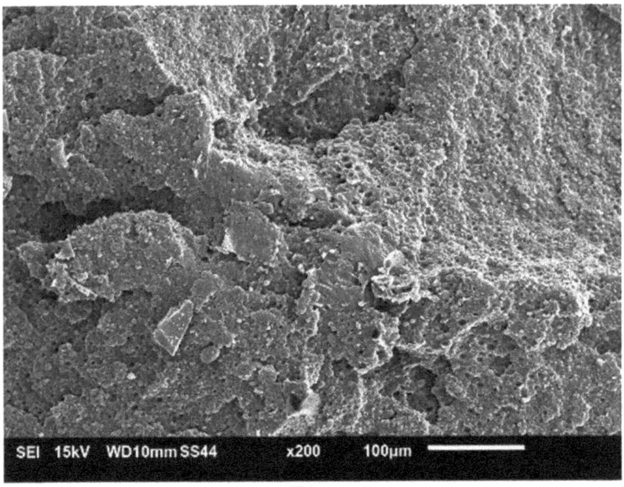

(B)

FIGURE 7.6 Comparison of scanning electron microscope images: (A) recycled polypropylene/polyethylene; (B) virgin polypropylene/polyethylene.

be used for NFC production. Agoua, Allognon-Houessou, Adjovi, and Togbedji (2013) reported NFC production from recycled PS which was obtained from packaging reinforced by sawdust filler.

Meanwhile, a study on date palm leaflets reinforced recycled PS composite was reported by Masri, Ounis, Sedira, Kaci, and Benchabane (2018). The effects of various fiber sizes and fiber loadings toward the NFC physical properties have been reported. The fiber size varies at

0.1–0.315, 0.315–0.5, and 0.5–1 mm, while the fiber loading varied at 70, 75, and 80 wt.%. It was found that the bulk density increased as the fiber size and fiber loading increased. The result trends were similar to the findings for pinus wood flour reinforced recycled PS reported by Poletto, Dettenborn, Zeni, and Zattera (2011).

7.4.2 Recycled polypropylene-based composites

PP is a well-known commodity plastic widely used in many applications since it has light weight, structural stability, dielectric vitality, and corrosion resistance as compared to traditional plastics (Khan, Hussain, & Mujtaba, 2014). Due to the high demands on PP, the estimate of PP waste was 14% of global waste (Sarker & Rashid, 2014). Thus, recycling PP for NFC production may reduce this statistic significantly.

In other findings, Ashori and Nourbakhsh (2009) reported the result of recycled newspaper reinforced recycled polypropylene (rPP/rNP) and recycled polyethylene (rPE/rNP) as NFCs. The effects of increasing fiber loadings revealed a drop in tensile strength for both types of matrices. The trend of dropping tensile strength with increasing fiber content is due to poor interfacial properties between hydrophilic and hydrophobic materials. However, the fiber loading of 70 wt.% showed the best tensile and flexural properties for both rPP and rPE composites on their strength and modulus. The increasing in fiber content from 50 to 85 wt.% also proved an increment in water absorptivity and thickness swelling for both types of composites. This might be due to the addition of fiber content, as adding more porous tubular structures accelerates the water penetration through capillary action and thus more water molecules reside in the composite structure.

Coconut (*Cocos nucrifera*) has become the fourth largest crop in Malaysia, after oil palm, rubber, and paddy in terms of cultivation area, with an estimated 109,185 hectares in 2007 (Sivapragasam, 2008). Among the by-products of coconut plantations, coconut shell has the potential to be used as a reinforcement agent for NFC production. Coconut shell is dried and crushed to form coconut shell powder (CSP). NFC made using CSP reinforced recycled polypropylene (CSP/rPP) composite has been described by Chun, Husseinsyah, and Azizi (2013). The composite was produced using a blade mixer at a temperature of 180°C and rotor speed of 50 rpm. The composites were produced with various CSP contents, 10, 20, 30, and 40 php. The comparison on SEM images (Fig. 7.7) of unmodified and modified CSP with ethanol revealed that the modified CSP promoted better compatibility with an rPP matrix as the modified CSP particles are embed inside the rPP matrix. However, the addition of CSP into the rPP matrix degraded the tensile strength even for modified CSP. Fig. 7.8 shows thermal scanning

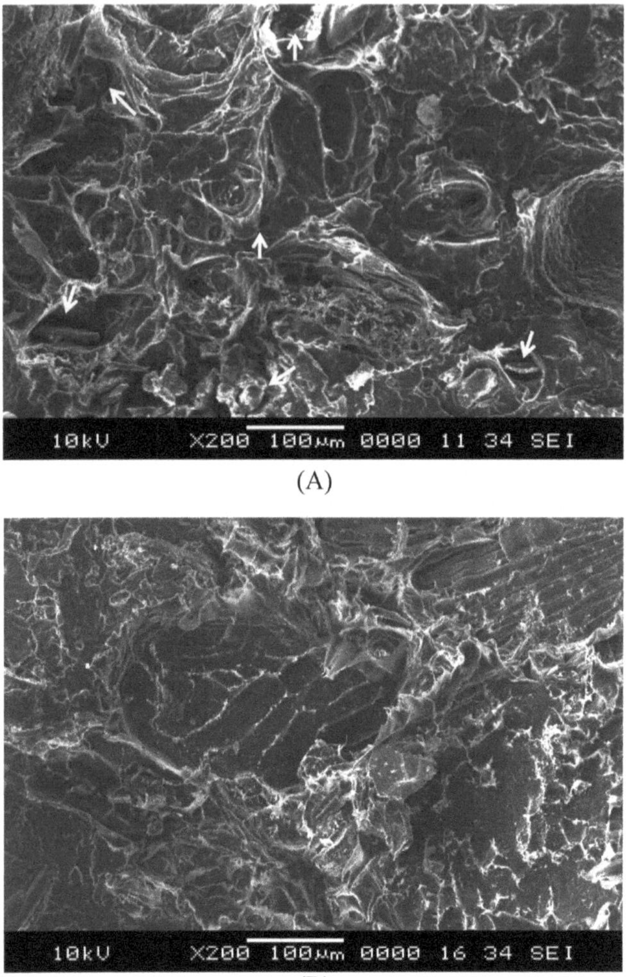

(B)

FIGURE 7.7 Scanning electron microscope images of 20 php coconut shell powder content: (A) unmodified; (B) ethanol modified (Chun et al., 2013).

through DSC which revealed that the addition of CSP content reduced the crystallinity of composites, which might be due to the presence of CSP hindering the crystallization of PP chains in composites. The modified CSP composites have higher crystallinity than unmodified CSP composites, indicating that modified CSP induces the migration and diffusion of the PP chain to the CSP surface.

In another scenario, the softwood lumber has been a key player in the automotive and building structure industries. In automotive industries, softwood lumber-based NFC is used for door inner panels, seat

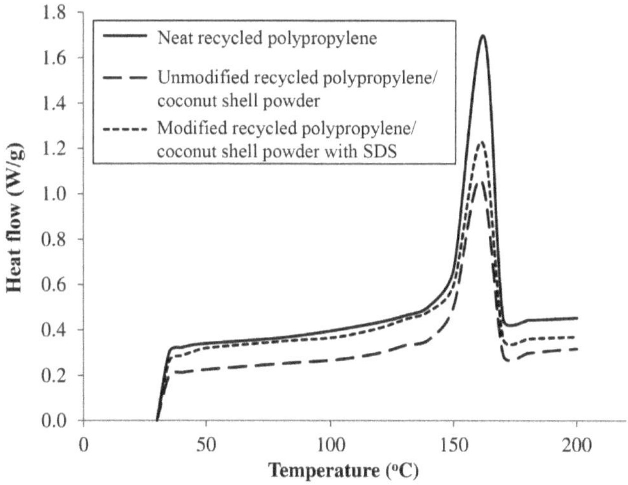

FIGURE 7.8 Differential scanning calorimetry thermograph of neat recycled polypropylene, unmodified, and modified coconut shell powder (Chun et al., 2013). SDS, Sodium dodecyl sulfate.

backs, and headrests. Meanwhile, the application of softwood lumber-based NFC in building structures can be seen in decking, cladding, and fencing (Homkhiew, Ratanawilai, & Thongruang, 2013). In addition, a durability study of NFC made by softwood lumber reinforced recycled PP (SL) through exposure to UV radiation was also investigated (Homkhiew, Ratanawilai, & Thongruang, 2014). Through SEM imaging, the effect of exposure to UV on the composite surface can be identified. The exposure caused surface cracks, indicating chain scission. There are three main stages of degradation, starting with the surface being eroded with many cavities, followed by an increase in the size and frequency of cavities, and finally the development of small cracks on the weathered surface. A composite with higher SL content (45 wt.%) exhibits strong swelling when exposed to water and moisture causes cracks to accumulate and contribute to deeper light penetration.

7.4.3 Recycled polyethylene-based composites

There are a huge amount of publications on recycled polyethylene-based natural fiber composites. An investigation into the impact of fiber size on composite performance was carried out by Yang, Leu, Yang, and Lo (2012). A recycled wood flour reinforced recycled HDPE plastic container was prepared in their study with three main fiber sizes: <125, 125–180, and 180–250 μm. The results showed that using a larger fiber size caused more water absorptivity and thickness swelling. This was because of the recycled matrix being unable to capsulate the entire wood particle during compounding. The uneven distribution of matrix, and fibers entangled with each other, also allows more water molecules to be absorbed. Furthermore, the effects of using larger fiber sizes also reduced the flexural and tensile strengths as compared to using a smaller size.

Differential scanning calorimetry (DSC) is a common method used for thermal properties characterization. A study on the effect of sugarcane fiber addition on the thermal transition of recycled HDPE was conducted by Ribeiro, Gandara, Moreno, and Saron (2017). The DSC results indicated the presence of sugarcane fiber increased the degree of crystallinity of material as the fiber content was increased. This is explained by the influence of fiber on the HDPE crystallinity, which is improved in the fiber to matrix interfaces, which later significantly affects the behavior of the mechanical properties such as elongation at break. However, at 20 wt.%, which was the highest fiber content used, a decrease in the degree of crystallinity was recorded. This might be due to the arrangement of fibers affecting the mobility of the polymeric chain and its crystallinity.

In addition, research on using recycled HDPE detergent bottles as an NFC matrix has been reported by Oza, Wang, and Lu (2011). As a comparison, virgin HDPE was also used for NFC production. Hemp fiber was used as the reinforcement agent with addition ratios of 10, 20, 30, and 40 wt.%. For thermal analysis, thermogravimetric analysis revealed the hemp reinforced recycled-based matrix composite required more heat and energy to decompose as compared with virgin-based matrix. The effect of variations of hemp fiber content (10, 20, 30, and 40 wt.%) indicated increasing flexural strength and modulus as the fiber content was increased. They claimed that the recycled matrix composite has good potential to be use for semistructural applications such as walls and short-span bridges.

A different formulation method was used by Lu and Oza (2013) in producing NFC from hemp fiber reinforced recycled HDPE detergent bottles and virgin HDPE composite. In the compounding process, they used the volume ratio instead of the weight ratio in which significant results were observed such as the recycled-based matrix having superior mechanical performances at 40 vol.% of fiber as compared with virgin-based matrix on the same formulation. This could be due to the recycled HDPE having less polymer branching, which provides stronger intermolecular forces.

NFC produced from rice husk reinforced post-consumer polyethylene has been prepared by Fávaro, Ganzerli, de Carvalho Neto, da Silva, and Radovanovic, (2010a), Fávaro, Lopes, Vieira de Carvalho Neto, Rogério de Santana, and Radovanovic (2010b). For this study, two types of matrices were used, one was untreated matrix, and the other was treated matrix. Prior to the composite production, post-consumer polyethylene (rPE) was placed in an acidic solution as the surface modification procedure. The composites produced with treated matrix showed an increment in Izod impact strength and tensile modulus as compared to the untreated matrix. The significance of using treated matrix included improved interfacial interaction and prevention of fiber pull-out condition. However, the treated matrix did not improve the tensile and flexural strengths or the flexural modulus. This behavior might be the effect of a wetting phase, in which the stresses in the fiber and matrix boundaries are concentrated around the reinforcement particle. Thus, the debonding at the boundary region might be caused by the weaken matrix and fiber interaction.

Furthermore, Sommerhuber, Welling, and Krause (2015) discovered the properties of NFC produced from recycled wood waste reinforced recycled high-density polyethylene (rPE). The composite from virgin materials (vPE) was also produced in order to compare the water absorption on recycled sourced composites. The recycled sourced composite showed high water absorption in a water absorption test, potentially due to incomplete encapsulation of wood fibers and the probable occurrence of wood fiber aggregates in the matrix. For tensile strength, the increasing fiber content, from 0 to 30 to 60 wt.%, resulted in linear dropping in the tensile strength for recycled-based composite compared to virgin-based. Meanwhile, the addition of 30 wt.% of fiber increased the flexural strength of recycled-based composite but then decreased when 60 wt.% of fiber was reached. The virgin-based composites showed a linear increment as the fiber contents increased. A morphological analysis also displayed the recycled-based composite with 60 wt.% had gaps generated between the fiber and matrix, proving a weak interface interaction.

An interesting observation on the morphological properties through SEM for the utilization of recycled plastics in NFCs has been reported by several scholars. The SEM morphological analysis of NFC manufactured from sisal fiber reinforced recycled HDPE composite was described by Fávaro et al. (2010a, 2010b). The recycled plastic containing HDPE obtained from the Brazilian recycling industry was reinforced with sisal fiber which lengths between 0.5–1.0 cm. Both matrix and fiber were passed through the pretreatment processes so that the comparison of treated and untreated samples could be conducted. The materials were extruded at 40 rpm rotor speed and injection molded at 180°C under 35 bar of pressure. Fig. 7.9A–C show the SEM images from several samples. The SEM analyses revealed the fiber pull-out occurred due to the fractured sample for the untreated samples. Meanwhile, both matrix-treated and fiber-treated samples showed better fiber to matrix phases adhesion and adherence. Despite the improvements for the treated samples on their morphological properties, Fig. 7.10 reveals the detected nonhomogeneous region on the fractured sample. This may be due to the increased mobility of the HDPE

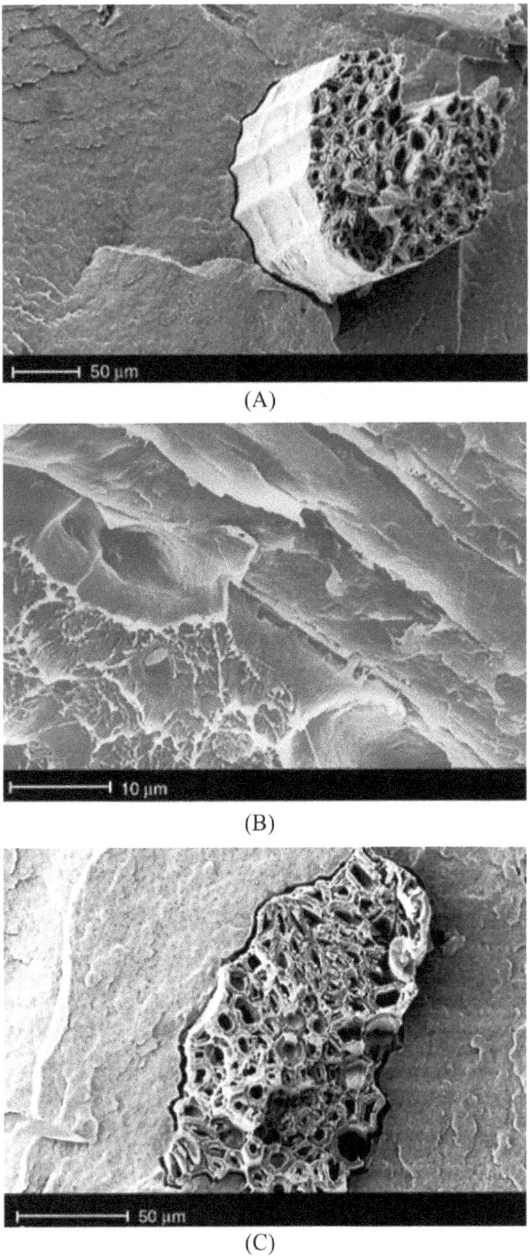

FIGURE 7.9 Scanning electron microscope images of sisal/high-density polyethylene fractured samples: (A) nontreated sample; (B) matrix-treated sample; (C) fiber-treated sample (Fávaro et al., 2010a, 2010b).

polymeric chains during the extrusion process that caused the HDPE polymer contacted with nontreated parts of the fibers (bulk). Thus, the treated matrix and fiber did not induce an improvement in the mechanical properties, especially for tensile modulus and flexural modulus.

7.5 Natural fiber reinforced recycled polymer blend composites

There are difficulties in separating recycled plastics that may contain various types of polymers, such as PP and PE. In order to have a better understanding of natural fiber reinforced recycled polymer blend composites, this subsection briefly reviews recent studies reported by several researchers on the outcomes of their study.

7.5.1 Recycled industrial and municipal plastics-based composites

Research into spruce wood particle fiber reinforced recycled plastic obtained from industrial and municipal wastes was reported by Turku, Keskisaari, Kärki, Puurtinen, and Marttila (2017). At fixed 54 wt.% of wood particles, it improved the mechanical properties of both types of recycled polymer blends. The use of fiber reinforced the blend morphology properties, thus assisting the stress transfer inside the composite. In order to obtain the water absorptivity, the water absorption test was conducted with the results showing a composite with municipal sourced polymer had the highest water absorption and thickness swelling because of water damaging the composite interphase and poor adhesion between

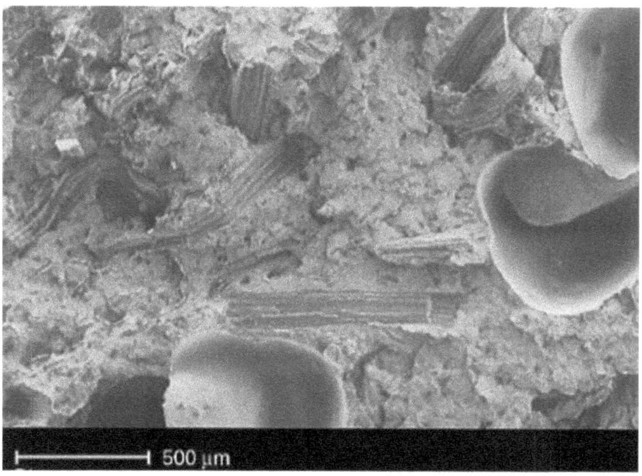

FIGURE 7.10 Scanning electron microscope image of the nonhomogeneous region from a treated fiber and matrix sample (Fávaro et al., 2010a, 2010b).

the fiber and matrix. Similar results were obtained by Moreno and Saron (2017). In their study, the NFCs were produced from polyethylene waste and recycled pine wood. In fiber-rich composite (>15 wt.%), the fiber improved the tensile strength and modulus, meanwhile it reduced the elongation at break. This was caused by the characteristics of natural fiber including lower strain compared to a thermoplastic matrix.

In addition, Turku, Keskisaari, Kärki, Puurtinen, and Marttila (2018) discovered the durability of NFC with the same materials with UV radiation and water immersion as the weathering conditionings. They found that water immersion strongly affected the mechanical properties of NFC. This was due to the degradation on interfacial interaction in the composite by the absorbed moisture. This reduced the stress transfer, thus lowering the mechanical properties. In addition, the water molecules caused fiber swelling which later contributed to stress in the matrix and microcracking growth. The microcracks later allowed the water molecules to migrate into the composite structure.

7.5.2 Recycled polypropylene/polyethylene-based composites

Taufiq, Ridzuan, and Mustafa (2018) reported the findings from the production of NFC made using kenaf fiber and recycled PP and polyethylene. The recycled PP and polyethylene were derived from rejected and unused disposable diapers, with an average ratio of 70:30 wt.% of PP:PE. The recycled plastics were compounded with kenaf fiber, which acted as the reinforcement. From their experimental works, the addition of kenaf fiber in recycled PP and PE, rPP/PE showed the optimum flexural strength and modulus at 60 wt.% of fiber (14 MPa and 1250 MPa, respectively) which improved the strength and modulus by 83% and 85%, respectively. This was due to the force direction which in the perpendicular direction contributed to better stress transfer and promoting of the stiffness of the structure. However, the results from the Izod impact tests showed there was no improvement when the kenaf fiber was added to rPP/PE. Since the direction of the load differs with the bending load, the decrement in the impact strength may be due to fiber to fiber contact causing fiber breakage to be the predominant failure mechanism. This later decreased the effective stress transfer between the fiber and matrix for the impact load. Other than that, the failure mechanisms found from the impact tests were matrix cracking and the presence of voids or air traps.

7.5.3 Recycled low-density polyethylene/high-density polyethylene/polypropylene-based composites

NFC made using date palm leaf fiber reinforced ternary recycled plastic blend was reported by Noorunnisa Khanam and AlMaadeed (2014). The recycled plastics were obtained from a plastic manufacturer in Qatar. These recycled plastics consisted of LDPE, HDPE, and PP. In their study, the LDPE/HDPE/PP blend was produced in the ratio of 20/40/40 wt.%. The results of the addition of date palm leaf fiber (20 wt.%) and additional maleic anhydride (1 and 2 wt.%) were revealed.

The tensile test showed the addition of date palm leaf fiber and coupling agent induced a slight increment in strength but more significantly on modulus. The addition of fiber reinforced the matrix structures, promoting the matrix to become more brittle and thus addressing the increase in modulus of rigidity. Furthermore, the addition of a coupling agent supported the improvement in tensile properties, especially in modulus, by improved the bonding between fiber and matrix, and obstructing stress propagation. Through the addition of a coupling agent, enhancement occurred on the interfacial adhesion between fiber and matrix. However, the addition of 2 wt.% of coupling was described as excess coupling agent since it caused a reduction in tensile properties. This may be due to polymer chain scission, where the increase in the number of short polymer chains leads to poor adhesion between the short polymer chains and the fibers. This can induce later void formation, as can be seen in Fig. 7.11.

7.5.4 Recycled high-density polyethylene/polyethylene terephthalate-based composites

The study of rice husk flour fiber reinforced recycled polymer blend obtained from a Malaysian local recycling plant has been reported by Chen, Salleh, Ab Ghani, Ahmad, and Gan (2015). The recycled plastic containing high-density polyethylene and polyethylene terephthalate was compounded to form an HDPE/PET blend. The effects of rice husk flour fiber loading variation (50, 60, 70, and 80 wt.%) on the water absorptivity characteristics were revealed in their study. In addition, a comparison of the water absorptivity for treated and untreated fibers was also presented. Since rice husk fiber has a hydrophilic nature, the water absorptivity is increased as the fiber loading is increased. This is because of the increment in the number of water residence sites in the form of free hydroxyl groups (–OH), which attracted water molecules through hydrogen bonding (+H).

Interestingly, the composites containing alkali-treated rice husk fiber showed higher water absorptivity compared to untreated ones. In their opinion, this could be the NaOH solution being attributed to the enlargement of the specific surface area of the fibers, thus resulting in fiber swelling after mercerization, which later induced the accessibility of water molecules to the reactive regions.

Furthermore, Chen & Ahmad, 2016 discovered the thermal properties of the same composite using DSC. They found that the increasing fiber

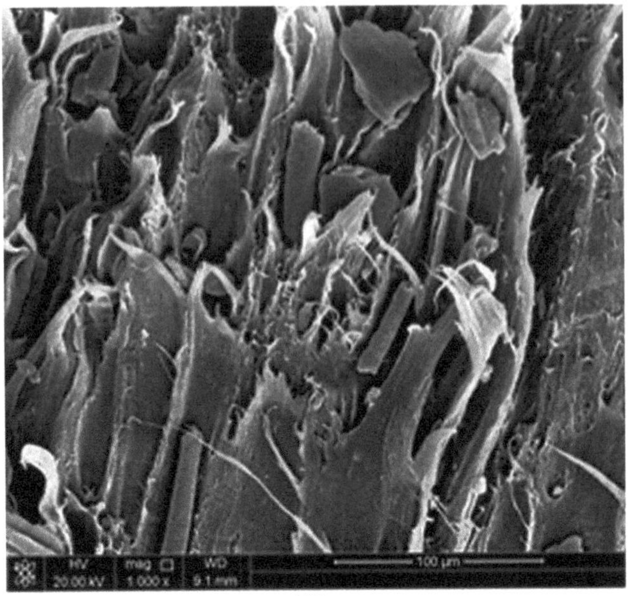

(A)

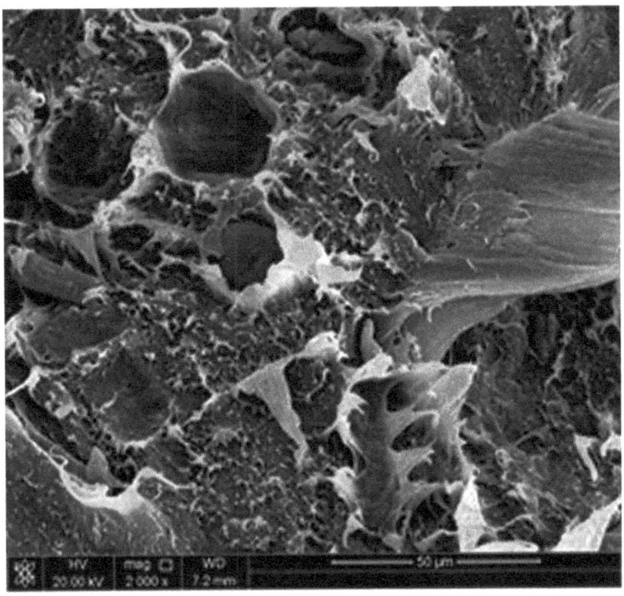

(B)

FIGURE 7.11 Comparison on SEM images of the composite: (A) addition of 1 wt.% coupling agent; (B) addition of 2 wt.% coupling agent (Noorunnisa Khanam & AlMaadeed, 2014).

loading content from 50 to 80 wt.% reduced the melting temperature of the composites. This was due to the incorporated rice husk fiber that caused restriction of the movement of the HDPE and PET chains from the crystallization process. In addition, the morphology properties for the composites with various fiber loadings were also revealed. As can be seen in Fig. 7.12, there are changes to the morphological properties as the rice husk fiber

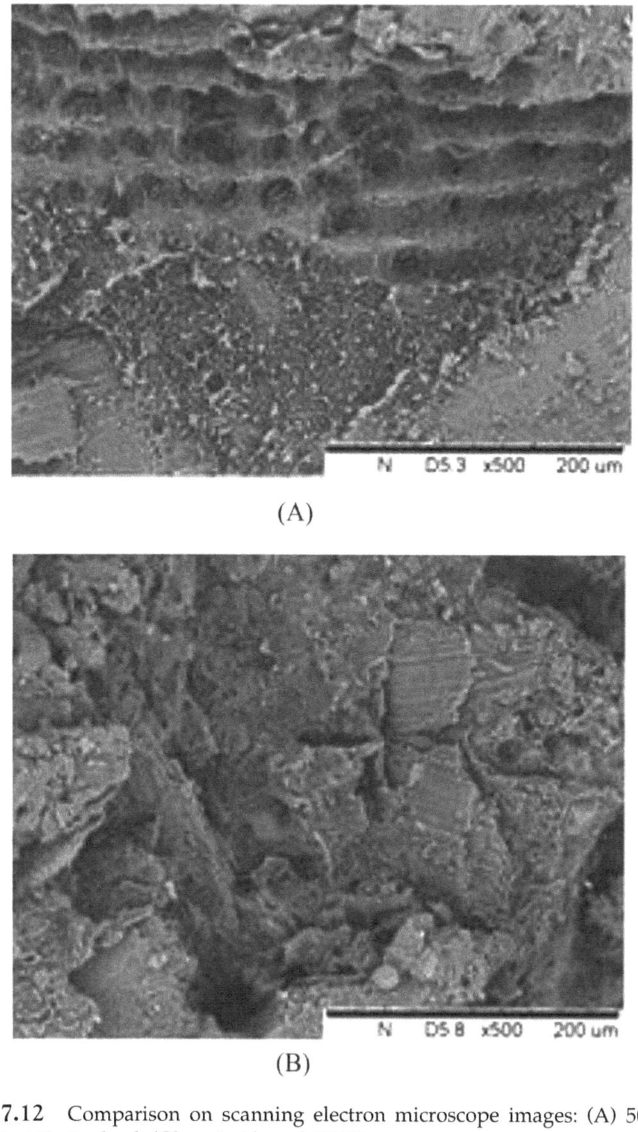

FIGURE 7.12 Comparison on scanning electron microscope images: (A) 50 wt.% rice husk (B) 80 wt.% rice husk (Chen & Ahmad, 2016).

content is increased. The composite with 50 wt.% fiber loading provided a perfect and strong adhesion between fiber and matrix, as shown in Fig. 7.12A. This indicates the adequate matrix and fiber to form good compounding quality, and good fiber–matrix interfacial bonding. However, the 80 wt.% fiber loading composite in Fig. 7.12B shows poorer fiber dispersion and the presence of clearer cavities in the surface morphologies. This resulted from the excessive fiber that led to insufficient adhesion between fiber and matrix, hence promoting fiber to fiber contact and fiber agglomeration dominated in the composites.

7.6 Conclusions

Based on the literature, NFC is an essential material with important roles in the daily lives of many people. The combined purposes of product sustainability and environmental preservation were the key factors driving the development of NFC. In another situation, the abundance of polymer waste that came from municipal and industrial wastes was observed as potential material for NFC production, as an alternative to using virgin synthetic polymer. Although the recycled polymer-based NFC development was at the laboratory scale, it has the potential to be used in mass production. Further studies into this recycled polymer-based NFC should be done to enable this new potential material to become the main alternative to conventional materials, thus improving the local economics and reducing the amount of polymer waste.

Acknowledgment

The authors wish to thank Universiti Teknikal Malaysia Melaka for providing support throughout this study.

References

Aberu, H., & Freire, M. (1995). Methoxyl content determination of lignins by 1H NMR. *Anais de Academia Brasileira de Ciencias*, *67*, 379.

Adhikary, K. B., Pang, S., & Staiger, M. P. (2008). Dimensional stability and mechanical behaviour of wood-plastic composites based on recycled and virgin high-density polyethylene (HDPE). *Composites Part B: Engineering*, *39*(5), 807–815.

Agoua, E., Allognon-Houessou, E., Adjovi, E., & Togbedji, B. (2013). Thermal conductivity of composites made of wastes of wood and expanded polystyrene. *Construction and Building Materials*, *41*, 557–562.

Ashori, A. (2008). Wood-plastic composites as promising green-composites for automotive industries!. *Bioresource Technology*, *99*(11), 4661–4667.

Ashori, A., & Nourbakhsh, A. (2009). Characteristics of wood-fiber plastic composites made of recycled materials. *Waste Management*, *29*(4), 1291–1295.

Aulin-Erdtman, G. (1958). Studies on ultraviolet absorption changes caused by modifications of chromophores with special reference to lignin chemistry. *Svensk Kemisk Tidskrift*, *70*, 145–156.

Azwa, Z. N., Yousif, B. F., Manalo, A. C., & Karunasena, W. (2013). A review on the degradability of polymeric composites based on natural fibres. *Materials and Design*, *47*, 424–442.

Bertin, S., & Robin, J. J. (2002). Study and characterization of virgin and recycled LDPE/PP blends. *European Polymer Journal*, *38*(11), 2255–2264.

Bledzki, A. K., & Gassan, J. (1999). Composites reinforced with cellulose based fibres. *Progress in Polymer Science*, *24*(2), 221–274.

Chan, C. M., Vandi, L.-J., Pratt, S., Halley, P., Richardson, D., Werker, A., & Laycock, B. (2018). Composites of wood and biodegradable thermoplastics: A review. *Polymer Reviews*, *58*(3), 444–494.

Chen, R. S., & Ahmad, S. (2016). Characterization of rice husk biofibre-reinforced thermoplastic blend biocomposite. In M. Poletto (Ed.), *Composites from Renewable and Sustainable Materials* (pp. 25–44). London: InTechOpen.

Chen, R. S., Salleh, M. N., Ab Ghani, M. H., Ahmad, S., & Gan, S. (2015). Biocomposites based on rice husk flour and recycled polymer blend: Effects of interfacial modification and high fibre loading. *BioResources*, *10*(4), 6872–6885.

Cheung, H. Y., Ho, M. P., Lau, K. T., Cardona, F., & Hui, D. (2009). Natural fibre-reinforced composites for bioengineering and environmental engineering applications. *Composites Part B: Engineering*, *40*(7), 655–663.

Chun, K. S., Husseinsyah, S., & Azizi, F. N. (2013). Characterization and properties of recycled polypropylene/coconut shell powder composites: Effect of sodium dodecyl sulfate modification. *Polymer-Plastics Technology and Engineering*, *52*(3), 287–294.

Djafari Petroudy, S. R. (2017). Physical and mechanical properties of natural fibers. In M. Fan, & F. Fu (Eds.), *Advanced high strength natural fibre composites in construction* (pp. 59–84). Duxford: Woodhead Publishing Limited.

Doherty, W. O., Mousavioun, P., & Fellows, C. M. (2011). Value-adding to cellulosic ethanol: lignin polymers. *Industrial Crops and Products*, *33*, 259–276.

Elanchezhian, C., Ramnath, B. V., Ramakrishnan, G., Rajendrakumar, M., Naveenkumar, V., & Saravanakumar, M. K. (2018). Review on mechanical properties of natural fiber composites. *Materials Today: Proceedings*, *5*(1), 1785–1790.

Fan, M., Ghaffar, S. H., & Fan, M. (2013). Structural analysis for lignin characteristics in biomass straw Structural analysis for lignin characteristics in biomass straw. *Biomass and Bioenergy*, *57*, 264–279.

Fávaro, S. L., Ganzerli, T. A., de Carvalho Neto, A. G. V., da Silva, O. R. R. F., & Radovanovic, E. (2010a). Chemical, morphological and mechanical analysis of sisal fiber-reinforced recycled high-density polyethylene composites. *Express Polymer Letters*, *4*(8), 465–473.

Fávaro, S. L., Lopes, M. S., Vieira de Carvalho Neto, A. G., Rogério de Santana, R., & Radovanovic, E. (2010b). Chemical, morphological, and mechanical analysis of rice husk/post-consumer polyethylene composites. *Composites Part A: Applied Science and Manufacturing*, *41*(1), 154–160.

Fraunhofer, J. A. Von, & Sichinat, W. J. (1992). Characterization of surgical suture materials using dynamic mechanical analysis. *Biomaterials*, *13*(10), 715–720.

Fuqua, M. A., Huo, S., & Ulven, C. A. (2012). Natural fiber reinforced composites. *Polymer Reviews*, *52*(3–4), 259–320.

Gurunathan, T., Mohanty, S., & Nayak, S. K. (2015). A review of the recent developments in biocomposites based on natural fibres and their application perspectives. *Composites: Part A*, *77*, 1–25.

Homkhiew, C., Ratanawilai, T., & Thongruang, W. (2013). Composites from recycled polypropylene and rubberwood flour: Effects of composition on mechanical properties. *Thermoplastic Composite Materials*, *28*(2), 179–194.

Homkhiew, C., Ratanawilai, T., & Thongruang, W. (2014). Effects of natural weathering on the properties of recycled polypropylene composites reinforced with rubberwood flour. *Industrial Crops and Products, 56*, 52–59.

Hori, R., & Wada, M. (2005). The thermal expansion of wood cellulose crystals. *Cellulose, 12*(5), 479–484.

Jao, D., Xue, Y., Medina, J., & Hu, X. (2017). Protein-based drug-delivery materials. *Materials, 10*(5), 1–24.

Kaya, A., & Kar, F. (2016). Properties of concrete containing waste expanded polystyrene and natural resin. *Construction and Building Materials, 105*, 572–578.

Khan, M. J. H., Hussain, M. A., & Mujtaba, I. M. (2014). Polypropylene production optimization in fluidized bed catalytic reactor (FBCR): Statistical modeling and pilot scale experimental validation. *Materials, 7*(4), 2440–2458.

Lu, N., & Oza, S. (2013). A comparative study of the mechanical properties of hemp fiber with virgin and recycled high density polyethylene matrix. *Composites Part B: Engineering, 45*(1), 1651–1656.

Masri, T., Ounis, H., Sedira, L., Kaci, A., & Benchabane, A. (2018). Characterization of new composite material based on date palm leaflets and expanded polystyrene wastes. *Construction and Building Materials, 164*, 410–418.

Miracle, D. B., Donaldson, S. L., Henry, S. D., Moosbrugger, C., Anton, G. J., Sanders, B. R., ... Scott Jr, W. (2001). ASM handbook: Composites Vol. 21, S. D. Henry, C. Moosbrugger, G. J. Anton, B. R. Sanders, N. Hrivnak, C. Terman, ... J. Williamson (Eds.). Materials Park: ASM International. Retrieved from <http://www.aeromat.com/content/ASM/StoreFiles/06781G_Frontmatter.pdf>.

Moreno, D. D. P., & Saron, C. (2017). Low-density polyethylene waste/recycled wood composites. *Composite Structures, 176*, 1152–1157.

Newman, R. H. (2008). Simulation of X-Ray diffractograms relevant to the purported polymorphs cellulose IV. *Cellulose, 15*(6), 769–778.

Noorunnisa Khanam, P., & AlMaadeed, M. A. (2014). Improvement of ternary recycled polymer blend reinforced with date palm fibre. *Materials and Design, 60*, 532–539.

Oza, S., Wang, R., & Lu, N. (2011). Thermal and mechanical properties of recycled high density polyethylene hemp fiber composites. *International Journal of Applied Science and Technology, 1*(5), 31–36.

Pandey, G., & Sreekumar, M. (2015). Development of high strength bio-composite material for light weight serial robot mechanisms. *Procedia Computer Science, 76*, 512–521. (Iris).

Poletto, M., Dettenborn, J., Zeni, M., & Zattera, A. J. (2011). Characterization of composites based on expanded polystyrene wastes and wood flour. *Waste Management, 31*(4), 779–784.

Radzi, A., Sapuan, S., & Jawaid, M. (2018). Mechanical and thermal performances of roselle fibre reinforced thermoplastic polyurethane composites. *Plastics Technology and Engineering, 57*(7), 601–608.

Ribeiro, G. L., Gandara, M., Moreno, D. D. P., & Saron, C. (2017). Low-density polyethylene/sugarcane fiber composites from recycled polymer and treated fiber by steam explosion. *Journal of Natural Fibers, 00*(00), 1–12.

Santucci, B. S., Maziero, P., Rabelo, S. C., Curvelo, A. A. S., & Pimenta, M. T. B. (2015). Autohydrolysis of hemicelluloses from sugarcane bagasse during hydrothermal pretreatment: A kinetic assessment. *Bioenergy Resources, 8*(4), 1778–1787.

Sapuan, S. M., Leenie, a, Harimi, M., & Beng, Y. K. (2006). Mechanical properties of woven banana fibre reinforced epoxy composites. *Materials & Design, 27*(8), 689–693.

Sarker, M., & Rashid, M. M. (2014). Polypropylene waste plastic conversion into fuel oil by using thermal degradation with fractional process. *American Journal of Environment, Energy and Power Research, 2*(3), 1–10. Retrieved from <www.ajeepr.com>.

Satyanarayana, K. G., Arizaga, G. G. C., & Wypych, F. (2009). Biodegradable composites based on lignocellulosic fibers—An overview. *Progress in Polymer Science (Oxford), 34*(9), 982–1021.

Selvaraju, S., & Ilaiyavel, S. (2011). Application of composites in marine industry. *Journal of Engineering Research and Studies*, *2*(2), 89–91.

Singha, A. S., & Thakur, V. K. (2008). Effect of fibre loading on urea-formaldehyde matrix based green composites. *Iranian Polymer Journal*, *17*(11), 861–873.

Sivapragasam, A. (2008). Coconut in Malaysia: Current developments and potential for re-vitalization. In *2nd International Plantation Industry Conference and Exhibition (IPICEX2008)*, (November), (pp. 1–9). Shah Alam: MARDI.

Sommerhuber, P. F., Welling, J., & Krause, A. (2015). Substitution potentials of recycled HDPE and wood particles from post-consumer packaging waste in wood-plastic composites. *Waste Management*, *46*, 76–85.

Stojanović, B., & Ivanović, L. (2013). Application of aluminium hybrid composites in automotive industry. *Technical Gazette*, *3651*, 247–251.

Taufiq, M. J., Mustafa, Z., & Mansor, M. R. (2017). Utilisation of recycled thermoplastics sourced from rejected-unused disposable diapers as polymer blends. *Journal of Mechanical Engineering and Sciences*, *11*(4), 3137–3143.

Taufiq, M. J., Ridzuan, M., & Mustafa, Z. (2018). Characterisation of wood plastic composite manufactured from kenaf fibre reinforced recycled-unused plastic blend. *Composite Structures*, *189*, 510–515.

Testova, L., Borrega, M., Tolonen, L. K., Penttilä, P. A., Serimaa, R., Larsson, P. T., & Sixta, H. (2014). Dissolving-grade birch pulps produced under various prehydrolysis intensities: Quality, structure and applications. *Cellulose*, *21*(3), 2007–2021.

Tian, G., Fu, Y., Zhuang, J., Wang, Z., & Li, Q. (2017). Separation of saccharides from prehydrolysis liquor of lignocellulose to upgrade dissolving pulp mill into biorefinery platform. *Bioresource Technology*, *237*, 122–125.

Turku, I., Kärki, T., Rinne, K., & Puurtinen, A. (2016). Characterization of plastic blends made from mixed plastics waste of different sources. *Waste Management & Research*, *35*(2), 200–206.

Turku, I., Keskisaari, A., Kärki, T., Puurtinen, A., & Marttila, P. (2017). Characterization of wood plastic composites manufactured from recycled plastic blends. *Composite Structures*, *161*, 469–476.

Turku, I., Keskisaari, A., Kärki, T., Puurtinen, A., & Marttila, P. (2018). Durability of wood plastic composites manufactured from reycled plastic. *Heliyon*, *4*, 1–30.

Wada, M. (2001). In situ observation of the crystalline transformation from cellulose IIII to Iβ. *Macromolecules*, *34*(10), 3271–3275.

Wada, M., Nishiyama, Y., & Langan, P. (2006). X-Ray structure of ammonia – cellulose I: New insights into the conversion of cellulose I to cellulose IIII. *Macromolecules*, *39*(8), 2947–2952.

Xu, J., Fu, Y., Tian, G., Li, Q., Liu, N., Qin, M., & Wang, Z. (2018). Mild and efficient extraction of hardwood hemicellulose using recyclable formic acid/water binary solvent. *Bioresource Technology*, *254*, 353–356.

Xu, F., Sun, J.-X., Sun, R., Fowler, P., & Baird, M. S. (2006). Comparative study of organosolv lignins from wheat straw. *Industrial Crops and Products*, *23*(2), 180–193.

Yang, T. H., Leu, S. Y., Yang, T. H., & Lo, S. F. (2012). Optimized material composition to improve the physical and mechanical properties of extruded wood-plastic composites (WPCs). *Construction and Building Materials*, *29*, 120–127.

Zhang, X. (2014). Chemical structure of natural polymer fibers. In *Fundamentals of fiber science* (pp. 53–64). Pennsylvania: DEStech Publications, Inc.

Zhang, Y.-Q. (2002). Applications of natural silk protein sericin in biomaterials. *Biotechnology Advances*, *20*(2), 91–100.

CHAPTER

8

Degradable composites: processes and applications

Zaleha Mustafa[1], *Siang Yee Chang*[1], *Siti Hajar Sheikh Md Fadzullah*[2], *Azlin Fazlina Osman*[3] *and Ismail Ibrahim*[3]

[1]Faculty of Manufacturing Engineering, Universiti Teknikal Malaysia Melaka, Melaka, Malaysia, [2]Centre for Advanced Research on Energy, Fakulti Kejuruteraan Mekanikal, Universiti Teknikal Malaysia Melaka, Melaka, Malaysia, [3]Center of Excellence Geopolymer and Green Technology (CEGeoGTech), Faculty of Chemical Engineering Technology, Universiti Malaysia Perlis (UniMAP), Perlis, Malaysia

8.1 Introduction

In the last few decades, reports have shown that with the increase in the global population there has been a significant rapid development in the manufacturing of polymer-based products either for domestic or high-end applications to meet demand owing to their mechanical properties, easy production process, low cost, and versatile applications (Quitadamo, Massardier, & Valente, 2017). Most of these products are either made from synthetic polymers or contain synthetic polymers as the major component. This results in an increase in the disposal of materials to landfill, which ultimately increases the burden on the environment and significantly contributes to both land and sea pollution (Pietrelli, Poeta, Battisti, & Sighicelli, 2017). Therefore the use of degradable materials, particularly from renewable outlets, is viewed as the potential solution to decrease the environmental issues that are imposed by non-degradable petroleum-based polymers (Alix, Colasse, Morvan,

Design for Sustainability
DOI: https://doi.org/10.1016/B978-0-12-819482-9.00017-4

Lebrun, & Marais, 2014; Gholampour & Ozbakkaloglu, 2020). One of the initiatives is to produce a totally green composite, by combining biofiber and resin from compostable polymers and natural sources. This composite often utilizes the waste from agricultural and forest feedstock and has the ability to degrade harmoniously with nature at the end of its life cycle (Gurunathan, Mohanty, & Nayak, 2015; Ramli, Fadzullah, & Mustafa, 2017).

8.2 Degradable composites system

In a composite system, there is a dispersed phase which is continuous and surrounds the other phase; this phase is termed the matrix. Meanwhile, the second constituent of a composite system which is embedded into the matrix to enhance or reinforce the mechanical properties of the matrix is referred to as the reinforcement.

Owing to the superior specific mechanical properties, composite materials have been widely used for more than half a century, in various applications demanding high performance. However, composite materials are mainly sourced from nonrenewable resources such as petroleum-based materials and remarkable amounts of these composite materials are introduced into the ecosystem as solid waste after their final use (Pietrelli et al., 2017). In the last few decades, high demand of the composite has resulting in increase of the composite waste in our environment, lead to major environmental issue.

With increase of environmental awareness, much attention has been diverted to the search for eco-friendly materials as part of the consideration in designing new products. This underscores a change from nonrenewable and nondegradable composite materials to renewable and easily degradable ones. In other words, degradable composite materials are the need of the moment as well as the future due to their compatibility with the environment. In this context, the focus is placed upon fully biodegradable composites in which both the matrix and reinforcement are derived from biodegradable resources, to be exact, biodegradable polymers as the matrix material and natural fibers as the second constituent.

8.2.1 Matrices

Though the matrix phase is not always present in greater quantities in a composite system, it plays a number of significant roles such as binding the reinforcing materials together, to transfer the stresses between reinforcing materials, protecting reinforcing materials from surface damage, and acting as a barrier to crack propagation.

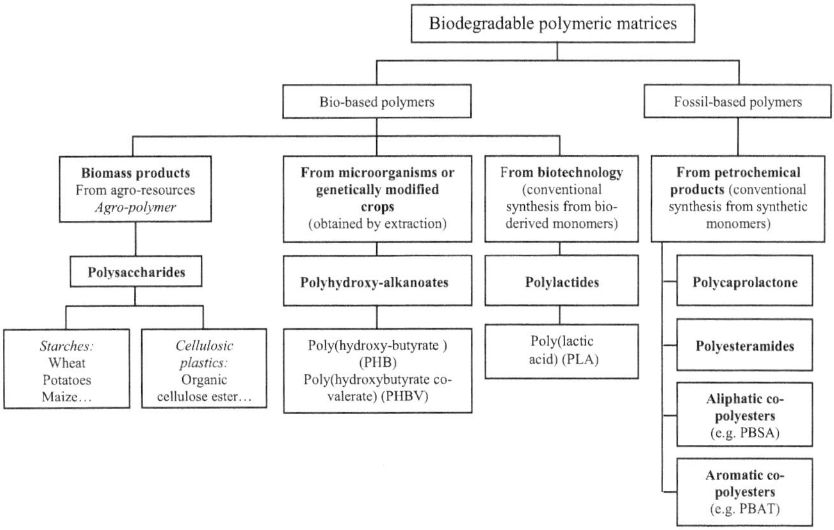

FIGURE 8.1 Classification of biodegradable polymers used as the matrix in degradable composites. *Source: Modified from Avérous, L. & Boquillon, N. (2004). Biocomposites based on plasticized starch: Thermal and mechanical behaviours. Carbohydrate Polymers, 56, 111–122.*

Biodegradable polymers used as the matrix can be classified into four families as illustrated in Fig. 8.1. The sources of the variety of polymeric materials range from renewable resources such as plants, bacteria, and nonrenewable petroleum. Accordingly, biodegradable polymers have been classified as bio-based or fossil-based depending on their origin. The first family is agro-polymers (e.g., polysaccharides) obtained from biomass by fractionation which make use of natural polymers with minor modification, whereas the other three families are polyesters originating from diverse sources. The second and third families represent polyesters sourced from biomass, respectively, through direct extraction from microorganisms or genetically modified plants and fermentation/polymerization from bio-based monomers (Avérous & Boquillon, 2004; Shen, Haufe, & Patel, 2009). Meanwhile, the fourth family is generally chemically synthesized from petrochemical processes.

Agro-polymers such as starch and cellulose derivatives are inexpensive, annually renewable materials derived from crops. The properties of agro-polymers can be regulated either through their natural mechanisms or chemical modification; for instance, by varying the ratio of amylose and amylopectin in starch as well as the degree of modification of glucose repeat unit hydroxyl groups in plant cellulose, the properties can be regulated easily and efficiently (Gross & Kalra, 2002). However, it is noted that the rigidity and water sensitivity of starches may be an adverse problem (Campilho, 2015; Yu, Dean, & Li, 2006) unless plasticizing them or blending them with biodegradable polyesters to increase flexibility and water resistance or down-the-drain disposal is desired (Gross & Kalra, 2002; Wu, 2003).

Cellulosic plastics, such as cellulose acetate (CA), cellulose acetate propionate, and cellulose acetate butyrate are bio-based polymers produced through the esterification or acetylation of cellulose. A variety of raw materials such as cotton, wood pulp, and sugarcane are used in making cellulose ester biopolymers (Mohanty, Wibowo, Misra, & Drzal, 2003). Eastman Chemical Company has successfully developed fully biodegradable CAs by reducing the acetyl substitution number down to 1.7 (Gross & Kalra, 2002). However, similar to starch plastics, CAs require plasticization for ease of processing. Several researchers have investigated cellulosic plastics as matrix polymer in degradable composites to form composite structures (Mohanty et al., 2003; Mohanty, Wibowo, Misra, & Drzal, 2004), films (Lu & Drzal, 2010), and filters (Janda, 2015).

PHAs, on the other hand, are produced directly by microbes throughfermentation of sugars or lipids (Gaceva et al., 2007). They can be accumulated to high levels in bacteria ($\sim$95% of the cellular dry weight). Their structures can be manipulated by genetic or physiological strategies. The widespread synthesis of polyhydroxyalkanoate (PHA) is associated with the corresponding abundance of microbes that produce PHA-degrading enzymes. PHAs with short side chains behave similarly to polypropylene (PP) (Gaceva et al., 2007), whereas PHAs with long side chains are elastomeric (Gross & Kalra, 2002; John & Thomas, 2008).

As for polylactic acid (PLA), the breakthrough in process technology in the 1980s has driven its extensive production and applicability in the market outside the medical field. The PLA crystallinity, crystallization rate, transparency, and degradation rate of finished products are conveniently controlled by the copolymerization of selected L- to D-isomer ratios of lactide or lactic acid (Gaceva et al., 2007). Upon disposal, PLA degrades primarily by hydrolysis, not microbial attack. Hence, even at high humidity, it is uncommon to encounter contamination of high-molecular-weight PLA by fungi, mold, or other microbes (Gross & Kalra, 2002). In contrast to agro-polymers, PHAs, and PLA, a number of biodegradable polyesters are also chemically synthesized through non-renewable crude oil resources, be it polycaprolactone (PCL), polyester amide (PEA), or another.

PCL is a linear aliphatic polyester which is generally prepared by the ring-opening polymerization of ε-caprolactone (Gaceva et al., 2007; Khatiwala, Shekhar, Aggarwal, & Mandal, 2008). There is a variety of molecular weights PCL commercially available by controlling the use of dry monomer and addition of a specific amount of active-hydrogen initiator (Piskin, 2002). Although low-molecular-weight PCL ranges from liquids to hard waxes, the higher molecular weight polymer is a strong, ductile polymer with excellent mechanical characteristics. It is a hydrophobic and semicrystalline polymer with the following thermal properties: melting temperature of 59°C–64°C and glass transition temperature of ~ −60°C (Khatiwala et al., 2008); hence, PCL is in a rubbery state at room temperature. Mechanically, it has a relatively low tensile strength (23 MPa) but good flexibility characterized by very high fracture strain (>700%) (di Franco, Cyras, Busalmen, Ruseckaite, & Vázquez, 2004; Pan et al., 2007; Rosa, Lopes, & Calil, 2005; Sen & Bhattacharya, 2000; Zhao et al., 2008). PCL has the unique characteristic of being miscible with almost all other polymers (polyethylene (PE), PP, polystyrene (PS), polymethyl methacrylate, polycarbonates, polysulfone, polyvinyl acetate, etc.). Degradation of PCL is through hydrolysis at the ester linkages at its backbone, but at a very slow rate (Edlund & Albertsson, 2002). Although PCL is an eco-friendly material, its widespread usage has been restricted due to its high cost and low melting temperature (Zhao et al., 2008). As a remedial measure, PCL is usually mixed with other low-cost biodegradable polymers such as starch (Chang, Ismail, & Ahsan, 2012; McKeen, 2017).

Undoubtedly, biodegradable polymers are very attractive in offering a possible solution to waste-disposal problems but these polymers tend to have a high price associated with them. Hence, applications of these polymers need to be found and taken on by manufacturers in order to consume sufficiently large quantities of these materials and drive the price down so that they can compete economically in the market (Blackburn, 2005). Fig. 8.2 shows the cost variation between traditional and biodegradable polymers.

8.2.2 Reinforcements

It is apparent that most biodegradable polymer blends possess a large range of properties but with lower strength properties as compared to commercially available nondegradable polymers; hence, this would suggest the incorporation of appropriate fibers to form composites. Like the matrix, the second constituent of degradable composites must be able to break down upon disposal and this is mainly accounted for by natural fibers. Natural fibers are mainly sourced from plants, animals, and minerals (Campilho, 2015; Ramamoorthy, Skrifvars, & Persson, 2015); however, the latter two have not been widely explored as reinforcing phases (Ramamoorthy et al., 2015). The following section, therefore, mainly deliberates on plant fibers.

Plant fibers, also known as lignocellulosic fibers, are usually referred to as wood fiber and agro-based leaf, bast, seed, straw, and fruit fibers. Typical examples are: (1) leaf: abaca, date palm, sisal, pineapple leaf, and henequen; (2) bast: flax, ramie, kenaf, hemp, jute and banana; (3) seed:

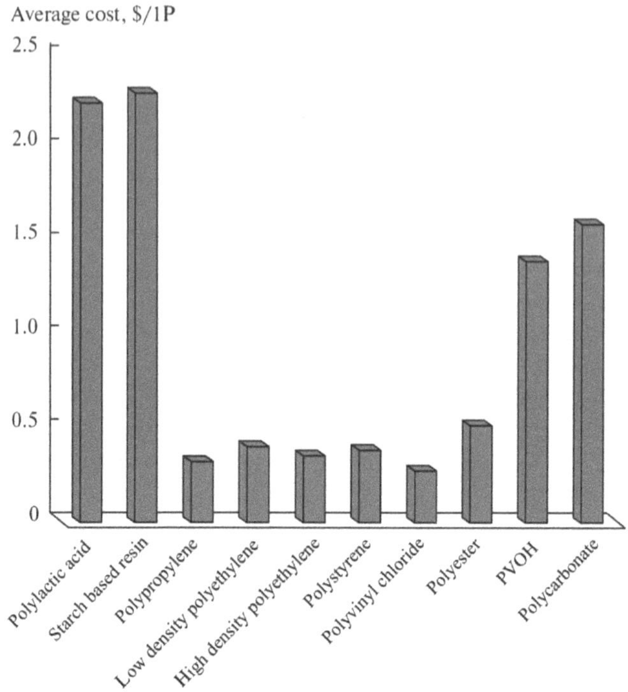

FIGURE 8.2 Cost comparison (in US$) between some biodegradable and traditional polymers (Hamad, Kaseem, Ko, & Deri, 2014). *PVOH*, Polyvinyl alcohol.

cotton; (4) straw: corn, rice, and wheat; (5) fruit: coir, kapok, oil palm, and sponge gourd; and (6) grass: bagasse, and bamboo (John & Thomas, 2008; Mohanty, Misra, & Hinrichsen, 2000; Ramamoorthy et al., 2015). These fibers often contribute greatly to the structural performance of the plant and, when used in polymeric composites, can provide significant reinforcement (Gaceva et al., 2007; John & Thomas, 2008).

The use of these lignocellulosic fibers in the preparation of composites has been done since historical times, with one of their applications in aircraft as early as the 1940s. However, due to their inferior properties and incomplete data on different lignocellulosic fibers compared with synthetic fibers, their applications were not explored extensively until the 1980s (Satyanarayana, Azrizaga, & Wypych, 2009). Thereafter there has been growing motivation to substitute synthetic fibers with lignocellulosic fibers in composite manufacturing, principally for the following reasons:

1. The presence of sources throughout the world, with some of them being available in abundance in the tropics and subtropics as agricultural waste products after cultivation (Bledzki & Gassan, 1999; John & Thomas, 2008).

2. Increasing environmental awareness, which has resulted in a greater focus on plant-derived fiber and crop-derived plastics as materials for the 21st century, simply due to their attractive aspects, such as CO_2 sequestration, independent of petroleum products, and value-added opportunities for agricultural industries (Bledzki & Gassan, 1999; John & Thomas, 2008).
3. The availability of complete data in regards to their structure and properties which depend greatly on the source of extraction, the age of the source, method of extraction, processing, etc. (Satyanarayana et al., 2009).
4. Lower cost and lightweight with comparable mechanical properties in comparison to synthetic fibers (Gaceva et al., 2007; John & Thomas, 2008) as illustrated in Table 8.1.

Despite the reasons stated above, there are several other merits in terms of technical aspects for using lignocellulosic fibers in composite fabrication, such as it being nonabrasive to machinery, lower processing energy requirement, and possessing fewer health hazards to the workforce (Sreekala, Kumaran, Joseph, Jacob, & Thomas, 2000). The economic and technological advantages of these lignocellulosic fibers may lead to sustainable development whereby incorporation of these fibers may increase the durability as well as the recyclability of the products produced (John & Thomas, 2008; Satyanarayana et al., 2009).

Also, it is also well-known that lignocellulosic fibers possess a few drawbacks, such as poor interfacial adhesion with the nonpolar/hydrophobic polymers, relatively low processing temperature, high moisture absorption, low microbial resistance, and susceptibility to rotting. All these disadvantages impair the efficiency of degradable composites, in terms of mechanical properties as well as dimensional stability. Nevertheless, it is quite clear that the advantages outweigh the disadvantages and most of the shortcomings have remedial measures in the form of chemical treatments (John & Thomas, 2008).

8.2.3 Celluloses

Lignocellulosic fiber is a three-dimensional polymeric composite, principally composed of cellulose (α-cellulose), hemicelluloses, and lignin. These basic components result in the physical properties of the fiber and give rise to its structure. However, a number of factors, in general, may influence the structure and chemical composition of the fiber, including climatic conditions, age, and the digestion process (Bledzki & Gassan, 1999).

Cellulose is an essential component of all lignocellulosic fibers. It is a natural linear polymer, having an average molecular weight of between

TABLE 8.1 Comparison of some properties and costs of lignocellulosic fibers and synthetic fibers (Biagiotti, Puglia, & Kenny Jose, 2004).

Fiber	Density (g/cm^3)	Diameter (μm)	Extension at break (%)	Tensile strength (MPa)	Young's modulus (GPa)	Specific modulus (GPa/cm^3)	Cost (Euro/kg)
Flax	1.40–1.50	40–620	2.7–3.2	343–1035	27–80	19–53	2.29–11.47
Jute	1.30–1.50	30–140	1.4–3.1	187–773	3–55	2–37	0.12–0.35
Abaca	1.5	17–21	10–12	980	72	481	0.81–0.92
Sisal	1.30–1.50	100–300	2.0–2.9	507–855	9.0–28.0	7–19	0.70–1.02
Kenaf	1.22–1.40	40–90	3.7–6.9	295–930	22–53	18–38	0.53–0.61
Ramie	1.5	40–60	3.6–3.8	400–938	44–128	29–85	1.44–2.40
Hemp	1.40–1.50	16–50	1.3–4.7	580–1110	3–90	2–60	0.57–1.73
Cotton	1.50–1.60	16–21	2.0–10.0	287–597	5.5–12.6	4–8	1.61–4.59
Banana	1.30–1.35	50–280	3–10	529–914	7.7–32.0	6–24	0.7–0.9
Henequen	1.49	20–500	3.0–5.0	430–580	10.1–16.3	7–11	0.38–0.67
Bagasse	0.55–1.25	200–400	0.9	20–290	2.7–17.0	5–14	0.15
Pineapple	1.52–1.56	200–8800	0.8–3.0	170–1627	6.21–82	4–53	0.36–0.72
E-glass	2.50–2.55	10–20	2.5	2000–3500	73.0	29	1.25
Aramide	1.4–1.45	12	3.3–3.7	3000–3150	63.0–67.0	45–48	7.2
Carbon	1.40–1.75	5.5–6.9	1.4–1.8	4000	230.0–240.0	164–171	12.0

130,000 and 190,000 (Rowell & Stout, 2007) with an average degree of polymerization (DP) of approximately 10,000. However, purification procedures usually reduce its DP to something in the order of 2500 (Bledzki & Gassan, 1999). Cellulose consists of D-anhydroglucose ($C_6H_{11}O_5$) repeating units joined by 1,4-β-D-glycosidic linkages at the C_1 and C_4 positions, with each repeating unit containing three hydroxyl groups. Crystallinity and physical properties of cellulose are rendered by these hydroxyl groups and their ability to form a hydrogen bond. The structure of cellulose is formed of slender rod-like crystalline microfibrils, which are resistant to strong alkaline (17.5%) but easily hydrolyzed by acid to water-soluble sugars (John & Thomas, 2008).

More importantly, the mechanical properties of lignocellulosic fiber depend greatly on its cellulose type (Bledzki & Gassan, 1999; Gaceva et al., 2007), because each type of cellulose has its own cell geometry, as illustrated in Fig. 8.3 and Table 8.2, and the geometrical conditions determine the mechanical properties (Bledzki & Gassan, 1999). For instance, the microfibril angle, β, as a measure of the cellulose fibril orientation in the secondary wall layer of a lignocellulosic fiber plays a crucial role in its mechanical properties. An increasing microfibril angle decreases cell wall stiffness but increases the strain to fracture (Burgert, 2006). Also, the elastic modulus of the crystalline regions (crystal modulus, E_1) of cellulose I is

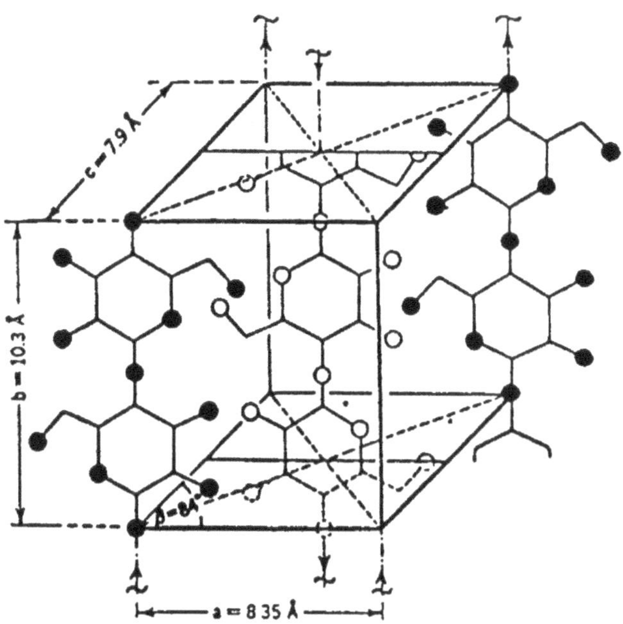

FIGURE 8.3 Designation of lattice parameters of elementary cell in cellulose (Bledzki & Gassan, 1999).

TABLE 8.2 Lattice parameters of elementary cells in different types of cellulose (Bledzki & Gassan, 1999).

Type	Source	Dimension (nm)			β (°)
		a	b	c	
Cellulose I	Cotton	0.821	1.030	0.790	83.3
Cellulose II	Cotton, mercerized	0.802	1.036	0.903	62.8
	Cotton, viscose	0.801	1.036	0.904	62.9
Cellulose III		0.774	1.030	0.990	58.0
Cellulose IV		0.812	1.030	0.799	90.0

138 GPa, which is comparable to those of high-performance fibers such as Kevlar, Technora, and liquid crystal polyesters (Ekonol, Vectran, Rodrum). This relatively high value is because of the rigid skeleton of glucopyranose rings and the intramolecular hydrogen bonds. Moreover, the maximum specimen modulus, Y_{max}, of cellulose I is also higher than those of glass or aluminum. These strongly suggest that lignocellulosic fiber at large possesses a potential ability as the reinforcement fiber replacing glass fiber (Nishino, Hirao, Kotera, Nakamae, & Inagaki, 2003).

On the other hand, hemicelluloses, which form a supportive matrix for cellulose microfibrils, are made up of a group of polysaccharides composed of a combination of 5- or 6-carbon ring sugars. The average molecular weight of the hemicelluloses is in the range of 24,000–27,000 Da (Rowell & Stout, 2007), with the DP at around 50–300. Hemicelluloses are not a form of cellulose, yet they remain associated with cellulose after lignin has been removed. Hemicelluloses are hydrophilic, soluble in alkaline, and easily hydrolyzed in acids (Bledzki & Gassan, 1999; John & Thomas, 2008). The third major constituent, lignin, is a complex, long-chained hydrocarbon polymer with both aliphatic and aromatic constituents. Like hemicelluloses, this amorphous compound gives rigidity to plants but is hydrophobic in nature. Lignin is found to contain five hydroxyl and five methoxyl groups per building unit. It is not hydrolyzed by acids, but soluble in hot alkaline and readily oxidized (John & Thomas, 2008).

Generally, lignocellulosic fiber has a structure of hollow cellulose fibrils held together by a hemicellulose and lignin matrix, as shown in Fig. 8.4. The hemicellulose molecules are hydrogen-bonded to cellulose and act as a cementing matrix between the cellulose microfibrils, forming the cellulose–hemicellulose network, which is thought to be the main structural component of the fiber cell. The stiffness of the cellulose/hemicelluloses composite is enhanced by the hydrophobic lignin network which acts as a coupling agent (John & Thomas, 2008).

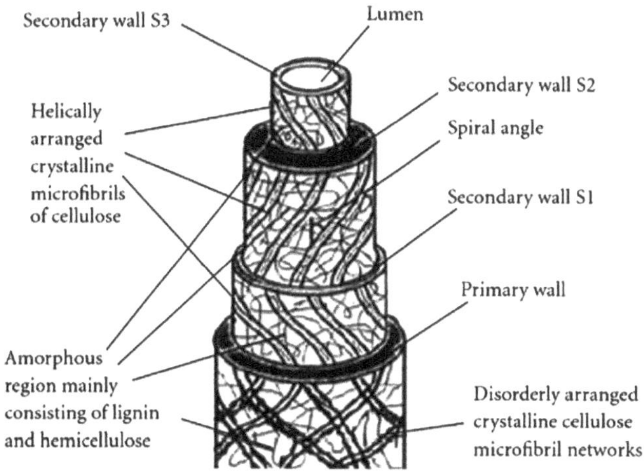

FIGURE 8.4 Structure of lignocellulosic fiber (John & Thomas, 2008).

TABLE 8.3 Infrared transmittance peaks (/cm) of untreated kenaf fiber (Sgriccia, Hawley, & Misra, 2008).

Bond type	Wave number (/cm)	Remarks
–CH– vibration	2900	Methyl and methylene groups in cellulose and hemicellulose
–C=O– stretching	1730	Acetyl group of hemicellulose
–C=C– stretching	1625	Presence of hemicellulose
–CO– stretching	1239	Acetyl group of lignin
–COOH	664	–COOH– bending peak

The presence of these three main components in lignocellulosic fiber can be verified through Fourier transform infrared spectroscopy (FTIR). Peaks for cellulose, hemicelluloses, and lignin of untreated kenaf fiber are detailed in Table 8.3.

8.3 Synthesis and fabrication process

Generally, the methods to fabricate the natural fiber composites depend very much on the type of polymer matrix materials. In addition,

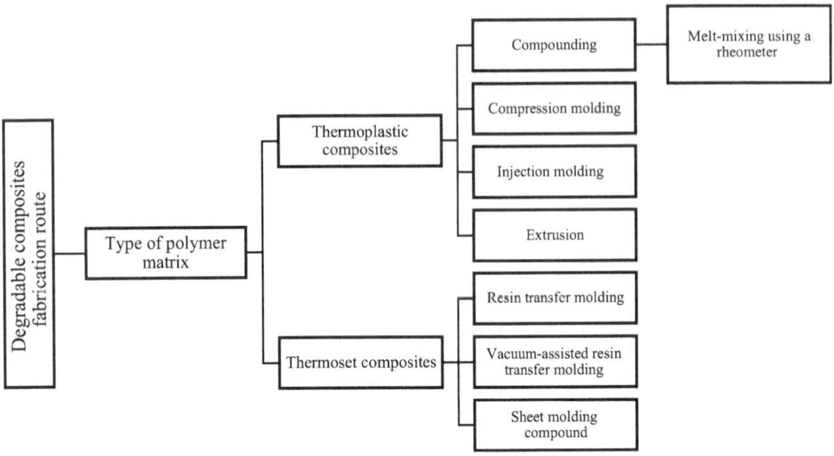

FIGURE 8.5 A schematic diagram showing the degradable composites fabrication route based on the polymer matrix (author owned).

the types of fiber used as well as their orientation also have a strong influence on the type of fabrication method suitable for the composites. Fig. 8.5 shows a schematic diagram of common fabrication routes for degradable composites, based on the polymer matrix, that is, either thermoplastic or thermoset polymer (Leong, Thitithanasarn, Yamada, & Hamada, 2013). The common fabrication processes for thermoplastic composites include compounding, compression molding, injection molding, and extrusion. For these types of fabrication processes, it was argued that composite properties vary in each technique due to the use of different processing pressures, temperatures, and speeds (Senthilkumar et al., 2018).

In addition, despite the relatively shorter cycle time as compared to those of the thermoset molding processes, one of the crucial processing parameters in fabricating the degradable natural fiber reinforced composites via injection molding is the processing temperature, that is typically below 220°C. Here, it is important to ensure that the processing temperature set for the composites does not exceed the maximum allowable temperature for the natural fiber and the polymer matrix, which is typically below 180°C for the natural fiber, and a minimum exposure time is set for the fiber in the barrel during the heating process.

As an example, in a review article by Sadashiv and Abasaheb (2019), it was summarized that pineapple leaf fiber (PALF) reinforced composite fabrication includes solution mixing, injection molding, a three-step processing consisting of melt-mixing, pre-pegging, and compression molding, two-roll milling, and extrusion using twin-screw extruder as some of the common fabrication methods for such materials.

Meanwhile, the thermoset composites are typically fabricated using resin transfer molding (RTM), vacuum-assisted resin transfer molding, or sheet molding compound. In addition, in a review on sisal fiber reinforced thermoset composites, Senthilkumar et al. (2018) summarized the fabrication method for such materials and included hand lay-up, manual molding, compression molding, the closed mold method, and also RTM as some of the commonly employed fabrication methods for these type of materials.

Nonetheless, despite the flexibility to fabricate various fiber architectures including mats in which the fiber is randomly oriented or woven fabrics of continuous natural fiber strands, Leong et al. (2013) highlighted that some of the main issues when selecting these processing routes in fabricating the thermoset-based biocomposites include the cost factor as well as relatively slow rate of curing, which is typically between 2–24 hours. In addition, they argued that post-processing may be required, which includes trimming and grinding, hence, such a fabrication route is more suited for large components such as those for the automotive industry (Leong et al., 2013).

The fabrication route for degradable polymer composites could also be classified based on the fiber architecture, which includes the fiber type, geometry, as well as length and orientation, which are some of the major contributing factors in the overall mechanical performance of degradable composites (Senthilkumar et al., 2018). For instance, in wood–plastic composites (WPC), production of composites with randomly distributed short fibers from wood is usually made via a compounding method. Chan et al. (2018) argued that the Brabender rheology mixers and twin-screw extruders are commonly used to melt mix the wood flour with biodegradable polymers at the laboratory scale. In their review, it was also highlighted that in a practical application, the processing of biodegradable composites should be handled with greater care since there is a risk of hydrolytic and also thermal degradation during processing requiring more care and attention to be taken in the processing (Chan et al., 2018).

Another review of natural fiber composites based on kenaf fiber reinforcement was reported by Tholibona et al. (2019) which summarized different fabrication routes for kenaf fiber reinforced composites, in which kenaf fibers were available in various geometries; these being continuous long fibers and discontinuous fibers in the form of yarn, woven, or nonwoven mat. Based on their summary, the fabrication routes for the unidirectional kenaf fiber reinforced composites include two-step compression molding which includes hand lay-up and hot press processing, press forming, wet lay-up, and cold press as well as microbraiding (Tholibona et al., 2019).

As an example, as shown in Fig. 8.6, Liu, Takagi, Osugi, and Yang (2012) proposed the use of a customized jig in ensuring full alignment of the fibers incorporated in the epoxy resin via RTM in fabricating a

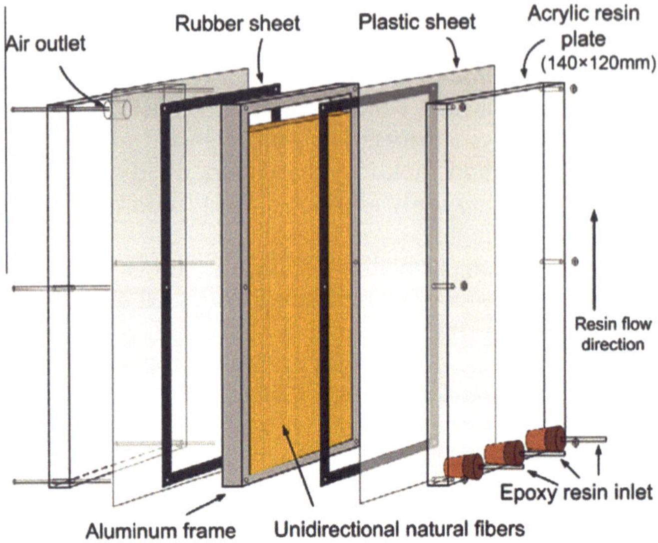

FIGURE 8.6 A customized mold used in fabricating unidirectional fiber composites via resin transfer molding (Liu et al., 2012).

unidirectional reinforced epoxy composite in which both abaca and bamboo are considered and characterized in terms of their transverse thermal conductivity (Liu et al., 2012).

8.4 Degradability of composites

The degradation of the biocomposite can be accelerated by many conditions such as moisture absorption, humidity, and pH of the soil. The ideal quality of a degradable composite is it will degrade at the end of its life cycle without harming the environment. Often, the degradation occurs during applications, particularly when the composite is incorporated with natural fibers as their reinforcement phase.

The natural fiber is hydrophilic in nature and often prone to moisture absorption either in a humid atmosphere and/or when immersed in water (Shubhra, Alam, & Beg, 2011; Taufiq, Mansor, & Mustafa, 2018). Hemicellulose contains low-molecular-weight polysaccharides consisting of hexoses, pentoses, and uronic acid residues (Fig. 8.4). Often a higher content of hemicellulose will lead to an increase in moisture absorption and biodegradation (Azwa, Yousif, Manalo, & Karunasena, 2013). Water transportation in polymer composite is expedited via three main mechanisms, which are diffusion (inside the matrix), defect within the matrix (pores, crack, or voids), or capillary action at the fiber–matrix interface

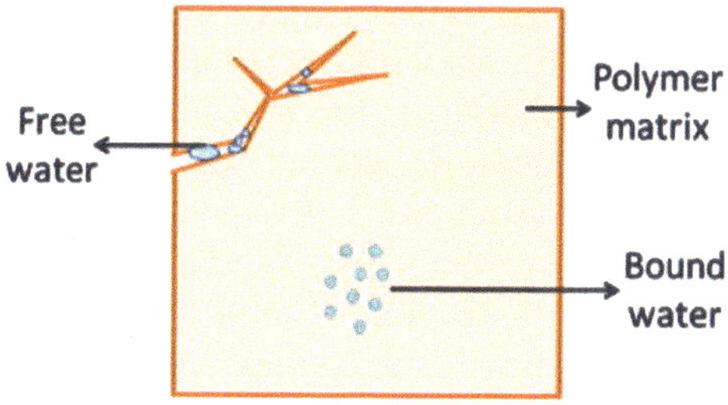

FIGURE 8.7 Water molecules in the polymer matrix. *Source: Reproduced with permission from Azwa, Z. N., Yousif, B. F., Manalo, A. C., & Karunasena, W. (2013). A review on the degradability of polymeric composites based on natural fibres. Materials & Design, 47, 424–442.*

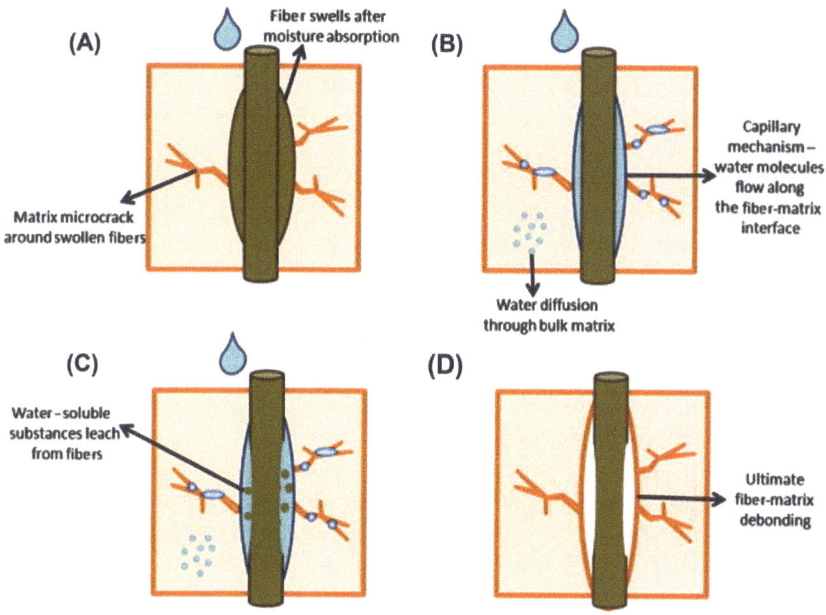

FIGURE 8.8 Effect of water on the fiber–matrix interface. *Source: Reproduced with permission from Azwa, Z. N., Yousif, B. F., Manalo, A. C., & Karunasena, W. (2013). A review on the degradability of polymeric composites based on natural fibres. Materials & Design, 47, 424–442.*

(Beg & Pickering, 2008). Water that is absorbed in a polymer is known as free water and bound water (Chen, Miao, & Ding, 2009). Free water moves independently through the pores, while bound water is attached to the polar groups of the polymer (Fig. 8.7). The overall effect of water absorption on the integrity of the polymer composite is summarized in Fig. 8.8. Upon exposure to moisture, the water molecule infiltrates and attaches to the hydrophilic region of the fiber creating intermolecular hydrogen bonding with the fiber. This reduces the interfacial adhesion of the fiber–matrix, leading to swelling of the cellulose fiber and resulting in microcracking at the interfaces. In turn, more water molecules are transported via the capillary and microcrack, reducing the free water with a higher amount of bound water present in the system, resulting in a high rate of leaching out of the water-soluble substances, which can be observed via debonding and disintegration of the composite's component that ultimately lead to failure of the composite. After a certain period of exposure, biological activities will take over aided by fungi to degrade the natural fibers (Dhakal, Zhang, & Richardson, 2007; Razali et al., 2018; Zulkafli et al., 2019).

Biodegradability of the composite can be measured using a soils burial test where weight loss is used as an indicator of the biodegradation process (Maran, Sivakumar, Thirugnanasambandham, & Sridhar, 2014). The ideal biocomposite should be able to be disposed of or decompose at the end of life without harming the environment with the aid of microorganisms. The biodegradation of the polymeric composite in soils can occur both aerobically and partly anaerobically, producing H_2O, CO_2, or CH_4 (Shah, Hasan, Hameed, & Ahmed, 2008). The enzymatic degradation of the polymer can occur via extracellular and intracellular degradation. Intracellular degradation occurs when an endogenous carbon source is hydrolyzed by the accumulating bacteria, while in extracellular degradation, both accumulating and other microorganisms utilize the exogenous carbon reservoir. Furthermore, enzymatic degradation can occur via a hydrolysis mechanism where enzymes will form an initial attachment onto the polymer substrate then subsequently catalyze the hydrolytic cleavage in the polymer chain (Shah et al., 2008).

Behera et al. (2012) studied the biodegradability of jute-reinforced soy-based composites in compost soil burial conditions for a period of 60 days. They reported that soy resin film experienced a weight loss of up to 92% compared to jute fabric (up to 64%). Higher weight losses in the soy-based film suggested that soil microbes tend to degrade the soy protein faster than cellulose and hemicellulose of the jute fiber, as well as confirming the biodegradable and environment-friendly nature of the composite.

8.5 Applications of degradable composites

8.5.1 Degradable composites for domestic use

The advancements of science and technology, as experienced in modern times, are responsible for the synthesis of petrochemical-based plastics for a wide variety of applications, including for packaging and domestic uses. Although these synthetic plastics have high mechanical and barrier properties, they are nonbiodegradable, thus bringing higher potential for major threats to both humans and the environment (Koushal, Sharma, Sharma, Sharma, & Sharma, 2014). These plastics will take millions of years to degrade under natural atmosphere and weathering conditions. During combustion, they can produce toxic by-products which pollute the atmosphere. Therefore the development of environmentally friendly yet sustainable packaging plastic has become a major research area for scientists and researchers (Balakrishnan, Gopi, & Thomas, 2018; Khan, Bilal Khan Niazi, Samin, & Jahan, 2017). Plastic film is the most commonly used material for packaging due to its aesthetic quality, low cost, easy to handle, flexibility, and lightweight. Plastic films which are used for domestic and food packaging are typically made from petroleum sources like PP, PS, and PE. The High-Level Expert Forum in 2009 reported that, by 2050, the need for food production would increase by 70% in order to feed the anticipated global population of 9 billion people (Food and Agricultural Organization of the United Nations FAO, 2009). Although, improving crop yield is one aspect of meeting this increased food supply need, packaging, shipping, and transportation are also important, especially with the record of substantial food losses to spoilage during storage and packaging.

In this context, food demands align with the packaging need, where millions of tons of food packaging made from plastics are produced each year (Hammond et al., 2015). Thin films, containers, and coatings are used to cover processed foods and farm produce to extend their shelf life. However, the majority of food industries still use petroleum-based plastics for the purposes of food packaging (Luijsterburg & Goossens, 2014; Piringer & Baner, 2008). Polyethylene terephthalate, PE, and PP are utilized to make a variety of films and containers for food, confectionery, and beverage products. As these synthetic plastics are nonbiodegradable in nature, waste disposal of food packaging plastics is causing increasing environmental issues to the global community every year (Abdel-Shafy & Mansour, 2018). Therefore, many environmental analysts believe that environmentally deployable plastic has a bright future in replacing the conventional plastics used by the food packaging industry.

The growing environmental awareness among people, the rise of the prepackaged disposable meal, and antipollution legislation have driven the development of bio-based plastics that could be derived from natural and renewable resources. Biopolymers from numerous resources like cellulose, chitin, and starch have been studied and researched for packaging applications to replace synthetic plastics (Mangaraj, Yadav, Bal, Dash, & Mahanti, 2019). The composite materials which are based on biopolymers as matrix or filler are gaining more attraction in current research due to the mechanical performance of these materials, which is always greater than their neat biopolymer counterparts. For instance, a study by Lani, Ngadi, Johari, and Jusoh (2014) showed that oil palm empty fruit bunch (OPEFB) nanocellulose reinforced polyvinyl alcohol/starch film showed better mechanical and water absorption properties as compared with neat starch. This improvement could be related to good compatibility between the starch matrix and the nanocellulose filler. Research performed by Saba, Jawaid, Paridah, and Alothman (2017) showed that the incorporation of 3% nano OPEFB filler into the epoxy resulted in an optimum improvement in the epoxy nanocomposite properties. This is due to the filler that was properly distributed and dispersed in the epoxy matrix.

Thermoplastic starch (TPS) is a biodegradable and renewable bio-based plastic which has great potential for the fabrication of edible and biodegradable film for general packaging and food packaging applications (Khan, et al., 2017). TPS can be utilized to manufacture any sorts of packaging, such as films, bags (for shopping or waste), and one-off products (e.g., catering equipment pieces). By adding filler or nanofiller, it is possible to obtain TPS biocomposite with improved mechanical and optical properties and lower sensitivity to vapor, demanding certain packaging applications. Typical mineral fillers such as clays, silica, talc, and calcium carbonate are usually used to improve the properties of TPS. Furthermore, organic fibers, such as cellulose fibers, particularly microcrystalline and nanofibers, are widely employed as reinforcing fillers for TPS (Balakrishnan, et al., 2018; Khan, et al., 2017). The filler can be dispersed in starch during the gelatinization extrusion process. The combination of TPS/filler and suitable plasticizer can produce a composite material with the required properties. Our research group has investigated TPS biocomposites for refrigerated food packaging applications using corn starch, bentonite, and cellulose as raw materials (Fig. 8.9). TPS biocomposites incorporating bentonite/cellulose (bent/cellulose) hybrid fillers were prepared by film casting. Ultrasonication was done to improve the filler dispersion in the TPS matrix and enhance the filler–matrix interactions. Ultrasonicated fillers resulted in greater improvement of the tensile properties when the samples were preconditioned at both an ambient temperature and 2°C. Upon preconditioning at 2°C, the TPS achieved its highest tensile strength and elongation at break when the ultrasonicated hybrid bent/cellulose in 80:20 ratio was added to its structure. These could be due to the effect of low-temperature interactions between the starch, plasticizers, and hybrid fillers. DSC analysis proved the changes in the crystallinity and dynamic characteristic of the TPS molecular chains which benefit in enhancing the biopolymer elongation at break and toughness, even when stored in refrigerated conditions. The work revealed that the TPS/bent/cellulose composite has the potential to be

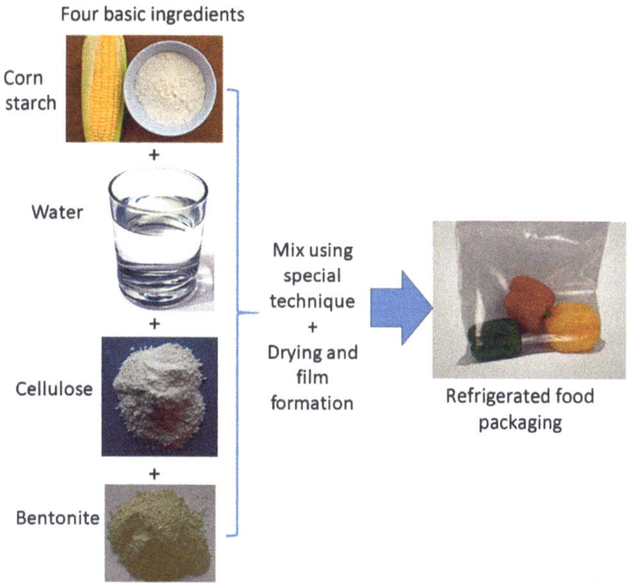

FIGURE 8.9 The production of thermoplastic starch biocomposite film for refrigerated food packaging applications (author owned).

further developed for refrigerated food packaging applications (Ismail, Osman, & Ping, 2019).

Cellulose-based biocomposites show high potential for packaging applications, in which the mechanical properties and biodegradability improvement of the bio-based packaging film are focused on saving the planet for future generations. With regard to this matter, researchers also investigated the potential of using both natural-based matrix and filler to form "green composite" in order to obtain fully biodegradable packaging films. All-cellulose composites (ACCs) can be deliberated as bio-based monocomponent or single-material composites due to them utilizing the same cellulose source as the matrix and reinforcement. The main benefit of this type of composite is that the chemical bonding or interface between the matrix and reinforcement can be improved (Huber et al., 2012). Research carried out by Ghaderi, Mousavi, Yousefi, and Labbafi (2014) on all-cellulose nanocomposite films made from bagasse nanocellulose showed the promising potential of the materials for the development of gas barrier and protective films. Other than bagasse-derived nanocellulose, researchers around the globe have investigated other types of lignocellulosic fiber such as *Nypa fruticans* (Govindan, Hussiensyah, Leng, & Amri, 2014) and OPEFB (Zailuddin, Husseinsyah, Hahary, & Ismail, 2017) for the production of biodegradable ACC films for packaging and protective films.

The production of palm oil in Malaysia is known to generate a great amount of biomass wastes, including like OPEFB. These wastes, if untreated, can eventually have an adverse effect on the environment. Several works have reported that these lignocellulose materials can be turned into valuable products, and thus, more research and studies are needed to explore and understand the potential uses of OPEFB (Chang, 2014; Lani, et al., 2014; Or, Putra, & Selamat, 2017). The motivation for research into OPEFB can be related to its characteristics such as abundance, availability, and efficiency to generate low composite materials production cost (Omoniyi, 2019). The cellulose dissolution and regeneration process could be an efficient method to produce green products from OPEFB that are not just natural but also cheap (Zailuddin, et al., 2017). The use of nano-sized cellulose (nanocellulose) that is synthesized from OPEFB can generate further advances in composite research as the unique characteristics of the "nanofiller" can be exploited to manufacture new nano-green materials with wider applications. In particular, the OPEFB nanocellulose special features such as high strength, lightweight, low cost, renewability, and abundantly available resources can bring great potential to the development of high-performance composite films. The high-strength biodegradable packaging film can potentially produce using the OPEFB as it turned into the "all-cellulose composite" (ACC) film. Our research group has produced biodegradable ACC film from OPEFB for potential use as fully biodegradable packaging plastic (Zailuddin, Osman, Husseinsyah, Ariffin, & Badrun, 2018). The films were prepared using OPEFB nanocellulose, microcrystalline cellulose (MCC), dimethylacetamide, and lithium chloride solvent system, whereby the partially dissolved cellulose was transformed into the matrix phase surrounding the remaining nondissolved fiber (Fig. 8.10). The film was formed upon drying at room temperature, thus saving energy and production cost. The application of acetic acid treatment on the OPEFB nanocellulose allowed a significant increase in the ACC film's mechanical properties. The film demonstrated a 279% increment in tensile strength when benchmarked with the untreated ACC film. This film also showed a high biodegradation rate after 3 months of a soil burial test. The findings of this research revealed the potential of the ACC to be used as biodegradable packaging film.

However, several concerns must be addressed prior to commercial use of bio-based food packaging materials. The degradation rates of the bio-plastic under various conditions, the potential for microbial growth, changes in mechanical properties during storage, and release of harmful compounds into the packaged food are among the factors to be critically analyzed. The packaging material must be composted completely in less than 200 days, leaving no toxic residue behind (Mitrus, Wojtowicz, & Moscicki, 2009). Furthermore, the chemical structure, molecular weights, crystallinity, and processing conditions of the bioplastics must be investigated because these

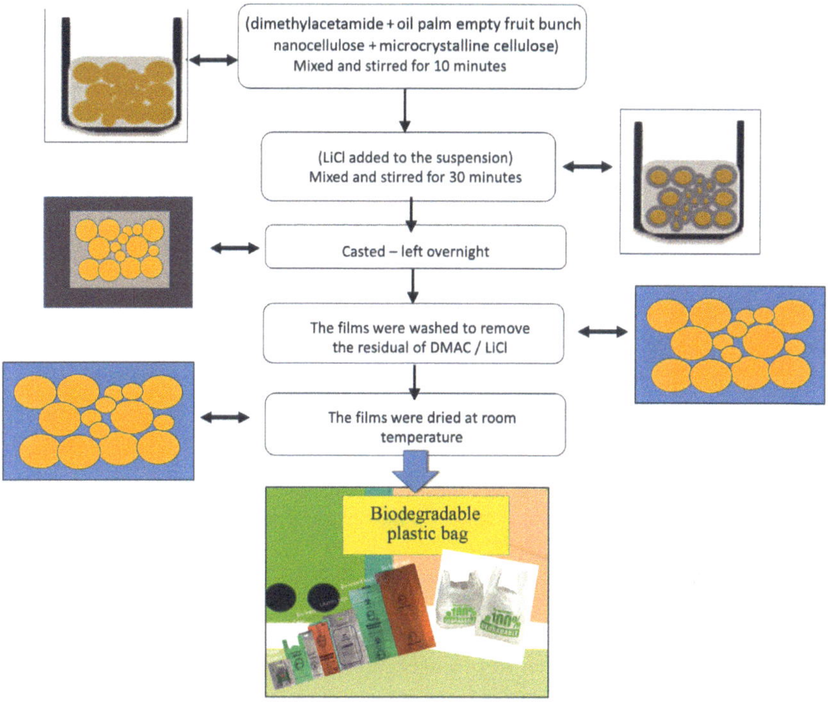

FIGURE 8.10 The production of oil palm empty fruit bunch-derived all-cellulose composite film for fully biodegradable plastic packaging (author owned).

factors can affect the physical and mechanical properties of the packaging material as the final product. The physical and mechanical properties required in packaging are generally dependent on the items to be packed and the conditions (or environments) in which the packages are stored (Hottle, Bilec, & Landis, 2013). Fundamental knowledge of the raw materials used for the development of the packaging material is therefore crucial.

8.5.2 Degradable composites for biomedical applications

A wealth of studies in bone tissue engineering have been accomplished over the past few decades, especially in developing porous biomaterials (Fu, Saiz, Rahaman, & Tomsia, 2011). The development of these synthetic biomaterials has gone through various stages from being inert to biocompatible, and now generative (Fig. 8.11) (Murugan & Ramakrishna, 2005). Third-generation biomaterials, which consist of biodegradable polymers reinforced with bioactive ceramics (bioglass (BG) or hydroxyapatite (HA)) phases, are becoming increasingly desirable as they mimic the natural structure of bone (Allo, Costa, Dixon, Mequanint, & Rizkalla, 2012).

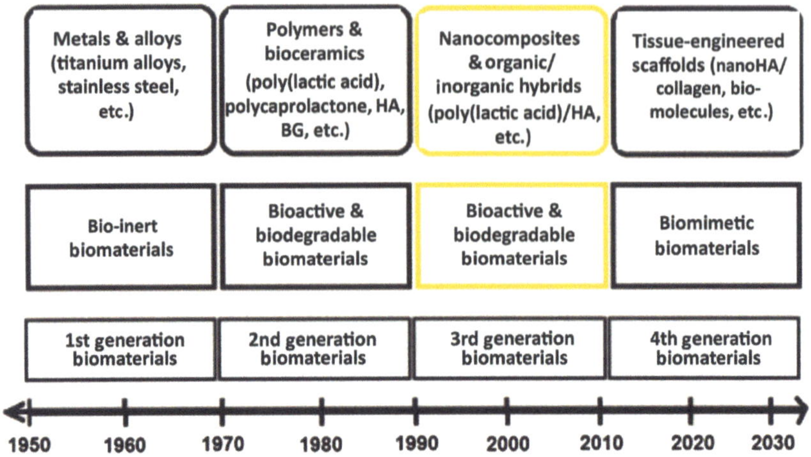

FIGURE 8.11 Evolution of synthetic biomaterials in bone tissue engineering. *Source: BG, bioglass; HA, hydroxyapatite. Adapted from Murugan, R., & Ramakrishna, S. (2005). Development of nanocomposites for bone grafting. Composites Science and Technology, 65(15-16), 2385–2406.*

Third-generation biomaterial in bone tissue engineering serves as an extracellular matrix for the stimulation of cell-driven tissue regeneration (Black et al., 2015). Therefore researchers are focusing on exploiting different combinations of materials in order to attain high biocompatibility and optimal mechanical properties of new composites with the desired porous structure (Bahril, Sahroni, & Mustafa, 2018; Hutmacher, 2001; Mustafa & Tanner, 2011). Although numerous preclinical trials with different animal models have shown positive results (Cancedda, Giannoni, & Mastrogiacomo, 2007), significant issues remain due to the lack of translation into clinical practice, including choice of biomaterial and in vitro preparation (Bao et al., 2013). This has motivated the search for new biomaterials for use in bone tissue engineering, including the biopolymers such as TPS.

Composite fabricated with biodegradable polymer together with bioactive filler, which exhibits similar characteristics to the natural inorganic–organic structures of bone, has been a topic of great interest due to its useful applications in bone tissue engineering. This is because the composite does not need to be removed from the body after implantation as the formation of new bone eventually replaces the implanted composite during degradation (Guarino, Raucci, Ronca, Cirillo, & Ambrosio, 2014). The use of TPS as the matrix material for a biocomposite scaffold was inspired by the natural origin of this material, its renewability, low cost, biodegradability, biocompatibility, and tailorable physical and mechanical properties (Niranjana Prabhu & Prashantha, 2018). The incorporation of bioglass as filler is anticipated to further enhance the mechanical properties and bioactivity of the host TPS to be more viable as a third-generation biomaterial (Fadzli, Roslinda, Zainuddin, & Ismail, 2016). With optimized formulation and processing parameters, the TPS/bioglass composite could potentially have the characteristics of the third-generation biomaterials which possess biodegradability, bioactivity, and feasibility adaptable for advanced tissue engineering applications. Adam and co-researchers focused on TPS/sol–gel-derived bioglass composite as a new biomaterial for bone tissue engineering (Adam, Osman, & Shamsudin, 2018). The composites were produced using a TPS/bioglass mass ratio of 80/20, with a combination of solvent casting and salt leaching techniques (Fig. 8.12). Tensile test results showed the addition of bioglass increased significantly the tensile strength and Young's modulus of the TPS/bioglass biocomposite. The modulus of all samples was higher than the requirement for cancellous bone (10–20 MPa). The SEM analysis proved the presence of a porous structure on the surface of all samples. XRD results confirmed the formation of a hydroxycarbonate apatite (HCA) layer on the surface of bioglass-containing samples, indicating the occurrence of surface reactions

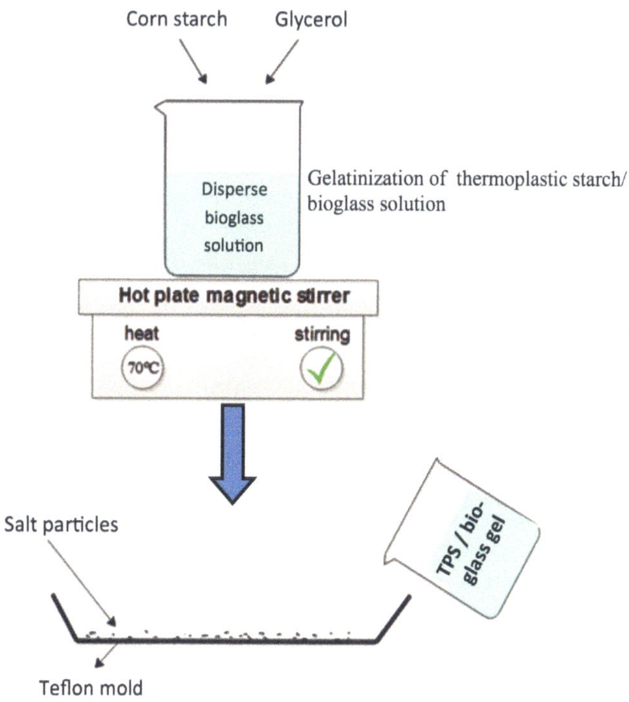

FIGURE 8.12 Production of thermoplastic starch/bioglass biocomposite for scaffold applications (author owned).

when the samples were immersed in simulated body fluid (SBF). Furthermore, the presence of a P-O stretch band in the FTIR spectrum between 1000 and 1150/cm and an Si-O-Si stretch band at 1000/cm also proved the bioactivity of the TPS/bioglass composite (Adam et al., 2018). These findings indicate the great potential of TPS/bioglass biocomposite for scaffold applications.

Utilizing the degradable polymers as the matrix in the production of porous scaffold may provide a comparable initial mechanical property to the bone. However polymeric scaffold becomes more brittle with the addition of a bioactive filler, therefore restricting the bioactive filler content to 20 or 40 vol.% (Othman et al., 2017; Othman et al., 2018; Tan et al, 2003). This, in turn, limits the osteointegration properties of the biocomposite. In order to overcome this issue, Tsuru, Munar, Ishikawa, and Othman (2013) combined two methods of preparation where the porous α-TCP powder was used as a matrix to form porous scaffold instead of polymeric materials. The porous scaffold was then coated with PCL to improve the mechanical properties of the composite structure. This allowed a higher content of bioactive materials to be utilized in the scaffold architecture. A similar approach was taken by Mustafa et al. (2018) where the porous scaffold was prepared by TCP as matrix and later coated with PHVB to improve their mechanical properties. The initial in vitro study showed that the ceramic scaffold provides good bioactivity in the SBF solutions as well as good initial mechanical properties that are comparable to human bone.

The idea of the utilizing degradable composite for bone fixation is that it would provide mechanical stability to support the fractured bone during the healing process and later safely degrade over time. Li et al. (2017) developed fully degradable 2D-braided magnesium alloy wires reinforced with PLA for an orthopedic implant that showed promising properties in terms of shear and impact strength. In the most up-to-date findings by Heimbach, Yu, and Wei (2020), utilizing degradable materials such as silk fibroin, HA, and PLA showed outstanding mechanical properties with flexural modulus and strength up to 21.1 GPa and 536 MPa, respectively, which is comparable to those of cortical bone. However, most of this study was carried out in static conditions. During the healing process, human bone is also subjected to cyclic conditions. Thus, while the degradability of the composite is desire to ensure removal of the implant once the healing has end, it must continuously provide sufficient mechanical support during the healing process to the host. Mustafa (2013), Mustafa and Tanner (2011), and Mustafa et al. (2018) studied the fatigue properties of fully degradable self-reinforced PLA fiber reinforced TCP composite under various loadings and it showed a great potential implant for bone fixation candidates (Bahril et al., 2018; Mustafa and Tanner, 2011; Mustafa, 2013).

Cellulose possesses good biocompatibility, high hydrophilicity, high surface area, and good mechanical strength that makes them a potential candidate for biomedical applications, such as biosensors and diagnostics, pharmaceuticals, cell culture and tissue engineering, antimicrobial agents, and wound dressings (Zhang et al., 2013). MCC is considered to be a partially depolymerized cellulose that is prepared by treating the cellulose which is acquired from the fibrous plant with a mineral acid. The noncrystalline part of the fiber plant is hydrolyzed during the acid hydrolysis process. This is to ensure the release of the cellulose crystals from the fiber (Rojas, Lopez, Guisao, & Ortiz, 2011). MCCs are widely used in the pharmaceutical industry, for example, as binders and fillers in medical tablets (Thoorens, Krier, Leclercq, Carlin, & Evrard, 2014).

In addition to plant- or vegetable-based fibers, animal-based fibers are also widely used natural fibers. For example, wool, mohair, and silk from animals are generally comprised of proteins and can be used as reinforcement materials in composites. Animal-based fibers can also be used as an option aside from plant-based fibers in the production of biomedical, biodegradable, and bioresorbable composite materials. This composite material is generally used for bioengineering and orthopedic applications (Sanjay, Arpitha, Naik, Gopalakrishna, & Yogesha, 2016).

The recent developments in materials science and chemistry in biomedical applications have led to the creation of various drug carrier systems and approaches. Cellulose-based drug delivery is one of the most successful systems according to the United States. For example, several antibiotics, HIV drugs, pain relievers, and flavonoids involve drug-delivery systems which are made of CA, whereas some oral drug-delivery systems are formulated with hydroxypropyl methylcellulose. Generally, cellulose is important as an excipient for drugs to allow good control of drug release, so that the right concentration of drug entering the human body can be ensured. Furthermore, due to its hydrophilicity (affinity toward water), this biopolymer can be cross-linked into hydrogels. Cellulose is nontoxic, biocompatible, and biodegradable, which mean it can pass through the human digestive system safely (Jorfi & Foster, 2015). Trovatti et al. (2011) used bacterial nanocellulose (BNC) membrane systems for the topical or transdermal release of lidocaine. A uniform drug distribution inside the BNC membranes indicates that the BNC membranes can produce a good release profile due to their good interactions with the lidocaine molecules. Müller et al. (2013) investigated the applicability of BNC as a drug-delivery system for proteins using serum albumin as a model drug. The results indicated that the freeze-dried BNC samples allowed a lower protein uptake as compared to native BNC samples. Furthermore, they showed that the biological activity and integrity of albumin was maintained during the loading and release processes. These works provide evidence that BNC can be an attractive and innovative biopolymer for a controlled drug-delivery system which possesses hydrophilicity, high biocompatibility, and controllable drug loading and release.

8.6 Conclusions

A degradable composite with attractive properties such as ease of accessibility, recyclability, and renewability provides intense competition to its nondegradable counterpart. Degradable composites are currently being used in packaging, building, transportation, and medical industries in an effort to reduce the greenhouse effect and ocean pollution to ensure a sustainable future. While some of the drawbacks make hinder it being considered for higher-end applications such as load-bearing or structural areas, effort from research works continue to address this drawback and will render their viability for wider applications in the future. An optimistic perspective would also see further development of degradable composites for bioabsorbable material, artificial skin, pharmaceutical and wound-healing applications, as well as lightweight materials.

Acknowledgments

The authors would like to thank the Universiti Teknikal Malaysia Melaka, Universiti Malaysia Perlis as well as family members for their support and encouragement.

References

Abdel-Shafy, H. I., & Mansour, M. S. (2018). Solid waste issue: Sources, composition, disposal, recycling, and valorization. *Egyptian Journal of Petroleum, 27*(4), 1275–1290.

Adam, S. N. F. S., Osman, A. F., & Shamsudin, R. (2018). Tensile properties, biodegradability and bioactivity of thermoplastic starch (TPS)/ bioglass composites for bone tissue engineering. *Sains Malaysiana, 47*(6), 1303–1310.

Alix, S., Colasse, L., Morvan, C., Lebrun, L., & Marais, S. (2014). Pressure impact of autoclave treatment on water sorption and pectin composition of flax cellulosic-fibres. *Carbohydrate Polymers, 102*, 21–29.

Allo, B. A., Costa, D. O., Dixon, S. J., Mequanint, K., & Rizkalla, A. S. (2012). Bioactive and biodegradable nanocomposites and hybrid biomaterials for bone regeneration. *Journal of Functional Biomaterials, 3*(2), 432–463.

Avérous, L., & Boquillon, N. (2004). Biocomposites based on plasticized starch: Thermal and mechanical behaviours. *Carbohydrate Polymers, 56*, 111–122.

Azwa, Z. N., Yousif, B. F., Manalo, A. C., & Karunasena, W. (2013). A review on the degradability of polymeric composites based on natural fibres. *Materials & Design, 47*, 424–442.

Bahril, N. H. S., Sahroni, T. R., & Mustafa, Z. (2018). Failure model for bioactive degradable composite. In *IOP Conference series: Earth and environmental science Vol. 195, No. 1*, p. 012030. IOP Publishing.

Balakrishnan, P., Gopi, S., & Thomas, S. (2018). UV resistant transparent bionanocomposite films based on potato starch/cellulose for sustainable packaging. *Starch-Stärke, 70*(1-2), 1700139.

Bao, C. L. M., Teo, E. Y., Chong, M. S., Liu, Y., Choolani, M., & Chan, J. K. (2013). *Advances in bone tissue engineering. Regenerative Medicine and Tissue Engineering*. IntechOpen.

Beg, M. D. H., & Pickering, K. L. (2008). Accelerated weathering of unbleached and bleached Kraft wood fibre reinforced polypropylene composites. *Polymer Degradation and Stability, 93*(10), 1939–1946.

Behera, A. K., Avancha, S., Basak, R. K., Sen, R., & Adhikari, B. (2012). Fabrication and characterizations of biodegradable jute reinforced soy based green composites. *Carbohydrate Polymers, 88*(1), 329–335.

Biagiotti, J., Puglia, D., & Kenny Jose, M. (2004). A review on natural fibre-based composites—Part 1: Structure, processing and properties of vegetables fibres. *Journal of Natural Fibers, 1*(2), 37–38.

Black, C. R., Goriainov, V., Gibbs, D., Kanczler, J., Tare, R. S., & Oreffo, R. O. (2015). Bone tissue engineering. *Current Molecular Biology Reports, 1*(3), 132–140.

Blackburn, R. S. (2005). Introduction. In R. S. Blackburn (Ed.), *Biodegradable and sustainable fibres*. Abington, CB: Woodhead Publishing Limited.

Bledzki, A. K., & Gassan, J. (1999). Composites reinforced with cellulose based fibres. *Progress in Polymer Science, 24*, 221–274.

Burgert, I. (2006). Exploring the micromechanical design of plant cell walls. *American Journal of Botany, 93*, 1391–1401.

Campilho, R. D. S. G. (2015). *Natural fiber composites*. FL: CRC Press Taylor & Francis Group.

Cancedda, R., Giannoni, P., & Mastrogiacomo, M. (2007). A tissue engineering approach to bone repair in large animal models and in clinical practice. *Biomaterials, 28*(29), 4240–4250.

Chan, C. M., Vandi, L., Pratt, S., Halley, P., Richardson, D., Werker, A., & Laycock, B. (2018). Composites of wood and biodegradable thermoplastics: A review. *Polymer Reviews, 58*(3), 444–494. Available from: https://doi.org/10.1080/15583724.2017.1380039.

Chang, S. H. (2014). An overview of empty fruit bunch from oil palm as feedstock for bio-oil production. *Biomass and Bioenergy, 62*, 174–181.

Chang, S. Y., Ismail, H., & Ahsan, Q. (2012). Effect of maleic anhydride on kenaf dust filled polycaprolactone/thermoplastic sago starch composites. *Bioresources, 7*(2), 1594–1616.

Chen, H., Miao, M., & Ding, X. (2009). Influence of moisture absorption on the interfacial strength of bamboo/vinyl ester composites. *Composites Part A: Applied Science and Manufacturing, 40*(12), 2013–2019.

Dhakal, H. N., Zhang, Z. Y., & Richardson, M. O. W. (2007). Effect of water absorption on the mechanical properties of hemp fibre reinforced unsaturated polyester composites. *Composites Science and Technology, 67*(7-8), 1674–1683.

di Franco, C. R., Cyras, V. P., Busalmen, J. P., Ruseckaite, R. A., & Vázquez, A. (2004). Degradation of polycaprolactone/starch blends and composites with sisal fibre. *Polymer Degradation and Stability, 86*, 95–103.

Edlund, U., & Albertsson, A.-C. (2002). Degradable polymer microspheres for controlled drug delivery. In A.-C. Albertsson (Ed.), *Degradable aliphatic polyesters*. NY: Springer-Verlag Berlin Heidelberg.

Fadzli, S. S. N., Roslinda, S., Zainuddin, F., & Ismail, H. (2016). Synthesis of sol-gel derived glass powder and in vitro bioactivity property tested in simulated body fluid. In *AIP conference proceedings, Vol. 1784, No. 1*, p. 040033. AIP Publishing.

Food and Agricultural Organization of the United Nations (FAO), (2009). High level expert forum-How to feed the world in 2050, 12-13 October, Rome, Italy.

Fu, Q., Saiz, E., Rahaman, M. N., & Tomsia, A. P. (2011). Bioactive glass scaffolds for bone tissue engineering: state of the art and future perspectives. *Materials Science and Engineering: C, 31*(7), 1245–1256.
Gaceva, G. B., Avella, M., Malinconico, M., Buzarovska, A., Grozdanoz, A., Gentile, G., & Errico, M. E. (2007). Natural fiber eco-composites. *Polymer Composite, 28*, 98–107.
Ghaderi, M., Mousavi, M., Yousefi, H., & Labbafi, M. (2014). All-cellulose nanocomposite film made from bagasse cellulose nanofibers for food packaging application. *Carbohydrate Polymers, 104*, 59–65.
Gholampour, A., & Ozbakkaloglu, T. (2020). A review of natural fiber composites: Properties, modification and processing techniques, characterization, applications. *Journal of Materials Science, 55*(3), 829–892.
Govindan, V., Hussiensyah, S., Leng, T. P., & Amri, F. (2014). Preparation and characterization of regenerated cellulose using ionic liquid. *Advances in Environmental Biology, 8*(8), 2620–2625.
Gross, R. A., & Kalra, B. (2002). Biodegradable polymers for the environment. *Science, 297*, 803–807.
Guarino, V., Raucci, M. G., Ronca, A., Cirillo, V., & Ambrosio, L. (2014). 5—Multifunctional scaffolds for bone regeneration. In K. Mallick (Ed.), *Woodhead publishing series in biomaterials, bone substitute biomaterials* (pp. 95–117). Woodhead Publishing.
Gurunathan, T., Mohanty, S., & Nayak, S. K. (2015). A review of the recent developments in biocomposites based on natural fibres and their application perspectives. *Composites Part A: Applied Science and Manufacturing, 77*, 1–25.
Hamad, K., Kaseem, M., Ko, Y. G., & Deri, F. (2014). Biodegradable polymer blends and composites: An overview. *Polymer Science Series A, 56*(6), 812–829.
Hammond, S. T., Brown, J. H., Burger, J. R., Flanagan, T. P., Fristoe, T. S., Mercado-Silva, N., & Okie, J. G. (2015). Food spoilage, storage, and transport: Implications for a sustainable future. *Bioscience, 65*(8), 758–768.
Heimbach, B., Yu, L., & Wei, M. (2020). In vitro behavior of silk fibroin-based composite resorbable bone fixation devices. *Materialia*, 100587.
Hottle, T. A., Bilec, M. M., & Landis, A. E. (2013). Sustainability assessments of bio-based polymers. *Polymer Degradation and Stability, 98*(9), 1898–1907.
Huber, T., Müssig, J., Curnow, O., Pang, S., Bickerton, S., & Staiger, M. P. (2012). A critical review of all-cellulose composites. *Journal of Materials Science, 47*(3), 1171–1186.
Hutmacher, D. W. (2001). Scaffold design and fabrication technologies for engineering tissues—State of the art and future perspectives. *Journal of Biomaterials Science, Polymer Edition, 12*(1), 107–124.
Ismail, I., Osman, A. F., & Ping, T. L. (2019). Effects of ultrasonication process on crystallinity and tear strength of thermoplastic starch/cellulose biocomposites. In *IOP Conference Series: Materials Science and Engineering* (Vol. 701, No. 1, p. 012045). Bristol: IOP Publishing.
Janda, K. (2015). Petroleum products. *Handbook of Material Biodegradation, Biodeterioration, and Biostablization*, 257–375.
John, M. J., & Thomas, S. (2008). Biofibres and biocomposites. *Carbohydrate Polymers, 71*, 343–364.
Jorfi, M., & Foster, E. J. (2015). Recent advances in nanocellulose for biomedical applications. *Journal of Applied Polymer Science, 132*(14).
Khan, B., Bilal Khan Niazi, M., Samin, G., & Jahan, Z. (2017). Thermoplastic starch: A possible biodegradable food packaging material—A review. *Journal of Food Process Engineering, 40*(3), e12447.
Khatiwala, V. K., Shekhar, N., Aggarwal, S., & Mandal, U. K. (2008). Biodegradationof poly(ε-caprolactone) (PCL) film by *Alcaligenes faecalis*. *Journal of Polymers and the Environment, 16*, 61–67.

Koushal, V., Sharma, R., Sharma, M., Sharma, R., & Sharma, V. (2014). Plastics: Issues challenges and remediation. *International Journal of Waste Resources*, *4*(1), 2–6.

Lani, N. S., Ngadi, N., Johari, A., & Jusoh, M. (2014). Isolation, characterization, and application of nanocellulose from oil palm empty fruit bunch fiber as nanocomposites. *Journal of Nanomaterials*, *2014*, 13.

Leong, Y. W., Thitithanasarn, S., Yamada, K., & Hamada, H. (2013). Compression and injection molding techniques for natural fiber composites. *Natural Fibre Composites: Materials, Processes and Applications*, 216–232. Available from: https://doi.org/10.1533/9780857099228.2.216.

Li, X., Chu, C., Zhou, L., Bai, J., Guo, C., Xue, F., & Chu, P. K. (2017). Fully degradable PLA-based composite reinforced with 2D-braided Mg wires for orthopedic implants. *Composites Science and Technology*, *142*, 180–188.

Liu, K., Takagi, H., Osugi, R., & Yang, Z. (2012). Effect of physicochemical structure of natural fiber on transverse thermal conductivity of unidirectional abaca/bamboo fiber composites. *Composites Part A: Applied Science and Manufacturing*, *43*(8), 1234–1241. Available from: https://doi.org/10.1016/j.compositesa.2012.02.020.

Lu, J., & Drzal, L. T. (2010). Microfibrillated cellulose/cellulose acetate composites: Effect of surface treatment. *Journal of Polymer Science Part B: Polymer Physics*, *48*(2), 153–161.

Luijsterburg, B., & Goossens, H. (2014). Assessment of plastic packaging waste: Material origin, methods, properties. *Resources, Conservation and Recycling*, *85*, 88–97.

Mangaraj, S., Yadav, A., Bal, L. M., Dash, S. K., & Mahanti, N. K. (2019). Application of biodegradable polymers in food packaging industry: A comprehensive review. *Journal of Packaging Technology and Research*, *3*(1), 77–96.

Maran, J. P., Sivakumar, V., Thirugnanasambandham, K., & Sridhar, R. (2014). Degradation behavior of biocomposites based on cassava starch buried under indoor soil conditions. *Carbohydrate Polymers*, *101*, 20–28.

McKeen, L. W. (2017). Renewable resource and biodegradable polymers. *Film Properties of Plastics and Elastomers*, 449–479.

Mitrus, M., Wojtowicz, A., & Moscicki, L. (2009). Biodegradable polymers and their practical utility. *Thermoplastic starch*, 1–33.

Mohanty, A. K., Misra, M., & Hinrichsen, G. (2000). Biofibres, biodegradable polymers and biocomposites: An overview. *Macromolecular Materials and Engineering*, *276/277*, 1–20.

Mohanty, A. K., Wibowo, A., Misra, M., & Drzal, L. T. (2003). Development of renewable resource-based cellulose acetate bioplastic: Effect of process engineering on the performance of cellulosic plastics. *Polymer Engineering & Science*, *43*(5), 1151–1161.

Mohanty, A. K., Wibowo, A., Misra, M., & Drzal, L. T. (2004). Effect of process engineering on the performance of natural fiber reinforced cellulose acetate biocomposites. *Composites Part A: Applied Science and Manufacturing*, *35*(3), 363–370.

Müller, A., Ni, Z., Hessler, N., Wesarg, F., Müller, F. A., Kralisch, D., & Fischer, D. (2013). The biopolymer bacterial nanocellulose as drug delivery system: investigation of drug loading and release using the model protein albumin. *Journal of Pharmaceutical Sciences*, *102*(2), 579–592.

Murugan, R., & Ramakrishna, S. (2005). Development of nanocomposites for bone grafting. *Composites Science and Technology*, *65*(15-16), 2385–2406.

Mustafa, Z., Ishak, N. F., Othman, R., Ahmad, N., Toibah, A. R., Fadzullah, S. S. M., & Tanner, K. E. (2018). In-vitro apatite growth on porous β-tricalcium phosphate scaffolds coated with PHBV. *Journal of Advanced Manufacturing Technology*, *12*(1), 135–146.

Mustafa, Z., & Tanner, K. E. (2011). Fatigue study of self reinforced PLA filled tricalcium phospate composite. In *Paper presented at the 24th European conference on biomaterials-Annual conference of the european society for biomaterials*.

Mustafa, Z. B. (2013). Multiaxial fatigue characterization of self-reinforced polylactic acid-calcium phosphate composite, Doctoral dissertation, University of Glasgow.

Niranjana Prabhu, T., & Prashantha, K. (2018). A review on present status and future challenges of starch based polymer films and their composites in food packaging applications. *Polymer Composites*, *39*(7), 2499–2522.

Nishino, T., Hirao, K., Kotera, M., Nakamae, K., & Inagaki, H. (2003). Kenaf reinforced biodegradable composite. *Composite Science and Technology*, *6*, 1281–1286.

Omoniyi, T. E. (2019). Potential of oil palm (*Elaeis guineensis*) empty fruit bunch fibres cement composites for building applications. *AgriEngineering*, *1*(2), 153–163.

Or, K. H., Putra, A., & Selamat, M. Z. (2017). Oil palm empty fruit bunch fibres as sustainable acoustic absorber. *Applied Acoustics*, *119*, 9–16.

Othman, R., Mustafa, Z., Ishak, N. F., Kien, P. T., Shamsudin, Z., Rosli, Z. M., & Noor, A. F. M. (2018). Intermediate phases formed during synthesis of β-tricalcium phosphate via wet precipitation and hydrothermal methods. *Journal of Advanced Research in Fluid Mechanics and Thermal Sciences*, *48*(2), 141–147.

Othman, R., Mustafa, Z., Mohd Juoi, J., Kien, P. T., Noor, M., & Fauzi, A. (2017). *Effect of calcination temperatures on phase transformation and stability of β-tricalcium phosphate powder synthesized by a wet precipitation method. Solid state phenomena* (pp. 132–135). Trans Tech Publications Ltd, Vol. 264.

Pan, P., Zhu, B., Dong, T., Serizawa, S., Iji, M., & Inoue, Y. (2007). Kenaf fibre/poly (ε-caprolactone) biocomposite with enhanced crystallization rate and mechanical properties. *Journal of Applied Polymer Science*, *107*, 3512–3519.

Pietrelli, L., Poeta, G., Battisti, C., & Sighicelli, M. (2017). Characterization of plastic beach debris finalized to its removal: a proposal for a recycling scheme. *Environmental Science and Pollution Research*, *24*(19), 16536–16542. Available from: https://doi.org/10.1007/s11356-017-9440-4.

Piringer, O. G., & Baner, A. L. (Eds.), (2008). *Plastic packaging materials for food: barrier function, mass transport, quality assurance, and legislation*. John Wiley & Sons.

Piskin, E. (2002). Biodegradable polymers in medicine. In G. Scott (Ed.), *Degradable polymers: Principles and applications* (2nd ed.). A.A. Dordrecht, Netherlands: Kluwer Academic Publishers.

Quitadamo, A., Massardier, V., & Valente, M. (2017). Oil-based and bio-derived thermoplastic polymer blends and composites. In L. Lucia, & A. Ayoub (Eds.), *Introduction to renewable biomaterials* (pp. 239–268).

Ramamoorthy, S. K., Skrifvars, M., & Persson, A. (2015). A review of natural fibers used in biocomposites: Plant, animal and regenerated cellulose fibers. *Polymer Reviews*, *55*(1), 107–162.

Ramli, S. N. R., Fadzullah, S. H. S. M., & Mustafa, Z. (2017). The effect of alkaline treatment and fiber length on pineapple leaf fiber reinforced poly lactic acid biocomposites. *Jurnal Teknologi*, *79*(5-2), 111–115.

Razali, A. S., Mustafa, Z., Shaaban, A., Fadzullah, S. H. S. M., Shamsudin, Z., Rahim, T. A., & Anjang, A. (2018). Mechanical degradation of pineapple leaf fibre reinforced polylactic acid green composite due to hydrolysis exposure. In *Proceedings of Mechanical Engineering Research Day 2018 (MERD)* (pp. 239–240). Melaka: Universiti Teknikal Malaysia. ISBN: 978-967-2145-20-2.

Rojas, J., Lopez, A., Guisao, S., & Ortiz, C. (2011). Evaluation of several microcrystalline celluloses obtained from agricultural by-products. *Journal of Advanced Pharmaceutical Technology & Research*, *2*(3), 144.

Rosa, D. S., Lopes, D. R., & Calil, M. R. (2005). Thermal properties and enzymatic degradation of blends of poly(ε-caprolactone) with starches. *Polymer Testing*, *24*, 756–761.

Rowell, R. M., & Stout, H. P. (2007). Jute and kenaf. In M. Lewin (Ed.), *Handbook of fiber chemistry* (*3rd ed.*, pp. 405–452). Boca Raton, FL: CRC Press.

Saba, N., Jawaid, M., Paridah, M. T., & Alothman, O. (2017). Physical, structural and thermomechanical properties of nano oil palm empty fruit bunch filler based epoxy nanocomposites. *Industrial Crops and Products, 108*, 840–843.

Sadashiv, S., & Abasaheb, S. (2019). Review on mechanical properties evaluation of pineapple leaf fibre (PALF) reinforced polymer composites. *Composites Part B, 174*, 106927. Available from: https://doi.org/10.1016/j.compositesb.2019.106927.

Sanjay, M. R., Arpitha, G. R., Naik, L. L., Gopalakrishna, K., & Yogesha, B. (2016). Applications of natural fibers and its composites: An overview. *Natural Resources, 7*(03), 108.

Satyanarayana, K. G., Azrizaga, G. C., & Wypych, F. (2009). Biodegradable composites based on lignocellulosic fibers—An overview. *Progress in Polymer Science, 34*, 982–1021.

Sen, A., & Bhattacharya, M. (2000). Residual stresses and density gradient in injection molded starch/synthetic polymer blends. *Polymer, 41*, 9177–9190.

Senthilkumar, K., Saba, N., Rajini, N., Chandrasekar, M., Jawaid, M., Siengchin, S., & Alotman, O. Y. (2018). Mechanical properties evaluation of sisal fibre reinforced polymer composites: A review. *Construction and Building Materials, 174*, 713–729. Available from: https://doi.org/10.1016/j.conbuildmat.2018.04.143.

Sgriccia, N., Hawley, M. C., & Misra, M. (2008). Characterization of natural fibre surfaces and natural fibre composites. *Composites: Part A, 39*, 1632–1637.

Shah, A. A., Hasan, F., Hameed, A., & Ahmed, S. (2008). Biological degradation of plastics: A comprehensive review. *Biotechnology Advances, 26*(3), 246–265.

Shen, L., Haufe, J., & Patel, M. K. (2009). *Product overview and market projection of emerging bio-based plastics PRO-BIP 2009. Report for European polysaccharide network of excellence (EPNOE) and European bioplastics (p. 243)*. Utrecht: EPNOE.

Shubhra, Q. T. H., Alam, A. K. M. M., & Beg, M. D. H. (2011). Mechanical and degradation characteristics of natural silk fiber reinforced gelatin composites. *Materials Letters, 65*, 333–336.

Sreekala, M. S., Kumaran, M. G., Joseph, S., Jacob, M., & Thomas, S. (2000). Oil palm fibre reinforced phenol formaldehyde composites: Influence of fibre surface modifications on the mechanical performance. *Applied Composite Materials, 7*, 295–329.

Tan, K. H., Chua, C. K., Leong, K. F., Cheah, C. M., Cheang, P., Bakar, M. A., & Cha, S. W. (2003). Scaffold development using selective laser sintering of polyetheretherketone–Hydroxyapatite biocomposite blends. *Biomaterials, 24*(18), 3115–3123.

Taufiq, M. J., Mansor, M. R., & Mustafa, Z. (2018). Influence of water absorptivity on kenaf fibre reinforced recycled-polymer composite properties. *Ecology, 12*(3), 220–233.

Tholibona, D., Tharazia, I., Sulonga, A. B., Muhamada, N., Ismaila, N. F., Radzia, M. K. F. M., & Huie, D. (2019). Kenaf fiber composites: A review on synthetic and biodegradable polymer matrix. *Journal of Kejuruteraan, 31*(1), 65–76.

Thoorens, G., Krier, F., Leclercq, B., Carlin, B., & Evrard, B. (2014). Microcrystalline cellulose, a direct compression binder in a quality by design environment—A review. *International Journal of Pharmaceutics, 473*(1-2), 64–72.

Trovatti, E., Silva, N. H., Duarte, I. F., Rosado, C. F., Almeida, I. F., Costa, P., & Neto, C. P. (2011). Biocellulose membranes as supports for dermal release of lidocaine. *Biomacromolecules, 12*(11), 4162–4168.

Tsuru, K., Munar, M., Ishikawa, K., & Othman, R. (2013). Mechanical behavior and cell response of PCL coated α-TCP foam for cancellous-type bone replacement. *Ceramics International, 39*(5), 5631–5637.

Wu, C. S. (2003). Physical properties and biodegradability of meleated polycaprolactone / starch composite. *Polymer Degradation and Stability, 80*, 127–134.

Yu, L., Dean, K., & Li, L. (2006). Polymer blends and composites from renewable resources. *Progress in Polymer Science (Oxford), 31*, 576–602.

Zailuddin, N. L. I., Husseinsyah, S., Hahary, F. N., & Ismail, H. (2017). Characterization and properties of treated oil palm empty fruit bunch regenerated cellulose biocomposite films with butyl methacrylate using ionic liquid. *Polymer-Plastics Technology and Engineering, 56*(2), 109–116.

Zailuddin, N.L.I., Osman, A.F., Husseinsyah, S., Ariffin, Z., Badrun, F.H. (2018). Mechanical properties and X-Ray diffraction of oil palm empty fruit bunch all-nanocellulose composite films. In Proceedings of the second international conference on the future of ASEAN (ICoFA) 2017–Vol. 2. (pp. 505–514). Springer, Singapore.

Zhang, Y., Nypelö, T., Salas, C., Arboleda, J., Hoeger, I. C., & Rojas, O. J. (2013). Cellulose nanofibrils. *Journal of Renewable Materials*, *1*(3), 195–211.

Zhao, Q., Tao, J., Yam, R. C. M., Mok, A. C. K., Li, R. K. Y., & Song, C. (2008). Biodegradation behavior of polycaprolactone/rice husk ecocomposites in simulated soil medium. *Polymer Degradation and Stability*, *93*, 1571–1576.

Zulkafli, N., Malingam, S. D., Fadzullah, S. H. S. M., Mustafa, Z., Zakaria, K. A., & Subramonian, S. (2019). Effect of water absorption on the mechanical properties of cross-ply hybrid pseudo-stem banana/glass fibre reinforced polypropylene composite. *Materials Research Express*, *6*(9), 095326.

CHAPTER

9

Thermoplastic starch as renewable plastics

R. Jumaidin and Z.A.S. Saidi

Fakulti Teknologi Kejuruteraan Mekanikal dan Pembuatan, Universiti Teknikal Malaysia Melaka, Hang Tuah Jaya, Durian Tunggal, Melaka, Malaysia

9.1 Introduction

Global population growth has dramatically increased the usage of different types of petroleum-based polymers for packaging materials that has then led to environmental problems because of their large quantities and inadequate disposal, whereby thermoplastics do not biodegrade for a very long time. These wastes can be classified as a type of material that cannot be broken down by water, sunlight, or ground soil, creating serious issues for wildlife, society, and the environment (Locklin, 2017). Moreover, another area of concern is air pollution, in addition to the produced greenhouse gas, methane, which can increase global temperatures. In addition to the air pollution and global warming, the use of nonbiodegradable packaging materials can cause flooding as they do not decompose, preventing proper drainage and clogging sewage systems. There is a severe threat to humanity and another living organisms if this pollution and the environmental issues are not resolved completely.

In recent years, thermoplastic starch (TPS)-based composites have been developed as environmentally friendly products that have created interest for packaging and material applications (Jumaidin, 2020). Aside from being degradable, starch is a promising renewable resource that has the ability to change or replace petroleum-based plastic packaging. However, TPS has some weaknesses or limitations in terms of high water and moisture absorption, and also low flexibility (Jumaidin, 2021).

Design for Sustainability
DOI: https://doi.org/10.1016/B978-0-12-819482-9.00005-8

To solve these issues, many fillers have been used to improve the properties of TPS, such as employing the addition of fibers (Bootklad & Kaewtatip, 2013). Among them, natural fibers are the most promising due to their suitable characteristics including being biodegradable, low density, low cost, recyclable, and highly abundant.

9.2 Biopolymers

Biopolymers are polymers that are produced by living organisms. They can be regarded as the most organic compound in the ecosphere that is a biodegradable chemical compound. They can chemically synthesize from biological starting materials which are sugars, fats, oils, and starch, and can be created by the system of biological creatures, micro-organisms, plants, and animals. The usage of plastics over metals and glass for packaging has continued to increase due to their good easy processability, low weight, and low cost, functional properties, and easy accessibility. The waste disposal problems can be solved and overcome to a considerable extent by using biopolymer-based materials (Garavand, Rouhi, Razavi, Cacciotti, & Mohammadi, 2017).

Currently, the use of biopolymers such as polylactic acid (PLA), silk, and chitosan has been progressively researched for medical applications. Biocompatibility and biodegradability, as essential properties of biopolymers, are great advantages and enhance their potential use in implantable medical applications. Since synthetic materials do not fulfill the needs of living systems, these new materials have great importance in medicine and medical areas. For bone applications, several examples of chitosan-based matrices have been studied. The use to chitosan mixed with another material has been established, in order to serve as osteoconductive matrices once chitosan matrices, per se, are mechanically weak (Rebelo, Fernandes, & Fangueiro, 2017). Table 9.1 shows the applications of chitosan in medical fields.

In general, there are three main types of biopolymers: starch, cellulose, and synthetic biopolymers.

TABLE 9.1 Applications of chitosan implants in medical fields (Rebelo et al., 2017).

Field of application	Implants
Cardiology	Heart valves—electrospun gelatin, chitosan, polyurethane
Dermatology	Gelatin, chitosan, hyaluronan scaffold for skin generation
Surgery	Nerve regeneration—chitosan, gelatin scaffold Liver—chitosan, gelatin scaffold
Ophthalmology	Contact lenses—chitosan, gelatin

9.2.1 Starch

Starch present in most green plants where carbohydrate is the largest portion in its composition. In microscopic view, it forms in individual granules which are dense, small, and insoluble in cold water. As starch exists in a granular form, it is stored at large concentrations in reproductive forms such as cereal grains (wheat, rice, maize, barley, rye, oats, millet, sorghum) (Soykeabkaew, Thanomsilp, & Suwantong, 2015). Starch is contained in glucose and can be released by melting and then be deformed thermoplastically. This polymer can be found in vegetables such as tapioca, corn, wheat, and potatoes, but is not present in animal tissues. This starch material is convenient for conventional plastic forming processes such as extruding and injection molding. Starch is classified as one of the most used biopolymers in manufacturing industries that can be used as a filler (Wiacek, 2015). Fig. 9.1 shows granules of native potato starch.

9.2.2 Cellulose

Cellulose is a polymer that can be obtained from natural resources such as cotton, wood, wheat, and corn. It is a polymer that has a natural long chain which performs an essential role in the indirect cycle of human food resources. This polymer has semisynthetic derivatives that are greatly needed in the cosmetics pharmaceuticals industries. Cellulose has two main groups, cellulose ethers and cellulose derivatives, with different mechanical and physicochemical properties. Those polymers are used in healthcare products and medicines. To give its bundled chains

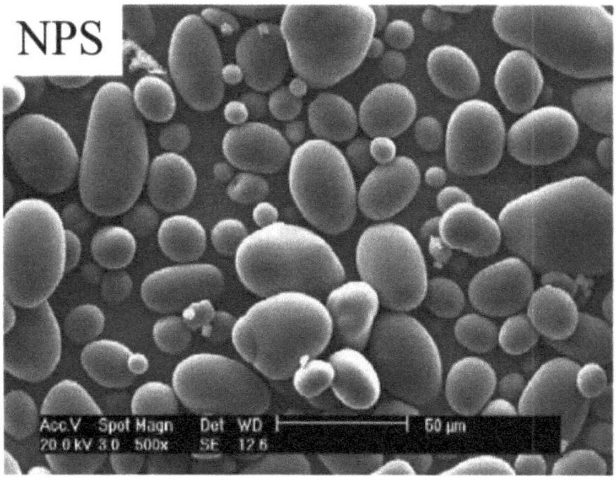

FIGURE 9.1 Native potato starch (NPS) granules (Wu, Geng, Chang, Yu, & Ma, 2009).

even more strength and stability, the chains are commonly packed in place resulting in hard and stable crystalline regions (Bogati, 2011).

9.2.3 Synthetic biopolymers

These materials can be made from chemicals and are widely used due to their durability at a lower cost than natural materials. Examples of materials made from plastic include polyethylene. These are called synthetic materials. Synthetic biopolymer-based nanocomposites can be applied to a scaffold for tissue engineering mainly due to their modifiable and biocompatible linear biodegradation. The saturated poly (α-hydroxyl esters), including PLA, polyglycolic acid, polylactic acid coglycolic acid, and polycaprolactone are customarily applied synthetic biopolymers in tissue engineering for three-dimensional scaffolds (Okamoto & John, 2013).

9.3 Natural fiber composites

Reinforcement is the composite part that imparts steadiness, strength, and the ability to carry a load (Chauhan & Chauhan, 2012). Natural fibers are capable chemical substances with the ability to replace fabrics, and also are similar products for energy conservation applications and have less weight. The purpose of the reinforcing natural fiber with natural-based resins and polymer composites is to make an application that can substitute current synthetic polymers or glass fiber reinforced materials in huge quantities (Sanjay, Arpitha, Naik, Gopalakrishna, & Yogesha, 2016). Jute, oil palm, sisal, kenaf, and banana are examples of dissimilar types of natural fiber plants that have been actively developed for the interior parts of automotives and aircraft. The lower process of natural fiber composites with high specific characteristics is attractive for different applications (Sanjay et al., 2016).

Researchers have progressed in using natural fibers as the reinforcements for polymer-based structural and cementitious materials over the last two decades. The used of straw combined with mud to construct walls in small towns has been practiced in China and Korea for over 100 years (Lau, Hung, Zhu, & Hui, 2018). Many automotive industries have already produced components with natural composites, mostly based on a polymer such as polypropylene and kenaf, sisal, or jute fibers. Automobile manufacturers in Germany such as Porsche, Opel, Mercedes-Benz, and Audi are examples of automotive industries that have used natural fiber composites in interior and exterior applications.

TABLE 9.2 Natural fiber composite applications (Mohammed, Ansari, Pua, Jawaid, & Islam, 2015).

Fiber	Application in the building, construction, and other industries
Hemp fiber	Infrastructure products, cordage, flexible materials, manufacture of pipes, electrical, and furniture
Palm oil fiber	Door frames for construction materials, windows, siding, roofing, fencing, and structural insulated panel building systems
Wood fiber	Window frames, decking, door shutters, panels, fencing, and railing systems
Sisal fiber	Shutting plates for infrastructure industry, panels, doors, roofing sheets, and also the manufacture of paper
Kenaf fiber	Mobile cases, packing materials, bags, soilless mixes, materials that absorb oils and liquids, and animal bedding
Cotton fiber	Furniture industry, textiles and yarn, and cordage
Jute fiber	Building panels, door shutters, roofing sheets, packaging, transport, chipboards, and geotextiles

The door trim panels for the Audi A2 midrange car were built from reinforcement of a mixed flax and sisal fiber with polyurethane (Sanjay et al., 2016). Table 9.2 shows some of the applications of natural fiber composites.

9.4 Starch-based composites

Starch has received growing interest from the scientific world due to its capacity to form bioplastics. It has been used as a starting point for a wide range of green biomaterials. Diverse methods have been applied to transform starch as a mean to improve the product properties and to develop the application range. Various kinds of starches which are deemed to be suitable as a matrix component of composites are highlighted along with their mechanical properties and characteristics. Starch-based plastics are very special due to their properties on biodegradable polymers gained from renewable plants as raw materials and are of increasing interest (Jumaidin, 2018). Starch is a common natural polymer that has biodegradable characteristics and can be converted into items using processing equipment after plasticization (Gupta, 2011). It is obtained from tubers, seeds, and roots of plants. It is distilled through processes of washing, sieving, wet grinding, and drying in which the properties of starch can be enhanced by different chemical and physical modifications such as blending, derivation, and graft copolymerization.

Biodegradable renewable materials have been identified as the most suitable for replacing single-use packaging materials. As a method to boost the microstructure, different approaches have been studied with the objective of solving these problems, such as the chemical modification of starches, combining with various biodegradable polymers, merging of natural fibers, and the inclusion of nanofillers (Soykeabkaew et al., 2015). Reinforcing commercially available TPS matrix with rapeseed fibers allowed the introduction of fully biodegradable composite materials. Espigulé et al. (2013) mentioned that when high reinforcement content was applied, the capacity of the TPS matrix to crystallize was lowered due to the relation with restrictions of the mobility of starch macromolecules caused by the reinforcement (Espigulé et al., 2013).

9.5 Polymerization of thermoplastic starch

The probability of changing starch into a thermoplastic material has created significant interest in the last decade. Starch can be changed into TPS when processing takes place at high temperature and shear along with a plasticizer (Jumaidin, 2019). The purpose of the plasticizer is to disturb the starch granules by shattering the starch bonds between the intermolecular and intramolecular hydrogen. The plasticizer and its chemical nature have a two-way effect on the physical properties of the processed starch. To begin with, it controls its destructuring and depolymerization. Second, it influences the final properties of the fabric, such as its glass movement temperature and modulus. As described by Roz, Carvalho, Gandini, and Curvelo (2006), TPS with high plasticizer content can produce phase separation, and these compositions are more susceptible to surrounding humidity as the plasticizer substance was expanded (Roz et al., 2006). Processing technology of starch plastics is one of the most important issues in starch processing. Adding the correct amount of plasticizer is a commonly used method to arrange plasticizer starches. Plasticizers are usually utilized to raise the flexibility of the film by reducing the bonding of intramolecular hydrogen with chains of polymers by expanding the intermolecular spacing. This development in expanded molecular mobility reduces the degree of crystallinity and also brings down the glass transition temperature (Zuo, Gu, Tan, & Zhang, 2015). Generally, the amount and type of plasticizer do not influence obviously the crystallinity of the processed materials (Roz et al., 2006). Thermoplastics are a polymeric material that can maintain its properties while being softened and solidified numerous times. Apart from being renewable and biodegradable, this material is recyclable, which increases its positive qualities.

There are several kinds of plasticizer used to create TPS, such as water which is mostly used as a fundamental plasticizer, second is glycerol, the most commonly used plasticizer after water, however, numerous categories are used including glycols, urea, acetamide, citric acid, sorbitol, and mannitol. Considering the physical properties of TPS, the selection of plasticizer is very important due to the essential factors for its utilization, which are biodegradability, processing, long-term stability, resistance to the environment, and also the final required properties (Ivanič, Jochec-Mošková, Janigová, & Chodák, 2017). It was found that sorbitol, which has a low power of plasticization, is a hexol that induces an antiplasticization effect, reduces the volume strain but raises the cohesion of the material. For glycerol is a proficient plasticizer, the portability of polymer chains is expanded, but any surplus of glycerol results in a phase separation, restricting the volume deformation. The characteristics of mannitol starch-based material not only consist of brittleness and breaks in deformation but are also able to exude and crystallize onto the surface of materials after several days of testing (Mikus et al., 2014). In recent, more studies have been carried out on TPS to improve its properties due to the minimal impact of this material to the environmental sustainability.

9.6 Thermoplastic cassava starch

Manihot esculenta, commonly called cassava, is the third most common food in tropical areas, and the fifth extensively grown starch crop in the world (Edhirej, Sapuan, Jawaid, & Zahari, 2017). From its extensive cultivation in tropical areas, CS is mostly used for food and another application throughout this nature due to its capability and large distribution in tropical areas. Starch has great potential as a natural thermoplastic resource to lead the comprehensive research using CS as a ground material for the biodegradability of thermoplastics. It has been reported that chitosan-coated starch films were made by coating CS film with chitosan solutions (Bangyekan, Aht-Ong, & Srikulkit, 2006). In this study, the starch film is coated by chitosan coating solutions from 1 to 4 wt.%. Some of the properties of these films, including physical and thermal properties, have led to improvements by coating chitosan solutions. The tensile strength, Young's modulus, and percent of elongation at break are greatly affected by the concentrations of glycerol. The tensile strength and Young's modulus were reduced by increasing the amount of glycerol.

Natural fibers are already used and studied as a reinforcement for polymer matrix composites due to their good properties and features, for instance, being inexpensive, lightweight, and renewable, and so can be used to solve the low mechanical strength characteristic of TPS. However, due to the diverse nature of hydrophilic fibers and hydrophobic matrices, their production in petroleum-based polymer was restricted, as they inhibited optimum surface adhesion and wetting between th matrix and fiber. Thus, natural fiber or filler that reinforced TPS was found to be the more promising way to improve and upgrade the properties of TPS while keeping its biodegradable and renewable qualities. Optimum fiber–matrix interaction can be achieved by this biocomposite because TPS has the same hydrophilic nature as natural fiber. Prachayawarakorn, Chaiwatyothin, Mueangta, and Hanchana (2013) mentioned that the mechanical properties of thermoplastic cassava starch (TPCS) and jute or kapok fibers substantially improved matrix reinforcement(Prachayawarakorn et al., 2013). The authors reported that expansion of the fiber contents seems to improve the strength of TPCS with jute fiber, and also TPCS and kapok fiber. Fig. 9.2 shows an scanning electron microscope image of TPCS/kapok fiber composites and Fig. 9.3 shows TPCS/jute fiber composites.

Moreover, using thermogravimetric analysis (TGA), by adding jute fibers, the TPCS matrix was increased, meanwhile by combining kapok fibers the thermal degradation temperature was reduced. This may be related due to jute fiber having less hydrophobic attributes compared with kapok fibers. Hence, the improvement of thermal stability following the incorporation of kapok fibers into TPCS were shown by lower weight loss of the material.

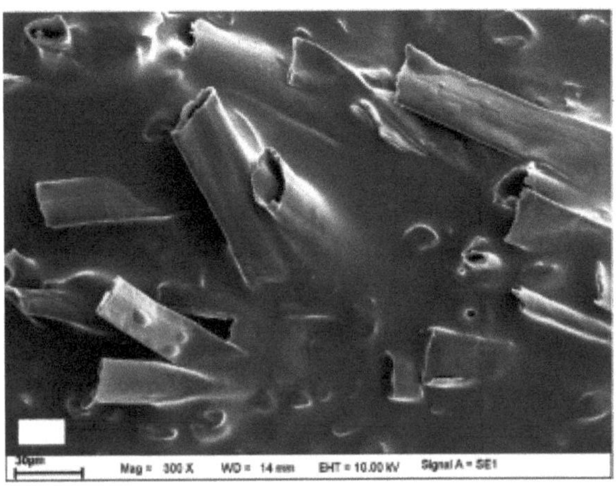

FIGURE 9.2 Scanning electron microscope image of thermoplastic cassava starch/kapok fiber (Prachayawarakorn et al., 2013).

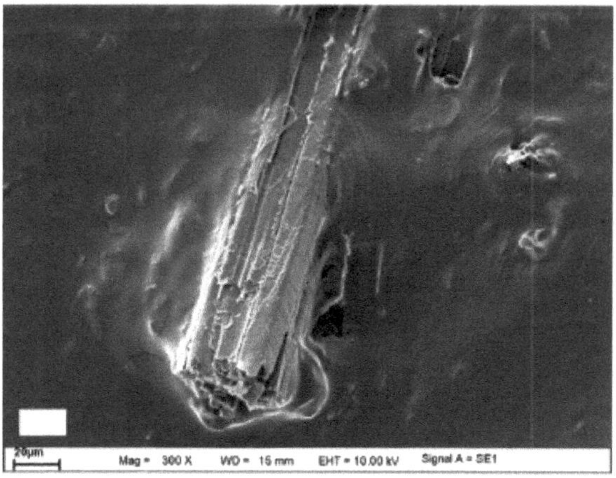

FIGURE 9.3 Scanning electron microscope image of thermoplastic cassava starch/jute fiber (Prachayawarakorn et al., 2013).

Another interesting study utilized cotton fiber reinforcement for CS biofoams (Sanhawong, Banhalee, Boonsang, & Kaewpirom, 2017). Cotton fiber and concentrated natural rubber latex (CNRL) were included in the biofoam products to overcome their two main weaknesses, which were susceptibility to moisture and a lack of flexibility. Natural rubber is a biodegradable polymer that presents excellent impact properties, high elongation, also high tensile strength. There are numerous formulations to prepare CS biofoam using a conventional compression molding machine. The homogeneous batter was obtained after all additives were mixed using a kitchen food mixer at room temperature for 5 minutes, including starch, cotton fiber, CNRL (premixed with an appropriate amount of vulcanization ingredients), and distilled water. Next, the homogeneous batter was placed in a compression mold and then poured into a $300 \times 300 \times 3$ mm^3 mold. The mold pressure was finalized at 1000 atm for 4 minutes and 220°C of temperature. Biofoam showed a denser cell structure, less porosity with increased density, and increased hydrophobicity with increasing CNRL content. With increasing hydrophobicity, better dimensional stability was achieved, as well as the moisture adsorption rate and moisture adsorption capacity of the biofoam being decreased. CNRL stimulates the thermal degradation of cotton fiber even though the use of CNRL did not affect the thermal degradation behavior of the biofoam. Suitable addition concentrations of CNRL led to better flexural properties of the biofoam, related to its cross-linked structure. The density of the biofoam increased after use of CNRL. The modulus rupture of the biofoam was enhanced with a process of either the reinforcing cotton fiber or the formed rubber network. The water content for biofoam products was significantly decreased due to the addition of CNRL to the starch batter because of its inherent hydrophobicity. When the biofoams were in contact with water, their dimensional stability was improved. The delayed biodegradation caused by the improved hydrophobicity of the biofoam occurred by hydrolysis.

In food industries, eggshell is produced in large quantities and is a major waste product. Eggshell is considered to be an excellent candidate to be used as a biofiller in the preparation of polymer composites as it is cheap, plentiful, has low density, and is environment-friendly and a renewable resource. A study on the effect of chicken eggshell on thermoplastic cassava starch has been carried out to investigate the properties as a comparison to commercial calcium carbonate (Bootklad & Kaewtatip, 2013). CS and glycerol were mixed together with calcium carbonate or eggshell at 10–50 wt.%, using polyethylene bags to drying the starch into. The preparation of mixed TPS and the composites was done prior to the use of a compression molding machine at 160°C for 10 minutes. A soil burial degradation test run was carried out in Hat Yai, Songkhla, Thailand, by burying samples in soil at 7 cm depth. The samples were gathered after 15 and 30 days and then cleaned using distilled water a number of times. An oven was then used to dry the samples at 105°C for 24 hours to calculate the loss of weight. The test indictaed that after 15 days of burial, the original weight for the matrix TPS loss was 71.55% and it was completely degraded following 30 days of burial. The weight loss of TPS is due to microorganism attack in the soil and the biodegradation process by moisture. Meanwhile, it has been found that the water absorption of the TPS and eggshell composites can be reduced by incorporation of calcium carbonate. This was attributed to the greater hydrophobic character of calcium carbonate than neat TPS matrix.

Natural fiber has become a possible material to replace petroleum-based products by combination with another material or alone to produce green composites. In order to make a hybrid composite, cassava bagasse (CB) with sugar palm fiber (SPF) was prepared by applying a casting method with CS to act as a plasticizer (Edhirej et al., 2017). Samples of 3.0 g were dissolved in 25 mL of distilled water and kept in preweighed centrifuge tubes for 25 minutes at 3000 rpm in order to conduct a water absorption test. The dispersion was combined and placed at room temperature for 1 hour after being kept in a centrifuge. The results showed that CB absorbed a higher amount of water, which signified that CB is much more hydrophilic than SPF (257.04%) which absorbed a lower amount of water (122.33%). Meanwhile, the moisture content was somewhat decreased when SPF and CB were included in the CS films. As a comparison of CS films with lower SPF content, the film with a greater amount of SPF had a lower moisture content. Table 9.3 gives a brief review of the previous studies using TPCS. TPS has great potential for many kinds of environmental friendly applications due to the suitable properties of the biopolymer matrix.

9.7 Thermoplastic rice starch

Rice is widely consumed as a staple food around the world, with over 500 million tons providing sustenance to many millions of people all over the world annually. Rice starch is a fundamental biopolymer, containing amylose and amylopectin, which are excellent raw materials for use in packaging materials. A study investigated waxy rice that was used as a biodegradable polymer using TPS. The cotton fiber was included as the fiber reinforcement into the TPS matrix to reduce water absorption and increase the mechanical properties of the TPS (Jutarat Prachayawarakorn, Ruttanabus, & Boonsom, 2011). The results showed that the TPRS sample had a greater stress maximum load and Young's modulus, however, decreasing strain at maximum load reinforcement because of the higher amylose content of rice starch created greater sturdiness than those of the TPWRS starch sample. The water absorption increased at the primary stage for TPRS and TPWRS samples without cotton fiber reinforcement, but due to the higher content of the more branched structure of amylopectin molecules, the TPRS was found to have more rapid water absorption. The percentage of water absorption for TPRS rose gradually and achieved its limit within 7 days, while for the TPWSR sample this was reached within 4 days.

As an alternative choice of fiber reinforcement, because of inadequate mechanical properties and elevated water absorption of TPS, thermoplastic rice starch (TPRS) was improved by reinforcing with natural silk protein fibers (Prachayawarakorn & Hwansanoet, 2012). In addition, silk fibers were added to the TPRS matrix bringing about a significant increase in the stress at maximum load and the Young's modulus. The interaction between the TPRS matrix and the silk reinforcement caused a reinforcing effect of the silk fibers. Meanwhile, the stress at maximum load and Young's modulus reduced the silk fibers because there was too high a content of lightweight silk fibers loaded into the TPRS matrix. Table 9.4 presents a review of TPS composites reinforced with rice starch. Apart from biodegradable packaging, the potential of this material has been also explored in other industries, such as agricultural and biomedical.

TABLE 9.3 Thermoplastic cassava starch composites.

Type of starch	Type of fiber	Type of testing	Main finding	Potential application	References
Cassava	Cotton fiber, natural rubber latex	Soil burial test	Better moisture resistance and dimensional stability	Packaging	Sanhawong et al. (2017)
Cassava	Rice straw	SEM, TEM, TGA, mechanical test, water drop test	Increased flexural strength and modulus by fourfold compared with the control	Nonfood packaging	Narkchamnan & Sakdaronnarong (2013)
Cassava	Carrageenan, low-density polyethylene composites, cotton fiber	IR spectroscopy, XRD, mechanical testing, water absorption, soil burial test	Increased the stress at maximum load and Young's modulus of the thermoplastic cassava starch (TPCS)/low-density polyethylene (LDPE) composites	Packaging	Prachayawarakorn & Pomdage (2014)
Cassava	Sugar palm	Water absorption, water content, chemical composition, thickness swelling, XRD	Increased the suitability of cassava starch composite films as environment-friendly food packaging	Bio-based product	Edhirej et al. (2017)
Cassava	Eggshell	SEM, water absorption, biodegradation in soil, thermal stability	Improved the water resistance and thermal stability of thermoplastic starch	Biodegradable material	Bootklad & Kaewtatip (2013)
Cassava	Jute, kapok	FT-IR spectroscopy, XRD, SEM, thermal properties, mechanical testing, water absorption	Increased the stress at maximum load and Young's modulus by the incorporation of both jute and kapok fibers	Packaging	Prachayawarakorn et al. (2013)

(Continued)

TABLE 9.3 (Continued)

Type of starch	Type of fiber	Type of testing	Main finding	Potential application	References
Cassava	Green coconut fiber	SEM, mechanical testing, water absorption, water sensitivity, moisture absorption	Improved the tensile properties of cassava starch, interfacial bonding between the matrix and fibers as well as the barrier to absorption caused by the fibers also improved	Packaging	Lomelí Ramírez et al. (2011)
Cassava	Pectin particles, cotton fiber	IR spectroscopy, XRD, SEM, mechanical tests, water absorption, thermal properties	Increased stress at highest load and Young's modulus of the TPCS/pectin particles and the TPCS/cotton fiber composites	Biodegradable polymer	Prachayawarakorn & Pattanasin (2016)
Cassava	Chitosan coating	XRD, physical and mechanical properties measurement, water absorption, WVTR, evaluation of mechanical properties	Increased tensile stress at maximum load and tensile modulus, and a reduction in percent elongation at break	Packaging	Bangyekan et al. (2006)

FT-IR, Fourier-transform infrared spectroscopy; *IR*, infrared spectroscopy; *SEM*, scanning electron microscopy; *TEM*, transmission electron microscope; *TGA*, thermogravimetric analysis; *TPCS*, thermoplastic cassava starch; *WVTR*, water vapor transmission; *XRD*, X-Ray diffraction.

TABLE 9.4 Thermoplastic rice starch composite.

Type of starch	Type of fiber	Type of testing	Main finding	Potential applications	References
Rice	Natural fiber	MFI, Fourier-transform infrared spectroscopy Fourier-transform, tensile test, enzymatic tests, SEM	Improved tensile properties	Biodegradable composite	Yew, Mohd Yusof, Mohd Ishak, & Ishiaku (2005)
Rice	Cotton	Tensile tests, SEM, water absorption, TGA, soil burial test	Development of tensile strength and Young's modulus	Biodegradable polymer	Prachayawarakorn, Sangnitidej, & Boonpasith (2010)
Rice	Silk protein fiber	IR spectroscopy, water absorption, SEM mechanical test	Increased the stress at maximum load and Young's modulus composite	Biodegradable material	Prachayawarakorn & Hwansanoet (2012)
Rice, waxy rice	Cotton	IR spectroscopy, XRD, water absorption, SEM, mechanical testing, soil burial test, TGA	Increased the stress at maximum load and Young's modulus of both TPRS and TPWRS	Biodegradable polymer	Prachayawarakorn et al. (2011)
Waxy rice	Agar, cotton fiber	SEM, tensile test, water absorption, color measurement, IR spectroscopy, tensile test, soil burial test, TGA, water uptake	Improved the tensile properties, thermal degradation temperature, and thermal stability of TPWRS	Packaging, agricultural, medical applications	Prachayawarakorn, Limsiriwong, Kongjindamunee, & Surakit (2011)

IR, Infrared spectroscopy; *MFI*, melt flow index; *SEM*, scanning electron microscopy; *TGA*, thermogravimetric analysis; *TPRS*, thermoplastic rice starch; *TPWRS*, thermoplastic waxy rice starch; *XRD*, X-Ray diffraction.

9.8 Thermoplastic cornstarch

Corn or maize is produced in greater quantities than wheat or rice and is regarded as a staple food in many parts of the world. Cornstarch is obtained from corn grain and is very versatile, easily modified, and is applicable in various industries such as food, pharmaceutical, and textile manufacturing. One of the ways to develop TPS thermoplastic properties, was corn starch that was used in experiments with numerous modifications being conducted. The modification of films based on thermoplastic corn starch (TPCRS) incorporation of chitosan and chitin were found by melt-mixing and thermocompression (Lopez et al., 2014). The film tensile strength was substantially improved with mixing of 10 g chitosan with 100 g starch to the TPS matrix. The film elongation at break was reduced by around 30% with the addition of 10 g chitosan. The interactions between chitosan or chitin and cornstarch reduce the accessibility of the hydrophilic groups. Hence, there is decreased water vapor transmission access for the matrix.

A characterization study of cornstarch using X-Ray powder diffraction and thermal analysis, along with characterization and processing of banana starch and sugarcane bagasse fiber composites, has been carried out (Guimarães, Wypych, Saul, Ramos, & Satyanarayana, 2010). The composites were made by compression molding of the matrix with banana and bagasse fibers presenting more identical composites with crude glycerin of cornstarch and glycerol as a plasticizer. TGA observation shows a positive thermal stability for making composites and had some cracks between the smooth and rough surface visualized by fractographic studies of 70 wt.% starch and 30% glycerol matrix indicating that the sample was ductile. The properties of strength for the matrix were improved by tensile testing of the composites. A higher quality of homogeneous composites was created from crude glycerin with the potential of better properties compared to material produced with commercial glycerol.

A mixture of organic polymer matrices and inorganic particles has been proved to develop the mechanical, physical, and thermal properties. According to Huang and Yu (2006), green composites were created through the blend extrusion method using TPCRS and activated montmorillonite (EMMT). Another aspect that needs to be focused on is the constancy of the thermal and water resistance properties of TPS with EMMT. These properties were also enhanced by being attributed with a huge aspect ratio of width to thickness of EMMT in TPS matrix.

Table 9.5 presents a brief introduction to TPCRS composites extracted from corn starch. Various actions have been conducted to resolve the problems created by plastic waste in order to produce a material that is environmentally friendly, especially a disposable material that can only

TABLE 9.5 Thermoplastic cornstarch composites.

Type of starch	Type of fiber	Type of testing	Main finding	Potential applications	References
Corn	Chitosan	XRD, DSC, SEM optical properties, WVP, tensile test, DMA	Increased tensile strength and elastic modulus, and decreased elongation at break	Packaging	Lopez et al. (2014)
Corn	Sisal	Tensile test, SEM, infrared spectrum analysis, biodegradability test, water absorption	Replacement for EPS as packing material, especially under large compression load (0.7–6 Mpa)	Plastic packing materials	Xie et al. (2018)
Corn	Date palm, flax fibers	Tensile test, SEM water uptake test, soil burial, TGA	Increased the composite static tensile and flexural mechanical properties (stiffness and strength). Thermal stability, water uptake, and biodegradation also improved	Biodegradable polymer	Ibrahim, Farag, Megahed, & Mehanny (2014)
Corn	Clay	SEM, XRD soil burial test, contact angle measurements	Improved properties in relation to TPS	Biodegradable material	Magalhães & Andrade (2009)
Corn	Banana, sugarcane bagasse	XRD, moisture content, thermal analysis, tensile test, SEM	Improvements in tensile properties	Biodegradable composite	Guimarães et al. (2010)

DMA, Dynamic mechanical analysis; *DSC*, differential scanning calorimetric; *EPS*, expandable polystyrene; *WVP*, water vapor permeability; *SEM*, scanning electron microscopy; *TGA*, thermogravimetric analysis; *TPS*, thermoplastic corn starch; *XRD*, X-Ray diffraction.

be used once. Many researchers have concentrated on petro-based commodity plastics that need to be substituted with a biodegradable material with competitive mechanical properties in a cost-effective manner.

9.9 Thermoplastic potato starch

Starch grains (leucoplasts) are extracted from the cells of root tubers of potato plants. The potatoes first needed to be crushed as a means of extracting the starch. The crushing releases starch grains from the damaged cells. A study by Wei, Wang, Xiao, Zheng, and Yang (2015) investigated the biodegradable of poly(butylene adipate-co-terephthalate) (PBAT) with TPS composite. First, they used the synthesized reactive compatibilizer which is a styrene-maleic anhydride glycidyl methacrylate (SMG), and the commercial compatibilizer. To change the part of PBAT in the PBAT/TPS blends, the second approach used modified PBAT (M-PBAT) with a greater molecular weight with a chain extension by reactive extrusion of PBAT. SEM images show an improvement dispersion of TPS when applying M-PBAT, leading to higher tensile strength and elongation at break of PBAT/TPS blends. Although the tensile strength was maintained at the same level or greater, in the presence of a compatibilizer (SMG), the elongation at break decreased although the tensile strength was constant at a similar level or slightly increased. Generally, the tensile and elongation results explain that it is suitable to use for various applications, as well as for disposable packaging and agricultural mulching films.

Table 9.6 summarizes the applications of TPS composites derived from potato starch. Potato starch has good promise as a renewable material with potential in packaging and agricultural industries.

TABLE 9.6 Thermoplastics of potato starch.

Type of starch	Type of fiber or polymer	Type of testing	Main finding	Potential applications	References
Potato, sweet potato, corn	Chitosan	DSC, TGA, SEM	Improved the thermal properties	Biodegradable composite	Gómez, Torres, Nakamatsu, & Arroyo (2006)
Potato	Clay	XRD, TEM, tensile test, thermal properties	Improved the mechanical properties by tensile and impact tests	Packaging application	Park et al. (2002)
Potato	Agar	FT-IR, SEM, water absorption, tensile test, WVP	Improved the microstructure of starch film, ameliorated mechanical properties, and WVP in a high-moisture environment	Food packaging	Wu et al. (2009)

(Continued)

TABLE 9.6 (Continued)

Type of starch	Type of fiber or polymer	Type of testing	Main finding	Potential applications	References
Potato, sweet potato, corn	Natural fibers (sisal, jute, and cabuya)	Tensile test, impact test, the fracture surface	Improved the mechanical properties by tensile and impact tests	Biodegradable composite	Torres, Arroyo & Gomez (2007)
Potato	PBAT	PBAT/TPS blends film, PBAT/TPS/SMG blends films, PBAT/M-PBAT/TPS blends films	Improved the mechanical properties by tensile strength and elongation	Disposable packaging, agricultural mulching film	Wei et al. (2015)
Potato	Corn, poly (vinyl alcohol)	TGA, SEM, trays testing	Improved the tensile properties	Biodegradable materials	Cinelli, Chiellini, Lawton, & Imam (2006)

DSC, Differential scanning calorimetry; *FT-IR*, Fourier-transform infrared spectroscopy; *PBAT*, Poly(butylene adipate-co-terephthalate); *SEM*, scanning electron microscopy; *SMG*, styrene-maleic anhydride glycidyl methacrylate; *TEM*, transmission electron microscopy; *TGA*, thermogravimetric analysis; *TPS*, thermoplastic starch; *WVP*, water vapor permeability; *XRD*, X-Ray diffraction.

9.10 Thermoplastics from various sources

The production of TPS was obtained through diverse starch types. Table 9.7 gives a summary of the TPS gained from many resources apart from the plants discussed above, and include mung bean, pea, sugar palm, and wheat. Many analyses of thermoplastics have been made from different kinds of starch with good properties and characteristic that are suitable for use in various applications besides being able to improve the product quality.

9.11 Conclusions

In general, TPS is considered to be a potential replacement for conventional synthetic plastic. Although TPS has advantages of low cost, biodegradable, renewable, and sustainable properties, it is not totally free of problems. A serious problem of TPS is the hydrophilic behavior which tends to absorb moisture from the environment, hence affecting the mechanical properties and dimensional stability of the material. Several modifications have been carried out on TPS including blending

TABLE 9.7 Thermoplastic starch from various sources.

Type of starch	Type of fiber	Type of testing	Main finding	Potential applications	References
Mung bean	Cotton, polyethylene	SEM, tensile test, water absorption, thermal properties, soil burial test	Increased the mechanical strength and rigidity of the films but water vapor permeability of the films decreased	Hard capsules for pharmaceutical applications	Prachayawarakorn, Hommanee, Phosee, & Chairapaksatien (2010)
Sugar palm	Seaweed	FT-IR, SEM, tensile test, flexural testing, impact test, TGA, density determination balance, moisture content, water absorption, thickness swelling, moisture absorption, water solubility, soil burial test, statistical analysis	Improved tensile and flexural properties accompanied by lower impact resistance	Biodegradable product	Jumaidin, Sapuan, Jawaid, Ishak, & Sahari (2017)
Wheat	Poly(lactic acid), methylenediphenyl diisocyanate	DMA, tensile test, water absorption	Improved the tensile properties, microstructure, and water absorption	Short-term disposable applications	Wang, Sun, & Seib (2002)
Pea	Citric acid	TGA, DMTA, WVP, mechanical testing	Improved the storage modulus, glass transition temperature, tensile strength, and water vapor barrier, but decreased thermal stability. CARS/TPS	Biodegradable plastics	Ma, Chang, Yu, & Stumborg (2009)

CARS, Citric acid-modified rice starch; *DMA*, dynamic mechanical analysis; *DMTA*, dynamic mechanical thermal analysis; *FT-IR*, Fourier-transform infrared; *SEM*, scanning electron microscopy; *TEM*, transmission electron microscopy; *TGA*, thermogravimetric analysis; *TPS*, thermoplastic starch; *WVP*, water vapor permeability.

with other polymers, reinforcing with natural fiber, and treatment to improve the functional characteristics of this material. The findings show that blending with other polymers, incorporation of fiber, and treatment of fiber to reduce the hydrophilic behavior of the material while improving the mechanical properties is feasible. However, various modifications and treatments have achieved various levels of success in improving the properties of TPS. In conclusion, TPS is a promising material to replace nonbiodegradable polymers, provided that proper modifications are carried out to achieve the desired characteristics of the final product.

Acknowledgment

The authors would like to thank Universiti Teknikal Malaysia Melaka and Ministry of Higher Education Malaysia for the financial support provided through the Fundamental Research Grant Scheme for Research Acculturation of Early Career Researchers (project number RACER/2019/FTKMP-CARE/F00413).

References

Bangyekan, C., Aht-Ong, D., & Srikulkit, K. (2006). Preparation and properties evaluation of chitosan-coated cassava starch films. *Carbohydrate Polymers, 63*, 61–71. Available from: https://doi.org/10.1016/j.carbpol.2005.07.032.

Bogati, D. R. (2011). Cellulose based biochemicals and their applications. *Degree Programme in Paper Technology, Bachelor, 37*.

Bootklad, M., & Kaewtatip, K. (2013). Biodegradation of thermoplastic starch/eggshell powder composites. *Carbohydrate Polymers, 97*(2), 315–320. Available from: https://doi.org/10.1016/j.carbpol.2013.05.030.

Chauhan, A., & Chauhan, P. (2012). Natural fiber reinforced composite: A concise review article. *Journal of Chemical Engineering & Process Technology, 03*(02), 2–4. Available from: https://doi.org/10.4172/2157-7048.1000132.

Cinelli, P., Chiellini, E., Lawton, J. W., & Imam, S. H. (2006). Foamed articles based on potato starch, corn fibers and poly(vinyl alcohol). *Polymer Degradation and Stability, 91*(5), 1147–1155. Available from: https://doi.org/10.1016/j.polymdegradstab.2005.07.001.

Edhirej, A., Sapuan, S. M., Jawaid, M., & Zahari, N. I. (2017). Cassava/sugar palm fiber reinforced cassava starch hybrid composites: Physical, thermal and structural properties. *International Journal of Biological Macromolecules, 101*, 75–83. Available from: https://doi.org/10.1016/j.ijbiomac.2017.03.045.

Espigulé, E., Puigvert, X., Vilaseca, F., Mendez, J. A., Mutjé, P., & Girones, J. (2013). Thermoplastic starch-based composites reinforced with rape fibers: Water uptake and thermomechanical properties. *BioResources, 8*(2), 2620–2630. Available from: https://doi.org/10.15376/biores.8.2.2620-2630.

Garavand, F., Rouhi, M., Razavi, S. H., Cacciotti, I., & Mohammadi, R. (2017). Improving the integrity of natural biopolymer films used in food packaging by crosslinking approach: A review. *International Journal of Biological Macromolecules, 104*, 687–707. Available from: https://doi.org/10.1016/j.ijbiomac.2017.06.093.

Gómez, C., Torres, F. G., Nakamatsu, J., & Arroyo, O. H. (2006). Thermal and structural analysis of natural fiber reinforced starch-based biocomposites. *International Journal of Polymeric Materials, 55*(11), 893–907. Available from: https://doi.org/10.1080/00914030500522547.

Guimarães, J. L., Wypych, F., Saul, C. K., Ramos, L. P., & Satyanarayana, K. G. (2010). Studies of the processing and characterization of corn starch and its composites with banana and sugarcane fibers from Brazil. *Carbohydrate Polymers, 80*(1), 130–138. Available from: https://doi.org/10.1016/j.carbpol.2009.11.002.

Gupta, K. M. (2011). CHAP 8 starch based composites for packaging applications. In Srikanth Pilla (Ed.), *Handbook of bioplastics and biocomposites engineering applications*, 189–266. Wiley. Available from: https://doi.org/10.1002/9781118203699.ch8.

Huang, M, & Yu, J (2006). Structure and properties of thermoplastic corn starch/montmorillonite biodegradable composites. *Journal of Applied Polymer Science, 99*, 170–176.

Ibrahim, H., Farag, M., Megahed, H., & Mehanny, S. (2014). Characteristics of starch-based biodegradable composites reinforced with date palm and flax fibers. *Carbohydrate Polymers, 101*, 11–19. Available from: https://doi.org/10.1016/j.carbpol.2013.08.051.

Ivanič, F., Jochec-Mošková, D., Janigová, I., & Chodák, I. (2017). Physical properties of starch plasticized by a mixture of plasticizers. *European Polymer Journal, 93*, 843–849. Available from: https://doi.org/10.1016/j.eurpolymj.2017.04.006.

Jumaidin, R., Sapuan, S. M., Jawaid, M., Ishak, M. R., & Sahari, J. (2017). Effect of seaweed on mechanical, thermal, and biodegradation properties of thermoplastic sugar palm starch/agar composites. *International Journal of Biological Macromolecules, 99*, 265–273. Available from: https://doi.org/10.1016/j.ijbiomac.2017.02.092.

Jumaidin, R, et al. (2018). Effect of Agar on Dynamic Mechanical Properties of Thermoplastic Sugar Palm Starch: Thermal Behavior. *ournal of Advanced Research in Fluid Mechanics and Thermal Sciences, 47*(1), 89–96.

Jumaidin, R, et al. (2019). Characteristics of cogon grass fibre reinforced thermoplastic cassava starch biocomposite: water absorption and physical properties. *Journal of Advanced Research in Fluid Mechanics and Thermal Sciences, 62*(1), 43–52.

Jumaidin, R, et al. (2020). Effect of cogon grass fibre on the thermal, mechanical and biodegradation properties of thermoplastic cassava starch biocomposite. *International Journal of Biological Macromolecules, 146*, 746–755. Available from: https://doi.org/10.1016/j.ijbiomac.2019.11.011.

Jumaidin, R, et al. (2021). Thermal degradation and mechanical characteristics of sugarcane bagasse reinforced biodegradable potato starch composites. *Journal of Advanced Research in Fluid Mechanics and Thermal Sciences, 78*(1), 157–166. Available from: https://doi.org/10.37934/arfmts.78.1.157166.

Lau, K., Hung, P., Zhu, M., & Hui, D. (2018). Properties of natural fibre composites for structural engineering applications. *Composites Part B, 136*(September 2017), 222–233. Available from: https://doi.org/10.1016/j.compositesb.2017.10.038.

Locklin, J. (2017). *Thermoplastic starch—Potato plastic*. 2–5.

Lomelí Ramírez, M. G., Satyanarayana, K. G., Iwakiri, S., De Muniz, G. B., Tanobe, V., & Flores-Sahagun, T. S. (2011). Study of the properties of biocomposites. Part I. Cassava starch-green coir fibers from Brazil. *Carbohydrate Polymers, 86*, 1712–1722. Available from: https://doi.org/10.1016/j.carbpol.2011.07.002.

Lopez, O., Garcia, M. A., Villar, M. A., Gentili, A., Rodriguez, M. S., & Albertengo, L. (2014). Thermo-compression of biodegradable thermoplastic corn starch films containing chitin and chitosan. *LWT-Food Science and Technology, 57*(1), 106–115. Available from: https://doi.org/10.1016/j.lwt.2014.01.024.

Ma, X., Chang, P. R., Yu, J., & Stumborg, M. (2009). Properties of biodegradable citric acid-modified granular starch/thermoplastic pea starch composites. *Carbohydrate Polymers, 75*, 1–8. Available from: https://doi.org/10.1016/j.carbpol.2008.05.020.

Magalhães, N. F., & Andrade, C. T. (2009). Thermoplastic corn starch/clay hybrids: Effect of clay type and content on physical properties. *Carbohydrate Polymers, 75*(4), 712–718. Available from: https://doi.org/10.1016/j.carbpol.2008.09.020.

Mikus, P. Y., Alix, S., Soulestin, J., Lacrampe, M. F., Krawczak, P., Coqueret, X., & Dole, P. (2014). Deformation mechanisms of plasticized starch materials. *Carbohydrate Polymers, 114*, 450–457. Available from: https://doi.org/10.1016/j.carbpol.2014.06.087.

Mohammed, L., Ansari, M. N. M., Pua, G., Jawaid, M., & Islam, M. S. (2015). A review on natural fiber reinforced polymer composite and its applications. *International Journal of Polymer Science, 2015*, 1–15. Available from: https://doi.org/10.1155/2015/243947.

Narkchamnan, S., & Sakdaronnarong, C. (2013). Thermo-molded biocomposite from cassava starch, natural fibers and lignin associated by laccase-mediator system. *Carbohydrate Polymers, 96*(1), 109–117. Available from: https://doi.org/10.1016/j.carbpol.2013.03.046.

Okamoto, M., & John, B. (2013). Synthetic biopolymer nanocomposites for tissue engineering scaffolds. *Progress in Polymer Science, 38*(10–11), 1487–1503. Available from: https://doi.org/10.1016/j.progpolymsci.2013.06.001.

Park, H. M., Li, X., Jin, C. Z., Park, C. Y., Cho, W. J., & Ha, C. S. (2002). Preparation and properties of biodegradable thermoplastic starch/clay hybrids. *Macromolecular Materials and Engineering, 287*(8), 553–558. Available from: https://doi.org/10.1002/1439-2054(20020801)287:8 < 553::AID-MAME553 > 3.0.CO;2-3.

Prachayawarakorn, J., & Hwansanoet, W. (2012). Effect of silk protein fibers on properties of thermoplastic rice starch. *Fibers and Polymers, 13*(5), 606–612. Available from: https://doi.org/10.1007/s12221-012-0606-x.

Prachayawarakorn, J., & Pattanasin, W. (2016). Effect of pectin particles and cotton fibers on properties of thermoplastic cassava starch composites. *Songklanakarin Journal of Science and Technology, 38*(2), 129–136.

Prachayawarakorn, J., & Pomdage, W. (2014). Effect of carrageenan on properties of biodegradable thermoplastic cassava starch/low-density polyethylene composites reinforced by cotton fibers. *Materials & Design, 61*, 264–269. Available from: https://doi.org/10.1016/j.matdes.2014.04.051.

Prachayawarakorn, J., Chaiwatyothin, S., Mueangta, S., & Hanchana, A. (2013). Effect of jute and kapok fibers on properties of thermoplastic cassava starch composites. *Materials and Design, 47*, 309–315. Available from: https://doi.org/10.1016/j.matdes.2012.12.012.

Prachayawarakorn, J., Hommanee, L., Phosee, D., & Chairapaksatien, P. (2010). Property improvement of thermoplastic mung bean starch using cotton fiber and low-density polyethylene. *Starch/Staerke, 62*(8), 435–443. Available from: https://doi.org/10.1002/star.201000002.

Prachayawarakorn, J., Limsiriwong, N., Kongjindamunee, R., & Surakit, S. (2011). Effect of agar and cotton fiber on properties of thermoplastic waxy rice starch composites. *Journal of Polymers and the Environment, 20*(1), 88–95. Available from: https://doi.org/10.1007/s10924-011-0371-8.

Prachayawarakorn, J., Ruttanabus, P., & Boonsom, P. (2011). Effect of cotton fiber contents and lengths on properties of thermoplastic starch composites prepared from rice and waxy rice starches. *Journal of Polymers and the Environment, 19*, 274–282. Available from: https://doi.org/10.1007/s10924-010-0273-1.

Prachayawarakorn, J., Sangnitidej, P., & Boonpasith, P. (2010). Properties of thermoplastic rice starch composites reinforced by cotton fiber or low-density polyethylene. *Carbohydrate Polymers, 81*(2), 425–433. Available from: https://doi.org/10.1016/j.carbpol.2010.02.041.

Rebelo, R., Fernandes, M., & Fangueiro, R. (2017). Biopolymers in medical implants: A brief review. *Procedia Engineering, 200*, 236–243. Available from: https://doi.org/10.1016/j.proeng.2017.07.034.

Roz, A., Carvalho, A., Gandini, A., & Curvelo, A. (2006). The effect of plasticizers on thermoplastic starch compositions obtained by melt processing. *Carbohydrate Polymers, 63*, 417–424. Available from: https://doi.org/10.1016/j.carbpol.2005.09.017.

Sanhawong, W., Banhalee, P., Boonsang, S., & Kaewpirom, S. (2017). Effect of concentrated natural rubber latex on the properties and degradation behavior of cotton-fiber-reinforced cassava starch biofoam. *Industrial Crops and Products, 108*, 756–766. Available from: https://doi.org/10.1016/j.indcrop.2017.07.046.

Sanjay, M. R., Arpitha, G. R., Naik, L. L., Gopalakrishna, K., & Yogesha, B. (2016). Applications of natural fibers and its composites: An overview. *Natural Resources, 07*(03), 108–114. Available from: https://doi.org/10.4236/nr.2016.73011.

Soykeabkaew, N., Thanomsilp, C., & Suwantong, O. (2015). A review: Starch-based composite foams. *Composites Part A: Applied Science and Manufacturing, 78*. Available from: https://doi.org/10.1016/j.compositesa.2015.08.014.

Torres, F. G., Arroyo, O. H., & Gomez, C. (2007). Processing and mechanical properties of natural fiber reinforced thermoplastic starch biocomposites. *Journal of Thermoplastic Composite Materials, 20*(2), 207–223. Available from: https://doi.org/10.1177/0892705707073945.

Wang, H., Sun, X., & Seib, P. (2002). Mechanical properties of Poly(lactic acid) and wheat starch blends with methylenediphenyl diisocyanate. *Journal of Applied Polymer Science, 84*(6), 1257–1262. Available from: https://doi.org/10.1002/app.10457.

Wei, D., Wang, H., Xiao, H., Zheng, A., & Yang, Y. (2015). Morphology and mechanical properties of poly(butylene adipate-co-terephthalate)/potato starch blends in the presence of synthesized reactive compatibilizer or modified poly(butylene adipate-co-terephthalate). *Carbohydrate Polymers, 123*, 275–282. Available from: https://doi.org/10.1016/j.carbpol.2015.01.058.

Wiacek, A. E. (2015). Effect of surface modification on starch biopolymer wettability. *Food Hydrocolloids, 48*, 228–237. Available from: https://doi.org/10.1016/j.foodhyd.2015.02.005.

Wu, Y., Geng, F., Chang, P. R., Yu, J., & Ma, X. (2009). Effect of agar on the microstructure and performance of potato starch film. *Carbohydrate Polymers, 76*(2), 299–304. Available from: https://doi.org/10.1016/j.carbpol.2008.10.031.

Xie, Q., Li, F., Li, J., Wang, L., Li, Y., Zhang, C., ... Chen, S. (2018). A new biodegradable sisal fiber–starch packing composite with nest structure. *Carbohydrate Polymers, 189*, 56–64. Available from: https://doi.org/10.1016/j.carbpol.2018.01.063.

Yew, G. H., Mohd Yusof, A. M., Mohd Ishak, Z. A., & Ishiaku, U. S. (2005). Water absorption and enzymatic degradation of poly(lactic acid)/rice starch composites. *Polymer Degradation and Stability, 90*(3), 488–500. Available from: https://doi.org/10.1016/j.polymdegradstab.2005.04.006.

Zuo, Y., Gu, J., Tan, H., & Zhang, Y. (2015). Thermoplastic starch prepared with different plasticizers: Relation between degree of plasticization and properties. *Journal of Wuhan University of Technology-Materials Science Edition, 30*(2), 423–428. Available from: https://doi.org/10.1007/s11595-015-1164-z.

CHAPTER

10

Composites leading to a clean and green future

I. Ahmad[1], *M.D. Ahmad*[2] *and N. Abdullah*[3]

[1]Algae and Biomass Research Laboratory, Malaysia-Japan International Institute of Technology, Universiti Teknologi Malaysia, Jalan Sultan Yahya Petra, Kuala Lumpur, Malaysia, [2]Department of Post-Harvest Engineering and Technology, Aligarh Muslim University, Aligarh, Uttar Pradesh, India, [3]UTM International, Level 8, Menara Razak, Universiti Teknologi Malaysia, Jalan Sultan Yahya Petra, Kuala Lumpur, Malaysia

10.1 Introduction

A composite is a material which is manufactured using two or more dissimilar materials that have superior properties to that of the individual materials separately. Composites are used as the interiors in automobiles, railway compartments, and ships, etc. They help in reducing sound/noise pollution by absorbing inner noise that affects human health as well as the economy and environment. Nowadays, natural fiber reinforcement composites using bamboo, jute, ramie, kenaf, etc. are used by researchers and engineers due to their low density or light weight and good mechanical strength, and are friendly to the ecosystem. They are used in the advancement of technologies and finding alternatives for electricity production. Solar and wind energy are clean and green sources of energy to generate electricity using solar panels and wind turbines, and both are made up of composite materials that reduce the need for coal which produces air pollution. Turbine blades are made up of hybrid composites like carbon/glass fiber that plays a crucial role in energy conversion. The strength-to-weight ratio is higher as compared with wood and metals and they also have corrosion resistance (Huda, Mohanty, Drzal, Misra, & Schut, 2004). Global food

Design for Sustainability
DOI: https://doi.org/10.1016/B978-0-12-819482-9.00014-9

production is rising with increase in the world's population, and therefore food waste (FW) and food packaging waste also increase including bagasse from sugar industries, peel and seed waste from beverage industries, and also vegetables and food grain waste like rice husk, bran, etc.

Packaging plays a vital role in food protection from the external environment by providing protection from pathogenic microbes, moisture, light, heat, and physical damage by mechanical shocks during transportation of food. It also helps in maintaining its quality attributes, increases its shelf life, and hence provides safety to the ultimate consumer. FW material like bagasse, peel, etc. are used as raw materials for synthesizing biodegradable composites, which is an excellent method for utilizing FW as well as reducing food packaging waste because $\cong$79% of solid waste is dumped in landfill. Food packaging materials are dependent on the type of food to be packed. The main material used these days for food packaging is plastic due to its excellent barrier properties to the external environment, however it is non-biodegradable and can take up to 400 years to degrade (Miller, 2005). Its use is forecasted to increase up to 250 million metric tons in the next 30 years, from a currently level of about 150 million metric tons annually. As a result of this soil pollution occurs and landfill leachate penetrates from the soil to ground water causing potable water pollution. Composites are used in safeguarding and maintaining natural resources and also protect human health by reducing the extraction of natural resources by achieving the properties of similar materials that are extracted because resources beneath the Earth's crust or land resources extraction can cause disasters like tsunamis, earthquakes, and landslides. Despite being both biodegradable and non biodegradable, composites are helpful to our environment. It saves resources by using these composites created from plant waste hence protecting environment. They have mostly been used as packaging materials for food that produces a high amount of solid waste which is a threat to humans and habitats presently and also in the future if it not controlled. A survey indicated that 20% of the total solid waste in developed countries comes from food. As the population of the world increases at a tremendous rate, it is forecast to reach 10 billion in 2050. It is therefore crucial to reduce solid waste, as well as reduce soil, water, and air pollution (Hammond, 2007).

10.2 Composite processing, types, and their environmental applications

As the dilemma related to the environment is a somewhat complex, environmental engineers and researchers are unfolding the solution with the help of biology and chemistry. Currently the point of convergence on which engineers meet are issues like disposal or recycle of solid waste, peoples' health, global warming, and controlling air and water pollution (Tara, Siddiqui, & Rathi, 2019). All are perilous to ecosystems, conclusively affect human health, and are often found to be serious. One of the rapidly increasing problems globally is water pollution (Chaudhry, Ahmed, Siddiqui, & Ahmed, 2016).

Water is the most important resource on the planet and is very useful for humans in many ways, however its quality is deteriorated rapidly with the advancement of modern technologies. The main sources include agriculture, landfill, industrial effluents, etc. Industrialization is the origin of the excessive water pollution caused by the release of dyes, harmful chemicals, and metal ions into water bodies (Siddiqui, Manzoor, Mohsin, & Chaudhry, 2019). Plastics are the most durable and are used in applications due to their good structure, mechanical strength, low cost, and easy availability. On the other hand, they are a rapidly growing sector that produces solid waste. Plastics originate from a petroleum-based polymer that is nonbiodegradable and often dumped in landfills. Concerned about the ecosystem, and important issues of biodegradability and sustainability of materials, presently researchers, scientists, and engineers are working together to produce many types of materials or combining different materials that are helpful in reducing environmental pollution and these materials are described as composites (Siddiqui & Chaudhry, 2017; Yu, Lee, & Kim, 2009). Composites are made up of two or more discontinuous phases called reinforcement or reinforcing material embedded into a continuous phase called a matrix to achieve superior properties of materials, which are not present in the individual components (Camargo, Satyanarayana, & Wypych, 2009).

10.3 Classification of composites

Composites can be classified based on their matrix and reinforcement. The matrix can be of various kinds including metal, polymer, and ceramic and reinforcement is provided by fiber, both particulate and structural. Current developments in this field of nanotechnology composites have resulted in the term nanocomposites. Nanocomposites are similar to conventional composites but their component dimensions are around 10^{-9} m, which makes them differ from traditional composites. Nanocomposites have a high surface to volume ratio because the particle size is fine and they have a high aspect ratio (Paliwal & Cherkaoui, 2012).

10.3.1 Classification of composites based on matrices

When materials are in fibrous form high strength is only achieved when the fibers are bonded with a suitable matrix. The matrix combined with fiber prevents abrasion, reduces blemishes, and provides strength to keep the fiber in place. The quality of the matrix depends on its ability to deform easily when a load is applied, transfer this load to fibers, and uniformly distribute stress concentration (Pankaj, Rav, Kumar, & Kumar, 2015). Fig. 10.1 summarizes three main classifications of composite matrices. Descriptions of each classification are given in the subsequent section.

10.3.1.1 *Classification based on metal matrices*

When compared with organic matrices, metal matrix composites have some attractive properties. These include that they can withstand elevated temperatures, and have toughness, erosion resistance, transverse strength, good electrical conductivity, high thermal conductivity, etc. However, the major drawback of metal matrix composites is their higher densities and another disadvantage is that fabrication of such types of composites requires high energy (Seshan, Guruprasad, Prabha, & Sudhakar, 1995).

10.3.1.2 *Classification based on polymer matrices*

Various types of polymeric materials include thermosetting and thermoplastics, such as epoxy, polyvinylchloride, polystyrenes, polycarbonate, unsaturated polyester, nylon, and also polymer blends that are combined with fibers, whiskers, small disks, or particles; glass and carbon are used as matrix materials for composites. High porosity and the high surface area of nanoparticles are the most attractive features of the polymer matrix, enhancing the physical and chemical properties of the composite. Two principles that should always be kept in mind while forming composites are that the interaction of the polymer matrix and reinforcement must be very efficient and the other is that the nanoparticle distribution in the matrix must be very effective (Ciprari, Jacob, & Tannenbaum, 2006).

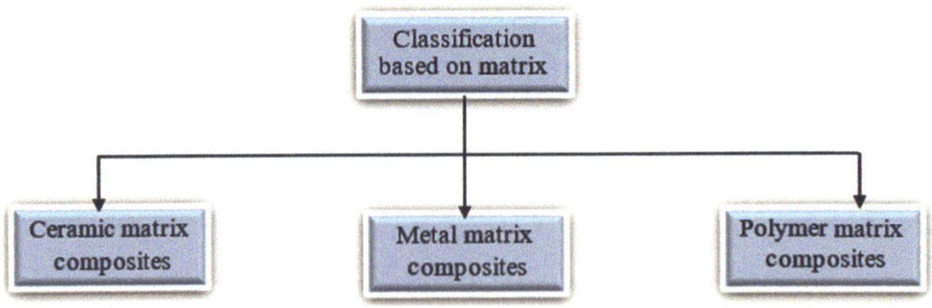

FIGURE 10.1 Classification of composites on matrix (Fatima, Rathi, Ahmad, & Chaudhry, 2019).

10.3.1.3 *Classification based on ceramic matrices*

The term ceramic comes from the Greek word "keramikos," which means pottery. Ceramics exhibit strong ionic and to some extent covalent bonding, and have crystalline nature. Ceramics have two forms being inorganic and nonmetallic hard materials such as alumina and silicon carbide (SC) and are advantageous in very high-temperature applications. Since ceramics have poor mechanical properties like tension and shear, most applications as reinforcements are in the particulate form, for example, zinc and calcium phosphate. Ceramic matrix composites are used in very high-temperature environments, these materials use a ceramic as the matrix and reinforce it with short fibers or whiskers, such as those made from SC and boron nitride (Camargo et al., 2009).

10.3.2 Classification of composites based on reinforcement

As mentioned earlier, composite materials can also be classified according to the type of reinforcement used in the formulation. Fig. 10.2 summarizes the classification of composites based on the type of reinforcement used. Details for each reinforcement classification are given in the subsequent sections.

10.3.2.1 *Fiber reinforced composites*

Composites having the dispersed phase is in the form of fibers are classified as whiskers, fibers and wires based on their diameter. Fibers are of glass fibers, silicon carbide fibers, high silica and quartz fibers,

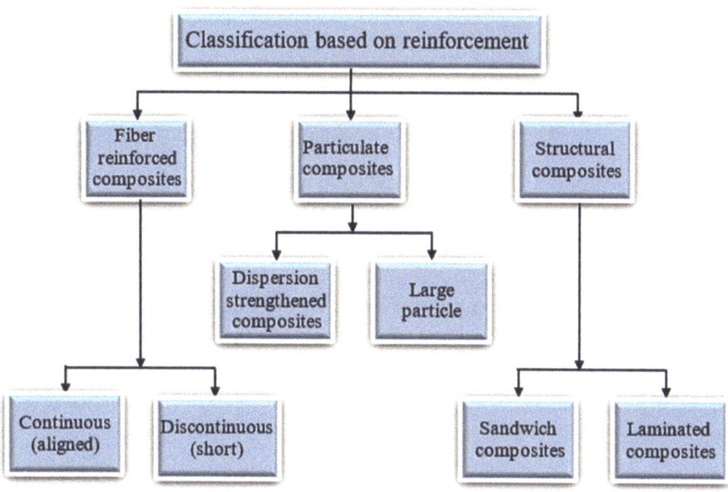

FIGURE 10.2 Classification based on reinforcement (Camargo et al., 2009).

alumina fibers, metal fibers and wires, graphite fibers, boron fibers, etc. The volume fraction varies from a few percent to as high as 70%. The use of fiber reinforcement is done to obtain high strength and high modulus.The properties of fiber augmented composites depend on its fiber length (short and long), orientations of fiber (continuous or discontinuous and aligned or randomly oriented), shape of fiber, composition, and application.

Hence it is necessary for the fibers to possess higher modulus than the matrix material so that the load is transferred to the fiber from the matrix more effectively. Reinforced composites are used for a number of industrial purposes because of their high specific strength and stiffness. Due to their excellent structural performance, the composites are also used in tribiological applications (Nakamura, Goda, Noda, & Ohgi, 2009).

10.3.2.2 *Particle reinforced composites*

In this type of composite, fine particles of sizes ranging from 0.01 to 0.14 μm are dispersed in a matrix, and as a result this material has very high strength and concentration varying from 1% to 15% by volume. To attain high strength and stiffness composites must have good bonding between the constituent materials, that is, the matrices and reinforcement. The cause of the good strength is their atomic level interactions. Reinforcement action helps to achieve good mechanical properties in composites. Composites made by dispersion of fibers have extremely tiny particles that are distributed to a great extent and well bonded in a matrix material, thus providing great strength. Composites with long and larger particles are not as well dispersed as smaller-particle composites (Fatima et al., 2019).

10.3.2.3 *Structural reinforced composites*

This type of composites ranges from those with 1-200 μm particles dispersed in the matrix. Wood is the best model for showing laminated composites. These are two-dimensional sheets of material containing a layer of materials bonded together. There are now laminated composites with complex structures, for example, metal–plastic laminates.

Methods of processing includes precoating or cladding and also adhesive bonding methods. Two sheets of materials are bonded with the help of an adhesive. These sheets are the sources of stiffness and strength in the composites which are composed of aluminum alloys, steel, and fiber reinforced plastics. The thickness of the sheets are decided so that they can withstand the stresses generated by the loading conditions. Applications of these composites are to cope with the stresses resulting from loading in various domestic areas such as roofs, walls, and floors (Fatima et al., 2019).

10.4 Recent advancements in composites

Composites that are gaining attention worldwide are green composites, because of their environmentally friendly nature and reduced use of petroleum-based materials, utilizing noncommercial and biodegradable resources through processing. They are employed in many engineering areas and have considerable economic significance. Green composites use natural resins or polymers for their processing. They are biodegradable materials which constitute both continuous and discontinuous phases with fully natural materials. Plant fibers are used to enhance the mechanical strength of composites. Commonly used polymers for processing of green composites include polylactic acid, polyhydroxybutyrate, and starch etc. Examples: Cellulose/SPC based resin, Kevlar/SPC based resin (Fatima et al., 2019). Presently researchers and engineers have been diverting attention from conventional synthetic fiber composites to natural fiber composites. These composites possess advantages like noncarcinogenic, weight ratio, and strength. The term natural fiber indicates a broad variety of plant, animal, or mineral fibers (Pankaj et al., 2015).

Composites have many advantages over individual materials as they are resistant to corrosion, chemical degradation, and heat. They are capable of coping with elevated temperatures, with carbon fiber composites being used in the wings of the space shuttle which encountered temperatures of around 2000°C. Apart from the mechanical properties they are also superior in properties such as heat sink, dent resistance, dimensional stability, and reliability against structural failure. Also, they have low cost and low density (Fatima et al., 2019).

Some of the advantages of the eco-friendly natural fiber reinforced thermoplastics are:

1. Biodegradability;
2. Lower greenhouse gas emissions;
3. Available throughout the world in an enormous variety of forms;
4. Employment creation in rural areas;
5. Development of a nonfood economy;
6. Efficient energy consumption;
7. Economic

The usage of agricultural materials for making composites provides a renewable resource, as well as evolution of a nonfood source for the economic development of rural areas. Furthermore, use of natural fibers in composites creates an overall CO_2 balance, as the amount of CO_2 taken up during their growth is matched (apart from the efforts necessary to grow and harvest the fibers) by the CO_2 released during their disposal.

Natural fibers are preferred over fibers which are based on petroleum resources because they reduce the generation of greenhouse gases considerably. The energy required to produce natural fibers is less as compared to glass fibers. The lower density of natural fibers gives them a niche over glass fibers as natural fibers result in fuel-efficient production of composite products, particularly in the automation industry, ultimately leading to a reduction of greenhouse gas emissions. The hemp-based natural fiber thermoplastic composites mat during their useful lifetime have a carbon sequestration and storage potential estimated to be about 325 kg carbon per metric tonne. A net carbon sequestration of 0.67 tonne/ha/year was estimated for a composite containing 65 wt.% hemp fiber. It has been found that replacing 30% glass fiber with 65% hemp fiber in thermoplastic composites produces a net energy saving of 50,000 MJ (about 3 tonne CO_2 emission) per tonne of thermoplastic. Also, by substituting 50% of the glass fiber with natural fiber in automotive applications, 3.07 million tonnes of carbon dioxide emissions and 1.9 million m^3 of crude oil can be saved (Sain & Panthapulakkal, 2004).

10.4.1 Biomaterials

Utilization of fat, oil, and grease (FOG) waste in highly valuable biopolymers and also biochemicals provide greater benefits (Carus, Carrez, Kaeb, & Venus, 2011). Recent research shows that used cooking oil UCO (which is highly reusable) and also free fatty acids (FFAs) have likely uses as an element for biopolymer synthesis such as nontoxic, biodegradable plastics which can alter plastics generated using petrochemical refining sources that are employed for various purposes (Ruiz, Kenny, Walsh, & O'Connor, 2014). Polyhydroxyalkanoates (PHAs) are the biopolymers synthesized using fermentation of bacteria which could be possibly used as a substitute for polymers produced as petroleum by-products (Babu, O'Connor, & Seeram, 2013). The important property of PHAs is that they are biocompatible with carbon dioxide (CO_2) and water (H_2O) using a various microorganisms. Recently, PHAs and its products have been employed in different disciplines such as agriculture, food, and biomedical sectors. PHA is variable depending on the different types of bacteria used in many renewable waste supply chains. The FFA content of grease trap waste (GTW) is below 15%, and for a sustainable supply chain it is required to minimize the impurities found in GTW and lower its moisture content. However, this is an expensive treatment that rises its process cost. Raw material value is very important in the production cost of PHA. Consequently, raw materials should be inexpensive to lower the overall manufacturing cost. By-products from biodiesel using the transesterification process include crude glycerol which is a feasible feedstock for value-added transformation into biochemicals or biopolymers (Luo, Ge, Cui, & Li, 2016).

A recent study has shown that bio-oil produced from waste cooking oil is employed as an asphalt modifier that increases the resistance capacity of surface pavement to thermal cracking and also minimizes the maintenance required (Sun, Yi, Huang, Feng, & Guo, 2016). A recent research shows the importance of GTW coming from restaurants acts as a binder used in the metal injection molding (MIM) process. MIM is a mass production process that produces small and complex parts. The process is associated with developing the supply chain from raw materials, that is, metal powder and various components, as a binder both mixed and injected in a mold forming the desired shapes (Ibrahim, Mohd Amin, Asmawi, & Mustafa, 2016).

10.5 Composites applied in different environmental applications

10.5.1 Composites for eco-friendly engineering applications

The earliest most probable use of composites was bricks made from dried grain stalks and mud in building construction. The oldest manmade composite materials were wattle and daub which were used about 6000 years ago. The integrated material that is used more than any other manmade building material around the globe is concrete (Shaffer, 1993). Most of the composites are fiber glasses, in which minor glass fibers are combined along with a polymeric material, usually an epoxy or polyester.

The glass fiber is hard, tough, and stiff but comparatively brittle. However, the polymeric materials are ductile and weak, but flexible. Therefore the resultant fiberglass composite has the features of being strong, stiff flexible, and ductile in nature (Jadhav, Pandit, Gayatri, Chavan, & Jadhav, 2019).

10.5.1.1 Composites employed in greenhouses

A greenhouse is a structure with walls constructed to modify climatic conditions according to the plants' requirements that are critical for cultivation of plants by creating transparent roofs. Various types of glasses are used for vegetables and flowers. A greenhouse can also contain equipment such as for cooling, lighting, and screening, which may be controlled with the help of a laptop to control the inside atmosphere for vegetable growth by controlling parameters like air temperature and relative humidity to increase production of a specific crop. The most important component in rising trends of greenhouses is ventilation. Cheaper composites with low reflectivity are less effective but economic advantages are gained with the use of antireflective glass (Choudhary, Suriyamoorthy, Moses, & Anandharamakrishnan, 2019).

Fiber reinforced polymer (FRP) composites have been developed as structural frameworks to construct greenhouse buildings. FRP composites are beneficial material alternatives and ideal for several kinds of frameworks. FRP composites possess various qualities like layout pliability, flame retardance, nonconductivity, low maintenance, inexpensive, UV resistance, and significant strength-to-weight ratio that make them suitable for the construction industry. The constituents of greenhouses are building components, such as square or circular tubes, stairways, angles, channel, I-section beams, gratings, and panels (Choudhary, Suriyamoorthy, Moses, & Anandharamakrishnan, 2019).

Furthermore, FRP has also been applied in the form of composite film for optical transmission of greenhouse buildings. When compared with other polymers, it is found that low-density polyethylene and linear low-density polyethylene (LLDPE) have superior properties for reducing environmental pollution (Myriounis, 2009). In general, LDPE is relevant for numerous materials such as trash bags and for greenhouse films, while LLDPE is excellent for farming purposes.

Both LDPE and LLDPE films are noticeable due to the fact that they are frequently used as covering to display a moderate window for light that is beneficial for plant life and used in photosynthesis. Thermal energy is frequently stored inside a greenhouse during the day hours and utilized at night, in winter this can help in providing constant development inside the greenhouse for vegetable cultivation (Dunne, Desai, & Sadiku, 2017).

10.5.1.2 Composites employed as sound absorbers

There has been an increase in the demand for reducing inside noise or sound in various transport vehicles such as railways, aerospace, and automobiles by effective use of higher sound absorption materials, which can absorb sounds in the elevated frequency range due to concerns with the safety and comfort of passengers. With the help of sound-absorbing materials such as mineral fibers, foam, and their composites, sandwich composites are widely used in aerospace, construction, and transportation and also in load-carrying parts due to their excellent strength-to-weight ratio (Thomas, 2010).

10.5.1.3 Composites employed in wind turbines for renewable energy

Composites play a vital role in utilization of renewable wind energy. Wind is an eco-friendly and clean source of energy. Composites used in the wind energy industry are mostly divided into two components: the nacelle and the blades. Turbine manufacturers now use hybrid composites of carbon or glass fiber. They are focusing on them because of their light weight, cost-effective, and eco-friendly nature for the construction of turbine blades (Choudhary, Suriyamoorthy, Moses, & Anandharamakrishnan, 2019). The application of composite materials in wind turbine blades is due to a number of physical (heat, humidity) and chemical (corrosion resistant) properties (Zhou, et al., 2014).

10.5.1.4 Composites employed in construction

Both bamboo and jute are natural fibers used for producing composites fabricated with the help of a polymer resin. They act as a substitute for wood and are helpful in reducing the use of wood in various domestic as well as commercial purposes such as roofing, ceiling, furniture, wardrobe, cupboard, and doors. Due to their organic low density some extraordinary biocomposite matrices can be manufactured that have excellent attributes (Das, Basak, Pandit, & Singha, 2018).

10.5.1.5 Composites employed in marine environments

Composite materials were introduced to marine environments from the late 1950s due to its various applications such as in sailing boats or ships, motor vessels, pontoons (flat-bottomed boats), and other buoyancy applications. The application of FRP composites to maritime crafts was initially pressed by a need for little, good, corrosion resistant long lasting naval boats. Many of these first applications are pushed by the necessity to overcome corrosion problems came across with steel or maybe environmental degradation suffered by wood.

A rising an element in boat construction will be performed by strength, lightness, durability and ease of manufacturing process. In spite of the new composites, FRP composites are increasingly prevalent for a long time to come in boat or ship production. Technology science and composite materials are in fact improving carbon nanotube and epoxy combined and merged by variety of various fresh composites (Choudhary, Suriyamoorthy, Moses, & Anandharamakrishnan, 2019).

10.5.1.6 Composites employed in aerospace

Numerous types of composites are applied in the aviation sector. For example, composites were initially used for defense aircraft in World War II. Presently they also are employed in the manufacture of private airplanes and modern commercial aircraft. The three most important types of reinforcement materials commonly used are glass fiber, aramid, and carbon fiber.

10.5.1.7 Composites employed in solar panels

The world's most abundant renewable source of power is solar energy. It is possibly surprising that this is looked at to replace the nonrenewable fossil fuels. Recent strategies are entirely dependent on semiconductor materials like photovoltaic and photocatalytic. Considerable work was based on the enhancement of semiconductor/biomolecular composites for changing sunlight to electricity and fuels. In fact, a broad variety of organic or inorganic nanocomposites are frequently used in making devices for phone systems to enhance their solar energy storage and for electrical charge transportation (Choudhary, Suriyamoorthy, Moses, & Anandharamakrishnan, 2019).

10.5.2 Composites developed from food wastes

FW is waste material rejected following human consumption that is later expelled, lost, deteriorated, or contaminated. The Food and Agriculture Organization specifies loss of food (FL) as any alteration in the quality, edibility, beneficial, and also eminence of food substances to stop it from being consumed by people. This definition describes the postharvest period from the farm to the end user (Gustavsson, Cederberg, Sonesson, Robert, & Alexander, 2011; Gustavsson, Cederberg, Sonesson, Van Otterdijk, & Meybeck, 2011).

According to the European Commission (EC) (2014), it has three groups which are:

- Food losses: Food products lost during the processing;
- Inevitable FW: These are food commodities that perish during the human consumption stage such as banana skins and fruit cores;
- Evitable FW: Those food products that could have been eaten, but were wasted during the human consumption period.

As compared with the various food production phases and distribution, Gustavsson, Cederberg, Sonesson, and Robert et al. (2011), Gustavsson, Cederberg, Sonesson, and Van Otterdijk et al. (2011) split FW into five sources where these losses generated are agricultural production, postharvest handling, storage, processing, transportation, and consumption (Gustavsson, Cederberg, Sonesson, & Robert et al., 2011; Gustavsson, Cederberg, Sonesson, & Van Otterdijk et al., 2011). Also, the FL was observed at all these stages including production, processing, and postharvest (PH) losses in the supply chain. Meanwhile, Parfitt et al. defined FW as the losses that happen between the trade and final consumption periods and its generation is connected to the conduct of retailers and ultimate consumers (Parfitt, Barthel, & Macnaughton, 2010).

FW is defined as "any food, and inedible parts of food, removed from the food supply chain to be disposed". These are comprised of compost, crops either cultivated or nonharvested, anaerobic utilization, production of bioenergy, cogeneration, carbonization, disposal to sewer, landfills, or that sent to water bodies. Some food materials suitable for consumption of people, but which does not reach the consumer, is described as FW. The organic materials produced for the non-food production chain are not considered as FW (Östergren et al., 2014). There is overlap between the definition of FL and FW. In Europe, FW consists of about 45% of the whole solid waste generated in municipal solid waste and approximately 55% generated from developing countries (Troschinetz & Mihelcic, 2009). Presently the methods used in FW organizations are feed for animals, mostly composed of organic fertilizer, anaerobic utilization, carbonization, and landfill (Adhikari, Barrington, Martinez, & King, 2009).

In developed countries there is slightly higher production of FW as compared to developing countries, which may be associated with the life style of the people in countries with high living standards usually resulting in higher FW generation (Lipinski et al., 2013). Consequently, this may result in the production of huge amounts of FW to meet the demands for food quality, such as a higher numbers of ingredients needed to enhance the quality of food. Moreover, consumers have also impacted the quantity of FW produced by retailers. Presently no method has been sustained for eco-friendly utilization of FW. Therefore, FW may be used in the production of biocomposite using the sustainable methods of FW management. This will assure an increase in economic value. Biocomposites made from different FW and their functions are described in Table 10.1.

10.5.2.1 Vegetable waste

Potato is the most highly consumed vegetable throughout the globe. A by-product of the potato is its peel which is produced during processing of the potato. Sugumaran et al. used potato peel powder as a filler material in polypropylene (PP) matrix in various proportions of potato peel powder ranging from 10% to 40% to produce biocomposite and inspected the physiochemical and thermodynamic properties.

The flexural strength of biocomposite was raised from 30% to 40%. With the inclusion of potato peel powder as a filler material, the flexural strength increased (Sugumaran et al., 2015). PepsiCo UK studied the feasibility of utilization of starch from potato peel for the development of crisp packaging material Rogols et al. invented an adhesive and binder material using potato peel for production of a packaging material. They stated that potato peel was employed in numerous applications in different fields such as paper manufacturing, tiling and paneling adhesives, in food industry binder is used for the development of composite food, meat analogue, vegetable analogue and in production of protein and fiber rich food product (Rogols, Sirovatka, & Widmaier, 2002).

TABLE 10.1 Different food wastes, their respective biocomposites, and functions.

Food waste	Type of biocomposite	Functions	References
Potato peel	Polypropylene—potato peel biocomposites	Enhanced physicomechanical and thermal properties	Sugumaran, Vimal, Kapur, & Narula (2015)
Potato peel	Potato peel powder—low-density polyethylene biocomposite	Enhanced tensile strength, flexural strength, and flexural modulus	Sugumaran, Prakash, Arora, Kapur, & Narula (2017)
Potato peel	Potato peel waste fermentation residue	Enhanced biodegradation rate	Wei, Liang, & McDonald (2015)
Potato peel	Cellulose nanocrystals	Improved mechanical and barrier properties	Chen, Lawton, Thompson, & Liu (2012)
Sugar beet	Biocomposite with polyhydroxybutyrate	Enhanced tensile modulus	Dufresne, Cavaille, & Vignon (1997)
Mango seed	Methyl cellulose	Adhesive agent	de los Santos-Villalobos et al. (2011)
Papaya seed	Papaya seed powder biocomposite	Biosorbents to remove heavy metals	Gilbert, Emmanuel, Adebanjo, & Olalere (2011)
Durian peel	Biocomposite	Insulator, stabilizer	Amid & Mirhosseini (2014)
Mangosteen peel	Composite	Biosorbents to remove heavy metals	Zein, Suhaili, Earnestly, & Munaf (2010)
Pineapple peel	Hydrogel and polyvinyl pyrrolindone composite	High absorption capacity of heavy metals	Bardiya, Somayaji, & Khanna (1996)
Jackfruit peel flour	Polyvinyl alcohol composite	Enhanced biodegradation rate	Ooi, Ismail, Abu Bakar, & Aziz (2011)
Jackfruit peel	Activated carbon	Enhanced rate of absorption of heavy metals	Inbaraj & Sulochana (2004)
Apple pomace	Films with polyvinyl alcohol	Enhanced thermal stability, decreased tensile strength	Gaikwad, Lee, & Lee (2016)

(Continued)

TABLE 10.1 (Continued)

Food waste	Type of biocomposite	Functions	References
Banana/ sisal	Polyester resin	Reduced the effective thermal conductivity	Idicula et al. (2006)
Coffee parchment and husk	Polyhydroxybutyrate composite	Enhanced thermal stability, tensile strength	Reis et al. (2015)
Tea mill waste	Polypropylene and high-density polyethylene	Increased the thermoplastic matrix, reduced the tensile strength, flexural strength, and impact strength of the produced composites	Dönmez Çavdar, Kalaycioğlu, & Mengeloğlu (2011)
Chicken feather	Polylactic acid, polybutyrate adipate terephthalate and PLA/ thermoplastic copolyester composite	Enhanced thermal and mechanical properties, wettability	Aranberri, Montes, Azcune, Rekondo, & Grande (2017)
Eggshell	Polypropylene biocomposite	Enhanced tensile and reinforcement properties	Toro, Quijada, Yazdani-Pedram, & Arias (2007)
Rice husk powder	Natural rubber latex foam	Reduced the soil burial taste	Ramasamy, Ismail, & Munusamy (2013a, 2013b)
Rice husk flour	Bio-flour filled biodegradable poly (butylene succinate)	Biocomposites as "green composites" or "eco-materials"	Yang, Kim, Park, Lee, & Hwang (2006)

In a further study, potato peel waste fermented residue fibers were utilized as filler and polyhydroxybutyrate (PHB) as matrix for the production of biocomposites (Wei et al., 2015). Dufresne et al. generated a film using cellulose microfibrils extracted from the pulp of sugar beet using mechanical and chemical treatments. Pectin is used as a binder between the cellulose microfibrils in the preparation of the film. They stated that the tensile strength was affected by the moisture content and pectin purity (Dufresne et al., 1997).

10.5.2.2 *Fruit waste*

Mango waste is divided into two main categories, seed and peel, generally ranging from 14% to 22% in the form of seed and 11%–18% in the form of peel (Mitra, Pathak, Lembisana Devi, & Chakraborty, 2010). Cellulose found in mango seed is the best source for the development of methyl cellulose and is used as an adhesive in various applications in food, civil construction, and pharmaceutical industries (de los Santos-Villalobos et al., 2011). Mango seed is also used to form malachite green dye which is used in food as a coloring agent, and also in the textile, paper, and acrylic industries (Franca, Oliveira, Saldanha, Santos, & Salum, 2010). Furthermore, it is used in the development of biosorbents to minimize heavy metal ions like chromium ions (de los Santos-Villalobos et al., 2011). Papaya waste is composed an equal amount of seed and peel at about 10%–20% (Parni & Verma, 2014). Papaya seeds have both food and nonfood applications, they are an excellent source of protein and fiber, and are appropriate for various food applications (Santos, Abreu, Freire, Queiroz, & Mendonça, 2014).

Papaya seed powder is also used in various dyes such as crystal violet (CV) and methylene blue (MB), but mostly for the development of biosorbents to minimize heavy metal ions such as lead and cadmium ions. Durian is a tropical fruit known for its nutritional value. The wastes generated from durian fruit are around 85% peel and 16% seed. Peel of durian fruit was utilized for the generation of different types of biocomposite with the use of pectin, polysaccharide gel, and fiber (Maran, 2015).

Khedari, Charoenvai, and Hirunlabh (2003) manufactured a particle board developed from durian fruit peel and used coconut coir which is an insulation material. Durian fruit seed was also used in the formation of a stabilizer to stabilize water in oil and vice versa (Amid & Mirhosseini, 2014). Ninety percent jackfruit waste was used in biofilm production, biosorbent, biohydrogen, and also for activated carbon, while 10% was used for extraction of pectins (Ooi et al., 2011). Also, jackfruit peel flour significantly increases degradability and reduces the tensile strength of polyvinyl alcohol (PVA). Jackfruit flour biocomposite and low-cost biosorbents are also prepared from jackfruit peel flour for removal of methylene (Hameed, 2009) and rhodamine (Jayarajan, Arunachalam, & Annadurai, 2011).

Apple pomace is the residue produced after extraction of juices and is used for the generation of active antioxidant films for food packaging material incorporation of PVA into various proportions ranging from 1, 5, 10, and 30% (w/w) of apple pomace (AP). The incorporation of AP increases the entire phenolic compounds content and antioxidant characteristics of packaging film material. The tensile strength of film is reduced as the AP percentage increases as does elongation at break, but the thickness is increased significantly (Choudhary, Suriyamoorthy, Moses, & Anandharamakrishnan, 2019).

Idicula et al. (2006) developed a biocomposite using banana, sisal, and fiber of pineapple leaf. The prepared biocomposite was utilized as the filler material for polyester resin. They studied the impact of the fibers used and the individual constituents of the composite on the thermal conductivity. Mixing of fiber like banana and sisal in the matrix of polyester reduced the biocomposite thermal conductivity. Polyester–glass composites have 50% higher thermal conductivity as compared with polyester−-pineapple leaf fiber (Idicula et al., 2006).

10.5.2.3 *Coffee and tea waste*

FW is produced from processing of tea and coffee and is mostly composed of lignocellulosic particles of lower products from coffee industries. Coffee parchment and husk are used as filler for the reinforcement with PHB to generate biocomposites. The impact of filler and its quantity affects the thermodynamics, mechanical strength, microstructure, and physical characteristics of the composite, resulting in significant tensile and izod strength increases as compared with pure polymer and with the incorporation of 20% coffee parchment, but with 10% coffee parchment no variation was found. The property of water absorption was found to be directly proportional to the coffee waste particle quality added for biocomposite production (Reis et al., 2015).

Donmez Cavdar et al. studied tea mill waste fibers (TMWF) used with a plastic PP matrix for the development of a thermoplastic composite and high-density polyethylene (HDPE). It was found that TMWF behaved as a lignocellulosic filler in biocomposite with the help of a single screw extruder. It was noted that the concentration of TMWF was directly proportional to the tensile and flexural strength, water absorption, and swelling thickness in the thermoplastic composite (Dönmez Çavdar et al., 2011).

10.5.2.4 *Animal-based food waste*

Aranberri and co-workers invented polymeric materials using chicken feather and various biopolymers for the fabrication of biocomposites, these are polylactic acid (PLA), polybutylene adipate terephthalate, and PLA/thermoplastic copolyester. A considerable effect has been found on the different kinds of eco-friendly materials through their physical, mechanical strength and thermodynamic characteristics. The biocomposite produced may be used for a panel component, flooring, and building structural materials. This is an eco-friendly method for utilization of CFs (Aranberri et al., 2017).

Eggshell nanopowder (ESNP) and protein extracted from soy were also used in the production of a hybrid nanocomposite. A ball mill was used for size reduction of the shell after sono-chemical treatment. The size of ESNP ranges from 20 to 40 nm as detected by transmission electron microscopy (TEM) and X-Ray diffraction (XRD).

10.5.2.5 Food grain waste

The major by-products or waste of rice industries are rice husk, bran, and germ. Rice husk flour (RHF) is used as a reinforcement combined with polybutylene succinate for preparation of biocomposites. According to Ramasamy et al., the rice husk powder characteristics, when mechanically filled with natural rubber latex (NRLF), the tensile strength developed increased at 2.5 pphr to 5, 7.5 and 10 pphr. Hardness of RHF increased as NRLF increased in bio-composites (Ramasamy et al., 2013a, 2013b).

10.5.3 Magnetic composites as environmental superadsorbents for dye separation

The land consists of 75% water, but only 3% is fresh water out of it and 1% is available for people to use (Ragunath, 2006). Water is crucial for the survival of all humans, animals, and plants. However, the supply from ground water required for humanity is reducing day by day. The major source of water pollution of most freshwater resources is industrialization. The soils and water bodies including lakes and rivers have become polluted with a rapid increment of diverse types of waste, extraction of crude oil and gasoline and also during refining in the petroleum industry, untreated waste or partially treated waste sent to municipal sewage, solid waste disposal at landfills generating leachate, toxic effluents from industries, discharge of nuclear reactors, and waters from modern agricultural activities like fertilizers and pesticides used to increase crop productivity (Dozier, 2005). Polluted water can put millions of lives in danger from water-related diseases. Fig. 10.3 shows the sources of water contamination.

In some conditions, water can be polluted naturally, however human influence is generally responsible for water pollutantion (Carneiro, Umbuzeiro, Oliveira, & Zanoni, 2010). One of the most significant pollutants include dyes which are generated from the wastes of the textile, paper, food, plastic, and cosmetic industries (Rafatullah, Sulaiman, Hashim, & Ahmad, 2010).

Dyes produced by many industries are toxic to such an extent that they are carcinogenic in nature (Mekkawy, Ali, & El-Zawahry, 1998; Oxspring, McMullan, Smyth, & Marchant, 1996). The production of dyes has reached enormous amounts, with more than 100,000 commercial dyes creating 7×10^5 tonnes of dyestuff are produced per year (McMullan et al., 2001; Pearce, Lloyd, & Guthrie, 2003). Consequently, a high amount of waste in the form of colored water is generated from these industries.

Presently, the removal of dyes from effluent has been a great challenge for environmental engineers. There are various types of techniques used for the removal of dyes from effluent, such as oxidation, adsorption, flocculation, and electrolysis (Crini, 2003; Wawrzkiewicz, 2012). Most of these are complicated, except adsorption, which is an

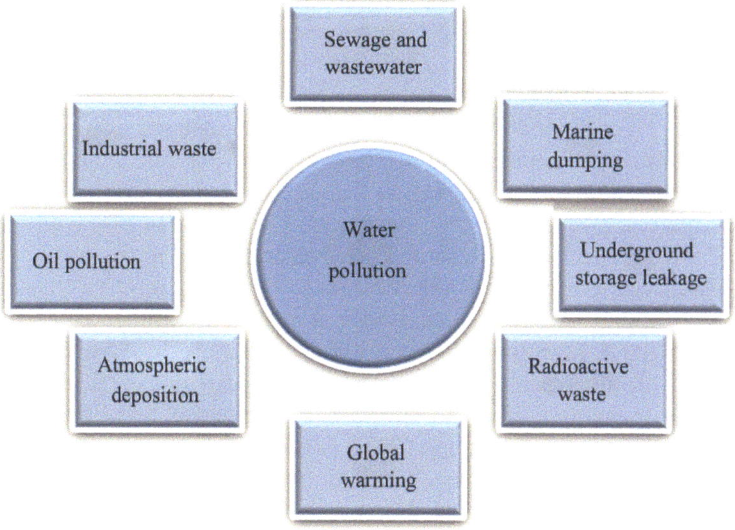

FIGURE 10.3 The sources of water contamination (Sivashankar, Sathya, Vasantharaj, & Sivasubramanian, 2014).

accessible, effective, and economical method (Shuang et al., 2012). Small adsorbents are used in wastewater treatment. The small size of the adsorbents increases their surface area, which increases the rate of adsorption, with the only problem being that it is difficult to separate the water after attaining saturated adsorption.

To overcome this complication, the technique of magnetic separation is adopted in wastewater treatment, because in this technique flocculent is not generated and the capacity of the discharge can be controlled within a short period of time (Luo & Zhang, 2009; Rocher, 2008). A magnetic separation method using MCs (magnetic composites) and MNCs (magnetic nanocomposites) has been achieved widely for use in the separation of dyes because of its extraordinary advantage in increasing the sequestration speed and ultimately improving the wastewater treatment efficiency (Qadri, Ganoe, & Haik, 2009).

In general, there are many types of materials used to produce industrial dyes. The colored organic compounds known as dyes are used to inculcate color into various substances such as paper, leather, fur, hair, drugs, cosmetics, waxes, greases, plastics, and textile materials (Amith, Sivanesan, Kannan, & Tapan, 2007; Monash & Pugazhenthi, 2009; Samanta & Agarwal, 2009). Dyes play a vital role in the textile industry, however considerable quantities are applied for coloring different types of commodities such as leather, paper, plastics, and crude oil products.

Generally, the dyes which are applied in the textile industries are basic dyes, acid dyes, reactive dyes, direct dyes, azo dyes, mordant dyes, vat dyes, disperse dyes, and sulfur dyes (Demirbas, 2009). Azo derivatives are the major group of dyes employed in these industries (Forgacs, Cserhati, & Oros, 2004). Dyes are broadly categorized into three classes: cationic, anionic, and nonionic dyes. Cationic dyes are basic in nature, while the anionic dyes include direct, acid, and reactive dyes (Mishra & Tripathy, 1993).

The dye compounds not only have harmful effects, but they also make the water aesthetically unpleasant. The dyes applicable to the textile industry include many structural varieties such as acidic, reactive, basic, disperse, azo, diazo, anthraquinone-based, and metal complex dyes (Banat, Nigam, Singh, & Marchant, 1996). Extreme exposure to dyes has caused various health-related issues such as increased heart rate, vomiting, shock, Heinz body formation, cyanosis, jaundice, quadriplegia, and tissue necrosis in humans (Vadivelan & Kumar, 2005). Metanil Yellow dyes have the tendency of causing tumors (Ramachandani, Das, Joshi, & Khanna, 1997) and instability of enzymes inside the human body (Das, Ramchandani, Upreti, & Khanna, 1997).

They can alter the expression of genes inside the body (Gupta, Sundarrajan, & Rao, 2003). Upon oral consumption they can cause toxic methamoglobinaemia and cyanosis, while exposure to the skin can result in allergic dermatitis (Chandro & Nagaraja, 1987; Hausen, 1994; Sachdeva et al., 1992). The oral consumption of dyes by animals can lead to the development of testicular lesions as a result of which the seminiferous tubules suffer injury and also a reduction in the rate of development of sperm (Tiwari, 2012).

The dye working mechanism can be explained through the sorption concept. Sorption is the process most extensively used for the removal of dyes and is efficiently applicable in wastewater treatment plants. Sorption process can be categorized into two major groups: (1) physisorption and (2) desorption. If the molecules which are adsorbed are physical in nature with a firm surface, then the adsorption is called physisorption, while if the forces of attraction are caused by chemical bonding then it is called desorption. Usually in physical adsorption Van der Waals forces of attraction are present between the adsorption molecule and the solid surface. When chemisorption is taking place then bonding forces will cause difficulty in separation of the adsorbate and the surface (Gupta & Suhas, 2009; Yadla, Sridevi, & Chandana Lakshmi, 2012).

The most significant aspect of an adsorbent is that the amount of adsorbate which is amalgamated on the adsorbent is calculated using adsorption isotherms. The adsorbent must have several essential features like the structure must be porous in nature, size should be compact, and the time required to attain adsorption equilibrium should be the minimum so that it can be used to separate the dye in a short time (Gupta & Suhas, 2009; Nguyen & Juang, 2013). The adsorbents which are available in nature are modified as MCs/MNCs which overcome the limitation of separation (Solis, Solis, Perez, Manjarrez, & Flores, 2012). The adsorbents can be easily separated from the solutions when imbuing with a magnet undergoing an external magnetic force (Pham et al., 2011; Slunjski, Cadee, & Tattersall, 2000; Yang, Yuan, Lu, & Cheng, 2009). Further exploration of different factors such as adsorbent and adsorbate concentrations, pH, activators and depressants, temperature, and agitation are sensible in determining the adsorption of dyes with incorporation of magnetic composites (Salleh, Mahmoud, Karim, & Idris, 2011). Fig. 10.4 gives an outline of the adsorption processes using MCs/MNCs.

There are many techniques used for the synthesis of MCs/MNCs. Co-precipitation is one of the methods used for the synthesis of magnetic nanoparticles (Drbohlavova et al., 2009; Wua et al., 2011; Zhao, Qiu, & Huang, 2008), and thermal decomposition (Daengsakul et al., 2009), solvothermal (Li, Wei, Liivat, Zhub, & Zhu, 2013a), microemulsion (Drmota, Drofenik, Koselj, & Žnidaršič, 2012), sono-chemical (Wu, He, Chen, Tang, & Nie, 2007), chemical vapor deposition (Morales et al., 2013), microwave assisting (Safarik, Horska, Pospiskova, Maderova, & Safarikova, 2013), polyol process (Karipoth, Thirumurugan, & Justin Joseyphus, 2013), hydrothermal (Nejati & Zabihi, 2012), photolysis (Nkosi, Mwakikunga, Sideras-Haddad, & Forbes, 2012), Sol–Gel (Xu et al., 2007), electrochemical (Starowicz et al., 2011), electrodeposition (Banerjee, Roy, Chen, & Chakravorty, 2000), laser pyrolysis methods (Sabino et al., 2004) have been investigated for synthesis of MNCs. Several critical issues have occurred during synthesizing of MNCs including distribution and control of particle size, crystallinity, shape, and alignment (Hyeon, 2003; Rao, Thomas, & Kulkarni, 2007).

These methods help in developing magnetic composites, that are steady at ambient temperature, of the proper size, uniform shape, high rank of monodispersity, nonaggregate, etc. Fig. 10.5 presents a classification of magnetic composite synthesizing techniques (Tiwari, 2012).

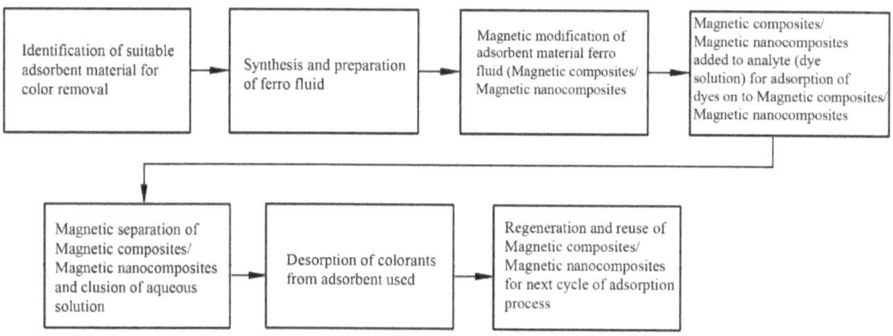

FIGURE 10.4 Outline of the adsorption process using magnetic composites/magnetic nanocomposites (Salleh et al., 2011).

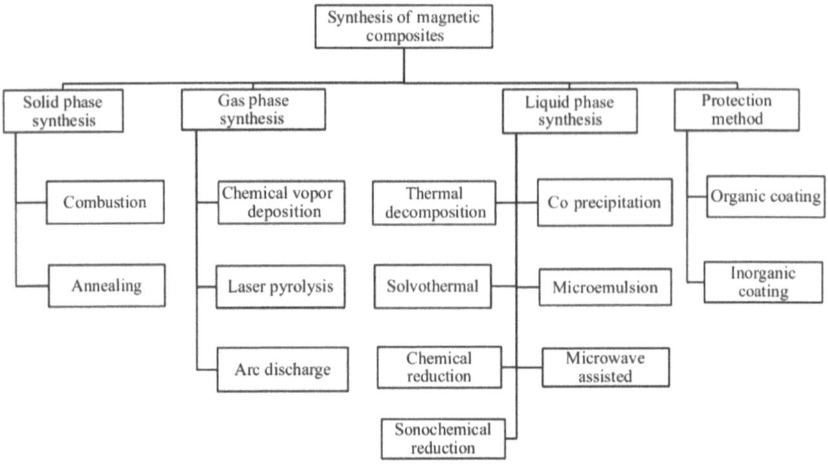

FIGURE 10.5 Techniques used in the synthesis of magnetic composites (Hyeon, 2003; Rao et al., 2007).

10.5.3.1 *Application of magnetic composites as dye adsorbents*

MCs are commonly utilized for the incorporation in the adsorption operation for removal of environmental contaminants. The most important benefit of MCs is that, less time is required for disposal of wastewater without generating pollutants to the environment. Commercially available magnetic elements are quite costly, and so they cannot be used for large scale (Safarik, Horska, Svobodova, & Safarikova, 2012). Magnetic modification of inexpensive adsorbents for highly beneficial materials is worthy of biological and ecological processes (Safarik & Safarikova, 2010).

The magnetic separation provides a desirable path for online separation, where particles with affinity to target species are merged with the heterogeneous solution. After integration with the mixture, the particles are marked, and the species are targeted. These particles are separated from the solution with the help of an externally applied magnetic force. Safarik and Safarikova (2010) employed ferrofluid modified peanut husks (FMMPH) for the removal of dissolved organic dyes from waste. This modified fluid was utilized as an absorbent to investigate the results with various dyes such as bismarck brown, acridine orange, crystal violet and safranin-O.

High adsorption was attained up to 95.3, 71.4, 80.9, and 86.1 mg/g for bismarck brown, acridine orange, crystal violet and safranin-O respectively. Ai, Zhou, and Jiang (2011) worked with a new adsorbent montmorillonite/$CoFe_2O_4$ composite and the magnetic sequestration in a simplistic, single-stage low-thermal refluxing passage. The characteristics, biotechnology, and magnetic attributes of the developed composite were distinguished by various instruments such as XRD, scanning electron microscope (SEM), and vibrating-sample magnetometer (VSM).

The adsorption qualities of the composite were investigated by applying an adsorbate such as MB.

The result indicated that the data obtained from adsorption isotherm was adapted with the Langmuir isotherm and followed the pseudo-second-order kinetic model. The adsorption mechanism was pH dependent. The silanol and aluminol categories on the margin are protonated at pH <4. Hence, the resulting adsorption quantity was proportional to the pH, and varies as variations in pH occurs. Yan, Li, Yang, Li, and Cheng (2013) developed biodegradable magnetic adsorbent utilizing glutamic acid altered chitosan with a single-stage process. The glutamic acid fabrication led to the modification of chitosan magnetic composite microspheres (CS-Glu-MCMs) and was distinguished by various instruments like Fourier transform infrared spectroscopy (FTIR), thermogravimetric analysis, VSM, SEM, and equilibrium swelling experiment. CS-Glu-MCM was utilized as an adsorbent for removal of cationic dyes, MB, CV, and light yellow 7GL (7GL), from aqueous solutions and the adsorption capacity was found to be 180 mg/g for MB, 375.4 mg/g for CV, and 217.3 mg/g for 7GL.

10.5.3.2 Application of magnetic nanocomposites as dye adsorbents

The importance of developing nanoparticles with incorporated magnetic characteristics and extreme adsorption quality for a good blend of compounds, is so that wastewater can be purified by this new treatment (Qu, Pedro, & Qilin Li, 2013). These specific magnetic nanoparticles are currently being investigated as a hopeful plan for identification and use in waste removal (Jung, Lee, & Shinkai, 2011).

Zhou, Jin, Liu, Liang, and Shang (2011) examined magnetic nanoparticles of chitosan modified by ethylenediamine (MNCME) for adsorption from aqueous solution of acid orange 7 (AO7) and acid orange 10 (AO10), and the highest capacity of the adsorption process was found at pH 4.0 for acid orange 7 (AO7) and pH 3.0 for acid orange 10 (AO10). Equilibrium experiments were found to be good with the Langmuir isotherm, and the highest adsorption of MNCME was found to be 3.47 mmol/g for AO7 at 298 K and 2.25 mmol/g for AO10. Singh, Gupta, Singh, and Sinha (2011) developed and distinguished nanocomposite adsorption of CV extracted from water, known as magnetic carbon-iron oxide nanocomposite (MCIONIC).

Four-factor central composite design combined with response surface modeling was employed to enhance CV removal from aqueous solution by applying MNCs and the Langmuir adsorption capacity was found to be 81.70 mg/g of the adsorbent. Konicki, Sibera, Mijowska, Lendzion-Bielun, and Narkiewicz (2013) synthesized magnetic $ZnFe_2O_4$ (MNZnFe)-incorporated microwave aided hydrothermal process for the separation of Acid Red 88 (AR88) from a dissolved water solution. The impact of various factors like initial concentration of the AR88, pH solution, and temperature having values of 10–56 mg/L, 3.2–10.7, and 20°C–60°C, respectively, were evaluated. The prepared magnetic $ZnFe_2O_4$ was characterized using the XRD, SEM, and FTIR methods. Thermodynamic parameters, $\Delta G°$, $\Delta H°$ and $\Delta S°$ indicate that the adsorption of AR88 onto MNZnFe was spontaneous and exothermic in nature. Desorption efficiency of AR88from MNZnFe in acetone and methanol was found to be 63% and 81%, respectively.

Yao et al. (2012) developed a magnetic Fe_3O_4 graphene composite (FGC) that was applied in dye separation of MB and CR from aqueous solution. The structure, surface, and magnetic attributes were examined with the help of SEM, TEM, and ED. The method of chemical deposition was applied, and Fe_3O_4 nanoparticles of 30 mm size were homogeneously dispersed onto graphene sheets. The maximum adsorption capacities of MB and CR on FGC were found to be 45.27 and 33.66 mg/g, respectively.

Nanotechnology is a diversified discipline which also has applications in the area of wastewater treatment (Mak & Chen, 2004). The use of various oxides of metals as adsorbent has been employed in individual or composite formation for the pioneering generation of nano-level environmental cleaning techniques such as Fe_2O_3, Fe_3O_4, TiO_2, and Al_2O_3 (Xiao et al., 2013; Yao et al., 2012).

There are several advantages of using MNCs over MCs for dye adsorbent applications. When comparing traditional natural or manmade adsorbents with macro size, the MNCs have an excellent advantage due of their reduced size, promising greater surface area, and elevated specific surface reactivity. Also, the MNCs can be developed for specific adsorbent separation by altering the functional class of the adsorbent and also by modifying the ionic magnitude of the solution and by changing the active sites of the MNCs (Debrassi et al., 2012; Inbaraj & Chen, 2011; Konicki et al., 2013; Zhou et al., 2011).

The recovery and regeneration of adsorbent have become simple with re-usability of nanocomposites for consecutive use, exhibiting an outstanding efficiency of 3–10 adsorption-desorption cycles (Wu, Feng, Wang, & Wang, 2013; Xia et al., 2013). The temperature required for the regeneration of MNCs is very low and the energy utilized in regeneration is greater than 50% of that of other technologies, low operatingpressure, low pressure drop, its performance has no restriction by orientation and resistance to vibration, that is, it has good flexibility and also durability, rapid adsorption speed, complete recovery, and it does not produce any contamination and hence is recovered efficiently (Deng et al., 2013; Mahmoodi, 2013; Mak & Chen, 2004; Singh et al., 2011; Wu et al., 2013; Xia et al., 2013; Xu, Zhang, Jiang, Reddy, & Yang, 2013; Yao et al., 2012).

The ratio of surface area to volume is high resulting in MNCs possessing of high adsorption quality compared to macroscopic adsorbent

MCs. The greater surface area generates high adsorption and hence comparatively low contact time. These MNCs are magnetic, hence they can be well seprated from aqueous solution by an external magnetic field. Finally, the unique qualities of MNCs and coexistence with recent treatment techniques have created a revolutionary tool in the field of wastewater treatment.

10.6 Conclusion

The demand for composite materials continues to rise because they can be more economical, durable, and lighter weight than traditional materials. They align well with the global trends of sustainability, energy efficiency, and water preservation. The properties of composites enable manufacturers to build sustainable attributes into projects or products that may increase energy efficiency, durability, and the use of materials with a lower environmental impact. The increasing demand for renewable energy, fuel-conserving transportation, and greener buildings opens a world of opportunity for the composites industry. Composites are complex materials. Reinforcement, resin, and additives can be combined in innumerable ways to provide the optimum combination of properties required for a specific application.

This chapter gave an extensive description of composites and their classifications based on the matrix and reinforcement material together with their practical applications. As composites are leading the progress to a clean and green future much emphasis is given to highlighting the environmental applications of composite materials in different areas, such as the environmental benefits of natural fiber reinforced thermoplastics. Various types of FW are utilized in the development of various innovative composite materials. The production of biocomposites from FW may be a sustainable technique for FW management and will assure an increase in its economic value and a reduction in the generation of greenhouse gases. When disposed FW decomposes to form methane, which is a greenhouse gas with a global warming potential 25 times greater than that of CO_2 over a 100-year time scale. Furthermore, the production of biomaterials from FOG in restaurant wastewaters is an interesting and novel prospect for environmental engineers.

Moreover, considering the prime importance and vast applications of dyes in the textile industry, the hazards associated with the usage of dyes should not be overlooked, hence this chapter focused on the adsorption of dyes using MCs and MNCs and their advantages. Hopefully, this chapter will provide an insight for readers to understand the environmental benefits and prospects of composite materials, leading to a clean and green future.

Acknowledgments

The authors would like to thank Malaysia-Japan International Institute of Technology, Universiti Teknologi, Malaysia, and Aligarh Muslim University for the support provided during the writing of this chapter. They also thank S.M. Sapuan and M.R. Mansor for the invitation to write this chapter.

References

Adhikari, B. K., Barrington, S., Martinez, J., & King, S. (2009). Effectiveness of three bulking agents for food waste composting. *Waste Management (New York, N.Y.)*, *29*(1), 197–203.

Ai, L., Zhou, Y., & Jiang, J. (2011). Removal of methylene blue from aqueous solution by montmorillonite/$CoFe_2O_4$ composite with magnetic separation performance. *Desalination*, *266*, 72–77.

Amid, B. T., & Mirhosseini, H. (2014). Stabilization of water in oil in water (W/O/W) emulsion using whey protein isolate-conjugated durian seed gum: Enhancement of interfacial activity through conjugation process. *Colloids and Surfaces B, Biointerfaces*, *113*, 107–114.

Amith, B., Sivanesan, D., Kannan, K., & Tapan, C. (2007). Kinetics of decolorization and biotransformation of Direct Black 38 by C. hominis and P. stutzeri. *Applied Microbiology and Biotechnology*, *74*, 1145–1152.

Aranberri, I., Montes, S., Azcune, I., Rekondo, A., & Grande, H. J. (2017). Fully biodegradable biocomposites with high chicken feather content. *Polymers*, *9*(11), 593.

Babu, R., O'Connor, K., & Seeram, R. (2013). Current progress on bio-based polymers and their future trends. *Progress in Biomaterials*, *2*(8).

Banat, I. M., Nigam, P., Singh, D., & Marchant, R. (1996). Microbial decolorization of textile-dye containing effluents: A review. *Bioresource Technology*, *58*, 217–227.

Banerjee, S., Roy, S., Chen, J. W., & Chakravorty, D. (2000). Magnetic properties of oxide-coated iron nanoparticles synthesized by electrodeposition. *Journal of Magnetism and Magnetic Materials*, *219*, 45–52.

Bardiya, N., Somayaji, D., & Khanna, S. (1996). Biomethanation of banana peel and pineapple waste. *Bioresource Technology*, *58*(1), 73–76.

Camargo, F. H. C., Satyanarayana, K. G., & Wypych, F. (2009). Nanocomposites: Synthesis, structure, properties and new application opportunities. *Materials Research*, *12*(1), 1–39.

Carneiro, P. A., Umbuzeiro, G. A., Oliveira, D. P., & Zanoni, M. V. B. (2010). Assessment ofwater contamination caused by a mutagenic textile effluent/dyehouse effluent bearing disperse dyes. *Journal of Hazardous Materials*, *174*, 694–699.

Carus, M., Carrez, D., Kaeb, H., & Venus, J. (2011). *Policy paper on bio-based economy in the EU: Level playing field for bio-based chemistry and materials* (pp. 04–18). Germany: Nova Institute.

Chandro, S. S., & Nagaraja, T. (1987). A food-poisoning outbreak with chemical dye. An investigation report. *Medical Journal Armed Forces India*, *43*, 291–300.

Chaudhry, S. A., Ahmed, M., Siddiqui, S. I., & Ahmed, S. (2016). Fe(III)-Sn(IV) mixed binary oxide-coated sand preparation and its use for the removal of As(III) and As(V) from water: Application of isotherm, kinetic and thermodynamics. *Journal of Molecular Liquids*, *224*, 431–441.

Choudhary, P., Suriyamoorthy, P., Moses, J. A., & Anandharamakrishnan, C. (2019). Bio-composites from food wastes. *Composites for Environmental Engineering*, 319–345.

Chen, D., Lawton, D., Thompson, M. R., & Liu, Q. (2012). Biocomposites reinforced with cellulose nanocrystals derived from potato peel waste. *Carbohydrate Polymers*, *90*(1), 709–716.

Ciprari, D., Jacob, K., & Tannenbaum, R. (2006). Characterization of polymer nanocomposite interphase and its impact on mechanical properties. *Macromolecules*, *39*(16), 6565–6573.

Crini, G. (2003). Studies on adsorption of dyes on beta-cyclodextrin polymer. *Bioresource Technology, 90*, 193–198.

Daengsakul, S., Mongkolkachit, C., Thomas, C., Siri, S., Thomas, I., Amornkitbamrung, V., & Maensiri, S. (2009). A simple thermal decomposition synthesis, magnetic properties, and cytotoxicity of La0.7Sr0.3MnO_3 nanoparticles. *Applied Physics A, 96*, 691–699.

Das, M., Ramchandani, S., Upreti, R. K., & Khanna, S. K. (1997). MetanilYellow: A biofunctional inducer of hepatic phase I and phase II xenoblastic-metabolising enzymes. *Food and Chemical Toxicology: An International Journal Published for the British Industrial Biological Research Association, 35*, 835–838.

Das, S., Basak, S., Pandit, P., & Singha, A. K. (2018). Hybrid bast fiber strengthened thermoset composites. *Thermoset Composites: Preparations, Properties and Applications, 38*, 112.

de los Santos-Villalobos, S., de-Folter, S., Délano-Frier, J. P., Gómez- Lim, M. A., Guzmán-Ortiz, D. A., Sánchez-García, P., & Peña-Cabriales, J. J. (2011). Critical aspects on the integral management of mango: Flowering, anthracnosis and industrial waste. *Revista Mexicana de Ciencias Agricolas, 2*(2), 221–234.

Debrassi, A., Correa, A. F., Baccarin, T., Nedelko, N., Slawska-Waniewska, A., Sobczak, K., ... Rodrigues, C. A. (2012). Removal of cationic dyes from aqueous solutions using N-benzyl-O-carboxymethylchitosan magnetic nanoparticles. *Chemical Engineering Journal, 183*, 284–293.

Demirbas. (2009). Agricultural based activated carbons for the removal of dyes from aqueous solutions: A review. *Journal of Hazardous Materials, 167*, 1–9.

Deng, J.-H., Zhang, X.-R., Zeng, G.-M., Gong, J.-L., Niu, Q.-Y., & Liang, J. (2013). Simultaneous removal of Cd (II) and ionic dyes from aqueous solution using magnetic graphene oxide nanocomposite as an adsorbent. *Chemical Engineering Journal, 226*, 189–200.

Dönmez Çavdar, A., Kalaycioğlu, H., & Mengeloğlu, F. (2011). Tea mill waste fibers filled thermoplastic composites: The effects of plastic type and fiber loading. *Journal of Reinforced Plastics and Composites, 30*(10), 833–844.

Dozier, M. C., (2005). What is water pollution, Water and Me Series, SCS-2005-02. In *Soil and crop sciences communications*. Texas A&M University System. <http://waterandme.tamu.edu/WaterPollution/waterpollution.pdf>.

Drbohlavova, J., Hrdy, R., Adam, V., Kizek, R., Schneeweiss, O., & Hubalek, J. (2009). Preparation and properties of various magnetic nanoparticles. *Sensors (Basel, Switzerland), 9*, 2352–2362.

Drmota, A., Drofenik, M., Koselj, J., & Žnidaršič, A. (2012). Microemulsion method for synthesis of magnetic oxide nanoparticles. In R. Najjar (Ed.), *Microemulsions—An introduction to properties and applications*. InTech, ISBN 978-953-51-0247-2.

Dufresne, A., Cavaille, J.-Y., & Vignon, M. R. (1997). Mechanical behavior of sheets prepared from sugar beet cellulose microfibrils. *Journal of Applied Polymer Science, 64*, 1185–1194.

Dunne, R., Desai, D., & Sadiku, R. (2017). Material characterization of blended sisal-kenaf composites with an ABS matrix. *Applied Acoustics, 125*, 184–193.

European Commission. (2014). *Food waste and its impacts: European week for waste reduction*. Belgium: European Commission.

Fatima, B., Rathi, G., Ahmad, R., & Chaudhry, S. A. (2019). Composites: Types, method of preparation and application as an emerging tool for environmental remediation. *Composites for Environmental Engineering*, 1–31.

Forgacs, E., Cserhati, T., & Oros, G. (2004). Removal of synthetic dyes from wastewaters: A review. *Environment International, 30*, 953–971.

Franca, A. S., Oliveira, L. S., Saldanha, S. A., Santos, P. I., & Salum, S. S. (2010). Malachite green adsorption by mango (*Mangifera indica* L.) seed husks: Kinetic, equilibrium and thermodynamic studies. *Desalination and Water Treatment, 19*(1–3), 241–248.

Gaikwad, K. K., Lee, J. Y., & Lee, Y. S. (2016). Development of polyvinyl alcohol and apple pomace bio-composite film with antioxidant properties for active food packaging application. *Journal of Food Science and Technology*, *53*(3), 1608–1619.

Gilbert, U. A., Emmanuel, I. U., Adebanjo, A. A., & Olalere, G. A. (2011). Biosorptive removal of Pb2 + and Cd2 + onto novel biosorbent: Defatted Carica papaya seeds. *Biomass Bioenergy*, *35*(7), 2517–2525.

Gupta, S., Sundarrajan, M., & Rao, K. V. K. (2003). Tumour promotion by metanil yellow and malachite green during rat hepatocarcinogenesis is associated with dysregulated expression of cell cycle regulatory proteins. *Teratogenesis, Carcinogenesis, and Mutagenesis* (Suppl. 1), 301–312.

Gupta, V. K., & Suhas. (2009). Application of low-cost adsorbents for dye removal—A review. *Journal of Environmental Management*, *90*, 2313–2342.

Gustavsson, J., Cederberg, C., Sonesson, U., Robert, V. O., Alexander, M., (2011). *Global food losses and food waste*, Dusseldorf, Germany.

Gustavsson, J., Cederberg, C., Sonesson, U., Van Otterdijk, R., & Meybeck. (2011). *A global food losses and food waste* (pp. 1–38). Rome: FAO.

Hameed, B. H. (2009). Removal of cationic dye from aqueous solution using jackfruit peel as non-conventional low-cost adsorbent. *Journal of Hazardous Materials*, *162*(1), 344–350.

Hammond, R. (2007). *The world in 2030: Summary and initial industry response*. Brussels, Belgium: Plastics Europe.

Hausen, B. M. (1994). A case of allergic contact dermatitis due to metanil yellow. *Contact Dermatitis*, *31*, 117–118.

Huda, M. S., Mohanty, A. K., Drzal, L. T., Misra, M., & Schut, E., (2004). Physico- mechanical properties of "Green" composites from poly(lactic acid) and cellulose fiber, *Proceedings of the 10th annual global plastics environmental conference (GPEC 2004)*, February 18 and 19, 2004, Detroit, MI, Paper # 11.

Hyeon, T. (2003). Chemical synthesis of magnetic nanoparticle. *Chemical Communications*, 927–934.

Ibrahim, M. H. I., Mohd Amin, A., Asmawi, R., & Mustafa, N. (2016). Influences of restaurant waste fats and oils (RWFO) from grease trap as binder on rheological and solvent extraction behavior in SS316L metal injection molding. *Metals*, *6*(2), 19.

Idicula, M., Boudenne, A., Umadevi, L., Ibos, L., Candau, Y., & Thomas, S. (2006). Thermophysical properties of natural fiber reinforced polyester composites. *Composites Science and Technology*, *66*(15), 2719–2725.

Inbaraj, B. S., & Sulochana, N. (2004). Carbonised jackfruit peel as an adsorbent for the removal of Cd (II) from aqueous solution. *Bioresource Technology*, *94*(1), 49–52.

Inbaraj, S., & Chen, B. H. (2011). Dye adsorption characteristics of magnetite nanoparticles coated with a biopolymer poly (_-glutamic acid). *Bioresource Technology*, *102*, 8868–8876.

Jadhav, A. C., Pandit, P., Gayatri, T. N., & Jadhav, N. C. (2019). Production of green composites from various sustainable raw materials. In S. Muthu (Ed.), *Green composites. Textile science and clothing technology* (pp. 1–24). Singapore: Springer.

Jayarajan, M., Arunachalam, R., & Annadurai, G. (2011). Agricultural wastes of jackfruit peel nano-porous adsorbent for removal of rhodamine dye. *Asian Journal of Applied Sciences*, *4*(3), 263–270.

Jung, J. H., Lee, J. H., & Shinkai, S. (2011). Functionalized magnetic nanoparticles as chemosensors and adsorbents for toxic metal ions in environmental and biological fields. *Chemical Society reviews*, *40*, 4464–4474.

Karipoth, P., Thirumurugan, A., & Justin Joseyphus, R. (2013). Synthesis and magnetic properties of flower-like FeCo particles through a one pot polyol process. *Journal of Colloid and Interface Science*, *404*, 49–55.

Khedari, J., Charoenvai, S., & Hirunlabh, J. (2003). New insulating particleboards from durian peel and coconut coir. *Building and Environment*, *38*(3), 435–441.

Konicki, W., Sibera, D., Mijowska, E., Lendzion-Bielun, Z., & Narkiewicz, U. (2013). Equilibrium and kinetic studies on acid dye Acid Red 88 adsorption by magnetic $ZnFe_2O_4$ spinel ferrite nanoparticles. *Journal of Colloid and Interface Science, 398*, 152–160.

Li, C., Wei, Y., Liivat, A., Zhub, Y., & Zhu, J. (2013a). Microwave-solvothermal synthesis of Fe_3O_4 magnetic nanoparticles. *Material Letters, 107*, 23–26.

Lipinski, B., Hanson, C., Lomax, J., Kitinoja, L., Waite, R., & Searchinger, T. (2013). *Reducing food loss and waste. World resource work paper, installment 2 of creating a sustainable food future* (21, pp. 1–41). Washington, DC: World Resources Institute.

Luo, X., Ge, X., Cui, S., & Li, Y. (2016). Value-added processing of crude glycerol into chemicals and polymers. *Bioresource Technology, 215*, 144–154.

Luo, X., & Zhang, L. (2009). High effective adsorption of organic dyes on magnetic cellulose beads entrapping activated carbon. *Journal of Hazardous Materials, 171*, 340–347.

Mahmoodi, N. M. (2013). Magnetic ferrite nanoparticle–alginate composite: synthesis, characterization and binary system dye removal. *Journal of the Taiwan Institute of Chemical Engineers, 44*, 322–330.

Mak, S. Y., & Chen, D.-H. (2004). Fast adsorption of methylene blue on polyacrylic acidbound iron oxide magnetic nanoparticles. *Dyes Pigments, 61*, 93–98.

Maran, J. P. (2015). Statistical optimization of aqueous extraction of pectin from waste durian rinds. *International Journal of Biological Macromolecules, 73*, 92–98.

McMullan, G., Meehan, C., Conneely, A., Kirby, N., Robinson, T., Nigam, P., … Smyth, W. F. (2001). Microbial decolourisation and degradation of textiles dyes. *Applied Microbiology and Biotechnology, 56*, 81–87.

Mekkawy, H. A., Ali, M. O., & El-Zawahry, A. M. (1998). Toxic effect of synthetic and natural food dyes on renal and hepatic functions in rats. *Toxicology Letters, 95*, 155–161.

Miller, R. (2005). *The landscape for biopolymers in packaging* (p. 2005) Heslington, York: National Non-Food Crops Centre (NNFCC) Publication.

Mishra, G., & Tripathy, M. (1993). A critical review of the treatment for decolorization of textile effluent. *Colourage, 40*, 35–38.

Mitra, S. K., Pathak, P. K., Lembisana Devi, H., & Chakraborty, I. (2010). Utilization of seed and peel of mango. *IX International Mango Symposium, 992*, 593–596.

Monash, P., & Pugazhenthi, G. (2009). Adsorption of crystal violet dye from aqueous solution using mesoporous materials synthesized at room temperature. *Adsorption, 15*, 390–405.

Morales, N. J., Goyanes, S., Chiliotte, C., Bekeris, V., Candal, R. J., & Rubiolo, G. H. (2013). One-step chemical vapor deposition synthesis of magnetic CNT–hercynite (FeAl2O4) hybrids with good aqueous colloidal stability. *Carbon, 61*, 515–524.

Myriounis, D. (2009). *Role of segregation and precipitates on interfacial strengthening mechanisms in metal matrix composites when subjected to thermo-mechanical processing*. Sheffield Hallam University, Research Archive.

Nakamura, R., Goda, K., Noda, J., & Ohgi, J. (2009). High temperature tensile properties and deep drawing of fully green composites. *eXPRESS Polymer Letters, 3*(1), 19–24.

Nejati, K., & Zabihi, R. (2012). Preparation and magnetic properties of nano size nickel ferrite particles using hydrothermal method. *Nejati Zabihi Chemistry Central Journal., 6*, 23.

Nguyen, T. A., & Juang, R. S. (2013). Treatment of waters and wastewaters containing sulfur dyes: A review. *Chemical Engineering Journal, 219*, 109–117.

Nkosi, S. S., Mwakikunga, B. W., Sideras-Haddad, E., & Forbes, A. (2012). Nanotechnology. *Science and Applications, 5*, 27–36.

Ooi, Z. X., Ismail, H., Abu Bakar, A., & Aziz, N. A. (2011). Effects of jackfruit waste flour on the properties of poly (vinyl alcohol) film. *Journal of Vinyl and Additive Technology, 17*(3), 198–208.

Östergren, K., Gustavsson, J., Bos-Brouwers, H., Timmermans, T., Hansen, O.-J., Møller, H., … Redlingshöfer, B. (2014). *FUSIONS definitional framework for food waste*. Sweden: EU FUSIONS.

Oxspring, D. A., McMullan, G., Smyth, W. F., & Marchant, R. (1996). Decolourisation and metabolism of the reactive textile dye, Remazol Black B, by an immobilized microbial consortium. *Biotechnology Letters*, *18*, 527–530.

Paliwal, B., & Cherkaoui, M. (2012). Estimation of anisotropic elastic properties of nanocomposites using atomistic-continuum interphase model. *International Journal of Solids and Structures*, *49*(18), 2424–2438.

Pankaj, T., Rav, B., Kumar, D., & Kumar, B. (2015). Study on mechanical behavior of sugarcane bagasse fiber reinforced polymer matrix composites. *International Journal of R&D in Engineering, Science and Management*, *1*(VII), 118–125, ISSN 2393-865X.

Parfitt, J., Barthel, M., & Macnaughton, S. (2010). Food waste within food supply chains: Quantification and potential for change to 2050. *Philosophical Transactions of the Royal Society London, B, Biological Science*, *365*(1554), 3065–3081.

Parni, B., & Verma, Y. (2014). Biochemical properties in peel, pulp and seeds of Carica papaya. *Plant Architecture*, *14*(1), 565–568.

Pearce, C. I., Lloyd, J. R., & Guthrie, J. T. (2003). The removal of color from textiles wastewater using whole bacterial cells: A review. *Dyes Pigments*, *58*, 179–196.

Pham, H. L., Do, H. M., Tran, D. L., Le, V. H., Nguyen, X. P., Nguyen, A. T., ... Vu, A. T. (2011). Magnetic nanoparticles: Study of magnetic heating and adsorption/desorption for biomedical and environmental applications. *International Journal of Nanotechnology*, *8*, 399–413.

Qadri, S., Ganoe, A., & Haik, Y. (2009). Removal and recovery of acridine orange from solutions by use of magnetic nanoparticles. *Journal of Hazardous Materials*, *169*, 318–323.

Qu, X., Pedro, J. J., & Qilin Li, A. (2013). Applications of nanotechnology in water and wastewater treatment. *Water Research*, *47*, 3931–3946.

Rafatullah, M., Sulaiman, O., Hashim, R., & Ahmad, A. (2010). Adsorption of methylene blue on low-cost sorbents: A review. *Journal of Hazardous Materials*, *177*, 70–80.

Ragunath, H. M. (2006). *Hydrology principles, analysis design*. India: New Age International (P) Limited.

Ramachandani, S., Das, M., Joshi, A., & Khanna, S. K. (1997). Effect of oral and parental administration of metanil yellow on some hepatic and intestinal biochemical parameters. *Journal of Applied Toxicology*, *17*, 85–91.

Ramasamy, S., Ismail, H., & Munusamy, Y. (2013a). Aqueous dispersion of rice husk powder as a compatible filler for natural rubber latex foam. *Advance Materials Research*, *626*, 530–536.

Ramasamy, S., Ismail, H., & Munusamy, Y. (2013b). Effect of rice husk powder on compression behavior and thermal stability of natural rubber latex foam. *Bioresources*, *8*(3), 4258–4269.

Rao, C. N. R., Thomas, P. J., & Kulkarni, G. U. (2007). *Nanocrystals: Synthesis, properties and applications, . Materials science* (Vol. 95, pp. 1–182). Berlin/Heidelberg/New York: Springer.

Reis, K. C., Pereira, L., Melo, I. C. N. A., Marconcini, J. M., Trugilho, P. F., & Tonoli, G. H. D. (2015). Particles of coffee wastes as reinforcement in polyhydroxybutyrate (PHB) based composites. *Materials Research*, *18*(3), 546–552.

Rocher, V. (2008). Removal of organic dyes by magnetic alginate beads. *Water Research*, *42* (4–5), 1290–1298.

Rogols, S., Sirovatka, D. M., Widmaier, R. G., (2002). *Packaging and structural materials comprising potato peel waste*. US Patent No. 6, 440, 204, U.S. Patent and Trademark Office, Washington, DC.

Ruiz, C., Kenny, S., Walsh, M., & O'Connor, K. (2014). *Production of biodegradable plastic by bacteria using waste resources. Science and solutions for a sustainable environment*. Ireland: University College Dublin, 11–12 Dec.

Sabino, V.-V., Morales, M. P., Bomati-Miguel, O., Bautista, C., Zhao, X., Bonville, P., ... Ferreiros, J. (2004). Colloidal dispersions of maghemite nanoparticles produced by laser pyrolysis with application as NMR contrast agents. *Journal of Physics D: Applied Physics*, *37*, 2054.

Sachdeva, S. M., Mani, K. V., Adval, S. K., Jalpota, V. P., Rasela, K. C., & Chadha, D. S. (1992). Acquired toxic methaemoglobinaemia. *The Journal of the Association of the Physicians of India*, *40*, 239–240.

Safarik, I., Horska, K., Pospiskova, K., Maderova, Z., & Safarikova, M. (2013). Microwave assisted synthesis of magnetically responsive composite materials. *IEEE Transactions on Magnetics*, *49*, 1.

Safarik, I., Horska, K., Svobodova, B., & Safarikova, M. (2012). Magnetically modified spent coffee grounds for dyes removal. *European Food Research and Technology*, *234*, 345–350.

Safarik, I., & Safarikova, M. (2010). Magnetic fluid modified peanut husks as an adsorbent for organic dyes removal. *Physics Procedia*, *9*, 274–278.

Sain, M., & Panthapulakkal, S. (2004). Green fiber thermoplastic composites. *Green composites: Polymer composites and the environment*, 181–183.

Salleh, M. A. M., Mahmoud, D. K., Karim, W. A. W. A., & Idris, A. (2011). Cationic and anionic dye adsorption by agricultural solid wastes: A comprehensive review. *Desalination*, *280*, 1–13.

Samanta, A. K., & Agarwal, P. (2009). Application of natural dyes on textiles. *Indian Journal of Fibre and Textile Research*, *34*, 384–399.

Santos, C. M. D., Abreu, C. M. P. D., Freire, J. M., Queiroz, E. D. R., & Mendonça, M. M. (2014). Chemical characterization of the flour of peel and seed from two papaya cultivars. *Food Science and Technology*, *34*(2), 353–357.

Seshan, S., Guruprasad, A., Prabha, M., & Sudhakar, A. (1995). Fiber-reinforced metal matrix composites-A review. *Journal of the Indian Institute of Science*, *76*, 1–14.

Shaffer, G. D. (1993). An archaeomagnetic study of a wattle and daub building collapse. *Journal of Field Archaeology*, *20*, 59–75.

Shuang, C. D., Li, P. H., Li, A. M., Zhou, Q., Zhang, M. C., & Zhou, Y. (2012). Quaternized magnetic microspheres for the efficient removal of reactive dyes. *Water Research*, *46*, 4417–4426.

Siddiqui, S. I., & Chaudhry, S. A. (2017). Removal of arsenic from water using nanocomposites. *A Review of Current Environmental Engineering*, *4*, 81–102.

Siddiqui, S. I., Manzoor, O., Mohsin, M., & Chaudhry, S. A. (2019). Nigella sativa seed based nanocomposite-MnO2/BC: An antibacterial material for photocatalytic degradation, and adsorptive removal of Methylene blue from water. *Environmental Research*, *171*, 328–340.

Singh, K. P., Gupta, S., Singh, A. K., & Sinha, S. (2011). Optimizing adsorption of crystal violet dye from water by magnetic nanocomposite using response surface modeling approach. *Journal of Hazardous Materials*, *186*, 1462–1473.

Sivashankar, R., Sathya, A., Vasantharaj, K., & Sivasubramanian, V. (2014). Magnetic composite an environmental super adsorbent for dye sequestration–A review. *Environmental Nanotechnology, Monitoring & Management*, *1*, 36–49.

Slunjski, M., Cadee, K., & Tattersall, J. (2000). MIEX *Resin water treatment process*. Netherlands: Aquatech Amsterdam.

Solis, M., Solis, A., Perez, H. I., Manjarrez, N., & Flores, M. (2012). Microbial decolouration of azo dyes: A review. *Process Biochemistry*, *47*, 1723–1748.

Starowicz, M., Starowicz, P., Zukrowski, J., Przewoznik, J., Lemanski, A., Kapusta, C., & Banas, J. (2011). Electrochemical synthesis of magnetic iron oxide nanoparticles with controlled size. *Journal of Nanoparticle Research*, *13*, 7167–7176.

Sugumaran, V., Prakash, S., Arora, A. K., Kapur, G. S., & Narula, A. K. (2017). Thermal cracking of potato-peel powder-polypropylene biocomposite and characterization of products—Pyrolysed oils and bio-char. *Journal of Analytical and Applied Pyrolysis*, *126*, 405–414.

Sugumaran, V., Vimal, K. K., Kapur, G. S., & Narula, A. K. (2015). Preparation and morphological, thermal, and physicomechanical properties of polypropylene-potato peel biocomposites. *Journal of Applied Polymer Science*, *132*, 34.

Sun, Z., Yi, J., Huang, Y., Feng, D., & Guo, C. (2016). Properties of asphalt binder modified by bio-oil derived from waste cooking oil. *Construction and Building Materials, 102*(Part 1), 496–504.

Tara, N., Siddiqui, S. I., Rathi, G., et al. (2019). Nano-engineered adsorbent for the removal of dyes from water: A review. *Curr. Anal. Chem., 15*, 1.

Thomas, S. (2010). *Surgical dressings and wound management*. Cardiff, South Wales: Medetec Publications.

Tiwari, H. (2012). Assessment of teratogenecity and embryo toxicity of dye wastewater untreated sludge from sanganer on swiss albino mice when administered during growth period of gestation. *Research & Development, 2*(1&2), 48–53.

Toro, P., Quijada, R., Yazdani-Pedram, M., & Arias, J. L. (2007). Eggshell, a new bio-filler for polypropylene composites. *Material Letters, 61*(22), 4347–4350.

Troschinetz, A. M., & Mihelcic, J. R. (2009). Sustainable recycling of municipal solid waste in developing countries. *Waste Management (New York, N.Y.), 29*(2), 915–923.

Vadivelan, V., & Kumar, K. V. (2005). Equilibrium, kinetics, mechanism, and process design for the sorption of methylene blue onto rice husk. *Journal of Colloids and Interface Science, 286*, 90–100.

Wawrzkiewicz, M. (2012). Anion exchange resins as effective sorbents for acidic dye removal from aqueous solutions and wastewaters. *Solvent Extraction and Ion Exchange, 30*, 507–523.

Wei, L., Liang, S., & McDonald, A. G. (2015). Thermophysical properties and biodegradation behavior of green composites made from polyhydroxybutyrate and potato peel waste fermentation residue. *Indian Crops Production, 69*, 91–103.

Wu, Q., Feng, C., Wang, C., & Wang, Z. (2013). A facile one-pot solvothermal method to produce superparamagnetic graphene–Fe3O_4 nanocomposite and its application in the removal of dye from aqueous solution. *Colloids and Surfaces B: Biointerfaces, 101*, 210–214.

Wu, W., He, Q., Chen, H., Tang, J., & Nie, L. (2007). Sonochemical synthesis, structure and magnetic properties of air-stable Fe3O_4/Au nanoparticles. *Nanotechnology, 18*, 145609.

Wua, S., Suna, A., Zhai, F., Wang, J., Xua, W., Zhang, Q., & Volinsky, A. A. (2011). Fe3O_4 magnetic nanoparticles synthesis from tailings by ultrasonic chemical co-precipitation. *Material Letters, 65*, 1882–1884.

Xia, Y., Dai, X., Huang, S., Tian, X., Yang, H., Li, Y., ... Zhao, M. (2013). Fast and highly efficient removal of methylene blue by a novel EDTAD-modified magnetic chitosan material. *Desalination and Water Treatment, 51*, 7586–7595.

Xiao, J. D., Qiu, L. G., Jiang, X., Zhu, Y. J., Ye, S., & Jiang, X. (2013). Magnetic porous carbons with high adsorption capacity synthesized by a microwave-enhanced high temperature ionothermal method from a Fe-based metal-organic framework. *Carbon, 59*, 372–382.

Xu, H., Zhang, Y., Jiang, Q., Reddy, N., & Yang, Y. (2013). Biodegradable hollow zein nanoparticles for removal of reactive dyes from wastewater. *Journal of Environmental Management, 125*, 33–40.

Xu, J., Yang, H., Fu, W., Du, K., Sui, Y., Chen, J., ... Zou, G. (2007). Preparation and magnetic properties of magnetite nanoparticles by sol–gel method. *Journal of Magnetism and Magnetic Materials, 309*, 307–311.

Yadla, S. V., Sridevi, V., & Chandana Lakshmi, M. V. V. (2012). A review on adsorption of heavy metals from aqueous solution. *Journal of Chemical, Biological, and Physical Sciences, 2*(3), 1585–1593.

Yan, H., Li, H., Yang, H., Li, A., & Cheng, R. (2013). Removal of various cationic dyes from aqueous solutions using a kind of fully biodegradable magnetic composite microsphere. *Chemical Engineering Journal, 223*, 402–411.

Yang, H., Yuan, B., Lu, Y. B., & Cheng, R. S. (2009). Preparation of magnetic chitosan microspheres and its applications in wastewater treatment. *Science in China Series B Chemistry, 52*, 249–256.

Yang, H. S., Kim, H. J., Park, H. J., Lee, B. J., & Hwang, T. S. (2006). Water absorption behavior and mechanical properties of lignocellulosic filler–polyolefin bio-composites. *Composite Structures, 72*(4), 429–437.

Yao, Y., Miao, S., Liu, S., Ma, L. P., Sun, H., & Wang, S. (2012). Synthesis, characterization, and adsorption properties of magnetic $Fe3O_4$@graphene nanocomposite. *Chemical Engineering Journal, 184*, 326–332.

Yu, W. R., Lee, B. H., & Kim, H. J. (2009). Fabrication of long and discontinuous naturalfibre reinforced polypropylene bio composites and their mechanical properties. *Fibres and Polymers, 10*(1), 83–90.

Zein, R., Suhaili, R., Earnestly, F., & Munaf, E. (2010). Removal of Pb (II), Cd (II) and Co (II) from aqueous solution using *Garcinia mangostana* L. fruit shell. *Journal of Hazardous Materials, 181*(1–3), 52–56.

Zhao, Y., Qiu, Z., & Huang, J. (2008). Preparation and analysis of $Fe3O_4$ magnetic nanoparticles used as targeted-drug carriers. *Chinese Journal of Chemical Engineering, 16*(3), 451–455.

Zhou, C.-G., Wang, S., Zhai, Y.-Y., Gao, Q., Luo, W.-J., & Xia, H. (2014). Highly efficient removal of Acid Red 18 from aqueous solution by magnetically retrievable chitosan/carbon nanotube: Batch study, isotherms, kinetics, and thermodynamics. *Chemical and Engineering Data, 59*, 39–51.

Zhou, L., Jin, J., Liu, Z., Liang, X., & Shang, C. (2011). Adsorption of acid dyes from aqueous solutions by the ethylenediamine-modified magnetic chitosan nanoparticles. *Journal of Hazardous Materials, 185*, 1045–1052.

Further reading

Ahmed, S., & Chaudhry, S. A. (Eds.), (2019). Composites for Environmental Engineering. John Wiley & Sons, Incorporated.

CHAPTER

11

A comprehensive review of natural fiber reinforced polymer biocomposites and their applications

S. Mohd Izwan[1], S.M. Sapuan[1,2], M.Y.M. Zuhri[1,2], A.R. Mohamed[3] and R.A. Ilyas[4,5]

[1]Advanced Engineering Materials and Composites Research Centre (AEMC), Department of Mechanical and Manufacturing Engineering, Faculty of Engineering, Universiti Putra Malaysia, Selangor, Malaysia, [2]Laboratory of Biocomposite Technology, Institute of Tropical Forestry and Forest Products (INTROP), Universiti Putra Malaysia, Selangor, Malaysia, [3]Department of Manufacturing and Material Engineering, Kulliyyah of Engineering, International Islamic University Malaysia, Kuala Lumpur, Malaysia, [4]School of Chemical and Energy Engineering, Faculty of Engineering, Universiti Teknologi Malaysia, Johor, Malaysia, [5]Centre for Advanced Composite Materials, Universiti Teknologi Malaysia, Johor, Malaysia

11.1 Introduction

The evolution in the material industries has created the word "composites" as a common term in many fields. Polymer matrix composites, one of the three main types of composites, have more extensive applications compared to the other two types of ceramic and metal matrix composites (Wang, Jiang, Zhou, Gou, & Hui, 2017). Although polymer composites are more likely to be used in nonstructural applications since their introduction several decades ago, the critical and structural

Design for Sustainability
DOI: https://doi.org/10.1016/B978-0-12-819482-9.00019-8

applications, such as those used in aircraft, now have approximately 50%–70% composite parts in the entire structure. This shows that composites have slowly gained the trust of transforming existing metals into lighter materials with comparable properties in vital structures and applications (Soutis, 2014; Zaaba & Ismail, 2019).

Concerned about the environmental issues and other aspects such as cost, material sustainability, and health, industrial practitioners in most fuel-based applications have been continuously searching for new ideas and materials to replace the high-cost, petroleum-based fibers in composites (Atiqah et al., 2019; Azammi et al., 2019; Friedrich & Abdulhakim, 2013). The introduction of natural fibers, which offer long-term continuous supply, are less hazardous to humans and the environment, and are also biodegradable, lightweight, and less expensive, bringing the hope of overcoming the disadvantages of the currently used synthetic composites (Alves et al., 2010; Atikah et al., 2019; Mazani, Sapuan, Sanyang, Atiqah, & Ilyas, 2019). Jute, kenaf, bamboo, pineapple, sugar palm, and bagasse are some examples of high-potential natural fibers reported to date (Ilyas, Sapuan, Atiqah, & Ibrahim et al., 2019; Ilyas, Sapuan, Ibrahim, Abral, & Ishak et al., 2019; Ilyas, Sapuan, Ibrahim, Atikah et al., 2019).

The idea of hybrid composites, which are a combination of two or more types of fibers in one matrix, has strengthened the potential of composites in various fields. Synthetic–synthetic hybrid composites can enhance the properties of existing synthetic composites to a higher level of advanced composites, while synthetic–natural and natural–natural hybrid composites provide an alternative toward greener and more environmentally friendly materials with comparable properties than wholly one-type natural composites (Jumaidin, Saidi, et al., 2019; Jumaidin, Ilyas, Saiful, Hussin, & Mastura, 2019; Jumaidin, Khiruddin, Saidi, Salit, & Ilyas, 2020; Sathishkumar, Naveen, & Satheeshkumar, 2014).

In polymer matrix composites, polymers such as epoxy, polyester, vinyl ester (VE), polypropylene (PP), polylactic acid (PLA), and polyethylene (PE) can be classified into two main categories—thermoplastics and thermosets (Shah, Sultan, Jawaid, Cardona, & Abu Talib, 2016). Table 11.1 shows the differences between these two types of polymers.

PP is one of the most commonly used thermoplastics in various applications including packaging for consumer products, plastic parts for various industries including the automotive industry, special devices like living hinges, and textiles (Jiang, Rui, & Chen, 2009). In an attempt to increase the potential of PP and its composites, researchers together with industrial practitioners are continuously putting efforts into ensuring that the properties of PP materials meet user demands. To date, numerous scientific studies have been reported in journals and presented in conferences, contributing to the development of PP composites (Asumani, Reid, & Paskaramoorthy, 2012; Hao, Zhao, Jiang, Yuan, & Chen, 2012; Maddah, 2016). This chapter reviews

TABLE 11.1 Differences between thermosets and thermoplastics (Sonnenfeld, Mendil-Jakani, Agogué, Nunez, & Beauchêne, 2017; Matweb, 2014; Szycher, 2012).

Properties	Thermoplastics	Thermosets
Strength	Less strong	Comparatively stronger
Nature	Moldable	Brittle
Recyclability	Recyclable	Nonrecyclable
Process	Reversible	Irreversible
Cost	Expensive	Cost-effective

the reported studies on the natural fiber/PP composites to describe their progress, which indirectly shows the gaps in the respective areas.

11.2 Extraction of natural fibers/different sources of natural fibers

Natural fibers used as reinforcement in polymer composites can be extracted from different parts of plants. Depending on the methods of extraction, fibers can be prepared in several forms such as long-continuous fibers, woven mat fibers, short-discontinuous fibers, and powdered micro-fiber or nano-sized fibers (Abral et al., 2019, 2020; Aisyah et al., 2019; Ilyas, Sapuan, Ishak, & Zainudin, 2018; Norizan, Nurazzi, Abdan, Ilyas, & Biofibers, 2020; Syafri et al., 2019; Zakikhani, Zahari, Sultan, & Majid, 2014). Fig. 11.1 shows the different parts of plants from which fibers are extracted to be used as reinforcement in polymer composites.

From Fig. 11.1, it can be seen that the most common source of reinforcement natural fiber extraction is the stems while the leaves are the least reported in earlier studies. Unlike manmade fibers, whose properties are tailored to be consistent, many factors can affect the properties of plant-based natural fibers as they grow naturally and need to be extracted in different ways either mechanically, chemically, or using a combination of both. Geographical factors of the plantation area, weather conditions, age of the plants during the harvesting period, and many other factors could affect the properties of natural fibers, even before the extraction methods are used to widen the variation in properties among fibers (Ilyas, Sapuan, Sanyang, Ishak, & Zainudin, 2018; Mahjoub, Yatim, Sam, & Hashemi, 2014).

Pineapple leaf fibers (PALF) can be extracted through a mechanical and retting process. A scrapping machine, made from a combination of a feed roller, leaf scratching roller, and serrated roller, is used to

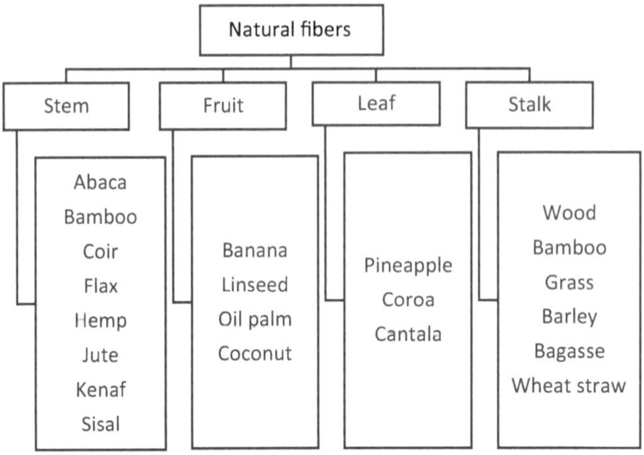

FIGURE 11.1 Reinforcement fibers extracted from different parts of plants (Azwa, Yousif, Manalo, & Karunasena, 2013; Sathishkumar et al., 2014).

mechanically break down the outer layer of fresh pineapple leaf as a preparation for the next retting process. The retting of PALF can be carried out using either tap water, chemicals, or natural pond water. To obtain fast retting reactions, the scratched PALF was immersed in a mixture of 0.5% urea or diammonium phosphate and substrate to liquor ratio of 1:20. After the PALF was loosened, it was segregated manually in pond water and dried in open air. The results suggested that fresh pineapple leaf can produce 2%–3% fibers (Asim et al., 2015).

Bamboo is known to have a very strong culm that is composed of longitudinal fibers stacked together in a lignin matrix. It is one of the hardest plant fibers that can be extracted as individual fiber bundles. Several methods of bamboo fiber extraction have been reported in previous studies, such as retting, crushing, grinding, rolling mill, degumming, and compression; all of which resulted in different forms of fibers with different ranges of properties as shown in Table 11.2. Steam explosion was found to produce bamboo fibers with better strength compared to the other extraction methods. To break down the plant cell wall, an autoclave was used to overheat the bamboo strips at 0.7–0.8 MPa and 175°C for 60 minutes. The steam was immediately released for 5 minutes to complete one cycle of this process. A minimum of nine cycles were carried out before the ash was removed from the fibers in the cleaning process. Hot water at 90°C–95°C with added soap was used to wash the fibers before the oven-dry process at 105°C for 24 hours (Zakikhani et al., 2014).

Similarly to other plant fibers, kenaf fibers can also be extracted through a retting and mechanical process. The decortication process is a mechanical

TABLE 11.2 Tensile properties of bamboo fibers extracted using different methods (Zakikhani et al., 2014).

Extraction procedure	Tensile strength (MPa)	Young's modulus (GPa)
Steam explosion	615–812	35.45
Rolling mill	270	–
Grinding	450–800	18–30
Retting	503	35.91
Crushing	420 ± 170	38.2 ± 16
Chemical	450	18
Alkaline	395 ± 155	26.1 ± 14.5
Chemical + compression	645; max: 1000	–
Chemical + roller mill	370; max: 480	–

extraction method that separates the bast from the pith of a kenaf stalk. In this method, the kenaf stalk is immersed in water for several days prior to the decortication process, either by hammer milling or roller milling. Compared to hammer milling, in which the kenaf stalk is beaten until the bast and the core are separated before they are passed through the meshes inside the hammer milling machine, the roller milling technique produces kenaf fibers with minimal damage. Fibers extracted using both of these techniques are then cleaned with running water, dried, and combed before further applications (Saba, Paridah, & Jawaid, 2015).

In addition to kenaf, flax and jute fibers are other types of bast fibers. Taking into account the general similarities of their microstructure, the extraction method applied to the kenaf fibers can directly be applied to jute and flax fibers. Various physical and chemical treatments have been done on these three plants to reduce the drawbacks of natural fibers in order for them to be comparable to synthetic fibers. In one study, commercial Duralin treatment was applied to flax fibers to reduce their moisture absorption and increase their strength. In this treatment, the flax was steamed at a high temperature above 160°C for 30 minutes in an autoclave prior to the drying process at 150°C for 2 hours. The extracted fibers, known as the Duralin flax, were found to absorb less moisture and exhibited higher and more uniform strength compared to untreated flax (Yan, Chouw, & Jayaraman, 2014).

Almost all plant fibers need to undergo several extraction processes and modifications before they are ready to be used as reinforcements in composites. To date, sugar palm fibers are the only fibers that can be used directly from the plant without any further modifications, except for general cleaning and drying processes (Ilyas et al., 2020; Ilyas, Sapuan, Ishak,

FIGURE 11.2 Sugar palm fibers.

Zainudin et al., 2018; Sanyang, Sapuan, Jawaid, Ishak, & Sahari, 2016; Sapuan et al., 2018). These plant fibers can be found on sugar palm trees along the length of the trunk, as shown in Fig. 11.2. The strength of the sugar palm fibers is reported to be different at different heights of the tree. Although sugar palm fibers can be directly used in composites, several chemical treatments have been applied to improve their compatibility in composites, thus enhancing their strength (Ishak et al., 2013).

It can be concluded that natural fibers extracted from different parts of plants exhibit different properties, which are not only caused by the different natural constituents controlled by nature, but also by different methods of extraction. The use of chemicals, heat, and pressure could indirectly change the structure of the extracted fibers along with their strength and other properties. Table 11.3 lists the tensile properties of plant fibers that have been studied previously.

11.3 Comparison between neat polypropylene and natural fiber/polypropylene composites

There are several reasons for introducing fibers or fillers in neat polymers such as PP, epoxy, and polyester. Depending on several factors,

TABLE 11.3 Tensile properties of different plant fibers (Jawaid & Abdul Khalil, 2011; Pickering, Efendy, & Le, 2016; Sinha, Kumar, & Bhattacharya, 2017).

Fiber	Tensile strength (MPa)	Tensile modulus (GPa)	Elongation at break (%)
Hemp	550–900	70	1.6
Ramie	400–1000	27–128	1.2–3.8
Flax	800–1500	60–80	1.2–1.6
Jute	400–800	10–30	1.8
Cotton	287–597	5.5–12.6	7–8
Sisal	600–700	38	2–3
Coconut	131–175	4–6	15–40
Kenaf	240–930	14–53	1.6
Bamboo	140–230	11–17	-
Pineapple	43–1627	60–82	-
Harakeke	440–990	14–33	-

the addition of fibers and fillers generally means the mechanical properties, electrical conductivity, and thermal stability of the polymer composites can be enhanced. Besides the improvements in the material properties, replacing a percentage of the polymers with fibers and fillers can reduce the dependency on petroleum-based materials, thus supporting the environmental protection.

In the early days of introducing fibers and fillers in polymers, synthetic materials such as glass and carbon were used. Today, natural fibers are actively reported to be used with various types of polymers including PP. However, a major concern regarding the weak fiber–matrix compatibility is continuously being voiced. It was found, in some studies, that the addition of untreated or raw natural fibers decreased the strength of the PP composites compared to neat PP.

A study found that the addition of 30% empty fruit bunch (EFB) and coir in their respective composites reduced the tensile strength from 42 MPa for neat PP to 25 MPa and 27 MPa for EFB/PP and coir/PP, respectively (Zainudin, Yan, Haniffah, Jawaid, & Alothman, 2014). Similar findings were reported in another study analyzing the addition of 40% wood flour of 100 mesh-size, in which the tensile strength of the neat PP was higher compared to the wood/PP composites, as shown in Table 11.4 (Zhang, Yang, Lin, & Tan, 2013). Introducing hybrid kenaf/sugar palm with a 3:7 ratio and a total of 10% fiber loading also decreased the tensile strength of the hybrid composites to 21.3 MPa compared to that of the neat PP at 31.6 MPa (Bachtiar, Siregar, Sulaiman, & Rejab, 2014).

TABLE 11.4 Comparison of tensile strengths between neat polypropylene and natural fiber/polypropylene composites.

Natural fiber (modification)	Tensile strength (MPa)		References
	Neat polypropylene	Natural fiber/ polypropylene composites (% fiber loading)	
Unmodified fibers and composites			
Empty fruit bunch	42.0	25.0 (30%)	Zainudin, Yan, Haniffah, Jawaid, & Alothman, 2014
Coir		27.0 (30%)	
Wood	31.8	29.1	Zhang et al. (2013)
Kenaf/sugar palm	31.6	21.3 (10%)	Bachtiar et al. (2014)
Kenaf	21.3	21.7 (10%)	Akhtar et al. (2016)
Banana	32.0	37.50 (10%)	Samal et al. (2009)
Pineapple	28.0	56.0 (25%)	Hoque et al. (2018)
Modified composites			
Sisal (chemical treatment)	Not broken	23.8	Jarukumjorn & Suppakarn (2009)
Palm (chemical treatment)	24.0	11.0 (8%)	Chollakup, Askanian, & Delor-Jestin (2017)
Pineapple (chemical treatment)		16.0 (6%)	
Banana/pineapple (chemical treatment)	27.0	24.0 (2%)	Rahman, Das, & Hasan (2018)
Bamboo (nanofillers)	21.9	21.2 (10%)	Rahman, Hamdan, Maryani, Hashim, & Islam (2014)
Bamboo (coupling agent)	32.2	35.2 (10%)	Chauhan, Aggarwal, Venkatesh, & Abhilash, 2018
Kenaf (coupling agent)	26.0	40.0 (10%)	Mirbagheri, Tajvidi, Hermanson, & Ghasemi, 2007

A different study found that the inclusion of 10% kenaf fibers did not significantly improve or reduce the tensile strength of the composites compared to neat PP. The tensile strengths of both neat PP and kenaf/PP composites showed similar values of 21.3 MPa and 21.7 MPa, respectively (Akhtar et al., 2016). A similar observation was found with the inclusion of 10% banana fibers, which had a tensile strength of

37.5 MPa, slightly higher than the neat PP with 32.03 MPa strength (Samal, Mohanty, & Nayak, 2009). However, a significant improvement was achieved when 25% pineapple fibers was added to PP, which increased the tensile strength from 28 MPa for neat PP to 56 MPa for pineapple/PP composites (Hoque, Hossain, Nahid, Bari, & Khan, 2018).

These studies provided a clear one-to-one comparison between neat PP and its composites, in which the natural fibers used were free from any chemical treatment, while the whole composites were free from any coupling agent during the fabrication process. Briefly, these preliminary comparisons indicate that introducing untreated raw natural fibers in PP does not significantly enhance the strength of the resulting composite. Therefore the use of a coupling agent, modified PP, or surface treatment on fibers needs to be further examined to study their effects on the natural fiber/PP composites. In addition, the percentage of fiber loading is also a key parameter to producing natural fiber/PP composites with comparable properties.

Table 11.4 lists a comparison of the tensile strengths between neat PP and natural fiber/PP composites reported in earlier studies. The comparison was limited to the lowest percentage of fiber loading used in the PP composites in each study (Akhtar et al., 2016; Bachtiar, Siregar, bin Sulaiman, & bin Mat Rejab, 2014; Hoque, Hossain, Nahid, Bari, & Khan, 2018; Samal, Mohanty, & Nayak, 2009; Zainudin, Yan, Haniffah, Jawaid, & Alothman, 2014; Zhang, Yang, Lin, & Tan, 2013). In several studies, a comparison was made between neat PP and modified composites with no analysis conducted on unmodified composites. In this discussion, the modified composites can be referred to as treated fibers, modified PP, or the use of coupling agents.

11.4 The effects of coupling agent and chemical treatment

As mentioned earlier, besides an optimum percentage of fiber loading and the size of fibers, the compatibility between hydrophilic natural fibers and hydrophobic polymers is one of researchers' concerns in achieving a high strength of the composites. Therefore, in some studies reported to date, the parameters of surface modification on the natural fibers, such as types and concentrations of chemicals used with different soaking times, were analyzed.

The use of 5 wt.% maleic anhydride-PP (MAH-PP) copolymer as a coupling agent in abaca/PP, jute/PP, and flax/PP enhanced the tensile strength of these composites compared to composites without the coupling agent. A comparison of 30% fiber loading between 25 mm fiber length of abaca-, jute-, and flax-reinforced PP composites showed a tensile strength of 45, 50, and 48 MPa with the use of MAPP compared to 33, 40, and 35 MPa without MAPP, respectively. The addition of MAH-PP also increased the Charpy impact strength for the abaca/MAH-PP composites at 20% and 30% fiber loadings, while the flexural strength improved at all percentages of fiber loading compared to the composites without MAH-PP (Bledzki, Mamun, & Faruk, 2007).

Another study indicated that a mixture of 40% wood flour, 57% PP, and 3% PP-g-MA compounded in a twin screw extruder at 200°C produced wood/PP composites with a higher tensile strength, flexural strength, and modulus compared to the composites without PP-g-MA at the same percentage of fiber loading (Zhang et al., 2013). A parallel trend was observed in different studies which found that the use of PP-g-MA in sisal/PP composites, made from 4 mm NaOH-treated sisal fiber, had a higher tensile strength of 28.66 MPa compared to those without PP-g-MA at 23.83 MPa. Except for the flexural modulus, the inclusion of MA-g-PP in the sisal/PP composites also increased its tensile modulus, flexural strength, and Izod impact strength (Jarukumjorn & Suppakarn, 2009).

Among the three different concentrations of MAPP used in the banana stem fiber (BSF)/PP composites, the 2% concentration marked the highest value for both tensile and flexural strengths and modulus compared to the 1% and 3% MAPP concentrations. However, the use of MAPP in all the BSF/PP composites regardless of its concentration improved the mechanical properties compared to those without MAPP (Samal et al., 2009).

It is known that the poor fiber–matrix compatibility is a result of the presence of hydroxyl and other polar groups on the fiber surface. Therefore, MAPP, which is composed of anhydride groups, can covalently bond with the hydroxyl groups of the natural fibers to form a stronger interfacial region. At the same time, a good stress transfer at the interface can be created from the chain entanglement between MAPP and PP in the composites (Pang, Ismail, & Bakar, 2018). These reasons are in good agreement with the results of the experimental analyses shown in Table 11.5, indicating that the use of MAPP significantly enhanced the mechanical properties of the natural fiber/PP composites.

Besides the use of MAPP as a coupling agent to promote better stress transfer and fiber–matrix adhesion, chemical treatments can be carried out on the natural fibers to eliminate several chemical compositions that weaken the interfacial adhesion, such as excess lignin and wax. In a study by Rahman et al., 5% sodium meta-periodate ($NaIO_4$) was used to treat bamboo fibers for 3 hours at 80°C, which led to enhancement of the mechanical properties from 21.5 MPa for untreated bamboo/clay/PP composites to 26 MPa for treated bamboo composites, both at 10% fiber loading (Rahman et al., 2014). The treatment conducted on the bamboo fibers

TABLE 11.5 Mechanical properties of natural fiber/polypropylene composites with and without coupling agent/chemical treatment.

Natural fiber	Percentage of fiber loading (%)	Coupling agent/chemical treatment	Natural fiber/polypropylene composites with (without) coupling agent/chemical treatment				References
			Tensile strength	Tensilemodulus	Flexural strength	Flexural modulus	
Abaca	30%	Maleic anhydride-PP	45 (33)	5.1 (4.3)	76 (56)	4.3 (4.0)	Bledzki, Mamun, & Faruk, 2007
Jute			50 (40)	4.8 (4.65)	74 (54)	4.5 (4.65)	
Flax			48 (35)	3.88 (3.8)	72 (55)	3.75 (3.8)	
Wood	40%	PP-g-MA	40.6 (29.1)	-	55.3 (44.8)	3.473 (2.883)	Zhang et al. (2013)
Sisal	30%	PP-g-MA	28.66 (23.83)	2.29 (2.11)	65.65 (60.88)	1.01 (3.91)	Jarukumjorn & Suppakarn (2009)
Banana	30%	MAPP (2%)	50 (45.25)	1.43 (0.985)	60.5 (48)	2.015 (1.815)	Samal et al. (2009)
Bamboo	10%	$NaIO_4$	26 (23.2)	2.4			Rahman et al. (2014)
Kenaf	20%	NaOH	22.048 (22.908)	1785.981 (1795.779)	43.284 (43.436)	1789.63 (1820.87)	Akhtar et al. (2016)
PAFL	45%	NaOH (7%)			–	–	Hoque et al. (2018)

improved both the tensile and flexural strengths and modulus of the bamboo/clay/PP composites at all percentages of fiber loading (Rahman et al., 2014; Zhang et al., 2018).

From Table 11.5, it can be seen that sodium hydroxide (NaOH) treatment on kenaf fibers at 6% concentration did not significantly affect the mechanical properties of kenaf/PP composites compared to untreated composites. The increase in the percentages of both the tensile and flexural strength and modulus were between 0% and less than 5%, implying that there was little improvement in the properties of the treated kenaf/PP composites at all percentages of fiber loading. Among all the treated kenaf/PP composites at 20% fiber loading marked the highest percentage improvement in the tensile strength by 3.9% (Akhtar et al., 2016).

However, for the PALF/PP composites, one study reported that a treatment with 7% NaOH for 24 hours resulted in a significant decrease in the tensile strength and modulus for the treated composites at all percentages of fiber loading (Hoque et al., 2018). The highest percentages of loss in the tensile strength and modulus were observed at 45% and 25% loading on the PALF/PP composites, respectively.

11.5 The influence of natural fiber loading on the composites

Different percentages of fiber loading were applied in different studies of natural fiber/PP composites. Overall, the range of fiber loading reported in earlier studies was from 10% to 60%, depending on the size of the fibers. These two parameters are interconnected to avoid a saturated mixture of fiber in the matrix during the fabrication process.

Varying the percentage of fiber loading from 20%, 30%, 35%, and 40% of untreated, 25 mm length abaca fibers in the abaca/PP composites continuously increased their tensile, flexural, and Charpy impact strengths. However, increasing the percentage further to 45% and 50% resulted in a loss of all the tensile, flexural, and impact strength. In a different study, using MAH-PP as a coupling agent, the tensile and flexural strengths of the abaca/PP composites were found to continue increasing in parallel with the increasing percentage of fiber loading from 20% to 50% without losing their strengths at higher percentages of fiber loading (Bledzki, Mamun, & Faruk, 2007).

A similar trend of increasing tensile strength with the increasing percentage of fiber loading was also observed in four other natural fiber/PP composites. Both untreated and NaOH-treated 40 mesh kenaf/PP composites, fabricated through the injection molding technique, showed increasing tensile and flexural strengths and modulus with the increase in the percentage of fiber loading from 10% to 40% (Akhtar et al., 2016).

Among all the mechanical properties analyzed on the untreated and NaOH-treated PALF/PP composites, only the tensile strength began to decrease at 45% fiber loading after continuously increasing from 25% to 40% fiber loading. Other properties, such as the tensile modulus, flexural strength, and modulus, increased continuously with the increase in the percentage of fiber loading from 25% to 45% (Hoque et al., 2018).

Similar findings were also reported for the banana/PP composites in another study (Samal et al., 2009). The tensile and flexural strengths increased with the increase in the percentage of fiber loading from 10% to 30%; however, both strengths began to drop at 40% fiber loading. In contrast, the tensile and flexural modulus of the untreated banana/PP composites increased continuously as the percentage of fiber loading was increased up to 40%, which was the highest percentage applied in the study.

A higher percentage of fiber loading in composites promotes high possibilities of fiber agglomeration in the matrix, which leads to a nonuniform stress transfer in the composites. The fiber agglomeration, due to the H-bonding between the fibers, results in the nonuniform distribution of fibers and the formation of voids, which later deteriorates the strength of the composites. However, increasing modulus explains the reinforcing effects in the composites.

A slightly different observation was reported in another study for bamboo/PP composites made from 180–225 μm mesh bamboo flour, in which increasing the fiber loading from 10% to 50% increased the mechanical properties of the composites (Chauhan, Aggarwal, Venkatesh, & Abhilash, 2018). The use of m-TMI as a coupling agent in the study was one of the factors that helped improve the mechanical properties of the bamboo/PP composites, although a higher percentage of fiber loading was applied.

Comparing the bamboo/PP composites analyzed in different studies showed various trends in the tensile and flexural strengths and modulus. Increasing the percentage of fiber loading from 10% to 25% decreased the tensile strength of the bamboo/PP composites, although two different approaches were applied: adding nanoclay as fillers and treating the bamboo fibers with sodium meta-periodate ($NaIO_4$). Although nanoclays can act as reinforcing fillers by the dispersion of its layers in the polymer matrix, a very small percentage in the composites with an increasing amount of hydrophilic natural fibers between the layers led to a nonsubstantial improvement in the mechanical properties of the natural fiber/PP composites (Rahman et al., 2014).

The three composite variations—untreated bamboo/PP, untreated bamboo/clay/PP, and treated bamboo/clay/PP—have similar behaviors in three other properties, that is, the tensile modulus, flexural strength, and flexural modulus. Their tensile modulus did not change substantially from 10% to 20% fiber loading but increased significantly at 25% fiber loading. The analysis of their flexural strength indicated that 20% fiber loading posed the highest value; however, increasing the fiber loading further to 25% decreased the flexural strength. Meanwhile, the flexural modulus had similar values at all percentages of fiber loading.

11.6 Various applications of natural fiber reinforced PP

PP is generally used in the interior components of vehicles. However, since a few years ago, researchers have suggested wider applications of PP based on its newly discovered potential and great properties. A study assessed the oxygen and water vapor transmission rates in PP integrated with less than 5 wt.% of zeolite and showed a strong prospective to be an alternate material in active plastic for industrial packaging (Toommee & Pratumpong, 2018). In a different study, excellent antibacterial activity was found on silver/PP nanocomposite films, thus proposing its use in active food packaging (Jo et al., 2018). Various process abilities of PP, which enable it to be transformed into blown films, sheets, flat films, and other different shapes, have widened its applications in the biomedical field (Gopanna, Rajan, Thomas, & Chavali, 2019). In addition, PP composites are also applicable in microwave and millimeter-wave communication systems based on the findings reported on the decreasing tendency of loss tangent with the increase in the frequency of PP–silicon composites. This indirectly opens up the potential uses of PP composites in the Internet of Things and 5G applications (Krupka et al., 2018). Milled carbon-fiber-reinforced PP composites were used as the conductive polymer composite materials in developing a bipolar plate for polymer electrolyte membrane fuel cells (Radzuan et al., 2019). Isotactic PP/hexagonal boron nitride composites have the ability to produce excellent dielectric properties for high-frequency applications (Takahashi, Imai, Kan, Hotta, & Ogawa, 2014). Compared to quasi-static properties, strain-rate-dependent mechanical properties are more reliable to be applied in designing various glass-fiber-reinforced PP composite-based structures (Kim, Kang, Kim, & Kim, 2019). A study concluded that 40% kenaf/PP composites with superior physical and mechanical properties may be used in a variety of applications such as automotive, sports, construction, animal bedding, and mass production industries (Akhtar et al., 2016). Furthermore, the addition of nanoclay in PP composites could enhance their mechanical properties for vehicle external wall panel applications (Uwa, Jamiru, Sadiku, Huan, & Mpofu, 2019). The hybrid cellulose–inorganic reinforcement composite approach leads to superior weight and cost savings and also environment benefits over inorganic reinforced composites (Reale Batista, Desireé, Drzal, & Mielewski, 2019). In a different study, blue-agave PP composites have been suggested as alternative materials for talc-filled PP components in the automotive interiors (Langhorst et al., 2018).

11.7 Conclusion

The use of different types of natural fibers as fillers in PP composites has been the focus of researchers recently. From a review of past literatures, it was found that some natural fibers, originating from the same part of plants, can be extracted using similar methods. For example, kenaf and jute, originating from the bast group, can be extracted using water, chemical, and enzymatic retting, besides a mechanical process using an extractor machine. The addition of untreated coir, EFB, wood flour, and kenaf/sugar palm fibers in PP composites decreased the tensile strength compared to neat PP.

The use of a coupling agent generally improved the mechanical properties of the composites compared to those without a coupling agent at the same fiber loading.

Coupling agents used in corresponding PP composites made from abaca, flax, and jute showed a higher tensile strength compared to unmodified abaca, flax, and jute fibers. Compared to the use of a coupling agent, chemical treatment of kenaf and bamboo fibers using sodium hydroxide did not significantly enhance the properties of the kenaf and bamboo composites at 10% and 20% fiber loading, respectively.

Acknowledgments

The authors are thankful to the University Putra Malaysia-Malaysia for supporting this research. The authors are especially indebted to the Ministry of Education for their sponsorship throughout this study.

References

Abral, H., Ariksa, J., Mahardika, M., Handayani, D., Aminah, I., Sandrawati, N., ... Ilyas, R. A. (2019). Highly transparent and antimicrobial PVA based bionanocomposites reinforced by ginger nanofiber. *Polymer Testing*, 106186.

Abral, H., Ariksa, J., Mahardika, M., Handayani, D., Aminah, I., Sandrawati, N., ... Ilyas, R. A. (2020). Transparent and antimicrobial cellulose film from ginger nanofiber. *Food Hydrocolloids*, *98*, 105266.

Aisyah, H. A., Paridah, M. T., Sapuan, S. M., Khalina, A., Berkalp, O. B., Lee, S. H., ... Ilyas., R. A. (2019). Thermal properties of woven kenaf/carbon fibre-reinforced epoxy hybrid composite panels. *International Journal of Polymer Science*, 1–8.

Akhtar, M., Niaz, A., Bakar Sulong, M. K. F., Radzi, N. F., Ismail, M. R., Raza, N. M., & Khan, M. A. (2016). Influence of alkaline treatment and fiber loading on the physical and mechanical properties of kenaf/polypropylene composites for variety of applications. *Progress in Natural Science: Materials International*, *26*(6), 657–664.

Alves, C., Ferrão, P. M. C., Silva, A. J., Reis, L. G., Freitas, M., Rodrigues, L. B., & Alves, D. E. (2010). Ecodesign of automotive components making use of natural jute fiber composites. *Journal of Cleaner Production*, *18*(4), 313–327.

Asim, M., Abdan, K., Jawaid, M., Nasir, M., Dashtizadeh, Z., Ishak, M. R., & Enamul Hoque, M. (2015). A review on pineapple leaves fibre and its composites. *International Journal of Polymer Science, 2015*, 1–16.

Asumani, O. M. L., Reid, R. G., & Paskaramoorthy, R. (2012). The effects of alkali-silane treatment on the tensile and flexural properties of short fibre non-woven kenaf reinforced polypropylene composites. *Composites Part A: Applied Science and Manufacturing, 43*(9), 1431–1440.

Atikah, M. S. N., Ilyas, R. A., Sapuan, S. M., Ishak, M. R., Zainudin, E. S., Ibrahim, R., ... Jumaidin, R. (2019). Degradation and physical properties of sugar palm starch/sugar palm nanofibrillated cellulose bionanocomposite. *Polimery, 64*(10), 27–36.

Atiqah, A., Jawaid, M., Sapuan, S. M., Ishak, M. R., Ansari, M. N. M., & Ilyas., R. A. (2019). Physical and thermal properties of treated sugar palm/glass fibre reinforced thermoplastic polyurethane hybrid composites. *Journal of Materials Research and Technology, 8*(5), 3726–3732.

Azammi, A. M. N., Ilyas, R. A., Sapuan, S. M., Ibrahim, R., Atikah, M. S. N., Asrofi, M., & Atiqah, A. (2019). *Characterization studies of biopolymeric matrix and cellulose fibres based composites related to functionalized fibre-matrix interface. Interfaces in particle and fibre reinforced composites- from macro to nano scales* (pp. 1–68.). London: Woodhead Publishing.

Azwa, Z. N., Yousif, B. F., Manalo, A. C., & Karunasena, W. (2013). A review on the degradability of polymeric composites based on natural fibres. *Materials and Design, 47*, 424–442.

Bachtiar, D., Siregar, J. P., bin Sulaiman, A. S., & bin Mat Rejab, M. R. (2014). Tensile properties of hybrid sugar palm/kenaf fibre reinforced polypropylene composites. *Applied Mechanics and Materials, 695*, 155–158.

Bledzki, A. K., Mamun, A. A., & Faruk, O. (2007). Abaca fibre reinforced PP composites and comparison with jute and flax fibre PP composites. *eXPRESS Polymer Letters, 1*(11), 755–762.

Chauhan, S. S., Aggarwal, P., Venkatesh, G. S., & Abhilash, R. M. (2018). Mechanical characterization, modelling and application. *International Journal of Plastics Technology, 22*(1), 85–103.

Chollakup, R., Askanian, H., & Delor-Jestin, F. (2017). Initial properties and ageing behaviour of pineapple leaf and palm fibre as reinforcement for polypropylene. *Journal of Thermoplastic Composite Materials, 30*(2), 174–195.

Friedrich, K., & Abdulhakim, A. A. (2013). Manufacturing aspects of advanced polymer composites for automotive applications. *Applied Composite Materials, 20*(2), 107–128.

Gopanna, A., Rajan, K. P., Thomas, S. P., & Chavali, M. (2019). *Polyethylene and polypropylene matrix composites for biomedical applications*. Elsevier Inc.

Hao, A., Zhao, H., Jiang, W., Yuan, L., & Chen, J. Y. (2012). Mechanical properties of kenaf/polypropylene nonwoven composites. *Journal of Polymers and the Environment, 20*(4), 959–966.

Hoque, M. B., Hossain, S., Nahid, A. M., Bari, S., & Khan, R. A. (2018). Fabrication and characterization of pineapple fiber-reinforced polypropylene based composites. *Nano Hybrids and Composites, 21*, 31–42.

Ilyas, R. A., Sapuan, S. M., Atiqah, A., Ibrahim, R., Abral, H., Ishak, M. R., ... Ya, H. (2019). Sugar palm (*Arenga pinnata* [Wurmb] Merr) starch films containing sugar palm nanofibrillated cellulose as reinforcement: Water barrier properties. *Polymer Composites*, 1–9.

Ilyas, R. A., Sapuan, S. M., Ibrahim, R., Abral, H., Ishak, M. R., Zainudin, E. S., ... Jumaidin, R. (2019). Effect of sugar palm nanofibrillated cellulose concentrations on morphological, mechanical and physical properties of biodegradable films based on agro-waste sugar palm (*Arenga pinnata* (Wurmb.) Merr) Starch. *Journal of Materials Research and Technology, 8*(5), 4819–4830.

Ilyas, R. A., Sapuan, S. M., Ibrahim, R., Abral, H., Ishak, M. R., Zainudin, E. S., ... Jumaidin, R. (2020). Thermal, biodegradability and water barrier properties of bio-nanocomposites based on plasticised sugar palm starch and nanofibrillated celluloses from sugar palm fibres. *Journal of Biobased Materials and Bioenergy, 14*, 1–13.

Ilyas, R. A., Sapuan, S. M., Ibrahim, R., Atikah, M. S. N., Atiqah, A., Ansari, M. N. M., & Norrahim, M. N. F. (2019). *Production, processes and modification of nanocrystalline cellulose from agro-waste: A review. Nanocrystalline materials*. IntechOpen.

Ilyas, R. A., Sapuan, S. M., Ishak, M. R., & Zainudin, E. S. (2018). Development and characterization of sugar palm nanocrystalline cellulose reinforced sugar palm starch bionanocomposites. *Carbohydrate Polymers, 202*, 186–202.

Ilyas, R. A., Sapuan, S. M., Ishak, M. R., Zainudin, E. S., & Atikah, M. S. N. (2018). *Characterization of sugar palm nanocellulose and its potential for reinforcement with a starch-based composite. Sugar palm biofibers, biopolymers, and biocomposites* (1st ed., pp. 189–220). Boca Raton, FL: CRC Press/Taylor & Francis Group.

Ilyas, R. A., Sapuan, S. M., Sanyang, M. L., Ishak, M. R., & Zainudin, E. S. (2018). Nanocrystalline cellulose as reinforcement for polymeric matrix nanocomposites and its potential applications: A review. *Current Analytical Chemistry, 14*(3), 203–225.

Ishak, M. R., Sapuan, S. M., Leman, Z., Rahman, M. Z. A., Anwar, U. M. K., & Siregar, J. P. (2013). Sugar palm (*Arenga pinnata*): Its fibres, polymers and composites. *Carbohydrate Polymers, 91*(2), 699–710.

Jarukumjorn, K., & Suppakarn, N. (2009). Effect of glass fiber hybridization on properties of sisal fiber–polypropylene composites. *Composites Part B, 40*(7), 623–627.

Jawaid, M., & Abdul Khalil, H. P. S. (2011). Cellulosic/synthetic fibre reinforced polymer hybrid composites: A review. *Carbohydrate Polymers, 86*(1), 1–18.

Jiang, X., Rui, Y., & Chen, G. (2009). Effect of single-mineral filler and hybrid-mineral filler additives on the properties of polypropylene composites. *Journal of Vinyl and Additive Technology, 21*(2), 129–133.

Jo, Y., Coralia, V. G., Ko, S., Lee, W., Hwa Shin, G., Chun Choi, J., ... Kim, J. T. (2018). Characterization and antibacterial properties of nanosilver-applied polyethylene and polypropylene composite films for food packaging applications. *Food Bioscience, 23*, 83–90.

Jumaidin, R., Ilyas, R. A., Saiful, M., Hussin, F., & Mastura, M. T. (2019). Water transport and physical properties of sugarcane bagasse fibre reinforced thermoplastic potato starch biocomposite. *Journal of Advanced Research in Fluid Mechanics and Thermal Sciences, 61*(2), 273–281.

Jumaidin, R., Khiruddin, M. A. A., Saidi, Z. A. S., Salit, M. S., & Ilyas, R. A. (2020). Effect of cogon grass fibre on the thermal, mechanical and biodegradation properties of thermoplastic cassava starch biocomposite. *International Journal of Biological Macromolecules, 146*, 746–755.

Jumaidin, R., Saidi, Z. A. S., Ilyas, R. A., Ahmad, M. N., Wahid, M. K., Yaakob, M. Y., ... Osman, M. H. (2019). Characteristics of cogon grass fibre reinforced thermoplastic cassava starch biocomposite: Water absorption and physical properties. *Journal of Advanced Research in Fluid Mechanics and Thermal Sciences, 62*(1), 43–52.

Kim, D. H., Kang, S. Y., Kim, H. J., & Kim, H. S. (2019). Strain rate dependent mechanical behavior of glass fiber reinforced polypropylene composites and its effect on the performance of automotive bumper beam structure. *Composites Part B: Engineering, 166*, 483–496.

Krupka, J., Shakhil, P. G., Arun, N. S., Ratheesh, R., Jantunen, H., Kim, H. T., & Sebastian, M. T. (2018). Low loss polypropylene-silicon composites for millimetre wave applications. *Materials Research Bulletin, 104*, 143–148.

Langhorst, A. E., Burkholder, J., Long, J., Thomas, R., Kiziltas, A., & Mielewski, D. (2018). Blue-agave fiber-reinforced polypropylene composites for automotive applications. *BioResources, 13*(1), 820–835.

Maddah, H. A. (2016). Polypropylene as a promising plastic: A review. *American Journal of Polymer Science, 6*(1), 1–11.

Mahjoub, R., Yatim, J. M., Sam, A. R. M., & Hashemi, S. H. (2014). Tensile properties of kenaf fiber due to various conditions of chemical fiber surface modifications. *Construction and Building Materials, 55*, 103–113.

Matweb, L. L. C. (2014). Online materials information resource. <http://www.matweb.com>

Mazani, N., Sapuan, S. M., Sanyang, M. L., Atiqah, A., & Ilyas, R. A. (2019). *Design and fabrication of a shoe shelf from kenaf fiber reinforced unsaturated polyester composites. In* Lignocellulose for Future Bioeconomy (pp. 315–332). Elsevier.

Mirbagheri, J., Tajvidi, M., Hermanson, J. C., & Ghasemi, I. (2007). Tensile properties of wood flour/kenaf fiber polypropylene hybrid composites. *Journal of Applied Polymer Science, 105*(5), 3054–3059.

Norizan, M., Nurazzi, K., Abdan, R. A., Ilyas., & Biofibers, S. P. (2020). Effect of fiber orientation and fiber loading on the mechanical and thermal properties of sugar palm yarn fiber reinforced unsaturated polyester resin composites. *Polimery, 65*(2), 34–43.

Pang, A. L., Ismail, H., & Bakar, A. A. (2018). Mechanical, morphological, and thermal properties of kenaf filled linear low-density polyethylene/poly(vinyl alcohol) composites: Effect of chemical treatment. *Journal of Vinyl and Additive Technology, 24*(Iii), E164–E171.

Pickering, K. L., Efendy, M. G. A., & Le, T. M. (2016). A review of recent developments in natural fibre composites and their mechanical performance. *Composites Part A: Applied Science and Manufacturing, 83*, 98–112.

Radzuan, M., Afiqah, N., Sulong, A. B., Somalu, M. R., Abdullah, A. T., Husaini, T., ... Rosli, M. I. (2019). Fibre orientation effect on polypropylene/milled carbon fiber composites in the presence of carbon nanotubes or graphene as a secondary filler: Application on PEM fuel cell bipolar plate. *International Journal of Hydrogen Energy, 44*(58), 30618–30626.

Rahman, M., Das, S., & Hasan, M. (2018). Mechanical properties of chemically treated banana and pineapple leaf fiber reinforced hybrid polypropylene composites. *Advances in Materials and Processing Technologies, 0698*, 1–11.

Rahman, R., Hamdan, S., Maryani, D., Hashim, A., & Islam, S. (2014). Bamboo fiber polypropylene composites: effect of fiber treatment and nano clay on mechanical and thermal properties. *Journal of Vinyl and Additive Technology*, 1–6.

Reale Batista, M., Desireé, L. T., Drzal, A. K., & Mielewski, D. (2019). Hybrid cellulose-inorganic reinforcement polypropylene composites: Lightweight materials for automotive applications. *Polymer Composites*, 1–16.

Saba, N., Paridah, M. T., & Jawaid, M. (2015). Mechanical properties of kenaf fibre reinforced polymer composite: A review. *Construction and Building Materials, 76*, 87–96.

Samal, S. K., Mohanty, S., & Nayak, S. K. (2009). Banana/glass fiber-reinforced polypropylene hybrid composites: Fabrication and performance evaluation. *Polymer-Plastics Technology and Engineering*, 37–41.

Sanyang, M. L., Sapuan, S. M., Jawaid, M., Ishak, M. R., & Sahari, J. (2016). Recent developments in sugar palm (*Arenga pinnata*) based biocomposites and their potential industrial applications: A review. *Renewable and Sustainable Energy Reviews, 54*, 533–549.

Sapuan, S. M., Ilyas, R. A., Ishak, M. R., Leman, Z., Huzaifah, M. R. M., Ammar, I. M., & Atikah, M. S. N. (2018). *Development of sugar palm-based products: A community project. Sugar palm biofibers, biopolymers, and biocomposites* (1st ed., pp. 245–266.). Boca Raton, FL: CRC Press/Taylor & Francis Group.

Sathishkumar, T. P., Naveen, J., & Satheeshkumar, S. (2014). Hybrid fiber reinforced polymer composites-A review. *Journal of Reinforced Plastics and Composites, 33*(5), 454–471.

Shah, A. U. M., Sultan, M. T. H., Jawaid, M., Cardona, F., & Abu Talib, A. R. (2016). A review on the tensile properties of bamboo fiber reinforced polymer composites. *BioResources, 11*(4), 10654–10676.

Sinha, A. K., Narang, H. K., & Bhattacharya, S. (2017). Mechanical properties of natural fibre polymer composites. *Journal of Polymer Engineering, 37*(9), 879–895.

Sonnenfeld, C., Mendil-Jakani, H., Agogué, R., Nunez, P., & Beauchêne, P. (2017). Thermoplastic/thermoset multilayer composites: A way to improve the impact damage tolerance of thermosetting resin matrix composites. *Composite Structures.*

Soutis, C. (2014). Introduction: engineering requirements for aerospace composite materials. In P. E. Irving, & C. Soutis (Eds.), *Polymer composites in the aerospace industry* (pp. 3–22). Duxford: Woodhead Publishing.

Syafri, E., Sudirman., Mashadi., Yulianti, E., Deswita., Asrofi, M., … Fudholi, A. (2019). Effect of sonication time on the thermal stability, moisture absorption, and biodegradation of water hyacinth (*Eichhornia crassipes*) nanocellulose-filled bengkuang (*Pachyrhizus erosus*) starch biocomposites. *Journal of Materials Research and Technology, 8*(6), 6223–6231.

Szycher, M. (2012). Elastomers. In *Handbook of polyurethanes,* (pp. 345–372). Boca Raton, FL: CRC Press.

Takahashi, S., Imai, Y., Kan, A., Hotta, Y., & Ogawa, H. (2014). Dielectric and thermal properties of isotactic polypropylene/hexagonal boron nitride composites for high-frequency applications. *Journal of Alloys and Compounds, 615,* 141–145.

Toommee, S., & Pratumpong, P. (2018). PEG-template for surface modification of zeolite: A convenient material to the design of polypropylene based composite for packaging films. *Results in Physics, 9,* 71–77.

Uwa, C. A., Jamiru, T., Sadiku, E. R., Huan, Z., & Mpofu, K. (2019). Polypropylene/nanoclay composite: A solution to refrigerated vehicles. *Procedia Manufacturing, 35,* 174–180.

Wang, X., Jiang, M., Zhou, Z., Gou, J., & Hui, D. (2017). 3D printing of polymer matrix composites: A review and prospective. *Composites Part B: Engineering.*

Yan, L., Chouw, N., & Jayaraman, K. (2014). Flax fibre and its composites-A review. *Composites Part B, 56,* 296–317.

Zaaba, N. F., & Ismail, H. (2019). Effects of natural weathering on the degradation of alkaline-treated peanut shell filled recycled polypropylene composites. *Journal of Vinyl and Additive Technology, 25*(1), 26–34.

Zainudin, E. S., Yan, L. H., Haniffah, W. H., Jawaid, M., & Alothman, O. Y. (2014). Effect of coir fiber loading on mechanical and morphological properties of oil palm fibers reinforced polypropylene composites. *Polymer Composites, 35*(7), 1418–1425.

Zakikhani, P., Zahari, R., Sultan, M. T. H., & Majid, D. L. (2014). Extraction and preparation of bamboo fibre-reinforced composites. *Materials and Design, 63,* 820–828.

Zhang, D., He, M., Luo, H., Qin, S., Yu, J., & Guo, J. (2018). Performance of long glass fiber-reinforced polypropylene composites at different injection temperature. *Journal of Vinyl and Additive Technology, 24*(3), 233–238.

Zhang, X., Yang, H., Lin, Z., & Tan, S. (2013). Polypropylene hybrid composites filled by wood flour and short glass fiber: effect of compatibilizer on structure and properties. *Journal of Thermoplastic Composite Materials,* 15–29.

CHAPTER

12

Electrospinning process for green polymeric nanomaterials

A.H. Nurfaizey[1,2], J. Jaafar[3], N.M. Mokhtar[4] and N.A.H. Nordin[5]

[1]Faculty of Mechanical Engineering, Universiti Teknikal Malaysia Melaka, Melaka, Malaysia, [2]Centre for Advanced Research on Energy, Universiti Teknikal Malaysia Melaka, Melaka, Malaysia, [3]Advanced Membrane Technology Research Centre (AMTEC), Universiti Teknologi Malaysia, Johor, Malaysia, [4]Faculty of Civil Engineering Technology, Universiti Malaysia Pahang, Pahang, Malaysia, [5]Chemical Engineering Department, Universiti Teknologi Petronas, Perak, Malaysia

12.1 Introduction

Nanomaterials are materials with one of their dimensions in the nanoscale, typically in the range of 1–100 nm. Nanomaterials are one of the many branches of nanotechnology. The United States National Nanotechnology Initiative defines "nanotechnology" as science, engineering, and technology of things at dimensions of less than 100 nm (Sargent, 2013). However, a broader definition of nanotechnology is commonly used in industry, which includes materials of less than 1000 nm as nanomaterials (Ramakrishna, Fujihara, Teo, Lim, & Ma, 2005). Most materials that are synthesized at the nanoscale exhibit superior properties in terms of strength, weight, durability, reactivity, electrical conductivity, etc., which have been proven beneficial for various applications (Sargent, 2013). Nowadays, nanotechnology has been well integrated in industries such as aerospace, automotives, chemicals, construction, cosmetics, electronics, energy, engineering, environment, food, household, medicine, military, security, sports, and textiles

Design for Sustainability
DOI: https://doi.org/10.1016/B978-0-12-819482-9.00015-0

(Correa, 2019; Dolez, 2015; Roco, 2011). It was reported that the global market for nanotechnology has reached USD1.05 billion in 2018 and the value is expected to reach USD2.23 billion by 2025 (Correa, 2019).

The birth of nanotechnology in the mid-20th century is often credited to a physicist named Richard Feynman who expressed his ideas of manipulating and controlling individual atoms and molecules for specific purposes. His ideas were presented during his talk "There's Plenty of Room at the Bottom" presented at the American Physical Society Meeting in 1959 (Feynman, 1959). It was only in 1974 when the word "nanotechnology" was first introduced by Norio Taniguchi during a conference (Nunes et al., 2019). Since then, nanotechnology has become a popular topic, especially in materials science as scientists attempt to synthesize and characterize new materials. The rapid growth of interest in nanotechnology was also fueled by relevant advances in imaging technologies at the time, leading to the invention of transmission electron microscopes and scanning tunneling microscopes.

Although the science and technology underpinning nanomaterials are fairly new, people have been manipulating the unique properties of nanomaterials for centuries. It was reported that samples of ancient Damascus steel swords dated back to 300 AD have been found to have mixtures of wire- and tube-like nanostructures with diameters of 40–50 nm (Reibold et al., 2009). It is believed that these nanostructures provide superior characteristics to the swords such as high strength and sharp cutting edges. Another example is a luster or thin metallic coating used on medieval pottery in the 13th century that was found to have a combination of a thin-layer (200–500 nm) film containing spherical metallic nanocrystals, and a thin (10–20 nm) metal-free outermost glassy layer (Pérez-Arantegui et al., 2004). This coating gives special attribution to the appearance of pottery products, which are well known for their impressive colors and iridescent looks.

Nanomaterials are normally classified according to their dimensionality, origin, chemical composition, or potential toxicity (Dolez, 2015). In this chapter, the classification of nanomaterials according to dimensionality is adopted. Using this approach, nanomaterials can be classified as zero-dimensional (0D) for materials in the form of particles and spheres, one-dimensional (1D) for fiber-like and tube-like materials, two-dimensional (2D) such as sheets and films, and three-dimensional (3D) for materials that have structures with length, width, and height such as crystals.

The methods of synthesizing nanomaterials can be generally divided into two: top-down and bottom-up processes. Bulk materials are reduced to nanoscale sizes using top-down methods by means of a physical, mechanical, or chemical approach. On the contrary, bottom-up methods assemble atoms or molecules of the materials into nanoscale structures using physical or chemical processes. The 0D nanoparticles can be produced using top-down processes such as crushing, grinding, and electro erosion; and bottom-up processes such as nucleation and growth of atoms. Nucleation and growth of atoms can also be used to produced 1D nanomaterials. An electrospinning process is another option for 1D nanomaterial production in the form of polymeric nanofibers. Two-dimensional nanocoating can be produced using physical or chemical vapor deposition, liquid processes, or chemical grafting. Finally, 3D nanomaterials can be produced using processes such as lithography, self-assembly, and phase inversion.

In this chapter, emphasis is given to the production of 1D and 2D polymeric nanofiber-based materials. Electrospinning is a method of producing 1D nanoscale polymeric fibers using electric charges (Long, Kamsom, Nurfaizey, Isa, & Masripan, 2017). However, the history of the electrospinning process can be traced back to the early 1900s (Cooley, 1900), although the process only became popular nearly a century later when it was introduced as an alternative method for synthesizing nanomaterials (Doshi & Reneker, 1995). To date, a wide range of electrospun fibers have been produced for various potential applications. Among others are polyethylene oxide (PEO), polyvinyl alcohol (PVA), polyacrylonitrile (PAN), polyamide (PA), polystyrene, polylactic acid (PLA), and polycaprolactone (PCL) (Huang, Zhang, Kotaki, & Ramakrishna, 2003; Munajat et al., 2017). These electrospun fibers have an average fiber diameter typically in the range of three to several hundreds of nanometers (Teo & Ramakrishna, 2006). Due to their nanoscale diameters, electrospun fiber mats offer a thin layer of porous medium with an excellent high surface area to volume ratio. This is particularly advantageous in many applications that require a high surface area, most notably in filtration applications (Roslan et al., 2018). Furthermore, electrospun fibers have also been proposed for other applications such as electronics, biomedicine, composites, sensors, tissue engineering, and wound dressing (Munajat et al., 2018; Teo, Inai, & Ramakrishna, 2011).

12.2 Electrospinning process

In its simplest form, an electrospinning apparatus consists of an electrically charged polymer source electrode that is positioned at a close proximity to a grounded collector electrode (Fig. 12.1). A continuous supply of polymer solution or melt is channeled through a pipette tip forming a pendant polymer droplet. When a high electrical potential is applied to the pendant drop, the shape of the droplet becomes distorted due to mutual repulsive forces within the polymer molecules and attractive forces toward the grounded collector. If the applied potential surpasses the minimum threshold, the acting electrical forces overcome the surface tension of the droplet causing a fine jet of polymer to be ejected

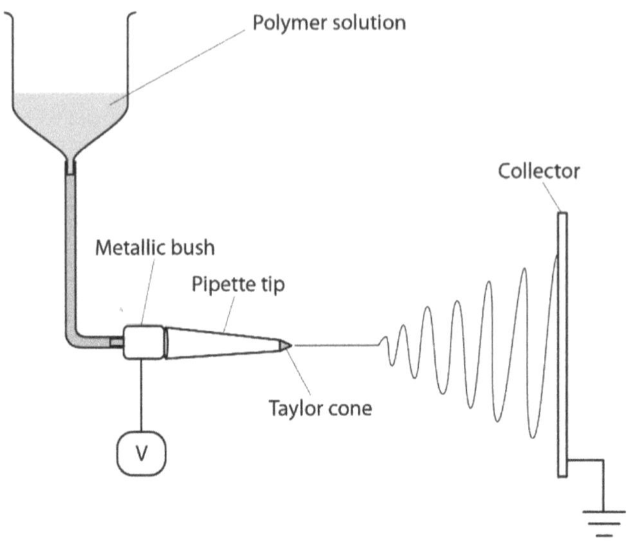

FIGURE 12.1 A typical electrospinning apparatus.

from the ruptured surface moving toward the grounded collector. The jet momentarily travels in a fairly straight trajectory, however as it goes further from the charged electrode, the electrical forces become weaker. The polymer jet starts to buckle and, beyond this point, it continues its journey to the grounded collector in a randomly spiral looping motion known as the "whipping instability" (Che Long, Nurfaizey, & Daud, 2016; Reneker, Yarin, Fong, & Koombhongse, 2000; Riboux, Marin, Loscertales, & Barrero, 2011). During flight, the wet polymer jet undergoes a solidification process as well as a fiber thinning process as the result of mechanical stretching of the fibers. Finally, thin and dry fibers are deposited on the collector electrode in the form of 2D randomly oriented fiber mats.

Due to the random nature of the deposition process in electrospinning it is difficult to tailor the deposited fibers for specific needs. Several attempts at controlling the deposition process in electrospinning have been reported. Controlled deposition in electrospinning was been used for several purposes such as to control the location of the deposition area in order to produce uniform coverage of electrospun fibers, to produce arrays of uniaxially aligned fibers, or to produce 3D nanofiber constructs.

A significant challenge in the synthesis of filtration membranes via electrospinning is in achieving a spatially uniform deposition of electrospun fibers. The problem of achieving a uniformly distributed collection of fibers is particularly pronounced in the case of a multispinneret system due to electrostatic repulsion between adjacent jets that are like-charged (Angammana & Jayaram, 2011; Kumar, Wei, Barry, Chen, & Mead, 2010). Various multispinneret setups with mechanically moving collectors or spinning heads have been described with the aim of producing uniform coverage of electrospun fibers (Filatov, Kirichenko, & Budyka, 2007). However, a major drawback of these methods based on mechanically moving components is that they are complex and require a higher level of supervision and maintenance of the apparatus. Uniaxially aligned fibers, on the other hand, are conventionally obtained by mechanical means such as rotating drums (Boland, Wnek, Simpson, Pawlowski, & Bowlin, 2001), moving spinning heads or collectors (Sun, Chang, Li, & Lin, 2006), or using patterned collectors or gap electrospinning (Li, Wang, & Xia, 2003). Near-field electrospinning (NFES) is another technique for producing aligned fibers. The main feature of this technique is the short distance between the spinneret and collector that is typically in the millimeter range (Zheng et al., 2010). Three-dimensional electrospun fiber structures can be produced using a melt electrospinning process, in which a polymer melt is used as the feed material rather than a polymer dissolved in solution. The absence of the whipping instability during melt electrospinning has been recognized as an advantage for controlled fiber collection. The capability of the process to direct-write a fairly complex 3D structure has been demonstrated previously (Brown, Dalton, & Hutmacher, 2011). However, since there was no whipping instability, thicker fibers are produced up to a few micrometers of diameter. This is due to less fiber stretching and thus less thinning of the fiber during the electrospinning process.

12.3 Green nanofibers using the electrospinning process

The International Union of Pure and Applied Chemistry defines "green materials" as materials that are synthesized from sustainable sources using energy-efficient processes with minimum carbon footprint (Green Materials, n.d.). The word "sustainable" refers to acts that could meet current needs without compromising the future generations (The United Nations ,1987). In materials science, green materials are defined as sustainable and environment-friendly materials that are newly synthesized or modified from existing materials, having superior or special properties (Ahmed & Hussain, 2018). In recent years, the global demand for biopolymers has increased significantly. The total production volume for bio-based polymers in 2019 was 3.8 million tonnes, which was equivalent to 1% of the production volume of conventional fossil-based polymers (nova-Institute, 2020). The growing demand was accelerated by concerns such as diminishing fossil resources, global warming, and increasing energy demand, triggering the needs for renewable and low carbon footprint alternative polymers (Mülhaupt, 2013).

In this section, emphasis is given to producing biopolymers and their bio-based derivative nanofiber structures using an electrospinning process. It is worth noting that there are several other methods to produce nano- to micro-size fibers such as self-assembly, phase separation, template synthesis, island in the sea, and melt and solution blowing (Teo et al., 2011). However, electrospinning is the most widely used method to produce polymeric fibers, mainly due to the simplicity, low cost, and versatility of the process.

In the literature, biopolymer-based electrospun fibers have been produced from a range of biopolymers and their derivatives for various potential applications, especially biomedical applications. This is particularly due to their unique properties that include biocompatibility, biodegradability, and nontoxicity. Clinical applications of these materials cause less side effects, which helps to improve patience compliance and drug stabilization. Biopolymer materials can be categorized into several main categories, that is, cellulose, chitin, lignin, protein, and synthetic biopolymers. Some of the example materials and their applications are described in the following sections.

12.3.1 Electrospinning of cellulose

Cellulose is a form of structural polysaccharide with linear chains of β-1,4-D-glucose monomers with a chemical formula of $(C_6H_{10}O_5)_n$. It is one of the most important polysaccharides besides starch, glycogen, and chitin. Cellulose is the most abundant of all naturally available polymers that forms the main component of plant cell wall structures. Cellulose is hydrophilic, white in color, and has no flavor or odor. Cellulose does not dissolve in water and most common organic solvents due to strong intramolecular hydrogen bonds (Kakoria & Sinha-Ray, 2018). This insolubility behavior makes it difficult to process and thus limits the potential applications of cellulose materials. However, cellulose has found its way into certain applications that require fairly strong fibrous materials. This includes paper industries, packaging, and textiles.

In one of the earliest recorded histories of the electrospinning process, cellulose was the first polymer to be used to produce polymeric fibers. In 1934 a US Patent US1975504 was filed titled *Process and Apparatus for Preparing Artificial Threads* (Formhals, 1934). The patent described the process and apparatus of producing artificial threads using cellulose and its derivative, cellulose acetate. The use of a mixture of acetone, acetic ether, benzene, and alcohol as the solvent was mentioned in the patent. A detailed description of the apparatus and parameter was given, including the application of electrical potential of between 5–10 kV. Due to the technological limitations at the time, no fiber size measurement was taken. However, the patent used the word "extremely fine threads" to describe the fine fibers obtained from the process. The word "electrospinning" was popularized only in 1995 by a group of researchers from The University of Akron (Doshi & Reneker, 1995). The process was considered as a new alternative for synthesizing nanoscale materials.

A decade after electrospinning was introduced, Kulpinski (2005) adapted an electrospinning process in an attempt to produce spruce cellulose electrospun fibers using a mixture of N-methylmorpholine-N-oxide and water as solvent. Several conditions of polymer dope solutions were prepared with concentrations of between 1–4 wt.%. The solutions were heated at 80°C–130°C to help reduce the viscosity to a level suitable for electrospinning with an applied voltage of 9–10 kV. The best electrospinning condition was found at a polymer concentration of 2 wt.%, solution temperature of 95°C–100°C, and applied voltage of 9.5 kV. Fairly smooth and uniform fibers were obtained, with an average fiber diameter of between 200–400 nm (Fig. 12.2). In another study, cellulose electrospun fibers were produce using a different solvent system. Cellulose solution was prepared by dissolving 3 wt.% of cellulose powders in a solvent system containing 8 wt.% of lithium chloride (LiCl) in N,N-dimethylacetamide (DMAc) (Kim, Frey, Marquez, & Joo, 2005). The electrospinning voltage was set at 15–25 kV and electrospinning was conducted at room temperature. Kim et al. (2005) reported that nonuniform fibers were produced at the collector due to the

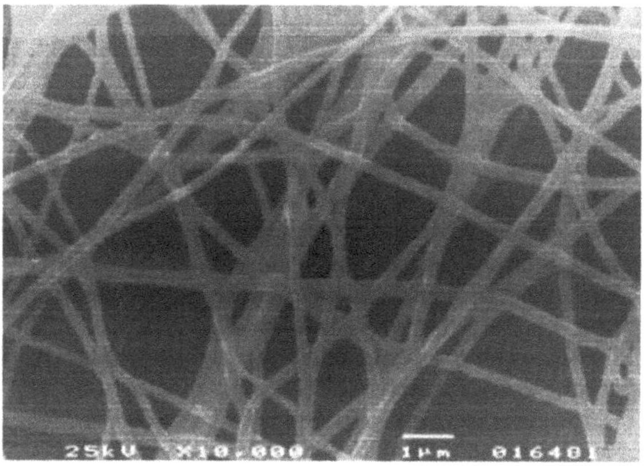

FIGURE 12.2 Scanning electron micrograph of cellulose electrospun nanofibers at × 10,000 magnification (Kulpinski, 2005).

presence of residual DMAc solvent. The authors concluded that by using an LiCl/DMAc system, a heated collector electrode (100°C–105°C) was required to improve the solvent evaporation and a post-electrospinning coagulation process with water was also required to remove LiCl salt. In this study, cellulose electrospun fibers with an average fiber diameter of between 150–500 nm were obtained.

An interesting finding was reported in a study by Ohkawa, Hayashi, Nishida, Yamamoto, and Ducreux (2009), who used a different polymer–solvent system, that is, cellulose in trifluoroacetic acid (TFA). The authors suggested that, as a solvent, TFA showed several characteristics that are crucial for electrospinning such as adequate charge capacity, high volatility, and the ability to dissolve the polymer with less intermolecular interactions. The study successfully produced cellulose electrospun fibers as small as 40 nm mean diameter.

Other than pure cellulose, several researchers have also used cellulose derivatives in their studies, such as cellulose acetate and ethyl cellulose. Cellulose acetate is the acetate ester of cellulose obtained by acetylation of cellulose (Falkiewicz-Dulik, Janda, Wypych, & Janda, 2015). Cellulose acetate is easier to dissolve in compared to pure cellulose due to its weaker intramolecular hydrogen bonds (Kakoria & Sinha-Ray, 2018). Cellulose acetate is highly soluble in acetone and other common organic solvents such as N,N-dimethylformamide (DMF) and tetrahydrofuran (THF). In a recent study, a solution of cellulose acetate in acetone was used to study the effects of processing parameters on fiber quality (Angel, Guo, Yan, Wang, & Kong, 2020). Angel et al. (2020) concluded that at a predetermined flow rate, needle size, airflow, and electrospinning distance, fiber diameter increased with increasing solution concentration. Another important derivative of cellulose is ethyl cellulose. In an attempt to study the effects of electrospinning parameters on the morphology of the fibers, Lim, Gwon, Pyo, and Nho (2010) prepared ethyl acetate ethyl acetate in trifluoroethanol at different concentrations. It was found that a concentration of more than 6 wt.% was required to produced smooth fibers with an average fiber diameter of several hundreds of nanometers.

12.3.2 Electrospinning of chitin and chitosan

Chitin is another form of structural polysaccharide with linear chains of β-1,4-N-acetyl-D-glucosamine monomers with a chemical formula of $(C_8H_{13}O_5N)_n$. On the other hand, chitosan is a derivation of chitin through a deacetylation process. Chitosan has a linear randomly distributed β-1–4-D-glucosamine (deacetylated unit) and N-acetyl-D-glucosamine (acetylated unit). Chitin is one of the most abundant biopolymers in the world. The major natural sources of chitin are crustaceans' shells, such as from crabs and shrimps (Rinaudo, 2006). Although chitin and chitosan are generally insoluble in distilled water and organic solvents, chitin and chitosan can be dissolved in certain inorganic solvents, acidic solutions, and ionic liquids (Roy et al., 2017). Chitin and chitosan have been used in tissue engineering, wound dressings, antimicrobial agents, antiaging cosmetics, and vaccine adjuvants (Elieh-Ali-Komi & Hamblin, 2016).

Due to their good biocompatibility and large surface area, chitin and chitosan electrospun fiber mats have been proposed for wound-dressing applications. Chitin powders were irradiated using gamma rays and later dissolved in hexafluoroisopropanol (HFIP) at varying concentrations between 1–6 wt.% (Min et al., 2004). Irradiation of chitin was carried out to enhance the solubility of the high-molecular-weight chitin in HFIP. Based on SEM examination, the best result was obtained when the solution concentration was at 5 wt.%, producing fibers that were mostly smooth with less irregularities (Fig. 12.3). The average fiber diameter was around 400 nm. Below or above 5 wt.%, the solution was either too thin or thick for electrospinning to sustain, producing irregular fibers with beads. Finally, the as-spun chitin fibers were deacetylated using NaOH aqueous solution to convert chitin fibers to chitosan fibers.

Studies on chitosan electrospun fibers using a number of acidic solvents were conducted by Ohkawa, Cha, Kim, Nishida, and Yamamoto (2004). Chitosan solutions were prepared separately in hydrochloric acid (HCl), acetic acid (AcOH), neat formic acid (FA), dichloroacetic acid (DCA), and trifluoroacetic acid (TFA). The authors noted that none of the acidic solvents were suitable for electrospinning except for TFA. Smooth homogeneous fibers were obtained using 8 wt.% chitosan in TFA

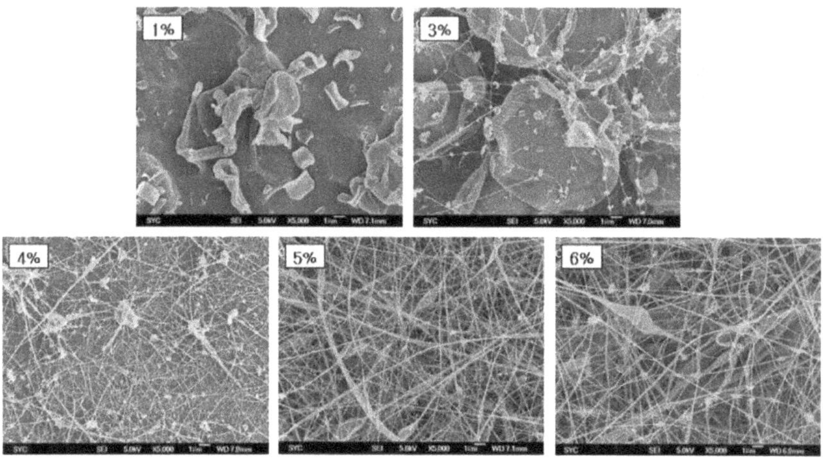

FIGURE 12.3 Scanning electron micrograph (×5000) of chitin electrospun nanofibers using different solution concentrations between 1–6 wt.% (Min et al., 2004).

with an average fiber diameter of 490 nm. It was concluded that electrospinning was possible using TFA as a solvent due to the ability of TFA to form salts with amino groups of chitosan and therefore reduce the rigid interactions between the molecules and increase the charge capacity of the solution (Ohkawa et al., 2004). Several other studies have also used different solvents in a number of attempts to produce chitin/chitosan electrospun fibers using ionic liquids of 1-ethyl-3-methylimidazolium [C_2C_1Im][OAc] and [C_2C_2Im][OAc] (Barber, Griggs, Bonner, & Rogers, 2013), methanesulfonic acid (MSA) (Skotak, Leonov, Larsen, Noriega, & Subramanian, 2008), and lithium chloride (LiCl)-DMAc (Kakoria & Sinha-Ray, 2018).

12.3.3 Electrospinning of lignin

The second most abundant natural polymer is lignin. Although lignin is also a plant-based polymer like cellulose, they are chemically and structurally different. Lignins are highly branched high-molecular-weight polymers with a three-dimensional molecular structure. The chemical formula of lignin is heterogeneous, complex, and varies between plant species. Molecular chains of lignin are formed mainly from three derivatives of benzene, also known as monolignols, that is, conifery alcohol, sinapyl alcohol, and para-coumaryl alcohol. Lignin provides structural support to plants by holding cells and microfibrils together to form cell walls. In industry, lignin has been used for various applications including composites, adhesives, films for energy storage devices, constructions, etc. (Agrawal, Kaushik, & Biswas, 2014; Mansor, Nurfaizey, Tamaldin, & Nordin, 2019).

One of the earliest attempts to produce lignin electrospun fibers was demonstrated by Lallave et al. (2007). The objective of the study was to produce carbon nanofibers by conversion of lignin electrospun fibers into carbon nanofibers through a pyrolysis process. The authors prepared solutions of Alcell lignin in ethanol with a mix ratio of 1:1 and viscosities between 350–400 cPs. A special electrospinning nozzle, which is often called a "coaxial spinneret," was used to cope with the high volatility of ethanol (Fig. 12.4). A coaxial spinneret allows two sources of different solutions to be electrospun simultaneously; the one at the center is the core material and the outer material is used to form a sheath layer. In the study, the lignin solution in ethanol was supplied through the center channel and pure ethanol was supplied through the outer channel. The purpose of supplying ethanol through the outer channel was to compensate for the high rate of solvent loss during the electrospinning process. The authors claimed that a stable electrospinning process was possible using this setup. Smooth and uniform lignin fibers were produced with an average fiber diameter between 400 and 2 μm. After stabilization and carbonization at 900°C, carbon nanofibers were produced with diameters as small as 200 nm.

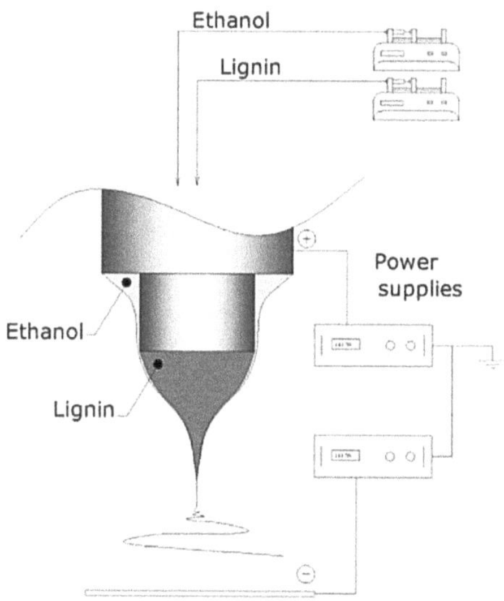

FIGURE 12.4 Schematic of a coaxial spinneret used by Lallave et al. (2007).

The same lignin in ethanol system was used by Ruiz-Rosas et al. (2010) in a similar attempt to produce carbon nanofibers from carbonization of lignin electrospun fibers. The authors prepared two types of lignin in ethanol solution: with and without the addition of platinum acetyl acetonate. The inclusion of platinum acetyl acetonate in the electrospinning solution was seen as an alternative method to introduce metal active catalyst material (such as platinum) onto the fiber surface. Prior to heat treatment, lignin electrospun fibers measuring between 800 nm and 3 μm were obtained. After stabilization and carbonization process, the average fiber diameter was reduced to below 1 μm (Ruiz-Rosas et al., 2010). A surface functionalization procedure of lignin electrospun fibers with phosphorus was also proposed. Phosphoric acid (H_3PO_4) was added into lignin in ethanol prior to electrospinning (García-Mateos, Berenguer, Valero-Romero, Rodríguez-Mirasol, & Cordero, 2018). After electrospinning, similar heat treatment procedures were implemented to produce carbon nanofibers.

12.3.4 Electrospinning of protein

Proteins are involved in vast biological structures and functions of living organisms such as ligaments, hair, fingernails, enzymes, and muscles (Torabizadeh, 2011). Proteins are large biopolymers consisting of long chains of amino acids with a basic formula of $RCH(NH_2)COOH$. In living organisms, the primary structure of a protein is determined by the sequence of different types of constituent amino acids. The unique structure will determine the specific biological function of the biopolymer. A number of industrial applications have been associated with protein engineering including biocatalysis in the food industry, and environmental, medical, and nanobiotechnology applications (Kapoor, Rafiq, & Sharma, 2017). One of the challenges in electrospinning protein is due to difficulties in finding solvents that can dissolve proteins without causing denaturation of the proteins (Kakoria & Sinha-Ray, 2018). In many electrospinning studies proteins were blended together with other biocompatible polymers as carriers.

Collagen-PEO electrospun fibers were produced by Huang, Nagapudi, Apkarian, and Chaikof (2001) using a blend of type I collagen (derived from tail tendons) in weak acid solution, PEO, and NaCL to final concentrations of 1–2 wt.%. The objective of the study was to demonstrate the feasibility of producing collagen–PEO electrospun fibers and to assess the influence of the processing parameters on fiber morphology. Based on SEM and TEM examinations, the optimal processing condition was found when solution concentration, applied voltage, and electrospinning distance were at 2 wt.%, 18 kV, and 15 cm, respectively. The process produced collagen-PEO electrospun fibers with an average fiber diameter of 100–150 nm (Huang et al., 2001). In another study, a blend of type I collagen from calf skin with PCL in 1,1,1,3,3,3-hexafluoro-2-propanol (HFP) was used to investigate the in vivo durability of electrospun collagen–PCL scaffolds (Tillman et al., 2009). The authors claimed that the collagen–PCL electrospun scaffolds could withstand a high degree of in vivo patency and structural integrity which are required for making scaffolds for endothelial and muscle cell growth.

Other than collagen, there were several other proteins that have been used in electrospinning for various biomedical applications. This includes silk, bovine serum albumin (BSA), and soy protein. For example, Jin, Fridrikh, Rutledge, and Kaplan (2002) produced blended *Bombyx mori*-PEO electrospun fibers for vascular graft scaffolds, Kowalczyk, Nowicka, Elbaum, and Kowalewski (2008) produced BSA-PEO electrospun fibers for biosensor applications, and Jiang et al. (2018) prepared serum albumin isolate (SPI)-polyamide 6 (PA6) with silver (Ag) nanoparticles for antibacterial filtrations.

12.3.5 Electrospinning of synthetic biopolymers

Ever since the introduction of synthetic petroleum-based polymers in the early to mid-1900s, the global demand for the "new materials" have continuously grown. According to the latest report by PlasticsEurope – The Facts (2019), global plastics production reached 359 million tonnes in 2018. Polyethylene (PE) and polypropylene were the two most produced polymers by production volume, followed by polyvinyl chloride, polyurethane, polyethylene terephthalate, and others (Beckman, 2018). However, growing concerns over recyclability and degradability of these polymers has led to new demands for greener replacement materials. Recent advances in biotechnology have allowed the exploration of synthetic bio-based polymers such as PLA and polyhydroxyalkanoate (PHA) that have been proven as potential replacements for their petroleum-based counterparts (Meraldo, 2016).

PLA is a synthetic polyester derived from renewable resources such as corn starch and tapioca roots. In an investigation into the electrospinnability of PLA polymer, Casasola, Thomas, Trybala, and Georgiadou (2014) prepared a solution of PLA in acetone at concentrations of 5–15 wt.%. Based on SEM examination, fibers with an average diameter of 757 nm were produced. However, there was formation of bead-like structures as can be seen in Fig. 12.5. In another study, biofunctionalization of electrospun fiber scaffolds was investigated using PLA as the supporting material for electrospinning of type I collagen (Hall Barrientos et al., 2017). PLA was chosen because of its biodegradability and biocompatibility. Electrospinning solution was prepared by mixing PLA in chloroform (CLF) and N,N-dimethylformamide (DMF) with type I collagen in hexafluoropropanol (HFP) to a final polymer concentration of 8 wt.%. The results of drug-release profiles showed that the collagen–PLA electrospun scaffolds demonstrated a sustained release behavior that would be useful for controlled drug-release applications (Hall Barrientos et al., 2017).

Another synthetic biopolymer that has been given much attention recently in biomedical applications is PHA. PHA is a linear polyester that can be made of many different types of hydroxyalkanoate monomers with various properties (Meraldo, 2016). PHA electrospun fiber scaffolds have been proposed for tissue engineering applications due to their good biological properties (Sanhueza et al., 2019). Other than PLA and PHA, polyethylene furanoate (Choi, Kim, Kim, & Kim, 2014) and

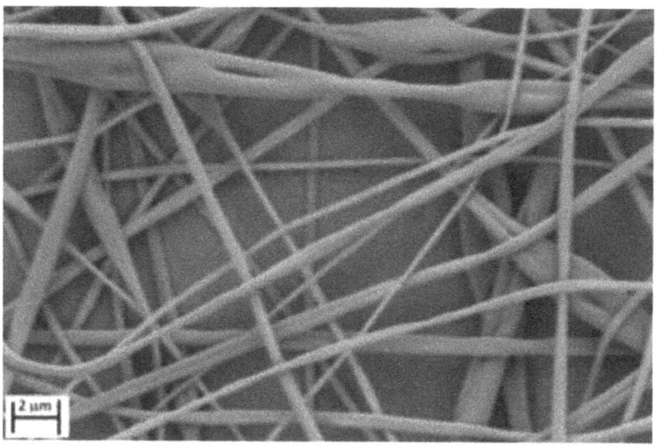

FIGURE 12.5 Scanning electron micrograph of polylactic acid electrospun fibers at × 10,000 magnification (Casasola et al., 2014).

polybutylene succinate electrospun fibers (Huang et al., 2018) have also been proposed for tissue engineering applications.

12.4 The advantages of the electrospinning process

As mentioned earlier, there are several other methods that have been introduced to produce submicron polymeric fibers. However, electrospinning stands out from the rest due to its simplicity and the low startup cost of the process. In electrospinning, the driving mechanism of the process is due to mutual repulsion of like electric charges. Hence, the prerequisite equipment for the electrospinning process is only a high-voltage DC power supply. Futhermore, a wide range of DC power supplies are commercially available for industrial and research applications. One of the leading manufacturers of these power supplies is Glassman High Voltage Inc. A typical high-voltage supply that is used in the electrospinning process costs only several thousand of dollars. This makes the electrospinning process more affordable and easier to setup compared with other nanofiber synthesizing methods.

Another big advantage of electrospinning is the versatility of the process. In the literature, a number of variations of the process have been described in order to obtain fibers that suit the intended applications. For example, several approaches have been described to obtain 2D uniaxially aligned fibers such as using a high-speed rotating collector (Lee, Yoon, & Kim, 2009), electrospinning onto a gap between two parallel collectors (Park & Yang, 2011), and electrospinning by means of direct writing onto a high speed x–y stage (Zheng et al., 2010). Studies have shown that cell growth rates were higher on scaffolds made of aligned fibers as compared with normal scaffolds (Shang, Yang, Cheng, Frank Walboomers, & Jansen, 2010). The versatility of the electrospinning process also allows the process to be upscaled to increase the fiber production rate. This is normally carried out using a multiple spinneret setup (SalehHudin, Mohamad, Mahadi, & Muhammad Afifi, 2018) and using a cylinder partly immersed in polymer solution bath also known as free surface electrospinning (Brettmann et al., 2012). Nowadays, there are several electrospinning machine manufacturers that are producing electrospinning machines which are capable of producing 2D electrospun fiber mats at mass production rates.

As mentioned earlier, the melt electrospinning process can be adapted to produce fairly complex 3D structures (Brown et al., 2011). This would be an advantage for applications that require scaffolds with good mechanical strength. In addition, the absence of solvent in melt electrospinning makes it a preferable process for making electrospun tissue scaffolds due to the lower risk of residual solvent toxicity (Dalton, Joergensen, Groll, & Moeller, 2008).

The processing parameters and characteristics of electrospinning offer a number of advantages for processing green biopolymer materials. In a typical electrospinning process, no heat is required during formation of the fibers. This is advantageous for processing materials that could be sensitive to heat such as proteins. Furthermore, the electrospinning process uses very minimal energy and produces minimal waste. Finally, a small-scale electrospinning process is normally carried out at room conditions and therefore, special controlled environments are not required. However, special attention may be required for air ventilation if toxic solvents are in use.

12.5 Current issues and recommendations

Despite the promising potential of biopolymer electrospun fibers, there are challenges that need to be addressed before these materials can be viably produced. One of the main issues is low solubility of biopolymers in common organic solvents. For example, the use of solvents such as 1,1,1,3,3,3-HFP, THF, FA, and CLF for dissolving proteins could cause denaturation of the molecule structures (Kakoria & Sinha-Ray, 2018). Furthermore, in studies that are intended for biomedical applications, toxicity from residual solvents remains a major concern as it will jeopardize the biocompatibility of electrospun scaffolds. The use of toxic and harmful solvents could also hinder large-scale production of these scaffolds, due to the high amount of solvent evaporation. For a large-scale operation, a special ventilation system would be required, and this could increase the production cost. In order to solve this issue, further exploration of the material–solvent system is required. Solvent-free electrospinning has been proposed by a group of researchers who claimed it could be one of the potential solutions to the problem (Zhang et al., 2017).

Due to the chaotic motion of the whipping instability during the fiber deposition process, a typical electrospinning process only produces 2D randomly oriented fiber mats in a round deposition area of several centimeters in diameter. Therefore it is not surprising that the one application in which electrospun fibers have achieved commercial success so far is for making small filtration media. There is a need to explore new controlled deposition methods for either producing more complex 3D structures or expanding the size of the deposition area. Recently, the author demonstrated a method to produce a 3D layer-by-layer construct consisting of uniaxilly aligned electrospun fibers (Fig. 12.6) (Nurfaizey et al., 2020). The method used a pair of electrically charged secondary electrodes to encourage fiber alignment during electrospinning.

A low production rate is also one of the major obstacles to the industrial application of electrospinning. Depending on the process parameters, such as solution flow rate and concentration, typical single spinneret electrospinning

FIGURE 12.6 Three-dimensional uniaxially aligned electrospun fiber construct produced using secondary electrode-assisted electrospinning (Nurfaizey et al., 2020).

will only produce a few tenths of a gram per hour (Ramakrishna et al., 2005). A commonly used technique for increasing the output of the electrospinning process is to increase the number of spinnerets by using a multispinneret spinning head (Zhou, Gong, & Porat, 2010). However, it is known that a multispinneret setup suffers from mutual Coulombic interactions between the like-charged jets. Thus, a multispinneret electrospinning setup will produce multiple deposition areas that do not freely overlay each other due to this mutual repulsion between jets. Therefore further studies on an effective upscaling method are required for mass production applications.

12.6 Conclusions

In this chapter, a comprehensive review of the electrospinning process for producing green nanomaterials was conducted. Due to simplicity, low cost, and versatility of the electrospinning process, the process has been proposed as an alternative for producing 1D nanomaterials in the form of polymeric nanofibers. The review includes several examples from relevant studies including the use of natural polymers such as cellulose, chitin, lignin, protein, and synthetic bio-based polymers such as PLA and PHA. The biocompatibility, biodegradability, and nontoxicity of these materials show promising potential for biomedical applications. However, several challenges remain such as the toxicity of residual solvents, control of the process, and low production output.

Acknowledgments

This work was variously supported by the Ministry of Higher Education of Malaysia, Universiti Teknikal Malaysia Melaka, and Universiti Teknologi Malaysia; through UTM Collaborative Research Grant (*Enhancement of Membrane Distillation Performance for Seawater Desalination Via Surface Functionalization*) and UTeM Short Term Grant (*Direct Electrospinning of Superhydrophobic Nanofibres on Ceramic Hollow Fiber Membrane for Seawater*

Desalination). Special thanks are expressed to the Advanced Materials Characterization Laboratory and Centre for Advanced Research on Energy of Universiti Teknikal Malaysia Melaka for their support.

References

Agrawal, A., Kaushik, N., & Biswas, S. (2014). Derivatives and applications of lignin—An insight. *The Scitech Journal*, 01.

Ahmed, S., & Hussain, C. M. (2018). *Green and sustainable advanced materials: Volume 1: Processing and characterization*. Hoboken, NJ: Scrivener Publishing LLC. Available from: https://doi.org/10.1002/9781119407089.

Angammana, C. J., & Jayaram, S. H. (2011). The effects of electric field on the multijet electrospinning process and fiber morphology. *IEEE Transactions on Industry Applications, 47*(2), 1028–1035. Available from: https://doi.org/10.1109/Tia.2010.2103392.

Angel, N., Guo, L., Yan, F., Wang, H., & Kong, L. (2020). Effect of processing parameters on the electrospinning of cellulose acetate studied by response surface methodology. *Journal of Agriculture and Food Research, 2,* 100015. Available from: https://doi.org/10.1016/j.jafr.2019.100015.

Barber, P. S., Griggs, C. S., Bonner, J. R., & Rogers, R. D. (2013). Electrospinning of chitin nanofibers directly from an ionic liquid extract of shrimp shells. *Green Chemistry, 15*(3), 601–607. Available from: https://doi.org/10.1039/c2gc36582k.

Beckman, E. (2018). *The world's plastic problem in numbers*. Retrieved from 03.03.2020. <https://www.weforum.org/agenda/2018/08/the-world-of-plastics-in-numbers>.

Boland, E. D., Wnek, G. E., Simpson, D. G., Pawlowski, K. J., & Bowlin, G. L. (2001). Tailoring tissue engineering scaffolds using electrostatic processing techniques: A study of poly (glycolic acid) electrospinning. *Journal of Macromolecular Science, Part A, 38*(12), 1231–1243.

Brettmann, B. K., Tsang, S., Forward, K. M., Rutledge, G. C., Myerson, A. S., & Trout, B. L. (2012). Free surface electrospinning of fibers containing microparticles. *Langmuir, 28* (25), 9714–9721. Available from: https://doi.org/10.1021/la301422x.

Brown, T. D., Dalton, P. D., & Hutmacher, D. W. (2011). Direct writing by way of melt electrospinning. *Advanced Materials, 23*(47), 5651–5657. Available from: https://doi.org/10.1002/adma.201103482.

Casasola, R., Thomas, N. L., Trybala, A., & Georgiadou, S. (2014). Electrospun poly lactic acid (PLA) fibres: Effect of different solvent systems on fibre morphology and diameter. *Polymer, 55*(18), 4728–4737. Available from: https://doi.org/10.1016/j.polymer.2014.06.032.

Che Long, F., Nurfaizey, A.H., & Daud, M.A. (2016). A preliminary study of greyscale intensity and deposited electrospun fibres using image analysis technique. In *Proceedings of Mechanical Engineering Research Day 2016* (pp. 165–166). Melaka: Centre for Advanced Research on Energy. https://doi.org/978967-025770-9.

Choi, H.-J., Kim, S. H., Kim, B.-J., & Kim, S. B. (2014). Fabrication of biomass based polyethylene furoate nanofiber by electrospinning. *Journal of the Korea Academia-Industrial Cooperation Society, 15*(6), 4024–4031. Available from: https://doi.org/10.5762/kais.2014.15.6.4024.

Cooley, J.F. (1900). *Apparatus for electrically dispersing fluids*. US Patent 692,631. US Patent 692, 631. United States. Retrieved from <http://www.google.com/patents/US692631>.

Correa, D. (2019). *Global nanotechnology market to reach $2.23 billion by 2025 at 10.5% CAGR: New study by allied market research*. Retrieved from 22.02.2020. <https://www.globenewswire.com/news-release/2019/11/19/1949526/0/en/Global-Nanotechnology-Market-to-Reach-2-23-Billion-by-2025-at-10-5-CAGR-New-Study-by-Allied-Market-Research.html>.

Dalton, P. D., Joergensen, N. T., Groll, J., & Moeller, M. (2008). Patterned melt electrospun substrates for tissue engineering. *Biomedical Materials, 3*(3). Available from: https://doi.org/10.1088/1748-6041/3/3/034109.

Dolez, P. I. (2015). *Nanomaterials definitions, classifications, and applications. Nanoengineering: Global approaches to health and safety issues* (pp. 3–40). Elsevier. Available from: https://doi.org/10.1016/B978-0-444-62747-6.00001-4.

Doshi, J., & Reneker, D. H. (1995). Electrospinning process and applications of electrospun fibers. *Journal of Electrostatics, 35*(2–3), 151–160. Available from: https://doi.org/10.1016/0304-3886(95)00041-8.

Elieh-Ali-Komi, D., & Hamblin, M. R. (2016). Chitin and chitosan: Production and application of versatile biomedical nanomaterials. *International Journal of Advanced Research, 4* (3), 411–427. <http://www.ncbi.nlm.nih.gov/pubmed/27819009>.

Falkiewicz-Dulik, M., Janda, K., Wypych, G., & Janda, K. (2015). Petroleum products. *Handbook of Material Biodegradation, Biodeterioration, and Biostablization*, 257–375. Available from: https://doi.org/10.1016/B978-1-895198-87-4.50010-4.

Feynman, R. P. (1959). There's plenty of room at the bottom. In *American Physical Society Meeting*. Reprinted in J. Microelectromechanical Systems. 1(1), 60–66 (1992).

Filatov, Y., Kirichenko, V., & Budyka, A. (2007). *Electrospinning of micro and nanofibers: Fundamentals and applications in separation and filtration processes*. Begell House.

Formhals, A. (1934). Process and apparatus for preparing artificial threads. United States Patent 1975504.

García-Mateos, F. J., Berenguer, R., Valero-Romero, M. J., Rodríguez-Mirasol, J., & Cordero, T. (2018). Phosphorus functionalization for the rapid preparation of highly nanoporous submicron-diameter carbon fibers by electrospinning of lignin solutions. *Journal of Materials Chemistry A, 6*(3), 1219–1233. Available from: https://doi.org/10.1039/c7ta08788h.

Green Materials. (n.d.). Retrieved from 25.02.2020. <https://iupac.org/materialschemistryedu/environmental/green-materials/>.

Hall Barrientos, I. J., Paladino, E., Szabó, P., Brozio, S., Hall, P. J., Oseghale, C. I., … Lamprou, D. A. (2017). Electrospun collagen-based nanofibres: A sustainable material for improved antibiotic utilisation in tissue engineering applications. *International Journal of Pharmaceutics, 531*(1), 67–79. Available from: https://doi.org/10.1016/j.ijpharm.2017.08.071.

Huang, A., Peng, X., Geng, L., Zhang, L., Huang, K., Chen, B., … Kuang, T. (2018). Electrospun poly (butylene succinate)/cellulose nanocrystals bio-nanocomposite scaffolds for tissue engineering: Preparation, characterization and in vitro evaluation. *Polymer Testing, 71*, 101–109. Available from: https://doi.org/10.1016/j.polymertesting.2018.08.027.

Huang, L., Nagapudi, K., Apkarian, P. R., & Chaikof, E. L. (2001). Engineered collagen-PEO nanofibers and fabrics. *Journal of Biomaterials Science, Polymer Edition, 12*(9), 979–993. Available from: https://doi.org/10.1163/156856201753252516.

Huang, Z. M., Zhang, Y. Z., Kotaki, M., & Ramakrishna, S. (2003). A review on polymer nanofibers by electrospining and their applications in nanocomposites. *Composites Science and TechnologyCompos. Sci. Technol., 63*(15), 2223–2253. Available from: https://doi.org/10.1016/S0266-3538(03)00178-7.

Jiang, Z., Zhang, H., Zhu, M., Lv, D., Yao, J., Xiong, R., & Huang, C. (2018). Electrospun soy-protein-based nanofibrous membranes for effective antimicrobial air filtration. *Journal of Applied Polymer Science, 135*(8), 45766. Available from: https://doi.org/10.1002/app.45766.

Jin, H. J., Fridrikh, S. V., Rutledge, G. C., & Kaplan, D. L. (2002). Electrospinning *Bombyx mori* silk with poly(ethylene oxide). *Biomacromolecules, 3*(6), 1233–1239. Available from: https://doi.org/10.1021/bm025581u.

Kakoria, A., & Sinha-Ray, S. (2018). A review on biopolymer-based fibers via electrospinning and solution blowing and their applications. *Fibers, 6*(3), 45. Available from: https://doi.org/10.3390/fib6030045.

Kapoor, S., Rafiq, A., & Sharma, S. (2017). Protein engineering and its applications in food industry. *Critical Reviews in Food Science and Nutrition, 57*(11), 2321–2329. Available from: https://doi.org/10.1080/10408398.2014.1000481.

Kim, C. W., Frey, M. W., Marquez, M., & Joo, Y. L. (2005). Preparation of submicron-scale, electrospun cellulose fibers via direct dissolution. *Journal of Polymer Science, Part B: Polymer Physics, 43*(13), 1673–1683. Available from: https://doi.org/10.1002/polb.20475.

Kowalczyk, T., Nowicka, A., Elbaum, D., & Kowalewski, T. A. (2008). Electrospinning of bovine serum albumin. Optimization and the use for production of biosensors. *Biomacromolecules, 9*(7), 2087–2090. Available from: https://doi.org/10.1021/bm800421s.

Kulpinski, P. (2005). Cellulose nanofibers prepared by the N-methylmorpholine-N-oxide method. *Journal of Applied Polymer Science, 98*(4), 1855–1859. Available from: https://doi.org/10.1002/app.22123.

Kumar, A., Wei, M., Barry, C., Chen, J., & Mead, J. (2010). Controlling fiber repulsion in multi-jet electrospinning for higher throughput. *Macromolecular Materials and Engineering, 295*(8), 701–708. Available from: https://doi.org/10.1002/mame.200900425.

Lallave, M., Bedia, J., Ruiz-Rosas, R., Rodríguez-Mirasol, J., Cordero, T., Otero, J. C., … Loscertales, I. G. (2007). Filled and hollow carbon nanofibers by coaxial electrospinning of Alcell lignin without binder polymers. *Advanced Materials, 19*(23), 4292–4296. Available from: https://doi.org/10.1002/adma.200700963.

Lee, H., Yoon, H., & Kim, G. (2009). Highly oriented electrospun polycaprolactone micro/nanofibers prepared by a field-controllable electrode and rotating collector. *Applied Physics A: Materials Science and Processing, 97*(3), 559–565. Available from: https://doi.org/10.1007/s00339-009-5371-3.

Li, D., Wang, Y. L., & Xia, Y. N. (2003). Electrospinning of polymeric and ceramic nanofibers as uniaxially aligned arrays. *Nano Letters, 3*(8), 1167–1171. Available from: https://doi.org/10.1021/Nl0344256.

Lim, Y.-M., Gwon, H.-J., Pyo, J., & Nho, Y.-C. (2010). *Preparation of cellulose-based nanofibers using electrospinning. Nanofibers*. InTech. Available from: https://doi.org/10.5772/8153.

Long, F., Kamsom, R., Nurfaizey, A., Isa, M., & Masripan, N. (2017). The influence of electrospinning distances on fibre diameter of poly (vinyl alcohol) electrospun nanofibres. In *Proceedings of Mechanical Engineering Research Day 2017* (pp. 377–378). Melaka, Malaysia: Centre for Advanced Research on Energy. https://doi.org/ISBN:978-967-0257-88-4.

Mansor, M. R., Nurfaizey, A. H., Tamaldin, N., & Nordin, M. N. A. (2019). Natural fiber polymer composites. *Biomass, biopolymer-based materials, and bioenergy* (pp. 203–224). Elsevier. Available from: https://doi.org/10.1016/b978-0-08-102426-3.00011-4.

Meraldo, A. (2016). Introduction to bio-based polymers. *Multilayer flexible packaging: second edition* (pp. 47–52). Elsevier Inc. Available from: https://doi.org/10.1016/B978-0-323-37100-1.00004-1.

Min, B. M., Lee, S. W., Lim, J. N., You, Y., Lee, T. S., Kang, P. H., & Park, W. H. (2004). Chitin and chitosan nanofibers: Electrospinning of chitin and deacetylation of chitin nanofibers. *Polymer, 45*(21), 7137–7142. Available from: https://doi.org/10.1016/j.polymer.2004.08.048.

Mülhaupt, R. (2013). Green polymer chemistry and bio-based plastics: Dreams and reality. *Macromolecular Chemistry and Physics, 214*(2), 159–174. Available from: https://doi.org/10.1002/macp.201200439.

Munajat, N. A., Nurfaizey, A. H., Bahar, A. A. M., You, K. Y., Fadzullah, S. H. S. M., & Omar, G. (2018). High-frequency dielectric analysis of carbon nanofibers from pan precursor at different pyrolysis temperatures. *Microwave and Optical Technology Letters, 60*(9), 2198–2204. Available from: https://doi.org/10.1002/mop.31326.

Munajat, N. A., Nurfaizey, A. H., Fadzullah, S. H. S. M., Omar, G., Jaafar, J., Roslan, N. S. A., & Husin, M. H. (2017). Fabrication and characterization of carbon nanofibres from polyacrylonitrile precursor. In Proceedings of Mechanical Engineering Research Day 2017, Centre for Advanced Research on Energy Melaka, Malaysia (pp. 362–363).

nova-Institute, (2020). *Press Release: The global bio-based polymer market 2019—A revised view on a turbulent and growing market*. Retrieved from 26.02.2020. <http://nova-institute.eu/press/?id = 164>.

Nunes, D., Pimentel, A., Santos, L., Barquinha, P., Pereira, L., Fortunato, E., & Martins, R. (2019). *Metal oxide nanostructures: Synthesis, properties, and applications* (pp. 1–328). Elsevier. Available from: https://doi.org/10.1016/C2016-0-01647-9.

Nurfaizey, A. H., Salim, M. A., Tamaldin, N., Nadlene, R., Kamarolzaman, A. A., & Tucker, N. (2020). Three-dimensional uniaxially aligned nanofibre construct using secondary electrode assisted gap electrospinning. *International Journal of Nanoelectronics and Materials, In Press.*

Ohkawa, K., Cha, D., Kim, H., Nishida, A., & Yamamoto, H. (2004). Electrospinning of chitosan. *Macromolecular Rapid Communications, 25*(18), 1600–1605. Available from: https://doi.org/10.1002/marc.200400253.

Ohkawa, K., Hayashi, S., Nishida, A., Yamamoto, H., & Ducreux, J. (2009). Preparation of pure cellulose nanofiber via electrospinning. *Textile Research Journal, 79*(15), 1396–1401. Available from: https://doi.org/10.1177/0040517508101455.

Park, S. H., & Yang, D. Y. (2011). Fabrication of aligned electrospun nanofibers by inclined gap method. *Journal of Applied Polymer Science, 120*(3), 1800–1807. Available from: https://doi.org/10.1002/app.33395.

Pérez-Arantegui, J., Molera, J., Larrea, A., Pradell, T., Vendrell-Saz, M., Borgia, I., ... Viti, C. (2004). Luster pottery from the thirteenth century to the sixteenth century: A nanostructured thin metallic film. *Journal of the American Ceramic Society, 84*(2), 442–446. Available from: https://doi.org/10.1111/j.1151-2916.2001.tb00674.x.

PlasticsEurope — The Facts (2019). *An analysis of European plastics production, demand and waste data.*

Ramakrishna, S., Fujihara, K., Teo, W. E., Lim, T. C., & Ma, Z. (2005). *An introduction to electrospinning and nanofibers*. Hackensack, NJ; London: World Scientific.

Reibold, M., Pätzke, N., Levin, A. A., Kochmann, W., Shakhverdova, I. P., Paufler, P., & Meyer, D. C. (2009). Structure of several historic blades at nanoscale. *Crystal Research and Technology, 44*(10), 1139–1146. Available from: https://doi.org/10.1002/crat.200900445.

Reneker, D. H., Yarin, A. L., Fong, H., & Koombhongse, S. (2000). Bending instability of electrically charged liquid jets of polymer solutions in electrospinning. *Journal of Applied Physics, 87*(9), 4531–4547. Available from: https://doi.org/10.1063/1.373532.

Riboux, G., Marin, A. G., Loscertales, I. G., & Barrero, A. (2011). Whipping instability characterization of an electrified visco-capillary jet. *Journal of Fluid Mechanics, 671*, 226–253. Available from: https://doi.org/10.1017/s0022112010005586.

Rinaudo, M. (2006). *Chitin and chitosan: Properties and applications. Progress in Polymer Science (Oxford)*. Pergamon. Available from: https://doi.org/10.1016/j.progpolymsci.2006.06.001.

Roco, M. C. (2011). The long view of nanotechnology development: The National Nanotechnology Initiative at 10 years. *Journal of Nanoparticle Research*. Available from: https://doi.org/10.1007/s11051-010-0192-z.

Roslan, N. S. A., Abdul Hamid, N., Md Isa, M. H., Muhammad, N., Mansor, M. R., & Abdul Munajat, N. (2018). Nylon electrospun nanofibre water filtration media for wastewater treatment. *Materials Research Express, 5*(10), 105010. Available from: https://doi.org/10.1088/2053-1591/aada94.

Roy, J. C., Salaün, F., Giraud, S., Ferri, A., Chen, G., & Guan, J. (2017). Solubility of chitin: Solvents, solution behaviors and their related mechanisms. *Solubility of polysaccharides*. InTech. Available from: https://doi.org/10.5772/intechopen.71385.

Ruiz-Rosas, R., Bedia, J., Lallave, M., Loscertales, I. G., Barrero, A., Rodríguez-Mirasol, J., & Cordero, T. (2010). The production of submicron diameter carbon fibers by the electrospinning of lignin. *Carbon, 48*(3), 696–705. Available from: https://doi.org/10.1016/j.carbon.2009.10.014.

SalehHudin, H. S., Mohamad, E. N., Mahadi, W. N. L., & Muhammad Afifi, A. (2018). Multiple-jet electrospinning methods for nanofiber processing: A review. *Materials and Manufacturing Processes, 33*(5), 479–498. Available from: https://doi.org/10.1080/10426914.2017.1388523.

Sanhueza, C., Acevedo, F., Rocha, S., Villegas, P., Seeger, M., & Navia, R. (2019). *Polyhydroxyalkanoates as biomaterial for electrospun scaffolds. International Journal of Biological Macromolecules.* Elsevier B.V. Available from: https://doi.org/10.1016/j.ijbiomac.2018.11.068.

Sargent, J. F., Jr (2013). *National nanotechnology initiative (NNI): Overview, reauthorization, and appropriations issues.* CreateSpace Independent Publishing Platform.

Shang, S., Yang, F., Cheng, X., Frank Walboomers, X., & Jansen, J. A. (2010). The effect of electrospun fibre alignment on the behaviour of rat periodontal ligament cells. *European Cells and Materials, 19,* 180–192. Available from: https://doi.org/10.22203/eCM.v019a18.

Skotak, M., Leonov, A. P., Larsen, G., Noriega, S., & Subramanian, A. (2008). Biocompatible and biodegradable ultrafine fibrillar scaffold materials for tissue engineering by facile grafting of L-lactide onto chitosan. *Biomacromolecules, 9*(7), 1902–1908. Available from: https://doi.org/10.1021/bm800158c.

Sun, D. H., Chang, C., Li, S., & Lin, L. W. (2006). Near-field electrospinning. *Nano Letters, 6*(4), 839–842. Available from: https://doi.org/10.1021/Nl0602701.

Teo, W. E., & Ramakrishna, S. (2006). A review on electrospinning design and nanofibre assemblies. *Nanotechnology, 17*(14), R89–R106. Available from: https://doi.org/10.1088/0957-4484/17/14/R01.

Teo, W.-E., Inai, R., & Ramakrishna, S. (2011). Technological advances in electrospinning of nanofibers. *Science and Technology of Advanced Materials, 12*(1), 013002. Available from: https://doi.org/10.1088/1468-6996/12/1/013002.

The United Nations (1987). Report of the world commission on environment and development: Our common future. Note by the Secretary-General. [New York] : UN, 4 Aug. 1987. https://digitallibrary.un.org/record/139811?ln = en.

Tillman, B. W., Yazdani, S. K., Lee, S. J., Geary, R. L., Atala, A., & Yoo, J. J. (2009). The in vivo stability of electrospun polycaprolactone-collagen scaffolds in vascular reconstruction. *Biomaterials, 30*(4), 583–588. Available from: https://doi.org/10.1016/j.biomaterials.2008.10.006.

Torabizadeh, H. (2011). All proteins have a basic molecular formula. World Academy of Science, Engineering and Technology, 54, 961–965. <http://www.expasy.org/cgi-bin/protparam>.

Zhang, B., Yan, X., He, H. W., Yu, M., Ning, X., & Long, Y. Z. (2017). Solvent-free electrospinning: Opportunities and challenges. *Polymer Chemistry*. Royal Society of Chemistry. Available from: https://doi.org/10.1039/c6py01898j.

Zheng, G. F., Li, W. W., Wang, X., Wu, D. Z., Sun, D. H., & Lin, L. W. (2010). Precision deposition of a nanofibre by near-field electrospinning. *Journal of Physics D-Applied Physics, 43*(41), 1–6. Available from: https://doi.org/10.1088/0022-3727/43/41/415501.

Zhou, F. L., Gong, R. H., & Porat, I. (2010). Needle and needleless electrospinning for nanofibers. *Journal of Applied Polymer Science, 115*(5), 2591–2598. Available from: https://doi.org/10.1002/app.31282.

CHAPTER

13

Recent development in kenaf (*Hibiscus cannabinus*)-based biocomposites and their potential industrial applications: A review

M.D. Hazrol[1], *S.M. Sapuan*[1,2], *M.Y.M. Zuhri*[1], *E.S. Zainudin*[1], *N.I.A. Wahab*[3] *and R.A. Ilyas*[4,5]

[1]Advanced Engineering Materials and Composites Research Centre (AEMC), Department of Mechanical and Manufacturing Engineering, Faculty of Engineering, Universiti Putra Malaysia, Selangor, Malaysia, [2]Laboratory of Biocomposite Technology, Institute of Tropical Forestry and Forest Products (INTROP), Universiti Putra Malaysia, Selangor, Malaysia, [3]Department of Electrical and Electronic Engineering, Universiti Putra Malaysia, Selangor, Malaysia, [4]School of Chemical and Energy Engineering, Faculty of Engineering, Universiti Teknologi Malaysia, Johor, Malaysia, [5]Centre for Advanced Composite Materials, Universiti Teknologi Malaysia, Johor, Malaysia

13.1 Introduction to natural fiber

The rapid growth since the industrial revolution and increasing interest in global warming issues have led to a proliferation of natural resources utilization for engineering products applications (Jumaidin, Ilyas, Saiful, Hussin, & Mastura, 2019; Jumaidin, Khiruddin, Asyul Sutan Saidi, Salit, & Ilyas, 2020; Jumaidin, Saidi, et al., 2019). A myriad of fiber crop varieties have been produced extensively by cultivation to meet the needs and preferences of society (Aziz & Ansell, 2004; Norizan, Abdan, Ilyas, & Biofibers, 2020;

Design for Sustainability
DOI: https://doi.org/10.1016/B978-0-12-819482-9.00007-1

Nurazzi, Khalina, Sapuan, & Ilyas, 2019; Nurazzi, Khalina, Sapuan, Ilyas, Rafiqah, et al., 2019). Scientific research into polymer science and engineering is playing a crucial role for the worldwide production of natural fibers and agro-waste-based products. The introduction of this technology is expected to produce lighter composites with lower costs compared to existing polymer composites reinforced by fiberglass. Natural fibers have lower density (1.2–1.6 g/cm^3) than glass fiber (2.4 g/cm^3), hence natural fiber-based composites are more lightweight than conventional composites (Huda, Drzal, Mohanty, & Misra, 2006). In addition, there are other advantages such as low wear of tools (Wambua, Ivens, & Verpoest, 2003), low density, cheaper cost, availability, and biodegradability (Nishino, Hirao, & Kotera, 2003). Bast fibers, cotton, jute, flax, kenaf, and sisal are the most commonly used natural plants in the above-mentioned applications (Kozlowski, Mieleniak, Helwig, & Przepiera, 1999).

Despite the abundance, availability, biodegradability, and affordability, it is well known that bioplastics developed from natural resources have certain disadvantages, especially in terms of their mechanical performance and water sensitivity compared to fossil-sourced plastics (Avérous & Boquillon, 2004). Therefore it is necessary to maintain and increase research efforts, particularly for locally available plants such as corn, sugar palm, and kenaf. Also, it is also worth aiming for reproducibility of those biodegradable plastics on an industrial scale while taking the precise functional requirements into consideration (Sapuan et al., 2018). In addition, most conventional building-oriented materials and constructional processes are energy-intensive to produce, and are responsible for a significant amount of landfill volume, and producing about 40% of total greenhouse gas emissions (Rhim, Park, & Ha, 2013). Due to population growth, an increase in the demand for conventional materials has resulted in a massive ecological footprint.

Natural fiber composites (NFCs) have been recently highlighted in various industrial applications and have been slowly replacing conventional materials based upon several factors (Ilyas & Sapuan, 2020). Proper material selection has become pivotal in engineering to attain successful and sustainable design, as well as customer satisfaction attributes (Al-Oqla, Salit, Ishak, & Aziz, 2015b). Moreover, the implementation of new materials in the industrial sector is generally limited by several constrains and limitations, such as the inherent relationship between the materials and their availability, cost, compatibility with the product design, machinability, recyclability, and performance in the final product form. This makes compromising these constraints, advantages, and disadvantages in selecting materials an intricate matter, where proper decisions have to be made concerning modern techniques such as optimization methods, informative decisions, and expert systems utilizing pairwise comparisons (Al-Oqla, Salit, Ishak, & Aziz, 2015a; Sapuan et al., 2011). In comparison to conventional composites, NFCs have greater specific strength and stiffness, better resistance to corrosion, greater fatigue strength and impact absorption capacities, recyclability, adaptability to hazardous environments, lower life-cycle costs, and nontoxicity (Faruk, Bledzki, Fink, & Sain, 2012). These properties are contributed by their constituents (fillers and polymers), particularly the natural fibers that have major advantages over traditional glass fibers. Such advantages include low cost, high/low energy recovery, good thermal and acoustical insulation characteristics, readily available, easily degradable, enhanced CO_2 sequestration, low dermal and respiratory irritation, and minimized tool wear in machining operations (Alves et al., 2010; Beckermann & Pickering, 2008; Faruk et al., 2012; Kalia, Avérous, Njuguna, Dufresne, & Cherian, 2011).

Many naturally occurring fibers are also used as composites but are most common in low-stress applications. Many fibers are obtained by processing waste from agriculture, industry, or end consumers (Bullions, Hoffman, Gillespie, Price-O'Brien, & Loos, 2006; Ilyas, Sapuan, Atiqah, et al., 2019; Ilyas, Sapuan, Ibrahim, Abral, et al., 2019a, 2019b; Ilyas, Sapuan, & Ishak, 2018; Ilyas, Sapuan, Ishak, & Zainudin, 2017, 2018a, 2018b, 2019; Ilyas et al., 2020; Ilyas, Sapuan, Ibrahim, et al., 2019). Some material engineers are attempting to manufacture safer and more environmentally friendly plastics. To alleviate the problem of nonbiodegradable plastics and the disposal issue of biomass waste, the development and manufacturing of environmentally friendly plastic as an alternative to long-lasting plastics is unavoidable (Atikah et al., 2019; Edhirej, Sapuan, Jawaid, & Zahari, 2017; Halimatul, Sapuan, Jawaid, Ishak, & Ilyas, 2019a, 2019b; Syafri et al., 2019). The development of environmentally friendly materials from natural renewable resources has reduced reliance on conventional plastics, which in turn helps to solve environmental pollution complications. Recently, there has been growing interest in the use of natural fibers and agricultural by-products to manufacture biodegradable plastics, such as cassava, kenaf, potato tubers, sugar palms, and maize (Ilyas, Sapuan, Ibrahim, et al., 2019).

The features and performance of products made from NFCs rely heavily on the properties of their individual components and their compatibility, as well as the polymer/filler interfacial characteristics that extend the possibilities of manufacturing various exciting new materials with completely new qualities (Abral, Ariksa, et al., 2019, 2020; Abral, Atmajaya, et al., 2020; Abral, Basri, et al., 2019; Al-Oqla & Sapuan, 2018). The use of natural fiber-enhanced polymer composites instead of synthetic fiber composites offers numerous long-term benefits from the perspective of sustainability, clean manufacturing, and infrastructure management (Al-Oqla & Sapuan, 2018). However, there is an output uncertainty associated with variation in the properties of natural fibers (Al-Oqla, Sapuan, Ishak, & Abdul Aziz, 2014; Alaaeddin, Sapuan, Zuhri, Zainudin, & Al-Oqla, 2019). This requires meticulous study in selecting the best conditions to achieve maximum performance of NFCs.

13.2 Classification of natural fibers

Natural fibers are classified according to their origin, whether obtained from plants, animals, or minerals (Asyraf, Ishak, Sapuan, Yidris, & Ilyas, 2020; Siregar, 2005). Natural plant fibers are extensively used, particularly as reinforcement material in biocomposites (Asyraf, Ishak, Sapuan, Yidris, Ilyas, et al., 2020; Asyraf, Ishak, Sapuan, Yidris, Shahroze, et al., 2020; Azammi et al., 2020). Plant fibers are further subdivided based on the sections of the plant they are obtained from (Lilholt & Lawther, 2000; Mohanty, Misra, & Drzal, 2005). Fig. 13.1 displays the different classes of plant fibers which include the foundation, grass fibers, leaves, beans, fruit, and stalks. The listed plant fibers are classified as nonwood fibers.

The use of nonwood fibers has been the subject of much recent research. Utilizing these natural fibers helps to conserve natural forests, as deforestation is an emerging environmental concern worth addressing. Heavy wood exploitation (i.e., timber) for building materials or wood–plastic composites contributes to deforestation, which in turn leads to a loss of biodiversity. For example, Malaysia had the highest rate of forest depletion in the world, with approximately 14.4% of forest cover lost in the 11-year period from 2000 to 2012 (Butler, 2013). Substituting to natural nonwood fibers is expected to be able to address ongoing forest degradation. Malaysia has vast types of underutilized natural fibers that can be used as alternatives to synthetic fibers. Those natural aboriginal fibers range from kenaf, coconut trunk fibers, sugar palm fibers, sugarcane, sago, pineapple leaf, cocoa pod husk to bunch of oil palm fruit, oil palm trunks, and many more. Nearly all of the natural fibers are ideal as raw materials for composites and other value-added products.

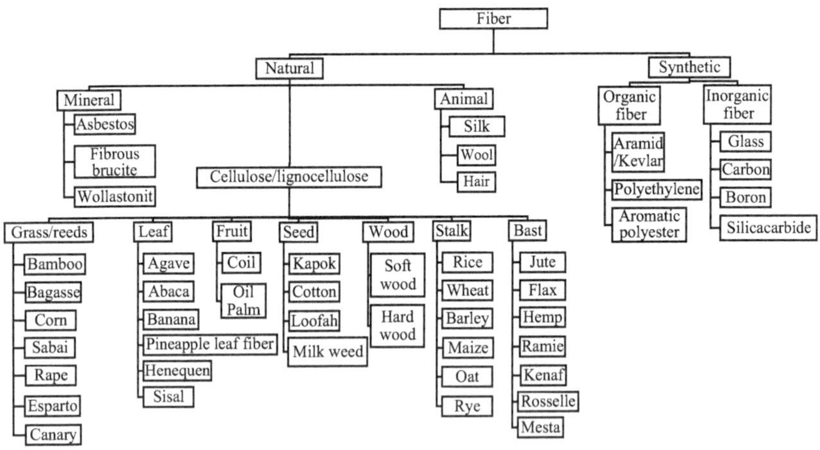

FIGURE 13.1 Classification of natural fibers.

13.3 Kenaf (*Hibiscus cannabinus* L.)

Kenaf is a commercially grown agricultural crop in Malaysia and has been used for various applications in other parts of the world. Kenaf is one of the most commonly harvested plants globally for its fibers, after cotton African (Saba, Paridah, & Jawaid, 2015). Aside from enhancing the strength of a product, kenaf is relatively cheaper than other types of fibers in the market and is readily available. The kenaf plant (*Hibiscus cannabinus*) belongs to the Hibiscus genus and the Malvaceae family akin to okra (*Abelmoschus esculentus* L. Moench) and cotton (*Gossypium* spp.) (Salleh et al., 2012). It is a wild dicotyledon plant native to subtropical and tropical African and Asian regions. The word kenaf originates from Persia and is an annual herbaceous plant that has fast-growing capability and is able to grow in hot weather, with an average fiber diameter of 67.6 mm (Mahjoub, Yatim, Mohd Sam, & Hashemi, 2014). Kenaf is a rugged, sturdy, and hard-wearing plant with a fibrous stalk that is resistant to insect damage and needs relatively little or no pesticide treatment (Elsaid, Dawood, Seracino, & Bobko, 2011). Kenaf is suitable for most soil types and needs minimal chemical treatment, a small amount of fertilizer, and a single herbicide treatment to improve the health and growth of the plant (Elsaid et al., 2011). Kenaf consists of three types of fibers: bast, core, and pith (Karimi, Tahir, Karimi, Dufresne, & Abdulkhani, 2014). The stalk is divided into two distinct parts: 35% bark and 65% core fibers (Abdul Khalil, Yusra, Bhat, & Jawaid, 2010). Kenaf has two distinct types of raw material which can be differentiated by their anatomical appearance in chemical compositions, which are bast and core fiber (Karimi et al., 2014). The pith consists entirely of parenchymatic cells, and is polygonal in form that is not usually prismatic (Paridah, Basher, & SaifulAzry, 2001).

According to researchers, kenaf bast fibers possess impressive mechanical properties which make them suitable for use as a substitute for glass fibers in polymer composites as reinforcing elements (Faruk et al., 2012). Kenaf fiber length, fiber composition, and fiber position affect the physical and mechanical properties of soy-based biocomposites reinforced by kenaf fiber (Liu, Drzal, Mohanty, & Misra, 2007). Kenaf fiber is commonly used in technical fields as a source of fiber for fiber reinforced polymer composite (Mahjoub et al., 2014). Kenaf has other salient characteristics, such as low feedstock, high biomass content, marginal demand for pesticides, and low crop rotation. The main benefits of crop rotation of kenaf are the contribution to weed control, resistance to drought, bringing diversity to dry land, and prevention of soybean cyst nematodes. In polymeric composites, kenaf fiber plays an important role in strengthening and reinforcing materials, which creates a product with more appealing properties. This allows for the replacement of artificial (synthetic) fibers such as glass and aramid with kenaf fibers for structural and nonstructural flexural applications (Yousif, Shalwan, Chin, & Ming, 2012).

13.4 Cultivation of kenaf

Kenaf was grown at a mean temperature of 22°C (condition A) and a heat index of 30°C (condition B) under two different environments. Miyagi Prefecture and Ehime Prefecture in Japan are the geographical locations of conditions A and B, respectively. The kenaf's height was assessed regularly. The impact of harvesting conditions on the kenaf fiber tensile strength was also investigated. In addition, the kenaf was divided into four parts (every 500 mm from the ground) to investigate the effect of fiber position on the plant (from root to tip) tensile strength and tensile properties.

13.5 Properties of kenaf fibers and kenaf fiber reinforced polymer composites

13.5.1 Chemical composition of kenaf fiber

In Malaysia, great efforts have been made to make kenaf an industrial crop since it can be converted into other commercially viable products through the national kenaf research and development program. Accordingly, under the 9th Malaysia Program (2006–2010), the government earmarked RM12 million for kenaf-based industry research and development to convert the plant into a commercially viable crop (Ishak, Leman, Sapuan, Edeerozey, & Othman, 2010). Countless studies have been performed across the nation to make kenaf an economically viable plant. Some of the research was conducted in an effort to explore the chemical and morphological characteristics of kenaf cultivated in Malaysia

TABLE 13.1 Composition of Everglades 71 kenaf fiber variety (Ashori, Harun, Raverty, & Yusoff, 2006).

Organic composition of fibers						**Inorganic (ash content %)**	**Fiber length average (mm)**	**Composition of whole stem (%)**
Fractions	Pentosan (%)	Klason lignin (%)	Acid-soluble lignin	Alpha-cellulose (%)	Hemicellulose (%)			
Core	20.7	19.8	2.3	46.1	29.7	1.6	0.72	65.7
Bast	13.4	13.4	13	56.4	26.2	2.2	2.48	34.3
Stem	19	17.9	2.1	48.7	28.1	1.8	–	–

and concluded that there are substantial similarities in the fractions of kenaf and wood, particularly for use in paper production.

As shown in Table 13.1, the kenaf variety explored by Ashori et al. (2006) was obtained from MARDI experimental area "Everglades 71." The results obtained from this study show that this variety had a high content of hemicellulose and alpha-cellulose in its bast fiber, which contributed greater composite strength and stability. The morphological data obtained suggested that a strong fiber–fiber hydrogen bond had the possibility of being developed due to the minor degradation of the pulp, an ideal property to enhance the strength of paper between the OH fiber-surface groups of polymeric materials.

13.6 Kenaf fibers as reinforcement for polymer composites

Cellulose and lignin are the principal components of natural fibers. Cellulose is responsible for providing mechanical support which depends on several aspects such as the fiber length, fiber loading or fiber fraction period, fiber aspect ratio, fiber orientation, and interfacial adhesion between the fiber–matrix (Rassiah & Megat Ahmad, 2013). Numerous researchers have reported studies of mechanical and other properties of kenaf fibres and kenaf fibre composites (Saba et al., 2015). This study, however, provides more data and more recent studies on the mechanical properties of kenaf composites and kenaf fiber reinforced polymer composites with up-to-date information to 2020. Among other natural fibers, kenaf fibers showed better and superior properties for reinforcement under varied flexural loading conditions in composites with different polymeric matrices. Table 13.2 highlights a list of recent works by various researchers on polymeric (thermoset or thermoplastic resin or biodegradable) reinforced kenaf fiber composites as well as its hybrid composites. Researchers have stated that the mechanical properties of kenaf fiber-enhanced composites differ

TABLE 13.2 Kenaf fiber reinforced polymer composites.

Reinforcement	Matrix	References
Kenaf fiber	HDPE	Salleh, Hassan, Yahya, and Azzahari (2014)
Treated and untreated kenaf	Epoxy	Yousif et al. (2012)
Kenaf fiber	Poly(furfuryl alcohol) bioresin	Khandelwal, Sahoo, Kumar, and Manik (2018)
Kenaf/fiber glass	Polyester	Ghani et al. (2012)

(*Continued*)

TABLE 13.2 (Continued)

Reinforcement	Matrix	References
Short fiber nonwoven kenaf	Polypropylene	Asumani, Reid, and Paskaramoorthy (2012)
Long kenaf/woven glass	Unsaturated polyester	Salleh et al. (2012)
Kenaf bast fiber	PP blended with TPNR and PP/EPDM	Anuar and Zuraida (2011)
Kenaf fiber	Polyurethane	Batouli, Zhu, Nar, and D'Souza (2014)
Kenaf fiber	Polylactide	Lee, Kim, Lee, Kim, and Dorgan (2009)
Kenaf	Poly(lactic acid)	Ochi (2008)
Kenaf-glass	Unsaturated polyester	Atiqah, Maleque, Jawaid, and Iqbal (2014)
Kenaf fiber and corn husk flour	Poly (lactic acid)	Kwon et al. (2014)
Nonwoven kenaf	Polypropylene	Hao, Zhao, and Chen (2013)
Kenaf fiber	Polypropylene	Shibata, Cao, and Fukumoto (2006)
Alkali-treated kenaf fiber	Poly(lactic acid)	Shukor, Hassan, Saiful Islam, Mokhtar, and Hasan (2014)
Kenaf fiber	Waste polypropylene	Suharty, Almanar, Sudirman, Dihardjo, and Astasari (2012)
Kenaf fiber	Cassava starch	Zainuddin, Ahmad, Kargarzadeh, Abdullah, and Dufresne (2013)
Kenaf/glass	Epoxy PBT epoxy	Davoodi et al. (2012)
Kenaf/glass	Epoxy	Davoodi et al. (2010)
Kenaf sheets	PLLA	Nishino, Hirao, and Kotera, (2003)
Kenaf fiber	Polypropylene	John, Bellmann, and Anandjiwala (2010)
Woven kenaf fiber	POM	Dan-Mallam, Abdullah, and Yusoff (2014)
Pultruded treated and untreated kenaf fiber	Polyester	Osman, Mazuki, Akil, Ishak, and Bakar (2013)
Kenaf fiber	Natural rubber	Nurul Aizan, Zainathul Akhmar, Mohd Muhiddin, Nor Hazwani, and Siti Sarah (2013)
Kenaf fiber	PS	Zheng, Xu, Kawai, and Kuroda (2012)

(*Continued*)

TABLE 13.2 (Continued)

Reinforcement	Matrix	References
Chemically treated kenaf fiber	Thermoplastic polyurethane	El-Shekeil, Sapuan, Khalina, Zainudin, and Al-Shuja'a (2012)
Unidirectional kenaf fiber	Epoxy	Abdullah, Alias, Jenal, Abdan, and Ali (2012)
Treated and untreated kenaf fiber	UPE	Ren, Qiu, Fifield, Simmons, and Li (2012)
Kenaf fiber and exfoliated graphite nanoplatelets	Polylactic acid	Han, Karevan, Bhuiyan, Park, and Kalaitzidou (2012)
Kenaf fiber and bacterial cellulose	PLA resin	Sukyai, Sriroth, Lee, and Hyun (2011)
Treated and untreated kenaf fiber	Epoxy resin	Mutasher, Poh, Than, and Law (2011)
Nonwoven kenaf fiber	Polypropylene	Hao, Zhao, Jiang, Yuan, and Chen (2012)
Kenaf fiber	Poly(vinyl chloride)/thermoplastic polyurethane poly-blend	El-Shekeil, Sapuan, Abdan, and Zainudin (2012)
Kenaf fiber	Thermoplastic polyurethane	Sapuan, Pua, El-Shekeil, and Al-Oqla (2013)
Kenaf fiber	HDPE	Salleh et al. (2014)
Kenaf fiber	PP	Bernard et al. (2011)
Kenaf fiber	Concrete	Elsaid et al. (2011)
Kenaf fiber	Epoxy	Xue, Du, Elder, Wang, and Zhang (2009)
Kenaf fiber	Polypropylene	Zampaloni et al. (2007)
Kenaf fibers	HDPE and PP	Meon, Othman, Husain, Remeli, and Syawal (2012)
Pultruded kenaf fiber	Unsaturated polyester	Mazuki, Akil, Safiee, Ishak, and Bakar (2011)
Treated banana/kenaf fibers	Polyester	Thiruchitrambalam, Alavudeen, Athijayamani, Venkateshwaran, and Perumal, (2009)
PALF and kenaf fiber	HDPE	Aji, Zainudin, Khalina, Sapuan, and Khairul (2011)
Kenaf fiber	PHBV and PBAT	Russo, Carfagna, Cimino, Acierno, and Persico (2013)

(Continued)

TABLE 13.2 (Continued)

Reinforcement	Matrix	References
Kenaf fiber	Thermoset composites	Sreenivasan, Sulaiman, Baharudin, Ariffin, and Abdan (2013)
Short kenaf fiber	HDPE	Verma and Shukla (2018)
Kenaf fibers	Cemented sand	Koutenaei, Choobbasti, and Kutanaei (2019)
Woven kenaf mat	Epoxy/bamboo	Chee, Jawaid, Sultan, Alothman, and Abdullah (2019)
Kenaf fibers	Magnesium hydroxide/ epoxy	Saba, Alothman, Almutairi, and Jawaid (2019)
Kenaf fiber	Sisal fibril/polyester	Nimanpure et al. (2019)
Kenaf fiber	*Azadirachta indica* blended epoxy	Arun Prakash and Viswanthan (2019)
Woven kenaf	Carbon fiber/epoxy	Aisyah et al. (2019)
Kenaf fiber	Oil palm/epoxy	Hanan, Jawaid, and Md Tahir (2020)
Kenaf fiber	Polypropylene	Asumani and Paskaramoorthy (2020)
Kenaf	Aloe vera fiber/PLA	Ramesh and Durga Prasad, (2020)
Kenaf fibers	Acrylic-based polyester	Salim et al. (2020)

HDPE, high-density polyethylene; *PALF*, pineapple leaf fiber; *PBAT*, polybutylenes adipate-co-terephthalate; *PHPV*, poly-3-hydroxybutyrate-co-3-hydroxyvalerate; *PLA*, polylactic acid; *PLLA*, poly-L-lactic acid; *POM*, polyoxymethylene; *PP*, polypropylene; *PP/EPDM*, polypropylene/ethylene–propylene–diene–monomer; *PS*, polystyrene; *TPU*, thermoplastic polyurethane; *TPNR*, thermoplastic natural rubber; *UP*, unsaturated polyester.

due to a great deal of variability in test methods used and the samples examined (Rassiah & Megat Ahmad, 2013).

13.7 Properties of kenaf fiber reinforced polymer composites

13.7.1 Physicomechanical properties

The kenaf variety described by Ashori et al. (2006) was collected from MARDI experimental field "Everglades 71." The evidence from this variety in Table 13.3 shows explicitly that it has a high content of helocellulose and alpha-cellulose in its bast fiber, which may lead to increased composite strength and stability. The morphology data obtained showed an indication of the possible formation of a strong fiber–fiber hydrogen bond by virtue of easy pulp collapse, a property

TABLE 13.3 Fiber characteristics of kenaf fractions (Ashori et al., 2006).

	Core			Bast		
Characteristics	Average[a]	Minimum and maximum	SD	Average[b]	Minimum and maximum	SD
Fiber length (L), mm	0.72	0.27–1.2	0.17	2.48	1.62–3.89	0.61
Fiber width (d), μm	31.88	10.69–55.84	8.64	24.24	12.17–47.85	6.92
Wall thickness (w), μm	4.08	0.58–11.50	1.68	4.85	16.5–12.68	1.57
Lumen width (l), μm	25.77	8.38–45.65	8.25	15.44	4.60–33.63	7.11
Runkel ratio (2w/l)	0.25	0.12–0.58	0.08	0.60	0.20–1.73	0.05
Felting factor (L/d)	24.36	10.15–69.32	10.43	104.83	52.93–183.10	29.86
Flexibility ratio (l/d)	80.03	63.40–89.24	4.88	63.87	36.60–83.25	9.39

[a]*Coefficient of suppleness.*
[b]*Arithmetic (number) average, not compensated for fiber mass.*

that is good for paper strength and can also benefit from effective strength between the fiber-surface OH-group and that of polymeric materials. Similarly, the fiber from bast has a wide length to diameter ratio (104:8) resulting in versatile fibers that are good for fiber bonding and wrapping, resulting in higher sheet, tear, and tensile strength properties used in natural fiber composites (NFCs).

Harun et al. (2009) also highlighted some of the findings obtained in Table 13.4 in the downstream field of ongoing work at Universiti Putra Malaysia on certain mechanical properties and had some accepted features, such as Ashori et al. (2006), on both chemical and biologically retted fibers. Most of the dimensions given by Ashori et al. (2006) for the variety Everglades 71 are also compatible with those of the kenaf variety (KB6 and V36) highlighted by Harun et al. (2009). Khalina et al. measured the tenacity of kenaf fibers and noted that chemically retted fibers (138.82 mN/tex) had a lower tenacity value than biologically retarded fibers (204.12 mN/tex). All of these studies have shown considerable prospects for commercial viability and increased technical application of Malaysian cultivated kenaf for the development of composite materials and other applications.

TABLE 13.4 Mechanical properties of chemically retted and biologically retted kenaf fiber (Harun et al., 2009).

	Biologically retted (volume fraction) (%)			Chemically retted (Volume fraction) (%)		
Mechanical properties	40	50	60	40	50	60
Tensile stress (MPa)	176.64	223.20	232.35	134.86	172.20	197.10
Tensile modulus (GPa)	18.04	23.51	13.99	15.37	19.56	21.19
Flexural stress (MPa)	199.33	225.27	248.10	191.97	197.45	214.45
Flexural modulus (GPa)	13.49	15.71	18.49	13.51	13.82	16.11
Impact resistance (kJ/m^2)	41.74	77.87	76.71	35.18	87.22	65.70

13.7.2 Thermal properties

The study of thermal analysis is another important characterization that must be considered in order to completely discern the overall behavior of composite reinforced kenaf fiber. There are three standard characterization methods that can be used for this purpose, namely differential calorimetric scanning (DSC), thermogravimetric analysis (TGA), and dynamic mechanical analysis (DMA). Several important parameters such as glass transition temperature (T_g), fusion temperature, crystalline degree, and oxidation can be determined from DSC analysis (Qu, Wirsén, & Albertsson, 2000). Meanwhile, the mass of the sample as a function of the temperature can be obtained from TGA calculation. Mass shifts usually occur during sublimation, evaporation, decomposition, and chemical reaction, and the material's magnetic or electrical transition, which is directly related to thermal stability (Julkapli & Akil, 2010). In particular, both DSC and TGA were performed by Julkapli and Akil (2010) on kenaf-filled chitosan biocomposites to determine their thermal properties. The heating scan was conducted twice for the DSC study and it was found that both samples displayed a strong endothermic peak during the first heating scan, consistent with water hydration. Meanwhile, it showed decreasing reduction of endothermic temperature for the second heating scan as the kenaf dust content was added to the chitosan film. The authors contended that this phenomenon was due to hydrogen bonding. However, with the rising kenaf fiber content, no strong evidence of changes in enthalpy values (DH) was observed. The TGA results showed that the introduction of kenaf dust in the chitosan film did not significantly alter its thermal stability. On the other hand, DMA is another flexible thermal tool, which can be controlled as a function of temperature and deformation to study the connection between the thermomechanical properties of the material. The measurement of DMA consists of observing a sample's time-dependent deformation activity x(t) under periodic often sinusoidal deformation force with very low F(t) amplitudes. For example, Young's modulus, E0 (storage modulus), and E00 (loss modulus), as well as the mechanical loss factor, and tan d (damping), can be calculated as functions of temperature and deformation (Mazuki et al., 2011). Nishino, Hirao, and Kotera, (2003) performed a DMA test on the kenaf surface, PLLA film, and its composites. They discovered that the kenaf/PLLA composites exhibited a higher dynamic storage module (E′) value than the kenaf sheet and PLLA film values. In addition, this high storage modulus was retained until the PLLA melting point ($T_m = 161°C$ from the DSC measurement). Although the tolerance was limited by the matrix resin's intrinsic property, the composite showed a relatively high thermal resistance. The principal tan d point, assigned as PLLA's glass transition, had changed to a higher temperature for the composite and lower absolute value. This indicated a close association between kenaf fiber and PLLA resin, focusing on the mechanical strengthening of kenaf fiber. Another important DMA finding was obtained by Mazuki et al. (2011) who explicitly studied the degradation of the dynamic mechanical properties of composites enhanced by kenaf fiber after immersion in various solutions. From their overall findings, it was concluded that after immersion in various solutions, the thermal mechanical properties of kenaf fiber-enhanced composite; tan d, storage modulus, and loss modulus decreased as illustrated in Figs. 13.2 and 13.3.

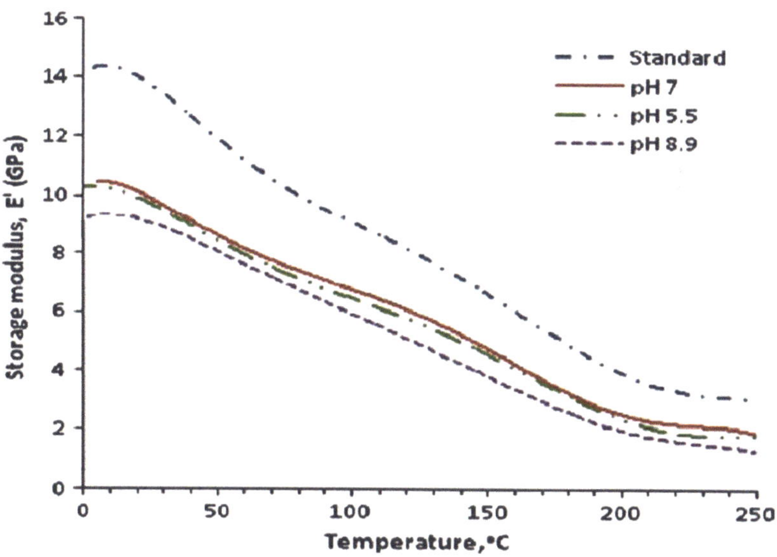

FIGURE 13.2 Variations of storage modulus, E0, for kenaf fiber reinforced composite after exposure to various solutions for 24 weeks (Akil et al., 2011).

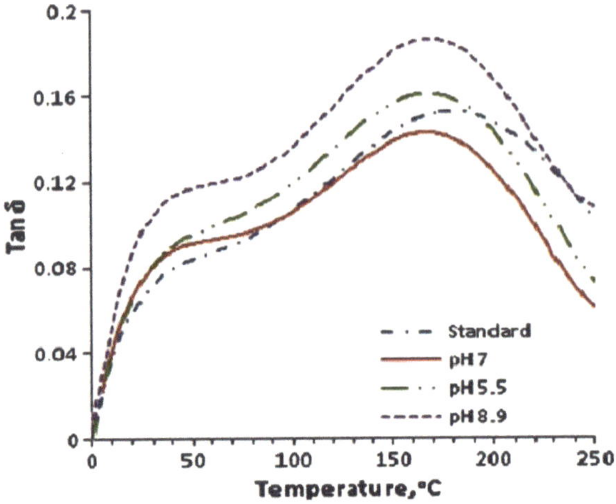

FIGURE 13.3 Variations of tan d for kenaf fiber reinforced composite after exposure to various solutions for 24 weeks (Akil et al., 2011).

Statistically, composites reinforced by kenaf fiber immersed in seawater (pH 8.9) showed the largest reduction, followed by acid rainwater (pH 5.5) and distilled water (pH 7). The reduction pattern reported for tan d is theoretically due to the increase in ductility caused by the restriction of polymer molecules movement after immersion for all tested samples (Gu, 2009). The addition of reinforcing fibers limits the mobility of the deformed polymer molecules and thus increased the values of the storage modules and decreased the viscoelastic gap between stress and strain. In addition, the reduction of storage modulus for immersed kenaf fiber reinforced composite was probably due to the damaged matrix, deteriorated interfacial adhesion, and bond strength between the matrix and fiber. The behaviors of the chemical chain mixture, the Van der Waals bonding, and the hydrogen bonding in the polymer's molecular constructure were responsible for the material's ability to withstand outside stress (Gu, 2009).

The hydrolysis and plasticization of the matrix would destroy the chemical combinations and bonding as water molecules reached the polymer. If a greater strain was caused in the event of stress, it would eventually lead to a decrease in the storage modulus. In addition, due to the increased mobility of polymer chains in composites with poor adhesion, the storage modulus of such materials was lower than that of weak interfacial bonding (Gérard, Perret, & Chabert, 1990). Instead of normal operation, the DMA system can also be controlled to determine the temperature of the heat deflection of kenaf fiber reinforced composites as shown by Liu et al. (2007). Temperature for heat deflection (HDT) is commonly defined as the temperature at which a material deflects by 0.25 mm under load application (0.45 MPa). According to Liu et al. (2007), in a scenario where modulus at room temperature is independent of the processing method, the heat deflection temperature of kenaf/soy-based plastic biocomposites showed a substantial increase of around 36°C. Contrary to expectations, the rate at which the modulus decreased with compressed molded samples temperature was lower than that observed for injected molded samples, as presented in Fig. 13.4. This change would cause the creep rate under a given load to decrease, resulting in higher HDT values. However, HDT values increased with better fiber quality (as well as fiber length), indicating the modulus' dependency on those variables. Taken together, they believed that improving the HDT of a composite by optimizing fiber size, fiber content, and processing methods could reveal new uses for biocomposites based on kenaf/soy plastics.

13.7.3 Water absorption properties

As discussed earlier, water absorption of natural fiber reinforced composites is a serious concern, in particular for their possible outdoor applications. The water absorption characteristic for a given composite system depends on the fiber material, fiber orientation, temperature, exposed surface area, fiber permeability, void material, and hydrophilicity of the individual components (Nosbi, Akil, Mohd Ishak, & Abu Bakar, 2010). A significant analysis and discussion of the subject were

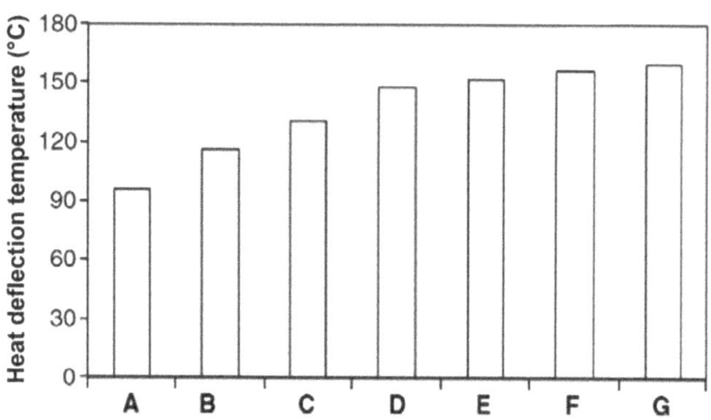

FIGURE 13.4 Heat deflection temperature of kenaf fiber reinforced soy-based biocomposites for (A) 30 wt.% kenaf fiber injection molded composite, (B) 33 wt.% 6 mm kenaf fiber compression molded composite, (C) 55 wt.% 2 mm kenaf fiber compression molded composite, (D) 56 wt.% 6 mm kenaf fiber compression molded composite, (E) 57 wt.% 25 mm kenaf fiber compression molded composite, (F) 53 wt.% 50 mm kenaf fiber compression molded composite, and (G) 54 wt.% long kenaf fiber compression molded composite (Akil et al., 2011).

presented by Nosbi et al. (2010), where it was described that the water absorption behavior of kenaf fiber reinforced unsaturated polyester composites was found to imitate the action of a Fickian, where it attained equilibrium at a particular time of immersion. More recent research from Mazuki et al. (2011) also showed a similar pattern had been documented. Interestingly, they found that the composite enhanced with kenaf fiber displayed a different magnitude of absorption characteristics under different solutions, as shown in Fig. 13.5.

From Fig. 13.5, it is evident that the moisture content increased as the immersion time was prolonged. A similar finding was reported by Doan, Brodowsky, and Mäder (2007). Nonetheless, the weight increase was not related to the time of immersion. The weight surged dramatically at the beginning of the curve (up to 17 days of immersion), showing rapid penetration of moisture into the composite material. A similar overall trend was observed for all the solutions considered in their analysis. They firmly believed that this phenomenon was due to penetration of water into the interface as the fiber swells, which activates capillary action. After 20 days of immersion, the water absorption rate decreased and eventually became saturated, starting from 260 days onwards. A noteworthy finding is that the reported weight gain of the specimen immersed in an acidic solution after 104 days of immersion (p50 h) was almost identical to that immersed in distilled water. Based on the solution's pH review, it was noticed that the neutralization phase occurred in the acidic solution, where the acidic solution's pH was nearly neutral (approximately pH 7) after 260 days of immersion compared to the initial pH of 3 at the initial phase of the study. As expected, the lowest weight gain in seawater was in the saturated condition (after 260 days of

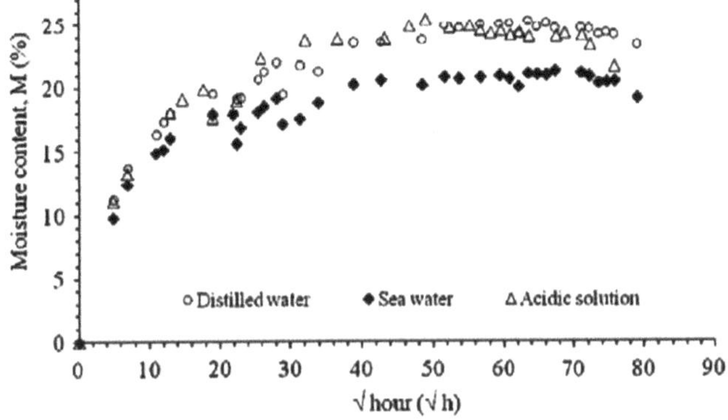

FIGURE 13.5 Water absorption content curves in distilled water, seawater, and acidic solution for compression specimens (Akil et al., 2011).

TABLE 13.5 Reverberation time of the reverberation room.

Centre frequency, f(Hz)	250	315	400	500	630	800	1 k
Rev time, T_{60}	11	11.7	11.5	9.7	8.8	7.4	7.8
Center frequency, f(Hz)	1.25 k	1.6 k	2 k	2.5 k	3.15 k	4 k	5 k
Rev time, T_{60}	6.6	5.7	5.6	4.8	4.4	3.3	2.9

immersion) of kenaf fiber reinforced composite. They assumed this was due to the presence of ionic salt in seawater, blocking the direction of diffusion after long-term exposure, thereby slowing down the process of absorption.

13.7.4 Sound absorption

Fiber sheets from industrial kenaf roll were used as the specimen for a random incidence sound absorption coefficient test (Lim, Putra, Nor, & Yaakob, 2018), with an expanding sheet of fiber of 0.43 m thickness. Forty-two kenaf sheets were obtained from a single fiber roll which in total provided 7.77 m^2 of absorption area. In accordance to BS EN ISO 354 standard procedure, the random incidence test was carried out in a reverberation room. The reverberation room had nonparallel surfaces with reflective walls, ceilings, and floors. The room volume was 171 m^3, and the empty space measured reverberation period is described in Table 13.5. The absorption coefficient was measured with center frequencies ranging from 250 Hz to 5 kHz in one-third-octave bands.

The finding showed that kenaf fibers had good sound absorption efficiency in both normal and random incidence. The result of an impedance tube test revealed that the normal-incidence sound absorption coefficient of kenaf fiber was found to be above 0.8 from 1.5 kHz onwards for specimens with 40 mm thickness and 93.5 kg/m^3 bulk density. Increasing the bulk density further increased the coefficient of sound absorption. Kenaf fibers would attain an absorption coefficient above 0.5 starting from 400 Hz, with an average value of 0.8 under the random incident test. Interestingly, the sound absorption output of kenaf fibers was comparable to the absorption level of synthetic rock wool (Lim et al., 2018).

13.8 Kenaf fiber reinforced hybrid polymer composites

Hybrid polymer composites can be produced by combining two fibers in one matrix or one fiber with two polymer mixtures (Asim et al., 2018). Hybrid composite properties were influence by fiber length, orientation, fiber/matrix interfacial bonding, fiber size, degree of fiber mixing, and arrangement of both fibers. Several researchers have produced composite hybrids by combining natural fibers with polymer matrices (Ibrahim, Sapuan, Zainudin, & Zuhri, 2020). In the case of kenaf-glass (KG) fibers, an unsaturated polyester hybrid composite that was produced by sheet molding compound (SMC) process exhibited higher tensile, better flexural and higher impact strength compared to treated kenaf fiber, according to the researchers (Ilyas et al., 2019). Kenaf with 15/15 v/v kenaf-glass fibers on 6% sodium hydroxide treatment using 3 hours mercerization method yielded stronger mechanical resistance to the composite (Mohd Nurazzi, Khalina, Sapuan, Dayang Laila, & Rahmah, 2017). The researchers concluded that kenaf fiber alone (30% volume fraction) or higher percentage (22.5% volume fraction) could not withstand a higher impact load due to its brittleness and poor toughness. According to the researchers, the dimensional stability and mechanical properties of composites are often affected by aging in humidity. In addition, work also had been carried out to depict the mechanical properties of kenaf fiber, which were found to depreciate due to the penetration of moisture into the kenaf/fiber glass polyester hybrid composite even in dissimilar environmental circumstances, including fresh water, rainwater, and seawater (acidic solutions) at room temperature from the initial phase of the test until the fourth week.

In other research work, researchers have effectively explored long kenaf/woven glass hybrid composites to study the pronounced impact of water absorption on mechanical properties (Ilyas, Sapuan, Ishak, & Zainudin, 2018c). A reduction of fracture were observed, presumably due to the effect of fiber orientation, fiber quality, exposed surface area, individual component hydrophilicity, void quality, and fiber permeability toward water absorption (Sreenivasan et al., 2013). Furthermore, the aging of humidity also produced undesirable effects on the composites' dimensional stability and fracture toughness properties. Work on a hybrid composite with banana also produced some promising results by treating fibers (banana/kenaf) for 30 minutes with 10% sodium lauryl sulfate (SLS) and 10% sodium hydroxide (NaOH). The SLS treatment for both the woven hybrid and random mix composite increased the flexural and tensile strength. The insight of the study showed that SLS treatment can provide better mechanical properties as it improved fiber–matrix adhesion, which in effect improved the mechanical strength more than the treated alkali. In addition, SLS treatment also cleaned the fibers on the surface and the amount of lignin in SLS fibers was found to be better than alkali-treated fibers (Osman et al., 2013). Scientists have investigated the mechanical properties of polyoxymethylene (POM) composite, reinforced by woven kenaf fiber after hybridization with polyethylene terephthalate fiber (PET) (Sohrab Azizi, Claudiane Ouellet-Plamondon, David, & Frechette, 2017). Hybridization with polyethylene terephthalate fiber (PET) fiber showed substantial improvements in terms of impact and tensile properties of the woven kenaf–POM composites. In another research, high-density polyethylene (HDPE) reinforced pineapple leaf fiber (PALF) and kenaf fiber showed no comparable improvement in flexural and tensile properties due to fiber aggregation of long fibers. The fiber agglomeration seemed to be a more important cause than the slow degradation of the fiber that caused mechanical properties to decline during internal mixing. Research has proved that the mechanical properties of thermoplastic glass mat (GMT) such as Young's modulus, tensile strength, flexural module, and flexural strength were close to hybrid kenaf/glass fiber epoxy composite (Davoodi et al., 2010).

13.9 Potential applications of kenaf reinforced polymer composites

Since the late 1950s, the use of kenaf fiber has attracted many studies, investigations, and projects to explore it as a potential national industrial crop. The utilization of this industrial crop expanded from historic and ancient uses as a cordage crop (sackcloth, rope, and twine) to modern uses including livestock feed, pulp and paper products, absorbents, electrical appliances, automotive panels, and building materials, enabling it to be a major potential industrial crop (Aisyah et al., 2019; Mazani, Sapuan, Sanyang, Atiqah, & Ilyas, 2019).

Natural fiber reinforced polymer composites or biocomposites have been a focus of interest for the research community for the past few decades (Azammi et al., 2020). These biocomposites are proposed to replace synthetic-based fibers because of their good attributes including low cost, low weight, biodegradability, high specific properties, less abrasiveness to equipment, environmentally friendly, sustainability, as well as renewability, hence producing a positive environmental impact (Ilyas & Sapuan, 2020). Amongst the natural fibers produced, kenaf fiber is considered to be one of the most important natural fibers and has received much consideration from the composite-based industry as a potential reinforcement for polymer composites. As with other natural fibers, kenaf fibers have lower thermal resistance compared with synthetic fibers. A study conducted by El-Shekeil, Sapuan, Jawaid, and Al-Shuja'a (2014) on reinforcement of kenaf fiber with a blended polymer of poly(vinyl chloride) (PVC)/thermoplastic polyurethane (TPU) (PVC/TPU/KF) composites showed reduction of impact strength, tensile strength, strain, and thermal stability with an increase in fiber loading. Bast kenaf fibers have higher mechanical properties compared to other parts of the kenaf plant. According to Anuar and Zuraida (2011), reinforcement of kenaf fibers with propylene polymer composites improved the tensile, stiffness, impact strength, and flexural properties of composites. Another study conducted by Ochi (2008) on the reinforcement of kenaf fibers and kenaf/PLA composites reported that the flexural and tensile strength of kenaf fiber/PLA polymer composites increased linearly with fiber contents of up to 50%. Likewise, a study conducted by Nishino, Hirao, and Kotera, (2003) on kenaf reinforced biodegradable composite observed that the reinforcement of kenaf fiber with poly-L-lactic acid (PLLA) improved the tensile strength (62 MPa) and Young's modulus (6.3 GPa) of the composites.

The reinforcement of kenaf fibers within the polymer matrix also resulted in thermal and charring stability improvement of polymer composites. In research work by Aisyah et al. (2019), incorporation of woven kenaf fiber and carbon fiber into the epoxy showed that a woven kenaf/carbon hybrid composite with higher kenaf fiber loading exhibited better thermal stability compared to other samples. Additionally, Nishino, Hirao, and Kotera, (2003) conducted dynamic mechanical analysis testing on the PLLA film, kenaf sheet, and kenaf/PLLA composites. It was observed that the kenaf reinforced PLLA composites displayed a higher value of dynamic storage modulus (E′) compared with the PLLA film and kenaf sheet.

Therefore, according to the above-mentioned benefits of kenaf fiber reinforced polymer composites, it can be deduced that this technology has huge potential to be used in various applications. Fig. 13.6 shows

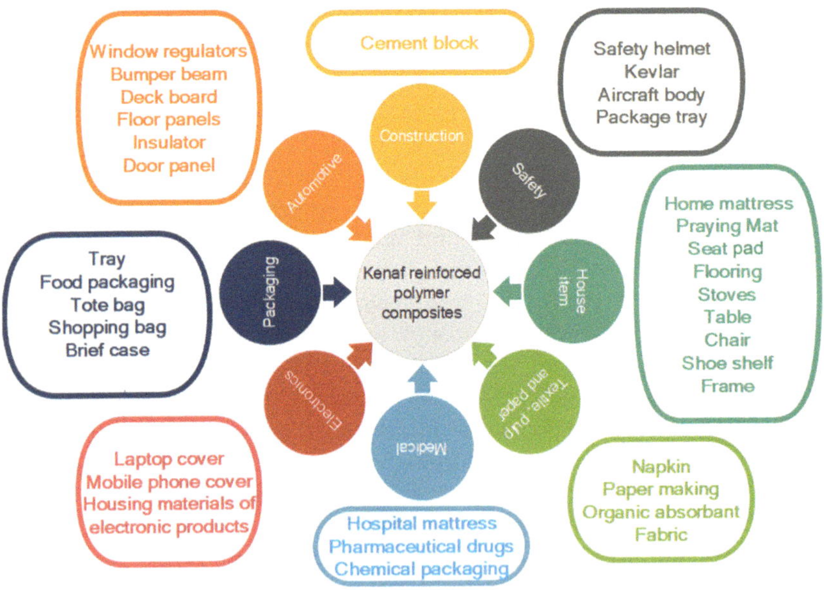

FIGURE 13.6 Applications of kenaf fiber reinforced polymer composites.

the potential applications of kenaf fiber reinforced polymer composites in various sectors including construction and housing, safety, household items, textiles, pulp and paper, decoration, electronics, packaging, wood panels, and automotive applications.

13.9.1 Housing materials of electronic products

Kenaf fiber also can be used as a housing material for electronic products. Kenaf fiber reinforced composites composed of 15%–20% fiber have been used for cell phone shells (Saba, Jawaid, Sultan, & Alothman, 2017). A study conducted by Serizawa, Inoue, and Iji (2006) on the development of high-performance bioplastics from polylactic acid (PLA) reinforced kenaf fiber showed significant improvement of modulus and heat resistance (distortion temperature under load). They concluded that the composite produced had comparable characteristics to current petroleum-based plastics used for housing materials of electronic products (Serizawa et al., 2006). Kenaf fiber reinforced polymer composites were also used for laptop covers, mobile phone covers, and housing materials for electronic products. According to Abdul Rahman, (2011), great success was noted in 2004, when the world's first and only kenaf fiberboard plant at Panasonic Electric Works Ltd. was established in Pahang, Malaysia. A total of 60,000 pieces of board were produced per month by the factory to be exported to Japan. These fiberboards became very well known to the consumers, especially among housing developers due to their eco-friendly properties, excellent strength, and lighter weight.

13.9.2 Safety equipment

Azmi, Sultan, Hamdan, Nor, and Jayakrishna (2018) conducted an experiment on the fabrication of bulletproof vests from kenaf fiber reinforced polymer composites using a hand-lay-up technique into seven layers of different configurations. The study was performed in order to analyze the impact of high-velocity and flexural properties on the novel composite materials as substitutes for existing insert panels for a bulletproof vest. The finding of this research addressed the disadvantages of conventional vests which are heavy and expensive. Based on the flexural properties of kenaf polymer composites, the third configuration was the best configuration compared to other samples, with a flexural modulus of 25.85 GPa.

The performance of a kenaf aero, flax aero, and Kabuto Aero SL cycling safety helmets in an oblique impact analysis with a simulated road surface was investigated by Abdullah (2016), as shown in Fig. 13.7. The findings indicated that the maximum acceptable impact velocity was 20 km/h (12.5 mph), which is equivalent to a weight of 300 g being dropped on the

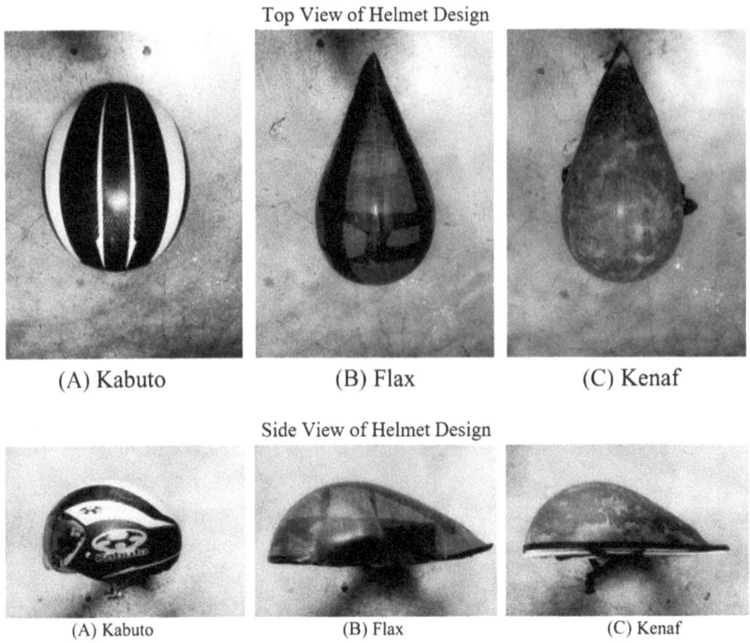

FIGURE 13.7 Top and side views of the helmet design. *Source: Reproduced with permission from Abdullah, M. N. (2016). Oblique impact analysis on kenaf (Hibiscus cannabinus) and flax (Linum usitatissimum) natural fiber sport cycling helmets. Movement, Health & Exercise, 5(1), 57–67.*

head. According to Abdullah (2016), the flax aero helmet and kenaf aero helmet showed 11.82% and 30.63% reductions in the resultant linear acceleration compared to a commercial cycling helmet, Kabuto Aero SL, respectively.

In another study conducted by Salman, Leman, Sultan, Ishak, and Cardona (2016) on the effect of stacking sequence and hybridization of plain woven kenaf/aramid reinforced polyvinyl butyral laminated hybrid composite under high-velocity ballistics, the ballistic limit gradually increased as the number of kenaf layers was increased in comparison to the aramid composite. Fig. 13.8 shows the ballistic hybrid helmet fabrication process. This study was conducted to reduce the reliance on ballistic resistance components such as aramid by fabricated novel helmets using a hot press technique. The reinforcement of plain woven kenaf and aramid reinforced polyvinyl butyral laminated hybrid composite led to the production of helmets that are cheap and easily available compared to conventional helmets. From their study, Salman et al. (2016) concluded that a positive effect on ballistic performance was observed for hybridization of kenaf fiber and aramid laminated composites. In addition, the Institute of Tropical Forestry and Forest Product (INTROP), Universiti Putra Malaysia fabricated a safety helmet from kenaf reinforced polymer composites, as shown in Fig. 13.9.

FIGURE 13.8 Ballistic hybrid helmet fabrication process. *Source: Reproduced with permission from Salman, S. D., Leman, Z., Sultan, M. T., Ishak, M. R., & Cardona, F. (2016). Ballistic impact resistance of plain woven. BioResources, 11(3), 7282–7295.*

FIGURE 13.9 Safety helmet from kenaf composites.

13.9.3 Food packaging and food containers

Kenaf reinforced polymer composites have been used widely in food packaging and food container industries. These polymer composites were extruded and cast into various types of packaging, including pharmaceutical and chemical packaging, electronics and electrical, disposal packaging, and shopping bags (Fig. 13.10), as well for commercial food storage packaging and containers (Fig. 13.11) (Saba, Paridah, Jawaid, Abdan, & Ibrahim, 2015).

FIGURE 13.10 Tote bag and shopping bag.

FIGURE 13.11 Kenaf composite plate and food container.

13.9.4 Construction and housing industry

Wood-based as well as metal-based products can be substituted by kenaf reinforced polymer composites which are extruded and molded into low cost and lightweight panels in construction and housing applications as an engineered material. Also, these composites can be used to make lightweight, strong, and fire-resistant cement blocks with great insulation properties. Currently, kenaf core blocks are utilized to construct solitary family homes and also multistory building blocks (Saba et al., 2015). The INTROP, Universiti Putra Malaysia fabricated flooring panels from kenaf/glass fiber reinforced polyester polymer composites, as shown in Fig. 13.12. INTROP had produced frames from kenaf fiber reinforced polypropylene composites, as shown in Fig. 13.13. Mazani et al. (2019) focused on designing and fabricating a shoe shelf using kenaf fiber reinforced unsaturated polyester composite, as shown in Fig. 13.14. These biocomposites have huge potential for making

FIGURE 13.12 Kenaf composite flooring.

FIGURE 13.13 Frame from kenaf fiber composites.

household furniture at a very low processing and manufacturing cost. It can serve as a good replacement for conventional plastic, wood, and metal furniture, where affordability, low cost of manufacture, light-weight, strength and environmental impact are priorities. Mahmood et al. conducted an experiment on designing and developing kenaf fiber reinforced polypropylene polymer composite chairs and table, as shown in Fig. 13.15. They concluded that the use of kenaf fiber reinforced polymer composite helps in reducing global warming, manufacturing cost, and the use of other synthetic materials. Kenaf has also been used in

FIGURE 13.14 Kenaf composites shoe shelf (Mazani et al., 2019).

FIGURE 13.15 Kenaf composite table and kenaf composite chair.

various applications including kenaf mattresses (Fig. 13.16), prayer mats, and seat pads (Fig. 13.17).

13.9.5 Textiles, pulp, and paper

Kenaf is a promising natural fiber for the textile, paperboard, and pulp and paper industries. Kenaf displays significant benefits compared to other natural fibers by possessing sufficient physical strength properties, short harvestable period, and nonchlorinated bleaching techniques. Kenaf core

FIGURE 13.16 Kenaf mattress.

FIGURE 13.17 Prayer mat and Econaf seat pad.

fibers can be potentially used as absorbents in paper products like hygienic pulp, tissues, and wipes. Nowadays, the whole stems are used to produce better strength and quality of pulp for the paper industry. According to Mossello et al. (2010), kenaf fiber is currently being explored widely in developed and developing countries as a feasible raw material for pulp and paper kenaf paper due to its better ink adherence, resistant to staining, strength, whiter color, and longer lasting compared to conventional tree paper.

13.9.6 Automotives

It is predicted that by 2020, the world car production rate per annum would reach about 76 million vehicles. EU End of Life Vehicles laws and regulations oblige automotive manufacturers to shift from the use of synthetic-based materials to agro-based materials. However, poor mechanical properties as well as limitations in the manufacturing process have limited the use agro-based material as semistructural and nonstructural automotive components. Therefore a study was conducted by Davoodi et al. (2010) to utilize hybrid kenaf/glass reinforced epoxy composites for a passenger car bumper beam as automotive structural components with modified SMC. The results showed that the hybridization of natural fiber with glass fiber improved the mechanical properties of hybrid composites compared to natural fibers alone (Davoodi et al., 2010). A company in Japan, Toyota Boshoku Corporation, became the first company to use kenaf as a raw bio-based material for their interior automotive components. The utilization of kenaf fiber reinforced biocomposites produced lighter weight seat back board and door trim, resulting in more efficient fuel consumption and a reduction of carbon dioxide (CO_2) gas emission. Table 13.6 shows the types of fibers used and their current applications. INTROP, Universiti Putra Malaysia, fabricated kenaf a composite automotive package tray (Fig. 13.18) and car window regulators (Fig. 13.19).

TABLE 13.6 Types of fiber and current applications (Hassan, Zulkifli, Ghazali, & Azhari, 2017).

Types of fiber	Application	Automotive manufacturer
Kenaf and flax mixture	Package trays and door panel inserts	Saturn L300s Opel Vectras
Kenaf and flax mixture	Package shelves	Lexus
Kenaf	Black surrounds for	BMW i3
Kenaf	Interior panel	Ford Escape
Kenaf	Interior panel	BMW

FIGURE 13.18 Kenaf composite automotive package tray.

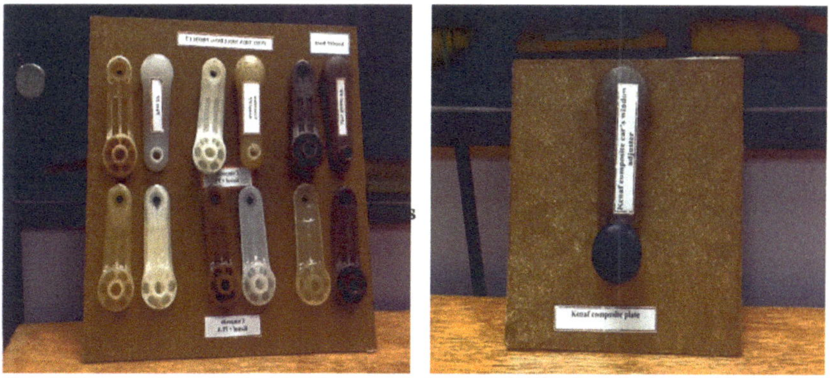

FIGURE 13.19 Kenaf composite car window regulators.

In order to make automotive vehicles more sustainable and environmentally friendly, BMW and Ford are fabricating material for automobile body parts from kenaf fiber. The first use of kenaf in Ford vehicles was in 2013, where it was used on Ford Escape model cars. In addition, the BMW i3 uses kenaf in its black surrounds.

13.10 Conclusions

Green and biodegradable materials are essential to achieve sustainable development and to protect the environment. With the existence of natural fibers and biopolymers for composite materials, the devastating environmental problems caused by petroleum-based materials can be eliminated or at least minimized. The development and manufacturing of these green composites will bring about major environmental changes, address the disposal of plastic waste, and reduce the carbon footprint of petroleum-based materials. Currently, the trend of utilizing bioresources as alternatives for nonbiodegradable synthetic materials has already started. The abundant supply and lower price of these green materials have earned them a lot of publicity in recent decades. However, the drawbacks and disadvantages of using natural fibers and biopolymers in composites need to be resolved by further research work. The combination of findings in this present work suggests that a number of mechanical tests and research works have verified the excellence of kenaf fiber's tensile strength and superior flexural strength which enable it to be used in a variety of applications such as in the automotive industry, lightweight construction materials, and other custom products such as yarns, fabrics, and clothes. Various tree components have been widely used for the production of a number of local items. Kenaf can compete with most natural fibers on the market such as coir, oil palm fiber, cotton, jute, and many other fibers due to its outstanding mechanical properties. Kenaf fiber is a good candidate for strengthening high-performance biodegradable polymer composites. The use of kenaf fiber in green composites will generally help: (1) minimize the negative environmental effect of synthetic polymers and fibers; (2) decrease the demand for reliance on petroleum products; and (3) turn kenaf plant into a sustainable industrial crop. This can therefore lead to better socioeconomic empowerment of rural people by increasing revenue and creating more employment opportunities. However, the huge opportunity to use kenaf fiber for various possible industrial applications in the composite industry has not been exploited widely. In fact, more advanced characterization of kenaf fibers and their biocomposites should be carried out in terms of research and development. Determining the barrier properties, feasibility, and moisture absorption of kenaf biocomposites are important for successful packaging applications. This is a new area of research and innovation that addresses certain issues that obstruct future industrial applications of kenaf fibers, biopolymers, and their composites.

Acknowledgment

The authors gratefully acknowledge Universiti Putra Malaysia for a Special Graduate Research Assistant (SGRA) Grant and for funding this project through Geran Putra Berimpak (GPB), UPM/800-3/3/1/GPB/2019/9679800. The authors also thank M.D. Damiri and Z.A. Maimon for their tremendous support provided during the writing of this chapter.

References

Abdul Khalil, H. P. S., Yusra, A. F. I., Bhat, A. H., & Jawaid, M. (2010). Cell wall ultrastructure, anatomy, lignin distribution, and chemical composition of Malaysian cultivated kenaf fiber. *Industrial Crops and Products, 31*(1), 113–121. Available from: https://doi.org/10.1016/j.indcrop.2009.09.008.

Abdul Rahman, K. (2011). Kenaf, a fibre for the future: the Harusmas experience. Available from: http://www.jeccomposites.com/knowledge/international-composites-news/kenaf-fibre-future-harusmas-experience.

Abdullah, A. H., Alias, S. K., Jenal, N., Abdan, K., & Ali, A. (2012). Fatigue behavior of kenaf fibre reinforced epoxy composites. *Engineering Journal, 16*(5), 105–114. Available from: https://doi.org/10.4186/ej.2012.16.5.105.

Abdullah, M. N. (2016). Oblique impact analysis on kenaf (*Hibiscus cannabinus*) and flax (*Linum usitatissimum*) natural fiber sport cycling helmets. *Movement, Health & Exercise, 5* (1), 57–67. Available from: https://doi.org/10.15282/mohe.v5i1.94.

Abral, H., Ariksa, J., Mahardika, M., Handayani, D., Aminah, I., Sandrawati, N., … Ilyas, R. A. (2019). Highly transparent and antimicrobial PVA based bionanocomposites reinforced by ginger nanofiber. *Polymer Testing* (October), 106186. Available from: https://doi.org/10.1016/j.polymertesting.2019.106186.

Abral, H., Ariksa, J., Mahardika, M., Handayani, D., Aminah, I., Sandrawati, N., ... Ilyas, R. A. (2020). Transparent and antimicrobial cellulose film from ginger nanofiber. *Food Hydrocolloids, 98*(August 2019), 105266. Available from: https://doi.org/10.1016/j.foodhyd.2019.105266.

Abral, H., Atmajaya, A., Mahardika, M., Hafizulhaq, F., Kadriadi., Handayani, D., ... Ilyas, R. A. (2020). Effect of ultrasonication duration of polyvinyl alcohol (PVA) gel on characterizations of PVA film. *Journal of Materials Research and Technology,* 1–10. Available from: https://doi.org/10.1016/j.jmrt.2019.12.078.

Abral, H., Basri, A., Muhammad, F., Fernando, Y., Hafizulhaq, F., Mahardika, M., ... Stephane, I. (2019). A simple method for improving the properties of the sago starch films prepared by using ultrasonication treatment. *Food Hydrocolloids, 93,* 276–283. Available from: https://doi.org/10.1016/j.foodhyd.2019.02.012.

Aisyah, H. A., Paridah, M. T., Sapuan, S. M., Khalina, A., Berkalp, O. B., Lee, S. H., ... Ilyas, R. A. (2019). Thermal properties of woven kenaf/carbon fibre-reinforced epoxy hybrid composite panels. *International Journal of Polymer Science, 2019*(December), 1–8. Available from: https://doi.org/10.1155/2019/5258621.

Aisyah, H. A., Paridah, M. T., Sapuan, S. M., Khalina, A., Berkalp, O. B., Wahab, M. S., ... Ramli, N. (2019). *Thermal and dynamic properties of woven kenaf/carbon fibre reinforced epoxy hybrid composite.* Available from: https://doi.org/10.20944/preprints201901.0264.v2.

Aji, I. S., Zainudin, E. S., Khalina, A., Sapuan, S. M., & Khairul, M. D. (2011). Studying the effect of fiber size and fiber loading on the mechanical properties of hybridized kenaf/PALF-reinforced HDPE composite. *Journal of Reinforced Plastics and Composites, 30*(6), 546–553. Available from: https://doi.org/10.1177/0731684411399141.

Akil, H. M., Omar, M. F., Mazuki, A. A. M., Safiee, S., Ishak, Z. A. M., & Bakar, A. A. (2011). Kenaf fiber reinforced composites: A review. *Materials and Design, 32*(8–9), 4107–4121. Available from: https://doi.org/10.1016/j.matdes.2011.04.008.

Alaaeddin, M. H., Sapuan, S. M., Zuhri, M. Y. M., Zainudin, E. S., & AL- Oqla, F. M. (2019). Physical and mechanical properties of polyvinylidene fluoride—Short sugar palm fiber nanocomposites. *Journal of Cleaner Production, 235,* 473–482. Available from: https://doi.org/10.1016/j.jclepro.2019.06.341.

Al-Oqla, F. M., Salit, M. S., Ishak, M. R., & Aziz, N. A. (2015a). A novel evaluation tool for enhancing the selection of natural fibers for polymeric composites based on fiber moisture content criterion. *BioResources, 10*(1), 299–312. Available from: https://doi.org/10.15376/biores.10.1.299-312.

Al-Oqla, F. M., Salit, M. S., Ishak, M. R., & Aziz, N. A. (2015b). Selecting natural fibers for bio-based materials with conflicting criteria. *American Journal of Applied Sciences, 12*(1), 64.

Al-Oqla, F. M., Sapuan, M. S., Ishak, M. R., & Abdul Aziz, N. (2014). Combined multi-criteria evaluation stage technique as an agro waste evaluation indicator for polymeric composites: Date palm fibers as a case study. *BioResources, 9*(3), 4608–4621. Available from: https://doi.org/10.15376/biores.9.3.4608-4621.

Al-Oqla, F. M., & Sapuan, S. M. (2018). Investigating the inherent characteristic/performance deterioration interactions of natural fibers in bio-composites for better utilization of resources. *Journal of Polymers and the Environment, 26*(3), 1290–1296. Available from: https://doi.org/10.1007/s10924-017-1028-z.

Alves, C., Ferrão, P. M. C., Silva, A. J., Reis, L. G., Freitas, M., Rodrigues, L. B., & Alves, D. E. (2010). Ecodesign of automotive components making use of natural jute fiber composites. *Journal of Cleaner Production, 18*(4), 313–327. Available from: https://doi.org/10.1016/j.jclepro.2009.10.022.

Anuar, H., & Zuraida, A. (2011). Improvement in mechanical properties of reinforced thermoplastic elastomer composite with kenaf bast fibre. *Composites Part B: Engineering, 42*(3), 462–465. Available from: https://doi.org/10.1016/j.compositesb.2010.12.013.

Arun Prakash, V. R., & Viswanthan, R. (2019). Fabrication and characterization of echinoidea spike particles and kenaf natural fibre-reinforced Azadirachta-Indica blended epoxy multi-hybrid bio composite. *Composites Part A: Applied Science and Manufacturing, 118*, 317–326. Available from: https://doi.org/10.1016/j.compositesa.2019.01.008.

Ashori, A., Harun, J., Raverty, W. D., & Yusoff, M. N. M. (2006). Chemical and morphological characteristics of Malaysian cultivated kenaf (*Hibiscus cannabinus*) fiber. *Polymer-Plastics Technology and Engineering, 45*(1), 131–134. Available from: https://doi.org/10.1080/03602550500373782.

Asim, M., Paridah, M. T., Saba, N., Jawaid, M., Alothman, O. Y., Nasir, M., & Almutairi, Z. (2018). Thermal, physical properties and flammability of silane treated kenaf/pineapple leaf fibres phenolic hybrid composites. *Composite Structures, 202*(February), 1330–1338. Available from: https://doi.org/10.1016/j.compstruct.2018.06.068.

Asumani, O., & Paskaramoorthy, R. (2020). Fatigue and impact strengths of kenaf fibre reinforced polypropylene composites: effects of fibre treatments. *Advanced Composite Materials*, 1–13. Available from: https://doi.org/10.1080/09243046.2020.1733308.

Asumani, O. M. L., Reid, R. G., & Paskaramoorthy, R. (2012). The effects of alkali–silane treatment on the tensile and flexural properties of short fibre non-woven kenaf reinforced polypropylene composites. *Composites Part A: Applied Science and Manufacturing, 43*(9), 1431–1440. Available from: https://doi.org/10.1016/j.compositesa.2012.04.007.

Asyraf, M. R. M., Ishak, M. R., Sapuan, S. M., Yidris, N., & Ilyas, R. A. (2020). Woods and composites cantilever beam: A comprehensive review of experimental and numerical creep methodologies. *Journal of Materials Research and Technology* (January). Available from: https://doi.org/10.1016/j.jmrt.2020.01.013.

Asyraf, M. R. M., Ishak, M. R., Sapuan, S. M., Yidris, N., Ilyas, R. A., Rafidah, M., & Razman, M. R. (2020). Evaluation of design and simulation of creep test rig for full-scale cross arm structure. *Advances in Civil Engineering*. Available from: https://doi.org/10.1155/2019/6980918.

Asyraf, M. R. M., Ishak, M. R., Sapuan, S. M., Yidris, N., Shahroze, R. M., Johari, A. N., . . . Ilyas, R. A. (2020). Creep test rig for cantilever beam: Fundamentals, prospects and present views. *Journal of Mechanical Engineering and Sciences, 14*(2), 6869–6887.

Atikah, M. S. N., Ilyas, R. A., Sapuan, S. M., Ishak, M. R., Zainudin, E. S., Ibrahim, R., . . . Jumaidin, R. (2019). Degradation and physical properties of sugar palm starch/sugar palm nanofibrillated cellulose bionanocomposite. *Polimery, 64*(10), 27–36. Available from: https://doi.org/10.14314/polimery.2019.10.5.

Atiqah, A., Maleque, M. A., Jawaid, M., & Iqbal, M. (2014). Development of kenaf-glass reinforced unsaturated polyester hybrid composite for structural applications. *Composites Part B: Engineering, 56*, 68–73. Available from: https://doi.org/10.1016/j.compositesb.2013.08.019.

Avérous, L., & Boquillon, N. (2004). Biocomposites based on plasticized starch: Thermal and mechanical behaviour. *Carbohydrate Polymers, 56*(2), 111–122. Available from: http://doi.org/10.1016/j.carbpol.2003.11.015.

Azammi, A. M. N., Ilyas, R. A., Sapuan, S. M., Ibrahim, R., Atikah, M. S. N., Asrofi, M., & Atiqah, A. (2020). Characterization studies of biopolymeric matrix and cellulose fibres based composites related to functionalized fibre–matrix interface. *Interfaces in particle and fibre reinforced composites* (1st ed., pp. 29–93). London: Elsevier. Available from: https://doi.org/10.1016/B978-0-08-102665-6.00003-0.

Aziz, S. H., & Ansell, M. P. (2004). The effect of alkalization and fibre alignment on the mechanical and thermal properties of kenaf and hemp bast fibre composites: Part 1—polyester resin matrix. *Composites Science and Technology, 64*(9), 1219–1230. Available from: https://doi.org/10.1016/j.compscitech.2003.10.001.

Azizi, S., Ouellet-Plamondon, C., David E., & Fréchette, M. F. (2017). Electrical and thermal properties of low-density polyethylene/graphene-like composite. *IEEE Conference on Electrical Insulation and Dielectric Phenomena (CEIDP)*, Fort Worth, TX (pp. 517–520).

Azmi, A. M. R., Sultan, M. T. H., Hamdan, A., Nor, A. F. M., & Jayakrishna, K. (2018). Flexural and impact properties of a new bulletproof vest insert plate design using kenaf fibre embedded with X-Ray films. *Materials Today: Proceedings*, *5*(5), 11193–11197. Available from: https://doi.org/10.1016/j.matpr.2018.01.143.

Batouli, S. M., Zhu, Y., Nar, M., & D'Souza, N. A. (2014). Environmental performance of kenaf-fiber reinforced polyurethane: A life cycle assessment approach. *Journal of Cleaner Production*, *66*, 164–173. Available from: https://doi.org/10.1016/j.jclepro.2013.11.064.

Beckermann, G. W., & Pickering, K. L. (2008). Engineering and evaluation of hemp fibre reinforced polypropylene composites: Fibre treatment and matrix modification. *Composites Part A: Applied Science and Manufacturing*, *39*(6), 979–988. Available from: https://doi.org/10.1016/j.compositesa.2008.03.010.

Bernard, M., Khalina, A., Ali, A., Janius, R., Faizal, M., Hasnah, K. S., & Sanuddin, A. B. (2011). The effect of processing parameters on the mechanical properties of kenaf fibre plastic composite. *Materials & Design*, *32*(2), 1039–1043. Available from: https://doi.org/10.1016/j.matdes.2010.07.014.

Bullions, T. A., Hoffman, D., Gillespie, R. A., Price-O'Brien, J., & Loos, A. C. (2006). Contributions of feather fibers and various cellulose fibers to the mechanical properties of polypropylene matrix composites. *Composites Science and Technology*, *66*(1), 102–114. Available from: https://doi.org/10.1016/j.compscitech.2005.03.017.

Butler, R. A. (2013). Malaysia has the world's highest deforestation rate, reveals Google forest map. Retrieved from <https://news.mongabay.com/2013/11/malaysia-has-the-worlds-highest-deforestation-rate-reveals-google-forest-map/>.

Chee, S. S., Jawaid, M., Sultan, M. T. H., Alothman, O. Y., & Abdullah, L. C. (2019). Thermomechanical and dynamic mechanical properties of bamboo/woven kenaf mat reinforced epoxy hybrid composites. *Composites Part B: Engineering*, *163*, 165–174. Available from: https://doi.org/10.1016/j.compositesb.2018.11.039.

Dan-Mallam, Y., Abdullah, M. Z., & Yusoff, P. S. M. M. (2014). The effect of hybridization on mechanical properties of woven kenaf fiber reinforced polyoxymethylene composite. *Polymer Composites*, *35*(10), 1900–1910. Available from: https://doi.org/10.1002/pc.22846.

Davoodi, M. M., Sapuan, S. M., Ahmad, D., Aidy, A., Khalina, A., & Jonoobi, M. (2012). Effect of polybutylene terephthalate (PBT) on impact property improvement of hybrid kenaf/glass epoxy composite. *Materials Letters*, *67*(1), 5–7. Available from: https://doi.org/10.1016/j.matlet.2011.08.101.

Davoodi, M. M., Sapuan, S. M., Ahmad, D., Ali, A., Khalina, A., & Jonoobi, M. (2010). Mechanical properties of hybrid kenaf/glass reinforced epoxy composite for passenger car bumper beam. *Materials & Design*, *31*(10), 4927–4932. Available from: https://doi.org/10.1016/j.matdes.2010.05.021.

Doan, T.-T.-L., Brodowsky, H., & Mäder, E. (2007). Jute fibre/polypropylene composites II. Thermal, hydrothermal and dynamic mechanical behaviour. *Composites Science and Technology*, *67*(13), 2707–2714. Available from: https://doi.org/10.1016/j.compscitech.2007.02.011.

Edhirej, A., Sapuan, S. M., Jawaid, M., & Zahari, N. I. (2017). Cassava/sugar palm fiber reinforced cassava starch hybrid composites: Physical, thermal and structural properties. *International Journal of Biological Macromolecules*, *101*, 75–83. Available from: https://doi.org/10.1016/j.ijbiomac.2017.03.045.

Elsaid, A., Dawood, M., Seracino, R., & Bobko, C. (2011). Mechanical properties of kenaf fiber reinforced concrete. *Construction and Building Materials*, *25*(4), 1991–2001. Available from: https://doi.org/10.1016/j.conbuildmat.2010.11.052.

El-Shekeil, Y. A., Sapuan, S. M., Abdan, K., & Zainudin, E. S. (2012). Influence of fiber content on the mechanical and thermal properties of Kenaf fiber reinforced thermoplastic polyurethane composites. *Materials & Design*, *40*, 299–303. Available from: https://doi.org/10.1016/j.matdes.2012.04.003.

El-Shekeil, Y. A., Sapuan, S. M., Jawaid, M., & Al-Shuja'a, O. M. (2014). Influence of fiber content on mechanical, morphological and thermal properties of kenaf fibers reinforced poly(vinyl chloride)/thermoplastic polyurethane poly-blend composites. *Materials and Design*, *58*, 130–135. Available from: https://doi.org/10.1016/j.matdes.2014.01.047.

El-Shekeil, Y. A., Sapuan, S. M., Khalina, A., Zainudin, E. S., & Al-Shuja'a, O. M. (2012). Influence of chemical treatment on the tensile properties of kenaf fiber reinforced thermoplastic polyurethane composite. *Express Polymer Letters*, *6*(12), 1032–1040. Available from: https://doi.org/10.3144/expresspolymlett.2012.108.

Faruk, O., Bledzki, A. K., Fink, H.-P., & Sain, M. (2012). Biocomposites reinforced with natural fibers: 2000–2010. *Progress in Polymer Science*, *37*(11), 1552–1596. Available from: https://doi.org/10.1016/j.progpolymsci.2012.04.003.

Gérard, J. F., Perret, P., & Chabert, B. (1990). Study of carbon/epoxy interface (or interphase): Effect of surface treatment of carbon fibers on the dynamic mechanical behavior of carbon/epoxy unidirectional composites. *Controlled interphases in composite materials* (pp. 449–456). Dordrecht: Springer Netherlands. Available from: https://doi.org/10.1007/978-94-011-7816-7_44.

Ghani, M. A. A., Salleh, Z., Hyie, K. M., Berhan, M. N., Taib, Y. M. D., & Bakri, M. A. I. (2012). Mechanical properties of kenaf/fiberglass polyester hybrid composite. *Procedia Engineering*, *41*, 1654–1659. Available from: https://doi.org/10.1016/j.proeng.2012.07.364.

Gu, H. (2009). Dynamic mechanical analysis of the seawater treated glass/polyester composites. *Materials & Design*, *30*(7), 2774–2777. Available from: https://doi.org/10.1016/j.matdes.2008.09.029.

Halimatul, M. J., Sapuan, S. M., Jawaid, M., Ishak, M. R., & Ilyas, R. A. (2019a). Effect of sago starch and plasticizer content on the properties of thermoplastic films: Mechanical testing and cyclic soaking-drying. *Polimery*, *64*(6), 32–41. Available from: https://doi.org/10.14314/polimery.2019.6.5.

Halimatul, M. J., Sapuan, S. M., Jawaid, M., Ishak, M. R., & Ilyas, R. A. (2019b). Water absorption and water solubility properties of sago starch biopolymer composite films filled with sugar palm particles. *Polimery*, *64*(9), 27–35. Available from: https://doi.org/10.14314/polimery.2019.9.4.

Han, S. O., Karevan, M., Bhuiyan, M. A., Park, J. H., & Kalaitzidou, K. (2012). Effect of exfoliated graphite nanoplatelets on the mechanical and viscoelastic properties of poly (lactic acid) biocomposites reinforced with kenaf fibers. *Journal of Materials Science*, *47* (8), 3535–3543. Available from: https://doi.org/10.1007/s10853-011-6199-8.

Hanan, F., Jawaid, M., & Md Tahir, P. (2020). Mechanical performance of oil palm/kenaf fiber-reinforced epoxy-based bilayer hybrid composites. *Journal of Natural Fibers*, *17*(2), 155–167. Available from: https://doi.org/10.1080/15440478.2018.1477083.

Hao, A., Zhao, H., & Chen, J. Y. (2013). Kenaf/polypropylene nonwoven composites: The influence of manufacturing conditions on mechanical, thermal, and acoustical performance. *Composites Part B: Engineering*, *54*, 44–51. Available from: https://doi.org/10.1016/j.compositesb.2013.04.065.

Hao, A., Zhao, H., Jiang, W., Yuan, L., & Chen, J. Y. (2012). Mechanical properties of kenaf/polypropylene nonwoven composites. *Journal of Polymers and the Environment*, *20*(4), 959–966. Available from: https://doi.org/10.1007/s10924-012-0484-8.

Hassan, F., Zulkifli, R., Ghazali, M. J., & Azhari, C. H. (2017). Kenaf fiber composite in automotive industry: An overview. *International Journal on Advanced Science, Engineering and Information Technology*, *7*(1), 315. Available from: https://doi.org/10.18517/ijaseit.7.1.1180.

Huda, M., Drzal, L., Mohanty, A., & Misra, M. (2006). Chopped glass and recycled newspaper as reinforcement fibers in injection molded poly(lactic acid) (PLA) composites: A comparative study. *Composites Science and Technology, 66*(11–12), 1813–1824. Available from: https://doi.org/10.1016/j.compscitech.2005.10.015.

Ibrahim, M. I. J., Sapuan, S. M., Zainudin, E. S., & Zuhri, M. Y. M. (2020). Preparation and characterization of cornhusk/sugar palm fiber reinforced cornstarch-based hybrid composites. *Journal of Materials Research and Technology, 9*(1), 200–211. Available from: https://doi.org/10.1016/j.jmrt.2019.10.045.

Ilyas, R. A., & Sapuan, S. M. (2020). The preparation methods and processing of natural fibre bio-polymer composites. *Current Organic Synthesis, 16*(8), 1068–1070. Available from: https://doi.org/10.2174/157017941608200120105616.

Ilyas, R. A., Sapuan, S. M., Atiqah, A., Ibrahim, R., Abral, H., Ishak, M. R., ... Ya, H. (2019). Sugar palm (*Arenga pinnata* [Wurmb.] Merr) starch films containing sugar palm nanofibrillated cellulose as reinforcement: Water barrier properties. *Polymer Composites* (July), 1–9. Available from: https://doi.org/10.1002/pc.25379.

Ilyas, R. A., Sapuan, S. M., Ibrahim, R., Abral, H., Ishak, M. R., Zainudin, E. S., ... Jumaidin, R. (2019a). Effect of sugar palm nanofibrillated cellulose concentrations on morphological, mechanical and physical properties of biodegradable films based on agro-waste sugar palm (*Arenga pinnata* (Wurmb.) Merr) starch. *Journal of Materials Research and Technology, 8*(5), 4819–4830. Available from: https://doi.org/10.1016/j.jmrt.2019.08.028.

Ilyas, R. A., Sapuan, S. M., Ibrahim, R., Abral, H., Ishak, M. R., Zainudin, E. S., ... Jumaidin, R. (2019b). Sugar palm (*Arenga pinnata* (Wurmb.) Merr) cellulosic fibre hierarchy: a comprehensive approach from macro to nano scale. *Journal of Materials Research and Technology, 8*(3), 2753–2766. Available from: https://doi.org/10.1016/j.jmrt.2019.04.011.

Ilyas, R. A., Sapuan, S. M., Ibrahim, R., Abral, H., Ishak, M. R., Zainudin, E. S., ... Jumaidin, R. (2020). Thermal, biodegradability and water barrier properties of bio-nanocomposites based on plasticised sugar palm starch and nanofibrillated celluloses from sugar palm fibres. *Journal of Biobased Materials and Bioenergy, 14*(2), 234–248. Available from: https://doi.org/10.1166/jbmb.2020.1951.

Ilyas, R. A., Sapuan, S. M., Ibrahim, R., Atikah, M. S. N., Atiqah, A., Ansari, M. N. M., & Norrahim, M. N. F. (2019). Production, processes and modification of nanocrystalline cellulose from agro-waste: A review. *Nanocrystalline Materials* (pp. 3–32). IntechOpen. Available from: https://doi.org/10.5772/intechopen.87001.

Ilyas, R. A., Sapuan, S. M., & Ishak, M. R. (2018). Isolation and characterization of nanocrystalline cellulose from sugar palm fibres (*Arenga pinnata*). *Carbohydrate Polymers, 181* (June 2017), 1038–1051. Available from: https://doi.org/10.1016/j.carbpol.2017.11.045.

Ilyas, R. A., Sapuan, S. M., Ishak, M. R., & Zainudin, E. S. (2017). Effect of delignification on the physical, thermal, chemical, and structural properties of sugar palm fibre. *BioResources, 12*(4), 8734–8754. Available from: https://doi.org/10.15376/biores.12.4.8734-8754.

Ilyas, R. A., Sapuan, S. M., Ishak, M. R., & Zainudin, E. S. (2018a). Sugar palm nanocrystalline cellulose reinforced sugar palm starch composite: Degradation and water-barrier properties. *IOP Conference Series: Materials Science and Engineering, 368*, article number 012006. Available from: https://doi.org/10.1088/1757-899X/368/1/012006.

Ilyas, R. A., Sapuan, S. M., Ishak, M. R., & Zainudin, E. S. (2018b). Development and characterization of sugar palm nanocrystalline cellulose reinforced sugar palm starch bionanocomposites. Development and characterization of sugar palm nanocrystalline cellulose reinforced sugar palm starch bionanocomposites. *Carbohydrate Polymers, 202* (September), 186–202. Available from: https://doi.org/10.1016/j.carbpol.2018.09.002.

Ilyas, R. A., Sapuan, S. M., Ishak, M. R., & Zainudin, E. S. (2018c). Water transport properties of bio-nanocomposites reinforced by sugar palm (*Arenga pinnata*) nanofibrillated cellulose. *Journal of Advanced Research in Fluid Mechanics and Thermal Sciences Journal, 51*(2), 234–246.

Ilyas, R. A., Sapuan, S. M., Ishak, M. R., & Zainudin, E. S. (2019). Sugar palm nanofibrillated cellulose (*Arenga pinnata* (Wurmb.) Merr): Effect of cycles on their yield, physic-chemical, morphological and thermal behavior. *International Journal of Biological Macromolecules, 123*, 379–388. Available from: https://doi.org/10.1016/j.ijbiomac.2018.11.124.

Ishak, M. R., Leman, Z., Sapuan, S. M., Edeerozey, aM. M., & Othman, I. S. (2010). Mechanical properties of kenaf bast and core fibre reinforced unsaturated polyester composites. *IOP Conference Series: Materials Science and Engineering, 11*, 012006. Available from: https://doi.org/10.1088/1757-899X/11/1/012006.

John, M. J., Bellmann, C., & Anandjiwala, R. D. (2010). Kenaf–polypropylene composites: Effect of amphiphilic coupling agent on surface properties of fibres and composites. *Carbohydrate Polymers, 82*(3), 549–554. Available from: https://doi.org/10.1016/j.carbpol.2010.05.015.

Julkapli, N. M., & Akil, H. M. (2010). Thermal properties of kenaf-filled chitosan biocomposites. *Polymer-Plastics Technology and Engineering, 49*(2), 147–153. Available from: https://doi.org/10.1080/03602550903284206.

Jumaidin, R., Ilyas, R. A., Saiful, M., Hussin, F., & Mastura, M. T. (2019). Water transport and physical properties of sugarcane bagasse fibre reinforced thermoplastic potato starch biocomposite. *Journal of Advanced Research in Fluid Mechanics and Thermal Sciences, 61*(2), 273–281.

Jumaidin, R., Khiruddin, M. A. A., Asyul Sutan Saidi, Z., Salit, M. S., & Ilyas, R. A. (2020). Effect of cogon grass fibre on the thermal, mechanical and biodegradation properties of thermoplastic cassava starch biocomposite. *International Journal of Biological Macromolecules, 146*, 746–755. Available from: https://doi.org/10.1016/j.ijbiomac.2019.11.011.

Jumaidin, R., Saidi, Z. A. S., Ilyas, R. A., Ahmad, M. N., Wahid, M. K., Yaakob, M. Y., … Osman, M. H. (2019). Characteristics of cogon grass fibre reinforced thermoplastic cassava starch biocomposite: Water absorption and physical properties. *Journal of Advanced Research in Fluid Mechanics and Thermal Sciences, 62*(1), 43–52.

Kalia, S., Avérous, L., Njuguna, J., Dufresne, A., & Cherian, B. M. (2011). Natural fibers, bio-and nanocomposites. *International Journal of Polymer Science, 2011*.

Karimi, S., Tahir, P. M., Karimi, A., Dufresne, A., & Abdulkhani, A. (2014). Kenaf bast cellulosic fibers hierarchy: A comprehensive approach from micro to nano. *Carbohydrate Polymers, 101*, 878–885. Available from: https://doi.org/10.1016/j.carbpol.2013.09.106.

Khandelwal, V., Sahoo, S. K., Kumar, A., & Manik, G. (2018). Electrically conductive green composites based on epoxidized linseed oil and polyaniline: An insight into electrical, thermal and mechanical properties. *Composites Part B: Engineering, 136*(July 2017), 149–157. Available from: https://doi.org/10.1016/j.compositesb.2017.10.030.

Koutenaei, R. Y., Choobbasti, A. J., & Kutanaei, S. S. (2019). Triaxial behaviour of a cemented sand reinforced with Kenaf fibres. *European Journal of Environmental and Civil Engineering*, 1–19. Available from: https://doi.org/10.1080/19648189.2019.1574607.

Kozlowski, R., Mieleniak, B., Helwig, M., & Przepiera, A. (1999). Flame resistant lignocellulosic-mineral composite particleboards. *Polymer Degradation and Stability, 64*(3), 523–528. Available from: https://doi.org/10.1016/S0141-3910(98)00145-1.

Kwon, H.-J., Sunthornvarabhas, J., Park, J.-W., Lee, J.-H., Kim, H.-J., Piyachomkwan, K., … Cho, D. (2014). Tensile properties of kenaf fiber and corn husk flour reinforced poly(lactic acid) hybrid bio-composites: Role of aspect ratio of natural fibers. *Composites Part B: Engineering, 56*, 232–237. Available from: https://doi.org/10.1016/j.compositesb.2013.08.003.

Lee, B.-H., Kim, H.-S., Lee, S., Kim, H.-J., & Dorgan, J. R. (2009). Bio-composites of kenaf fibers in polylactide: Role of improved interfacial adhesion in the carding process. *Composites Science and Technology, 69*(15–16), 2573–2579. Available from: https://doi.org/10.1016/j.compscitech.2009.07.015.

Lilholt, H., & Lawther, J. M. (2000). *Natural organic fibers. Comprehensive composite materials* (pp. 303–325). Elsevier. Available from: https://doi.org/10.1016/B0-08-042993-9/00048-6.

Lim, Z. Y., Putra, A., Nor, M. J. M., & Yaakob, M. Y. (2018). Sound absorption performance of natural kenaf fibres. *Applied Acoustics, 130,* 107–114. Available from: https://doi.org/10.1016/j.apacoust.2017.09.012.

Liu, W., Drzal, L. T., Mohanty, A. K., & Misra, M. (2007). Influence of processing methods and fiber length on physical properties of kenaf fiber reinforced soy based biocomposites. *Composites Part B: Engineering, 38*(3), 352–359. Available from: https://doi.org/10.1016/j.compositesb.2006.05.003.

Mahjoub, R., Yatim, J. M., Mohd Sam, A. R., & Hashemi, S. H. (2014). Tensile properties of kenaf fiber due to various conditions of chemical fiber surface modifications. *Construction and Building Materials, 55,* 103–113. Available from: https://doi.org/10.1016/j.conbuildmat.2014.01.036.

Mazani, N., Sapuan, S. M., Sanyang, M. L., Atiqah, A., & Ilyas, R. A. (2019). *Design and fabrication of a shoe shelf from kenaf fiber reinforced unsaturated polyester composites. Lignocellulose for future bioeconomy* (pp. 315–332). Elsevier. Available from: https://doi.org/10.1016/B978-0-12-816354-2.00017-7.

Mazuki, A. A. M., Akil, H. M., Safiee, S., Ishak, Z. A. M., & Bakar, A. A. (2011). Degradation of dynamic mechanical properties of pultruded kenaf fiber reinforced composites after immersion in various solutions. *Composites Part B: Engineering, 42*(1), 71–76. Available from: https://doi.org/10.1016/j.compositesb.2010.08.004.

Meon, M. S., Othman, M. F., Husain, H., Remeli, M. F., & Syawal, M. S. M. (2012). Improving tensile properties of kenaf fibers treated with sodium hydroxide. *Procedia Engineering, 41,* 1587–1592. Available from: https://doi.org/10.1016/j.proeng.2012.07.354.

Mohanty, A. K., Misra, M., & Drzal, L. T. (2005). *Natural fibers, biopolymers, and biocomposites.* Boca Raton: CRC Press.

Mohd Nurazzi, N., Khalina, A., Sapuan, S. M., Dayang Laila, A. H. A. M., Rahmah, M., & Hanafee, Z. (2017). A review: Fibres, polymer matrices and composites. *Pertanika Journal of Science & Technology, 25*(4), 1085–1102.

Mossello, A. A., Harun, J., Tahir, P. M., Resalati, H., Ibrahim, R., Fallah Shamsi, S. R., & Mohmamed, A. Z. (2010). A review of literatures related to using kenaf for pulp production (beating, fractionation, and recycled fiber). *Modern Applied Science, 4*(9). Available from: https://doi.org/10.5539/mas.v4n9p21.

Mutasher, S. A., Poh, A., Than, A. M., & Law, J. (2011). The effect of alkali treatment mechanical properties of kenaf fiber epoxy composite. *Key Engineering Materials, 471–472,* 191–196. Available from: https://doi.org/10.4028/www.scientific.net/KEM.471-472.191.

Nimanpure, S., Hashmi, S. A. R., Kumar, R., Bhargaw, H. N., Kumar, R., Nair, P., & Naik, A. (2019). Mechanical, electrical, and thermal analysis of sisal fibril/kenaf fiber hybrid polyester composites. *Polymer Composites, 40*(2), 664–676. Available from: https://doi.org/10.1002/pc.24706.

Nishino, T., Hirao, K., & Kotera, M. (2003). Kenaf reinforced biodegradable composite. Composites Science and Technology, *63,* 1281–1286. Available from: https://doi.org/10.1016/S0266-3538(03)00099-X.

Norizan, M. N., Abdan, K., Ilyas, R. A., & Biofibers, S. P. (2020). Effect of fiber orientation and fiber loading on the mechanical and thermal properties of sugar palm yarn fiber reinforced unsaturated polyester resin composites. *Polimery, 65*(2), 34–43. Available from: https://doi.org/10.14314/polimery.2020.2.5.

Nosbi, N., Akil, H. M., Mohd Ishak, Z. A., & Abu Bakar, A. (2010). Degradation of compressive properties of pultruded kenaf fiber reinforced composites after immersion in various solutions. *Materials & Design, 31*(10), 4960–4964. Available from: https://doi.org/10.1016/j.matdes.2010.04.037.

Nurazzi, N. M., Khalina, A., Sapuan, S. M., & Ilyas, R. A. (2019). Mechanical properties of sugar palm yarn/woven glass fiber reinforced unsaturated polyester composites : Effect of fiber loadings and alkaline treatment. *Polimery*, *64*(10), 12–22. Available from: https://doi.org/10.14314/polimery.2019.10.3.

Nurazzi, N. M., Khalina, A., Sapuan, S. M., Ilyas, R. A., Rafiqah, S. A., & Hanafee, Z. M. (2019). Thermal properties of treated sugar palm yarn/glass fiber reinforced unsaturated polyester hybrid composites. *Journal of Materials Research and Technology* (December). Available from: https://doi.org/10.1016/j.jmrt.2019.11.086.

Nurul Aizan, M. Z., Zainathul Akhmar, S. A. S., Mohd Muhiddin, A., Nor Hazwani, Z., & Siti Sarah, J. (2013). Study on cure characteristics and mechanical behaviours of kenaf fibre reinforced natural rubber composites. *Advanced Materials Research*, *812*, 66–72. Available from: https://doi.org/10.4028/www.scientific.net/AMR.812.66.

Ochi, S. (2008). Mechanical properties of kenaf fibers and kenaf/PLA composites. *Mechanics of Materials*, *40*(4–5), 446–452. Available from: https://doi.org/10.1016/j.mechmat.2007.10.006.

Osman, M. R., Mazuki, A. A. M., Akil, H. M., Ishak, Z. A. M., & Bakar, A. A. (2013). Effect of chemical treatment on the mechanical properties of pultruded kenaf fibre reinforced polyester composites. *Key Engineering Materials*, *594–595*, 691–695. Available from: https://doi.org/10.4028/www.scientific.net/KEM.594-595.691.

Paridah, M. T., Basher, A. B., SaifulAzry, S., & Z, A. (2001). Retting process of some bast plant fibers and its effect on fiber quality: a review. *Bio Resour*, *2011*(6), 5260–5281.

Qu, X., Wirsén, A., & Albertsson, A.-C. (2000). Effect of lactic/glycolic acid side chains on the thermal degradation kinetics of chitosan derivatives. *Polymer*, *41*(13), 4841–4847. Available from: https://doi.org/10.1016/S0032-3861(99)00704-1.

Ramesh, P., & Durga P. (2020). Characterization of kenaf/aloevera fiber reinforced pla-hybrid biocomposite. In *Advances in Applied Mechanical Engineering*, Singapore: Springer. Available from: https://doi.org/10.1007/978-981-15-1201-8_113.

Rassiah, K., & Megat Ahmad, M. M. H. (2013). A review on mechanical properties of bamboo fiber reinforced polymer composite. *Australian Journal of Basic and Applied Sciences*, *7*(8), 247–253.

Ren, X., Qiu, R., Fifield, L. S., Simmons, K. L., & Li, K. (2012). Effects of surface treatments on mechanical properties and water resistance of kenaf fiber-reinforced unsaturated polyester composites. *Journal of Adhesion Science and Technology*, *26*(18–19), 2277–2289. Available from: https://doi.org/10.1163/156856111X623087.

Rhim, J. W., Park, H. M., & Ha, C. S. (2013). Bio-nanocomposites for food packaging applications. *Progress in Polymer Science*, *38*(10–11), 1629–1652. Available from: https://doi.org/10.1016/j.progpolymsci.2013.05.008.

Russo, P., Carfagna, C., Cimino, F., Acierno, D., & Persico, P. (2013). Biodegradable composites reinforced with kenaf fibers: Thermal, mechanical, and morphological issues. *Advances in Polymer Technology*, *32*(S1), E313–E322. Available from: https://doi.org/10.1002/adv.21282.

Saba, N., Alothman, O. Y., Almutairi, Z., & Jawaid, M. (2019). Magnesium hydroxide reinforced kenaf fibers/epoxy hybrid composites: Mechanical and thermomechanical properties. *Construction and Building Materials*, *201*, 138–148. Available from: https://doi.org/10.1016/j.conbuildmat.2018.12.182.

Saba, N., Jawaid, M., Sultan, M. T. H., & Alothman, O. Y. (2017). Green biocomposites for structural applications. *Green Energy and Technology*. Available from: https://doi.org/10.1007/978-3-319-49382-4_1.

Saba, N., Paridah, M. T., & Jawaid, M. (2015). Mechanical properties of kenaf fibre reinforced polymer composite: A review. *Construction and Building Materials*, *76*, 87–96. Available from: https://doi.org/10.1016/j.conbuildmat.2014.11.043.

Saba, N., Paridah, M. T., Jawaid, M., Abdan, K., & Ibrahim, N. A. (2015). Potential utilization of kenaf biomass in different applications. *Agricultural biomass based potential materials* (pp. 1–34). Cham: Springer International Publishing. Available from: https://doi.org/10.1007/978-3-319-13847-3_1.

Salim, M. S., Ariawan, D., Ahmad Rasyid, M. F., Mat Taib, R., Ahmad Thirmizir, M. Z., & Mohd Ishak, Z. A. (2020). Accelerated weathering and water absorption behavior of kenaf fiber reinforced acrylic based polyester composites. *Frontiers in Materials, 7*. Available from: https://doi.org/10.3389/fmats.2020.00026.

Salleh, F. M., Hassan, A., Yahya, R., & Azzahari, A. D. (2014). Effects of extrusion temperature on the rheological, dynamic mechanical and tensile properties of kenaf fiber/HDPE composites. *Composites Part B: Engineering, 58*, 259–266. Available from: https://doi.org/10.1016/j.compositesb.2013.10.068.

Salleh, Z., Taib, Y. M., Hyie, K. M., Mihat, M., Berhan, M. N., & Ghani, M. A. A. (2012). Fracture toughness investigation on long kenaf/woven glass hybrid composite due to water absorption effect. *Procedia Engineering, 41*, 1667–1673. Available from: https://doi.org/10.1016/j.proeng.2012.07.366.

Salman, S. D., Leman, Z., Sultan, M. T., Ishak, M. R., & Cardona, F. (2016). Ballistic impact resistance of plain woven. *BioResources, 11*(3), 7282–7295.

Sapuan, S. M., Ilyas, R. A., Ishak, M. R., Leman, Z., Huzaifah, M. R. M., Ammar, I. M., & Atikah, M. S. N. (2018). *Development of sugar palm–based products: A community project. Sugar palm biofibers, biopolymers, and biocomposites* (1st ed., pp. 245–266). Boca Raton, FL: CRC Press/Taylor & Francis Group. Available from: https://doi.org/10.1201/9780429443923-12.

Sapuan, S. M., Mun, N. K., Hambali, A., Lok, H. Y., Fairuz, A. M., & Ishak, M. R. (2011). Prototype expert system for material selection of polymeric composite automotive dashboard. *International Journal of Physical Sciences, 6*(25), 5988–5995. Available from: https://doi.org/10.5897/IJPS11.939.

Sapuan, S. M., Pua, F., El-Shekeil, Y. A., & AL-Oqla, F. M. (2013). Mechanical properties of soil buried kenaf fibre reinforced thermoplastic polyurethane composites. *Materials & Design, 50*, 467–470. Available from: https://doi.org/10.1016/j.matdes.2013.03.013.

Serizawa, S., Inoue, K., & Iji, M. (2006). Kenaf-fiber-reinforced poly(lactic acid) used for electronic products. *Journal of Applied Polymer Science, 100*(1), 618–624. Available from: https://doi.org/10.1002/app.23377.

Shibata, S., Cao, Y., & Fukumoto, I. (2006). Lightweight laminate composites made from kenaf and polypropylene fibres. *Polymer Testing, 25*(2), 142–148. Available from: https://doi.org/10.1016/j.polymertesting.2005.11.007.

Shukor, F., Hassan, A., Saiful Islam, M., Mokhtar, M., & Hasan, M. (2014). Effect of ammonium polyphosphate on flame retardancy, thermal stability and mechanical properties of alkali treated kenaf fiber filled PLA biocomposites. *Materials & Design (1980-2015), 54*, 425–429. Available from: https://doi.org/10.1016/j.matdes.2013.07.095.

Siregar, J.P. (2005). Tensile and flexural properties of Arenga pinnata filament (IJUK filament) reinforced epoxy, Master of Science Thesis, Universiti Putra Malaysia (January).

Sreenivasan, S., Sulaiman, S., Baharudin, B.T.H.T., Ariffin, M.K.A., & Abdan, K. (2013). Recent developments of kenaf fibre reinforced thermoset composites: review, *17*, 2–11. Available from: https://doi.org/10.1179/1432891713Z.0000000000312.

Suharty, N. S., Almanar, I. P., Sudirman., Dihardjo, K., & Astasari, N. (2012). Flammability, biodegradability and mechanical properties of bio-composites waste polypropylene/kenaf fiber containing nano $CaCO_3$ with diammonium phosphate. *Procedia Chemistry, 4*, 282–287. Available from: https://doi.org/10.1016/j.proche.2012.06.039.

Sukyai, P., Sriroth, K. R., Lee, B. H., & Hyun, J. K. (2011). The effect of bacterial cellulose on the mechanical and thermal expansion properties of kenaf/polylactic acid. *composites. Applied Mechanics and Materials, 117–119*, 1343–1351. Available from: https://doi.org/10.4028/www.scientific.net/AMM.117-119.1343.

Syafri, E., Sudirman., Mashadi., Yulianti, E., Deswita., Asrofi, M., ... Fudholi, A. (2019). Effect of sonication time on the thermal stability, moisture absorption, and biodegradation of water hyacinth (*Eichhornia crassipes*) nanocellulose-filled bengkuang (*Pachyrhizus erosus*) starch biocomposites. *Journal of Materials Research and Technology, 8*(6), 6223–6231. Available from: https://doi.org/10.1016/j.jmrt.2019.10.016.

Thiruchitrambalam, M., Alavudeen, A., Athijayamani, A., Venkateshwaran, N., & Perumal, A. E. (2009). Improving mechanical properties of banana/kenaf polyester hybrid composites using sodium laulryl sulfate treatment. *Materials Physics and Mechanics, 8*, 165–173.

Verma, R., & Shukla, M. (2018). Characterization of mechanical properties of short kenaf fiber-HDPE green composites. *Materials Today: Proceedings, 5*(2), 3257–3264. Available from: https://doi.org/10.1016/j.matpr.2017.11.567.

Wambua, P., Ivens, J., & Verpoest, I. (2003). Natural fibres: Can they replace glass in fibre reinforced plastics? *Composites Science and Technology, 63*(9), 1259–1264. Available from: https://doi.org/10.1016/S0266-3538(03)00096-4.

Xue, Y., Du, Y., Elder, S., Wang, K., & Zhang, J. (2009). Temperature and loading rate effects on tensile properties of kenaf bast fiber bundles and composites. *Composites Part B: Engineering, 40*(3), 189–196. Available from: https://doi.org/10.1016/j.compositesb.2008.11.009.

Yousif, B. F., Shalwan, A., Chin, C. W., & Ming, K. C. (2012). Flexural properties of treated and untreated kenaf/epoxy composites. *Materials & Design, 40*, 378–385. Available from: https://doi.org/10.1016/j.matdes.2012.04.017.

Zainuddin, S. Y. Z., Ahmad, I., Kargarzadeh, H., Abdullah, I., & Dufresne, A. (2013). Potential of using multiscale kenaf fibers as reinforcing filler in cassava starch-kenaf biocomposites. *Carbohydrate Polymers, 92*(2), 2299–2305. Available from: https://doi.org/10.1016/j.carbpol.2012.11.106.

Zampaloni, M., Pourboghrat, F., Yankovich, S. A., Rodgers, B. N., Moore, J., Drzal, L. T., ... Misra, M. (2007). Kenaf natural fiber reinforced polypropylene composites: A discussion on manufacturing problems and solutions. *Composites Part A: Applied Science and Manufacturing, 38*(6), 1569–1580. Available from: https://doi.org/10.1016/j.compositesa.2007.01.001.

Zheng, C., Xu, Y. Y., Kawai, T., & Kuroda, S. (2012). Effect of polymeric coupling agent on mechanical properties of kenaf fiber/polystyrene composites. *Applied Mechanics and Materials, 268–270*, 127–133. Available from: https://doi.org/10.4028/www.scientific.net/AMM.268-270.127.

PART III

Manufacturing processes

CHAPTER

14

Recent progress on fused filament fabrication research: sustainable materials and processing parameters

S.N.M Farhan Han[1], *M.T. Mastura*[2,3] *and Muhd Ridzuan Mansor*[1]

[1]Faculty of Mechanical Engineering, Universiti Teknikal Malaysia Melaka, Melaka, Malaysia, [2]Faculty of Mechanical and Manufacturing Engineering Technology, Universiti Teknikal Malaysia Melaka, Melaka, Malaysia, [3]Centre for Advanced Research on Energy, Universiti Teknikal Malaysia Melaka, Melaka, Malaysia

14.1 Introduction

Additive manufacturing (AM) is defined as the process of joining materials to make objects from 3D model data, usually layer upon layer, as opposed to subtractive manufacturing methodologies, such as traditional machining (Bikas, Stavropoulos, & Chryssoluris, 2016). AM technology has rapidly grown and introduced fabrication of conceptual and functional prototypes which can shorten the production development process (Mellor, Hao, & Zhang, 2014). Many technologies encompass AM including stereolithography, fused filament fabrication (FFF), multijet modeling, selective laser sintering (SLS), selective laser melting (SLM), electron beam melting, laminated object manufacturing, laser metal deposition, and 3D printing (3DP) (Ford & Despeisse, 2016; Guo & Leu, 2013; Wong & Hernandez, 2012). The improvement in AM

Design for Sustainability
DOI: https://doi.org/10.1016/B978-0-12-819482-9.00003-4

technologies resulted in AM being widely used in various fields and being used in direct manufacturing industries (Ford & Despeisse, 2016). Three-dimensional printing has been used by architects and designers in producing functional prototypes, whereas the abilities of AM to produce customized products in small quantities is specifically useful for the biomedical field (Ngo, Kashani, Imbalzano, Nguyen, & Hui, 2018), especially in producing a variety of medical implants from CT-imaged tissue replicas (Stansburry & Idacavage, 2016). Other than that, AM also contributes to defense support services for the Royal Navy's platforms (Busachi et al., 2017). The applications of AM in defense support services are to provide platforms with the ability to sustain their system, recover its ability after damage, and dramatically collapse the supply chain (Busachi, Erkoyuncu, Colegrove, Martina, & Ding, 2015). In addition, AM can be used in the manufacture of body armor kits and special tools for military requirements (Busachi et al., 2016). Guo and Leu (2013) stated that AM is known to have the ability to produce freeform fabrication of geometrically complex parts and produce prototype parts with desired material properties for investigation and testing. In addition, AM can minimize the usage of materials, which leads to reducing wastage, lowering the cost of manufacturing, and time reduction (Berman, 2012; Wong & Hernandez, 2012).

The process of AM can be divided into three categories starting with the state of the material, which can be liquid-based, powder-based, or solid-based, as shown in Table 14.1. These three types of starting

TABLE 14.1 Starting state of materials (Ford & Despeisse, 2016; Guo & Leu, 2013; Wong & Hernandez, 2012).

Starting state of material	Additive manufacturing
Liquid	SLA
	MJM
Powder	SLS
	SLM
	EBM
	LMD
	3DP
Solid	FDM
	LOM

EBM, Electron beam melting; *FDM*, fused deposition modeling; *LMD*, laser metal deposition; *LOM*, laminated object manufacturing; *MJM*, multijet modeling; *SLA*, stereolithography; *SLM*, selective laser melting; *SLS*, selective laser sintering; *3DP*, 3D printing.

material state need to be considered for AM to determine the ability to produce the feedstock and it is easy to determine the specific AM process with appropriate materials (Bourell et al., 2017). There are various types of material that can be processed by AM technologies other than polymers, including metal matrix composites, ceramic matrix composites, and polymer composites. Composite filaments can produce different product functionalities and the advantages of AM have increased the use of this technology for applications such as advanced metals and alloys used for aerospace since 3DP consumes less time, is easy, and has less cost compared to traditional processes. Equally, defense industries have used metal components in AM technologies as it is able to produce functional components with complex shape and high-specification requirements (Ding, Pan, Cuiuri, & Li, 2015), whereas ceramics are used in producing scaffolds and concrete is the main material employed in the AM of buildings (Vaezi, Seitz, & Yang, 2013). However, not all AM technologies can use any material randomly, as stated by Ford and Despeisse (2016), since each AM has a specific material that can be processed to produce the output. In addition, the limitation of material availability for 3DP is the main drawback in utilizing these technologies. Thus, the development of materials is needed to widen the utilization of AM technologies.

This chapter presents a review of the development and optimization of composite materials used in FFF to investigate the improvement of mechanical properties on composite materials compared to virgin thermoplastic polymers. In addition, a review of the parameters of FFF technology is included to identify the effects of parameter alterations on the mechanical properties of materials.

14.2 Fused filament fabrication

Fused deposition modeling (FDM) was developed by Scott Cramp in 1988 and was founded by Stratasys Inc. in 1989. Stratasys Inc. developed most of the company's products based on FDM technology and commercialized FDM in the early 1990s (Chua, Leong, & Lim, 2003; Crump, 1992). FDM is also known as FFF, or plastic jet printing (PJP) by Stratasys Inc. (Šafka et al., 2016). FFF is one of the AM methods to melt or soften materials to produce layers of material other than SLM and SLS (Bikas et al., 2016). In Stratasys' patented process stated by Crump (1992), a computer-aided design (CAD) software is used to create a graphic model of the conceptual design and saved in formatted files of standard triangulation language or initial graphics exchange specification. Then, the files are imported into the workstation to generate supports using InsightTM software or Catalyst® software. The CAD files

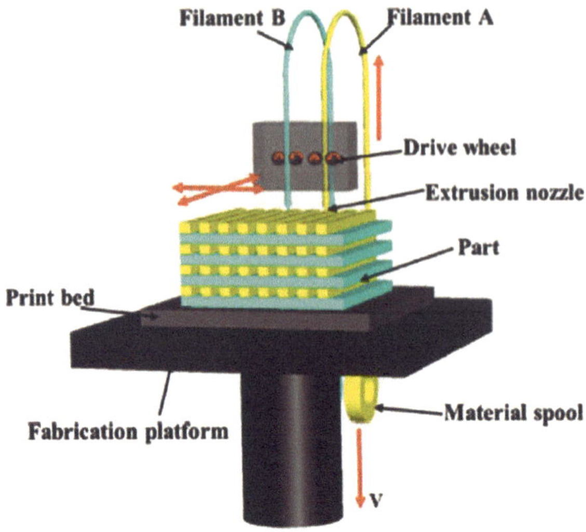

FIGURE 14.1 Schematic of an FFF setup (Wang et al., 2017).

are then sliced into horizontal layers after being oriented for the optimum build position and other necessary support structures. Finally, the tool paths of the build process are generated and downloaded to the FFF machine.

A schematic of an FFF setup is shown in Fig. 14.1. FFF printers work by being fed spooled filament of a thermoplastic polymer into a liquefier which is assisted by a pinch-feed mechanism (Gibson, Rosen, & Stucker, 2010). A three-dimensional object can be produced by successive layer upon layer of materials where the filament is melted inside the liquefier at a temperature above its melting point and pushed at the nozzle die by a solid upstream filament (Carneiro, Silva, & Gomes, 2015). The upstream filament or incoming solid filament acts as a plunger to extrude material in the form of a molten bead, also referred to as roads (Cerneels, Voet, Ivens, & Kruth, 2013). The extruded materials lies on a platform (Torres, Cole, Owji, & DeMastry Gordon, 2016) where layers are fused together and solidify into the final parts (Wang, Jiang, Zhou, Gou, & Hui, 2017). An FFF machine with a dual-tip mechanism dispensed two material filaments where the primary modeler material used to produce the model geometry and secondary material, or release material is used to produce the support structures (Chua & Leong, 2014). The secondary material forms a bond with the primary modeler material and can be washed away upon completion of 3D models. There is a control model that combines all parameters together to support the working principle of FFF. The parameters consist of filament feeding rate, extrusion width, linear plotting speed, and layer thickness (Carneiro et al., 2015).

FFF is one of the most popular AM technologies, especially for engineering applications in producing conceptual models, prototypes, and engineering components. Industries that have utilized FFF include aerospace (Kumar & Nair, 2017), automotive (Page, 2018), medical (Wang et al., 2017), and defense (Rathee, Srivastava, Maheshwari, & Siddiquee, 2017). FFF was initially a suitable manufacturing technology used for home and small businesses. Many types of FFF model can be found and the evaluation of FFF technology itself in terms of performance, cost, and other likely criteria involves build time, dimensional accuracy, material usage, and waste (Roberson, Espalin, & Wicker, 2013; Stanek et al., 2012). In addition, the qualities of FFF such as simple fabrication process (Masood & Song, 2004), fabricate geometrically complex shape (MacDonald & Wicker, 2016), less expensive machine (Wong & Hernandez, 2012), and cost-effectiveness (Cooper, 2001; Mohan, Senthil, Vinodh, & Jayanth, 2017; Noorani, 2006) in producing 3D objects with good resolution have increased the development of FFF technology.

The availability of feedstock for the FFF technique is the main limitation for industrial applications as FFF mainly uses thermoplastic polymer. As FFF technologies have become well developed, the range of filament has become wider with various choices of composite materials. Materials used include ceramic matrix composites, metal matrix composites, natural fiber composites, and synthetic fiber composites. However, there remain deficiencies when it comes to feedstock material for FFF. The vital specific requirement for FFF is the temperature of the material, where the melting point of the material should not be too high or too low as the heating element of a commercially available FFF can operate at around 300°C (Salem Bala & Bin Wahab, 2016). Thus, only thermoplastic polymers and some other materials with low melting points are suitable for FFF technology.

14.2.1 Process of product design and manufacturing process

The prototypes produced by FFF can vary significantly in quality, depending on the material properties, machine specifications, operation parameters, and part geometries (Villalpando, Eiliat, & Urbanic, 2014). The quality of prototypes can be determined by conducting mechanical testing to investigate the strength of prototypes, in addition, surface finishing and dimensional accuracy are vital in ensuring the quality of prototypes (Costa, Duarte, & Covas, 2017). The proper selection of process parameters is the key to success for the AM process in terms of product quality and mechanical properties (Masood, 1996). Mohamed, Masood, and Bhowmik (2015) stated that the optimum process for FFF is important to ensure the quality of products and improve dimensional accuracy, also, waste and large amounts of scrap can be reduced and the

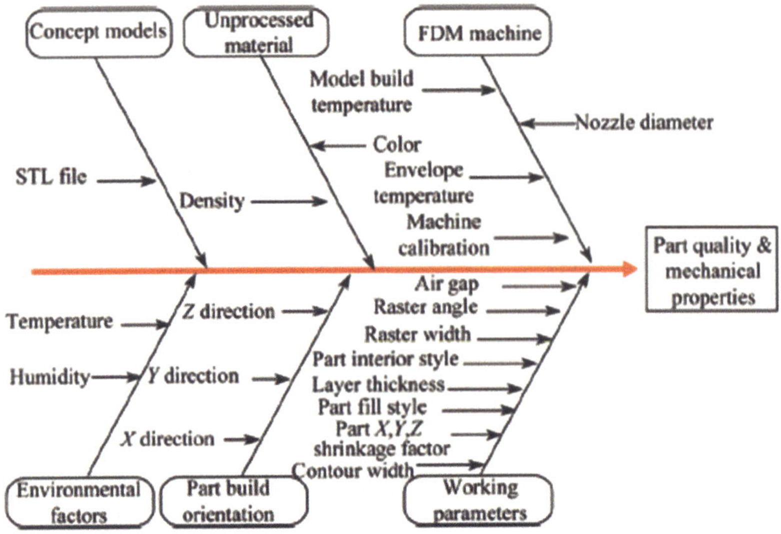

FIGURE 14.2 FFF process parameters (Mohamed et al., 2015).

production time decreased. The parameters of FFF are the build orientation, layer thickness, air gap, raster angle, raster width, contour width, contour-to-contour air gap- and perime ter-to-raster air gap (Croccolo, De Agostinis, & Olmi, 2013; Mohamed et al., 2015; Sood, Ohdar, & Mahapatra, 2010; Wang et al., 2017) as shown in Fig. 14.2.

The drawbacks to FFF technology include resisted accuracy, slow process, and unpredictable shrinkage (Chua & Leong, 2014). Agag, Koga, and Takeichi (2001) stated that thermoplastic materials tend to shrink during the cooling process, resulting in wrapping of products. Shrinkage and distortions of printed samples are common, but this can be overcome by adjusting the FFF process parameters. Other than that, poor surface quality, weak mechanical properties, layer-by-layer appearances, and limited thermoplastics materials are the main disadvantages of FFF (Chohan, Singh, Boparai, Penna, & Fraternali, 2017). In addition, the optimum process for FFF is difficult to determine due to the complexity of FFF where it affects the material properties (Mohamed et al., 2015).

14.2.2 Fused filament fabrication as a green processing method in design for sustainability

The design for sustainability is crucial in determining the social impact and usage of FFF on the environmental. The sustainability conditions that apply for FFF include the type of materials used and the efficiency of the manufacturing process (Medellin-Castillo & Zaragoza-Siqueiros, 2019).

One of the advantages of FFF technology is the ability to process recyclable materials in producing 3D objects. The used of recyclable materials, less usage of toxic materials, and the efficiency of the manufacturing process enable the production of items with longer life and reduced environmental effects (Medellin-Castillo & Zaragoza-Siqueiros, 2019). However, the characteristics of materials have to be considered to obtain energy-saving materials and produce less waste. The printing parameters used for the printing process affect the energy consumption when utilizing FFF. These parameters include the type of material, printing speed, thickness of layers, and complexity of object designs (Griffiths, Howarth, De Almeida-Rowbotham, Rees, & Kerton, 2016; Watson & Taminger, 2018). Freitas, Almeida, Bártolo, and Bártolo (2016) and Watson and Taminger (2018) agreed that the parameters of FFF such as building orientation, internal filling, and complex geometry affect the production time and energy consumption. Materials with high temperature, lower printing speed, thicker layers of objects, and high percentage of internal filling have disadvantage of increasing the energy consumption (Suárez & Domínguez, 2020). Therefore it is necessary to optimize the design of objects manufactured by FFF technology to reduce the energy consumption and minimize the harmful effects on the environment.

14.3 Materials of fused filament fabrication

FFF widely used feedstock included thermoplastic filaments such as acrylonitrile butadiene styrene (ABS), polylactic acid (PLA), polycarbonate (PC), polypropylene (PP), and polyethylene (PE) when FFF was first introduced (Mohan, Senthil, Vinodh, & Jayanth, 2017; Wong & Hernandez, 2012). Polymer matrices can be divided into three classes—thermosets, thermoplastics, and rubbers. There are many different types of polymers for each class. Epoxy, polyamide, and polyester are types of thermosets, whereas PP, PLA, ABS, and PE are examples of thermoplastic matrices. Thermoplastic polymer matrices have been used as reinforced material to improve the mechanical properties of composites. Thermoplastics are preferred to thermosets due to their ability to be recycled, low cost, and short processing time as curing is not needed (Baghaei, Skrifvars, & Berglin, 2013). In addition, thermoplastics have been found to have a high specific strength, high stiffness, and resistance to water absorption (Ning, Pillay, & Vaidya, 2009). Thermoplastics were selected at the best choice due to their low melting temperature (Wang et al., 2017) and less expensive materials (Torres et al., 2016). Across the timeline and technology, composite filaments started to be developed with the advantages of FFF attracting researchers to investigate the possibilities and abilities of custom-made composite filaments Grujovic et al. (2017).

14.3.1 Synthetic fiber reinforced polymer composites

In 2015, Hull et al. designed the process variables to investigate the effects of extrusion on carbon fiber reinforced ABS composite filaments. The process variables during extrusion are carbon fiber content, extrusion temperature, and nozzle size as shown in Table 14.2.

Significant effects of filament changes during the extrusion process included filament diameter, expansion percentage, and extrusion rate. It was concluded that as the carbon fiber content was increased, variability of filament diameter decreased. Next, as the settings of the extrusion temperature increased, the filament diameter decreased since the fluidity of melted thermoplastics increased. Meanwhile, a 1.75 mm nozzle gives a near zero expansion percentage and produced a higher extrusion rate. Thus, a design with well-controlled process parameters for carbon fiber reinforced ABS composites for FFF was successful. Fig. 14.3 illustrates the extruder used to fabricate the composite filaments.

Tambrallimath, Keshavamurthy, Saravanbavan, Kumar, and Kumar (2019) developed graphene reinforced PC-ABS composite filaments for FFF application with varying percentages of graphene from 0.1 to 0.25 vol.%. In this study, the parameter process of extrusion temperature was designed using Taguchi's L_{25} orthogonal array to achieve a filament with 1.75 mm diameter. Then, SEM and elemental mapping were used to analyze the dispersion characteristics of graphene-filled composites. Fig. 14.4 shows the composite filaments with varied percentages of graphene. They concluded that the temperature of the nozzle plays an important role in achieving 1.75 mm filament diameter. The range of the nozzle temperature used was 254°C–262°C, as a higher graphene content in composites

TABLE 14.2 Extrusion conditions (Hull et al., 2015).

	Process parameters		
Extrusion condition	**Carbon fiber content (%)**	**Extrusion temperature (°C)**	**Nozzle size (mm)**
1	0	220	3
2	5	220	3
3	10	220	3
4	15	220	3
5	5	190	3
6	5	200	3
7	5	210	3
8	5	190	1.75

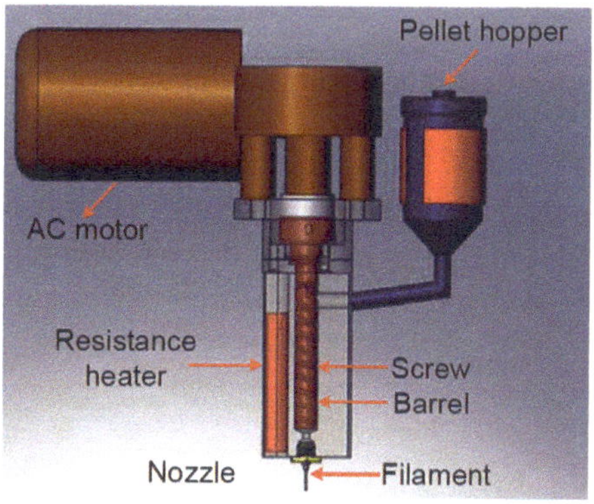

FIGURE 14.3 An illustration of the extruder (Hull et al., 2015).

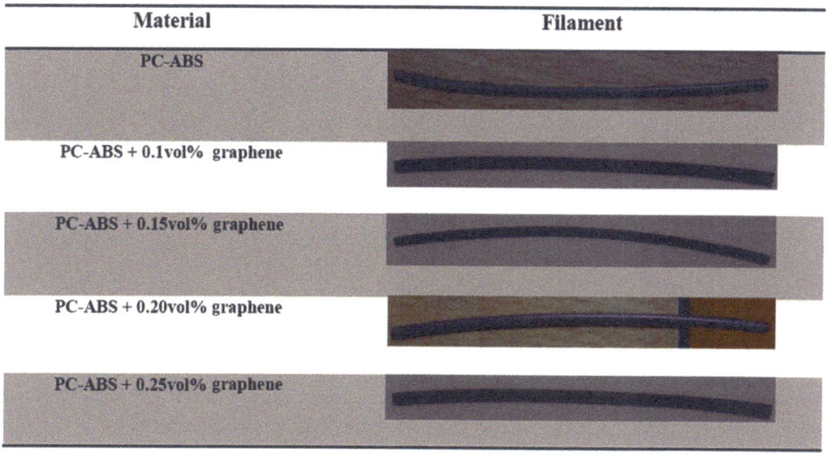

FIGURE 14.4 Composite filaments with various percentages of graphene content (Tambrallimath, Keshavamurthy, Saravanbavan, Kumar, & Kumar, 2019).

requires higher nozzle temperature. Results from SEM and elemental mapping illustrated that the dispersion of filler material increased with a higher graphene content. In addition, good interaction between the filler and polymer was found, with no agglomeration between particles.

14.3.2 Natural fiber reinforced polymer composites

Natural fiber composites are well developed as they are environmentally friendly materials and various types of natural fiber have been investigated for use in composites including kenaf, hemp, banana, and flax (Pankaj, Kant, Jawalkar, & Kumar, 2017). The characteristics of

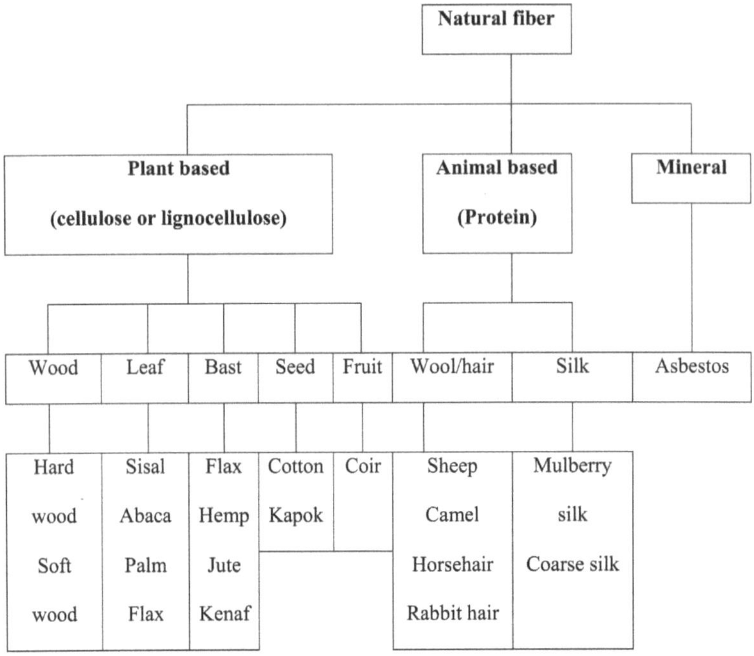

FIGURE 14.5 Classifications of natural fiber (Ahmad et al., 2015; Ho et al., 2012).

natural fibers such as high strength, low cost, and biodegradable are some of the reasons they have become alternatives to synthetic fiber and are widely used in industries due to their lightweight and environmental properties (Ahmad, Choi, & Park, 2015). Moreover, Ahmad et al. (2015) stated that natural fiber can be used as a reinforcement for thermosets or thermoplastics instead of synthetic fiber. Fig. 14.5 shows the classification of natural fiber.

Researchers have conducted studies into natural fiber reinforced thermoplastic composites such as banana (Alavudeen, Rajini, Karthikeyan, Thiruchitrambalam, & Venkateshwaren, 2015), bamboo (Ochi, 2015), sugar palm (Mastura, Sapuan, Mansor, & Nuraini, 2016), corn (Luo et al., 2017) hemp (Stoof & Pickering, 2018), wood (Kariz, Sernek, Obućina, & Kuzman, 2018), and kenaf (Dashtizadeh, Khalina, Cardona, & Lee, 2019; Han, Taha, Mansor, & Aminallah, 2019).

In 2018, Kariz et al. studied wood composite filaments for FFF application and focused on the physical and mechanical properties of 3D-printed composites. Various percentages of wood content from 10% to 50% were used and a single-screw extruder was used to extrude the filament with a final diameter between 1.45 and 1.75 mm. Based on the experimental results, they found that an increasing wood content decreased the density of filaments. In addition, filament with higher wood content often clogged the nozzle as the flow of composites at the nozzle during the printing process did not evenly melt, thus resulting in poor quality of printed parts. The printing process parameters such as printing speed influence the printed samples, by affecting the deposited materials connection between layers. Table 14.3 shows the 3D-printed parts of composite filaments.

Results from mechanical testing found that 10% wood fiber content gives the highest tensile strength value, then, the tensile strength value decreased as the percentage of wood content increased. The higher content of wood influenced the tensile strength of composites as the presence of porous areas and voids increases. This condition would affect the printing process of filaments, where the nozzle has a higher chance of becoming clogged and irregular flow of material through the nozzle would affect the quality of samples. Thus, they suggested that by adding the additives, and controlling the temperature and extrusion speed parameters the surface quality of 3D-printed samples can be improved.

14.3.3 Metal reinforced polymer composites

Wang, Castles, and Grant (2015) studied the compatibility of a polymer–ferrite composite filament for FFF. The development of the product design involved three stages: selection of nontoxic polymer matrix; produced composite filaments; and optimization of FFF parameters. A stripline technique was used to measure the permittivity and permeability of directly printed blocks and diamond-like lattice structure. Fig. 14.6 shows the diamond-like lattice structure. From the experimental results, higher fractions of ferrite produced brittle filament, which was very difficult for the feeding and printing processes. Also, the permittivity and permeability of composites increased with increasing ferrite fraction. They concluded that the uniformity of diameter, nozzle diameter, and printing parameters is crucial to obtaining successful printing of complex structures.

Carrico and Leang (2017) presented 3D-printed single degree of freedom and multi-degree of freedom for ionic polymer–metal composite (IPMC) actuators and demonstrated their use on soft-robotic devices. FFF technology was used to create the IPMC structure due to its ability in terms of shape, size, and time compared with the existing manufacturing process. The design process involved in creating IPMC structures included development of filament, a custom-designed 3D printer, and the printing process. Fig. 14.7 illustrates the creating of a precursor filament and Fig. 14.8 illustrates the custom-made 3D printer. Finally, they successfully demonstrated the ability of FFF technology on soft-robotic devices and future works will focus on the printing process of IPMC actuators.

TABLE 14.3 Three-dimensional (3D) printed parts of wood composite filaments (Kariz et al., 2018).

Filament (wood content)	3D-printed part		
	Filament (70 ×)	Surface (20 ×)	Edge (40 ×)
0%			
10%			
20%			

30%			
40%			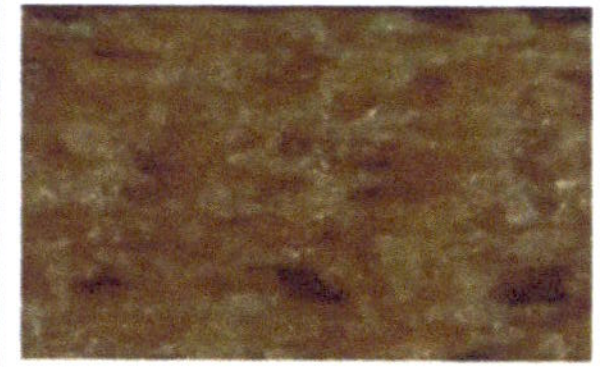
50%			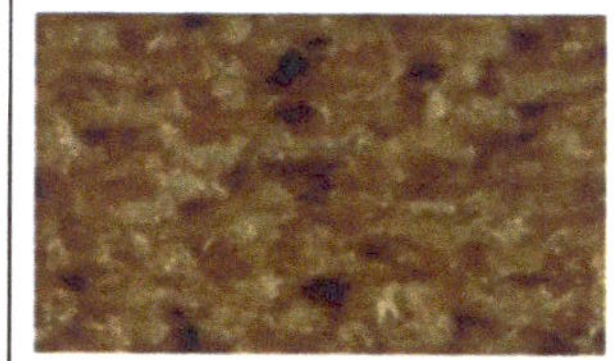

FIGURE 14.6 Diamond-like lattice structure (Wang et al., 2015).

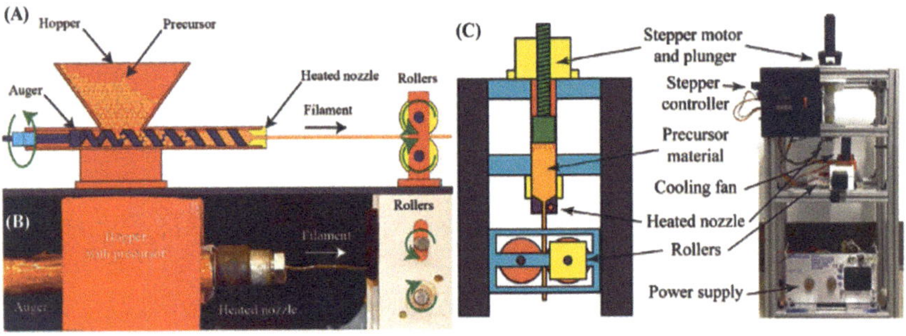

FIGURE 14.7 The designs of (A) extruder, (B) auger-based, and (C) plunger-style filament design (Carrico & Leang, 2017).

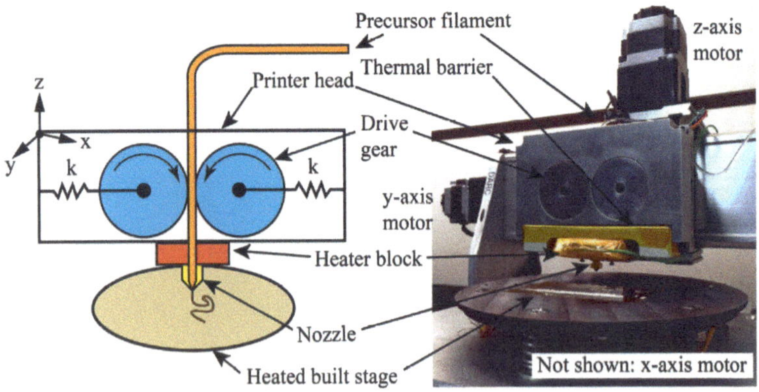

FIGURE 14.8 Custom-made 3D printer (Carrico & Leang, 2017).

14.3.4 Ceramic reinforced polymer composites

In 2012, Drummer et al. studied the processing conditions of PLA and beta tricalcium phosphate (β-TCP) for FFF in terms of thermal, mechanical, and microscopic properties. Three different processing temperatures were used to investigate the influence of temperature on the mechanical properties—215°C, 225°C, and 235°C. The experimental results showed that 225°C nozzle temperature has the best mechanical performance. They found that there was a connection between processing temperature and morphology of samples. As a higher nozzle temperature is used, the parts have a higher degree of crystallinity. Hence, the recrystallization is measured as the temperature of nozzle is increased, where the adjacent filament are heated again by the neighboring filament during the printing process.

Singh and Singh (2014) developed a nylon matrix with different proportions of aluminum oxide (Al_2O_3) as an additive to study the tensile strength of new composite filaments for an FFF machine. The Taguchi L9 technique was applied to design the final control log of the experiment. Twenty-seven experiments were conducted based on the control log of experimentation as shown in Table 14.4. S/N ratio analysis was used to analyze the tensile strength with respect to barrel temperature, screw speed, and volume of Al_2O_3. They successfully developed new composite filaments with an existing FFF process parameter setup, and concluded that new composite filaments give good mechanical properties as compared to commercialized filament. Five percent Al_2O_3 mixed at 180°C and 35 rpm of screw head speed gives the highest tensile strength. In addition, they successfully reduces the cost and time for the development of new composites for FFF applications.

TABLE 14.4 Control log of experimentation (Singh & Singh, 2014).

S. no.	Barrel temperature (°C)	Screw speed (RPM)	Vol.% of Al_2O_3 in nylon
1	180	25	5
2	180	30	7.5
3	180	35	10
4	185	25	7.5
5	185	30	10
6	185	35	5
7	190	25	10
8	190	30	5
9	190	35	7.5

14.3.5 Biocompatible material

Biomaterial is a combination of natural and synthetic materials that can be used in the treatment or replacement of any tissue organ or function of the body. Biocompatible materials are currently well developed as the potential of AM technologies in replicating structural features of biological materials in synthetic components (Studart, 2016). Bose, Ke, Sahasrabudhe, and Bandyopadhyay (2018) indicated that the advantages of advanced manufacturing technique such as reduced cost, minimal inventory, and maximum performance in vivo are the reason that it is constantly being explored to produce biocompatible materials. In addition, ceramic composite parts from FFF are able to control the pore size and pore porosity. The development process for ceramic composite parts is illustrated in Fig. 14.9.

Wang et al. (2016) stated that the design of biomaterial microarchitectures has received increasing attention. Materials widely used for biomaterials include bioceramics, metallic biomaterials, natural biopolymers, synthetic biopolymers, and polymer–ceramic composites (Bose, Ke, Sahasrabudhe, & Bandyopadhyay, 2018).

Bioceramic materials play an important role as biomaterials even though there are some drawbacks to bioceramics such as the melting point being too high for them to melt under normal heating points (Bose, Ke, Sahasrabudhe, & Bandyopadhyay, 2018). The applications of bioceramics include bones and teeth as the advantages of FFF include producing small parts with complex design and low cost. A number of authors have considered the effects of geometrical pattern for bone tissue engineering fabricated by FFF (Cavo & Scaglione, 2016; Korpela et al., 2013). Steffens et al. (2016) investigated the interaction between MSCs with poly(ε-caprolactone) biomaterials for obtaining scaffolds through AM as it is possible to print a wide range of materials with complex geometries and accuracies such as scaffolds. In 2018, Maia, Carvalho, Oliveira, and Reis (2018) stated that 3D technology allows

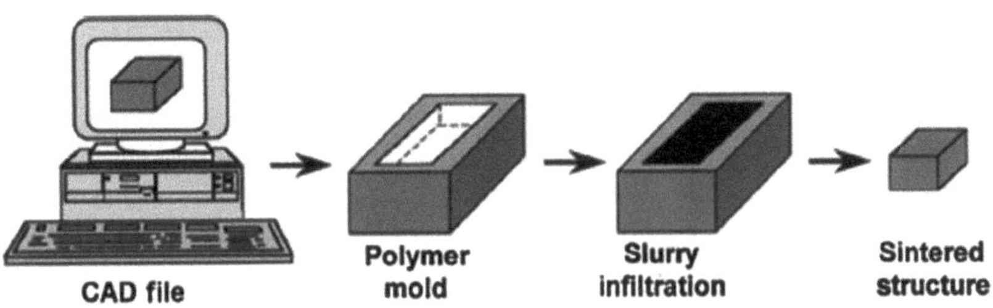

FIGURE 14.9 The process of making porous ceramic structures (Bose, Ke, Sahasrabudhe, & Bandyopadhyay, 2018).

fabrication with high control over spatial resolution, shape, and mechanical properties for osteochondral regeneration.

14.4 Applications and challenges of composite material filaments

Based on the analysis and experimental results from a literature review, FFF is known to produce conceptual models, prototypes, and currently can produce end-user products. Industries that utilize FFF include the aerospace, automotive, biomedical, and energy fields. Based on a literature review, metal/polymer composites are used in 3D-printed circuits and electromagnetic structures for shielding. In addition, a variety of ceramic and metallic materials such as silicon nitrate, Al_2O_3, and hydroxyapatite are applied in structural, electroceramic, and bioceramic fields. Ahlhelm et al. (2016) stated that ceramic is used in biomedical applications for its biocompatible, high corrosion resistance, and flexibility. Haq, Wahab, Saidin, and Jaimi (2014) stated that FFF technologies are potential processes to manufacture complex shapes and high-strength performance carbon fiber reinforced composites for complex aircraft structures.

The mechanical properties of composite material filaments are not limited to materials only, the process parameters of FFF are also important. According to Šafka et al. (2016), the mechanical properties of composite material filaments can be improved or decreased with the addition of reinforcement. Also, the addition of nanomaterials can improve both mechanical and thermal properties as some FFF printed products have poor properties (Salem Bala & Bin Wahab, 2016). Nanomaterials such as carbon nanotubes, nanoceramic particles, and metal nanoparticles have been used in previous research and have been proved to improve the properties of the final parts (Haq et al., 2014). Parandoush and Lin (2017) concluded that the development of fiber reinforced composites using FFF can strengthen the mechanical properties of 3D-printed parts. However, fiber orientation, bonding between reinforcement and matrix, and void formation are challenges faced in the 3DP of composites (Parandoush & Lin, 2017; Wang et al., 2016). Therefore the particle size distribution and filling rate of composite material should be controlled to avoid blocked paths in nozzles due to the limited cross-sections. Navarrete, Hidalgo-Salazar, Nunez, and Arciniegas (2017) stated that the optimum processing temperature is important to ensure the composite filament can print parts in FFF technologies.

In addition, researchers are aware of the working parameters of FFF. Thus, many experimental analyses is on going to obtain the optimum process parameters since different types and compositions of reinforcement and matrix require different parameters. Galantucci, Bodi, Kacani, and Lavecchia (2015), Lieneke, Denzer, Adam, and Zimmer (2016), Nuñez, Rivas, García-Plaza, Beamud, and Sanz-Lobera (2015), Raut, Jatti, Khedkar, and Singh (2014), Qattawi, Alrawi, and Guzman (2017) investigated the process parameters and agreed that the composite filament mechanical properties depend on the FFF processing parameters. The building direction, layer height, extrusion parameters, and dimensional accuracy influence the mechanical properties of composite material filaments and there can be a challenge to obtain the optimum parameters for each composition of reinforcement and matrix, and this area needs further research work. On the other hand, SEM is used to analyze the bonding interference and void content of composite filaments. Based on a literature review, it was proved that the addition of metal powder significantly improved the thermal conductivity of metal–polymer composite filaments, while the presence of Al and Al_2O_3 in nylon6 matrix gives high stiffness and stability when operating at a high sliding velocity and long term. The addition of natural fiber with polymer can significantly improve the mechanical properties of composite filaments, and the advantages of natural fiber include environmental-friendly, low cost, and potential to replace synthetic fiber reinforced composites have resulted in it being widely investigated.

14.5 Conclusions

This chapter presents a review of research works carried out on the development of new composite materials and optimization of FFF process parameters. The FFF manufacturing process is currently widely used in industries as it can produce complex geometric shapes, requires less time, and reduces waste. New composite material filaments developed by researchers are being used to investigate the mechanical properties and quality of printed parts using FFF. Hence, studies on process parameters for FFF have been conducted to obtain the desirable performance strength of printed parts. Overall, the literature review has revealed that the current research into composite materials used as feedstock filaments for FFF is looking for a wide range of materials with availability and optimum process parameters for FFF.

Acknowledgment

The authors would like to acknowledge Universiti Teknikal Malaysia Melaka and Ministry of Higher Education, Malaysia for financial support through a Fundamental Research Grant Scheme (FRGS) with grant number of FRGS/1/2020/TK0/UTEM/02/26.

References

Agag, T., Koga, T., & Takeichi, T. (2001). Studies on thermal and mechanical properties of polyimide–clay nanocomposites. *Polymer, 42*(8), 3399–3408.

Ahlhelm, M., Günther, P., Scheithauer, U., Schwarzer, E., Günther, A., Slawik, T., … Michaelis, A. (2016). Innovative and novel manufacturing methods of ceramics and metal-ceramic composites for biomedical applications. *Journal of the European Ceramic Society, 36*(12), 2883–2888.

Ahmad, F., Choi, H. S., & Park, M. K. (2015). A review: Natural fiber composites selection in view of mechanical, light weight, and economic properties. *Macromolecular Materials and Engineering, 300*(1), 10–24.

Alavudeen, A., Rajini, N., Karthikeyan, S., Thiruchitrambalam, M., & Venkateshwaren, N. (2015). Mechanical properties of banana/kenaf fiber-reinforced hybrid polyester composites: Effect of woven fabric and random orientation. *Materials & Design, 66*, 246–257.

Baghaei, B., Skrifvars, M., & Berglin, L. (2013). Manufacture and characterisation of thermoplastic composites made from PLA/hemp co-wrapped hybrid yarn prepregs. *Composites Part A: Applied Science and Manufacturing, 50*, 93–101.

Berman, B. (2012). 3-D printing: The new industrial revolution. *Business Horizons, 55*(2), 155–162.

Bikas, H., Stavropoulos, P., & Chryssoluris, G. (2016). Additive manufacturing methods and modelling approaches: A critical review. *The International Journal of Advanced Manufacturing Technology, 83*(1–4), 389–405.

Bose, S., Ke, D., Sahasrabudhe, H., & Bandyopadhyay, A. (2018). Additive manufacturing of biomaterials. *Progress in Materials Science, 93*, 45–111.

Bourell, D., Kruth, J. P., Leu, M., Levy, G., Rosden, D., Beese, A. M., & Clare, A. (2017). Materials for additive manufacturing. *CIRP Annals, 66*(2), 659–681.

Busachi, A., Erkoyuncu, J., Colegrove, P., Drake, R., Watts, C., & Martina, F. (2016). Defining next-generation additive manufacturing applications for the ministry of defence (MoD). *Procedia CIRP, 55*, 302–307.

Busachi, A., Erkoyuncu, J., Colegrove, P., Drake, R., Watts, C., & Wilding, S. (2017). Additive manufacturing applications in defence support services: Current practices and framework for implementation. *International Journal of System Assurance Engineering and Management*, 1–18.

Busachi, A., Erkoyuncu, J., Colegrove, P., Martina, F., & Ding, J. (2015). Designing a WAAM based manufacturing system for defence applications. *Procedia CIRP, 37*, 48–53.

Carneiro, O. S., Silva, A. F., & Gomes, R. (2015). Fused deposition modelling with polypropylene. *Materials & Design, 83*, 768–776.

Carrico, J. D., & Leang, K. K. (2017). Fused filament 3D printing of ionic polymer-metal composites for soft robotics. *In Electroactive polymer actuators and devices*. International Society for Optics and Photonics, 1016301.

Cavo, M., & Scaglione, S. (2016). Scaffold microstructure effects on functional and mechanical performance: Integration of theoretical and experimental approaches for bone tissue engineering applications. *Materials Science and Engineering: C, 68*, 872–879.

Cerneels, J., Voet, A., Ivens, J., & Kruth, J. P. (2013). Additive manufacturing of thermoplastic composites. In: *Composites Week@*. Leuven: KU Leuven, 1–7.

Chohan, J. S., Singh, R., Boparai, K. S., Penna, R., & Fraternali, F. (2017). Dimensional accuracy analysis of coupled fused deposition modelling and vapour smoothing operations for biomedical applications. *Composites Part B: Engineering, 117*, 138–149.

Chua, C. K., & Leong, K. F. (2014). *3D printing and additive manufacturing: Principles and applications (with companion media pack) of rapid prototyping* (4th ed.). World Scientific Publishing Company.

Chua, C. K., Leong, K. F., & Lim, C. S. (2003). *Rapid prototyping: Principles and applications.* Singapore: World Scientific.
Cooper, K. G. (2001). *Rapid prototyping technology*. New York: Marcel Dekker.
Costa, S. F., Duarte, F. M., & Covas, J. A. (2017). Estimation of filament temperature and adhesion development in fused deposition techniques. *Journal of Materials Processing Technology, 245*, 167–179.
Croccolo, D., De Agostinis, M., & Olmi, G. (2013). Experimental characterization and analytical modelling of the mechanical behaviour of fused deposition processed parts made of ABS-M30. *Computational Materials Science, 79*, 506–518.
Crump, S. (1992). Apparatus and method for creating three-dimensional objects, US patent 5121329.
Dashtizadeh, Z., Khalina, A., Cardona, F., & Lee, C. H. (2019). Mechanical characteristics of green composites of short kenaf bast fiber reinforced in cardanol. *Advances in Materials Science and Engineering.*
Ding, D., Pan, Z., Cuiuri, D., & Li, H. (2015). A practical path planning methodology for wire and arc additive manufacturing of thin-walled structures. *Robotics and Computer-Integrated Manufacturing, 34*, 8–19.
Drummer, D., Cifuentes-Cuéllar, S., & Rietzel, D. (2012). Suitability of PLA/TCP for fused deposition modelling. *Rapid Prototyping Journal, 18*(6), 500–507.
Ford, S., & Despeisse, M. (2016). Additive manufacturing and sustainability: An exploratory study of the advantages and challenges. *Journal of Cleaner Production, 137*, 1573–1587.
Freitas, D., Almeida, H. A., Bártolo, H., & Bártolo, P. J. (2016). Sustainability in extrusion-based additive manufacturing technologies. *Progress in Additive Manufacturing, 1*(1-2), 65–78.
Galantucci, L. M., Bodi, I., Kacani, J., & Lavecchia, F. (2015). Analysis of dimensional performance for a 3D open-source printer based on fused deposition modelling technique. *Procedia CIRP, 28*, 82–87.
Gibson, I., Rosen, D. W., & Stucker, B. (2010). Design for additive manufacturing. *Additive Manufacturing Technologies*, 299–332.
Griffiths, C. A., Howarth, J., De Almeida-Rowbotham, G., Rees, A., & Kerton, R. (2016). A design of experiments approach for the optimisation of energy and waste during the production of parts manufactured by 3D printing. *Journal of Cleaner Production, 139*, 74–85.
Grujovic, N., Zivic, F., Zivkovic, M., Sljivic, M., Radovanovic, A., Bukvic, L., … Sindjelic, A. (2017). Custom design of furniture elements by fused filament fabrication. *Proceedings of the Institution of Mechanical Engineers, Part C: Journal of Mechanical Engineering Science, 231*(1), 88–95.
Guo, N., & Leu, M. C. (2013). Additive manufacturing: Technology, applications and research needs. *Frontiers of Mechanical Engineering, 8*(3), 215–243.
Han, S. N. M. F., Taha, M. M., Mansor, M. R., & Aminallah, M. M. (2019). Thermal degradation study of acrylonitrile butadiene styrene (ABS) composites for FFF. In: *Proceedings of Mechanical Engineering Research Day*, Melaka: Centre for Advanced Research on Energy, 3–4.
Haq, R. H. A., Wahab, B., Saidin, M., & Jaimi, N. I. (2014). Fabrication process of polymer nano-composite filament for fused deposition modeling. *Applied Mechanics and Materials, 465*, 8–12.
Ho, M. P., Wang, H., Lee, J. H., Ho, C. K., Lau, K. T., Leng, J., & Hui, D. (2012). Critical factors on manufacturing processes of natural fibre composites. *Composites Part B: Engineering, 43*(8), 3549–3562.
Hull, E., Grove, W., Zhang, M., Song, X., Pei, Z.J., & Cong, W. (2015). Effects of process variables on extrusion of carbon fiber reinforced ABS filament for additive manufacturing. In: *ASME 2015 International Manufacturing Science and Engineering Conference.* American Society of Mechanical Engineers Digital Collection.

Kariz, M., Sernek, M., Obućina, M., & Kuzman, M. K. (2018). Effect of wood content in FDM filament on properties of 3D printed parts. *Materials Today Communications, 14*, 135–140.

Korpela, J., Kokkari, A., Korhonen, H., Malin, M., Närhi, T., & Seppälä, J. (2013). Biodegradable and bioactive porous scaffold structures prepared using fused deposition modelling. *Journal of Biomedical Materials Research Part B: Applied Biomaterials, 101*(4), 610–619.

Kumar, L. J., & Nair, C. K. (2017). Current trends of additive manufacturing in the aerospace industry. *Advances in 3D Printing & Additive Manufacturing Technologies*, 39–54.

Lieneke, T., Denzer, V., Adam, G. A., & Zimmer, D. (2016). Dimensional tolerances for additive manufacturing: Experimental investigation for fused deposition modelling. *Procedia CIRP, 43*, 286–291.

Luo, Z., Li, P., Cai, D., Chen, Q., Qin, P., Tan, T., & Cao, H. (2017). Comparison of performances of corn fiber plastic composites made from different parts of corn stalk. *Industrial crops and products, 95*, 521–527.

MacDonald, E., & Wicker, R. (2016). Multiprocess 3D printing for increasing component functionality. *Science, 353*(6307), 2093.

Maia, F. R., Carvalho, M. R., Oliveira, J. M., & Reis, R. L. (2018). Tissue engineering strategies for osteochondral repair. *Osteochondral Tissue Engineering*, 353–371.

Masood, S. H. (1996). Intelligent rapid prototyping with fused deposition modelling. *Rapid Prototyping Journal, 2*(1), 24–33.

Masood, S. H., & Song, W. Q. (2004). Development of new metal/polymer materials for rapid tooling using fused deposition modelling. *Materials & Design, 25*(7), 587–594.

Mastura, M. T., Sapuan, S. M., Mansor, M. R., & Nuraini, A. A. (2016). Application of hybrid sugar palm-filled polyurethane composites in conceptual design of an automotive anti-roll bar. *Journal of Engineering and Technology, 7*(1), 102–113.

Medellin-Castillo, H. I., & Zaragoza-Siqueiros, J. (2019). Design and manufacturing strategies for fused deposition modelling in additive manufacturing: A review. *Chinese Journal of Mechanical Engineering, 32*(1), 53.

Mellor, S., Hao, L., & Zhang, D. (2014). Additive manufacturing: A framework for implementation. *International Journal of Production Economics, 149*, 194–201.

Mohamed, O. A., Masood, S. H., & Bhowmik, J. L. (2015). Optimization of fused deposition modelling process parameters: A review of current research and future prospect. *Advances in Manufacturing, 3*(1), 42–53.

Mohan, N., Senthil, P., Vinodh, S., & Jayanth, N. (2017). A review on composite materials and process parameters optimisation for the fused deposition modelling process. *Virtual and Physical Prototyping, 12*(1), 47–59.

Navarrete, J. I. M., Hidalgo-Salazar, M. A., Nunez, E. E., & Arciniegas, A. J. R. (2017). Thermal and mechanical behaviour of bio-composites using additive manufacturing. *International Journal on Interactive Design and Manufacturing*, 1–10.

Ngo, T. D., Kashani, A., Imbalzano, G., Nguyen, K. T., & Hui, D. (2018). Additive manufacturing (3D printing): A review of materials, methods, applications and challenges. *Composites Part B: Engineering*.

Ning, H., Pillay, S., & Vaidya, U. K. (2009). Design and development of thermoplastic composite roof door for mass transit bus. *Materials & Design, 30*(4), 983–991.

Noorani, R. (2006). *Rapid prototyping: Principles and applications*. John Wiley and Sons.

Nuñez, P. J., Rivas, A., García-Plaza, E., Beamud, E., & Sanz-Lobera, A. (2015). Dimensional and surface texture characterization in fused deposition modelling (FDM) with ABS plus. *Procedia engineering, 132*, 856–863.

Ochi, S. (2015). Flexural properties of long bamboo fiber/PLA composites. *Open Journal of Composite Materials, 5*(03), 70.

Page, T. (2018). A survey of the use of additive fabrication in component replacement and customised automotive modifications. *International Journal of Manufacturing, Materials, and Mechanical Engineering, 8*(4), 23–24.

Pankaj, K., Kant, S., Jawalkar, C. S., & Kumar, S. (2017). A review on processing and mechanical performance of natural fiber composites. *Journal of Industrial Engineering and Advances*, *2*(2), 1–8.

Parandoush, P., & Lin, D. (2017). A review on additive manufacturing of polymer-fiber composites. *Composite Structures*, *117*(1), 36–53.

Qattawi, A., Alrawi, B., & Guzman, A. (2017). Experimental optimization of fused deposition modelling processing parameters: A design-for-manufacturing approach. *Procedia Manufacturing*, *10*, 791–803.

Rathee, S., Srivastava, M., Maheshwari, S., & Siddiquee, A. N. (2017). Effect of varying spatial orientations on build time requirements for FDM process: A case study. *Defence technology*, *13*(2), 92–100.

Raut, S., Jatti, V. S., Khedkar, N. K., & Singh, T. P. (2014). Investigation of the effect of built orientation on mechanical properties and total cost of FDM parts. *Procedia Materials Science*, *6*, 1625–1630.

Roberson, D. A., Espalin, D., & Wicker, R. B. (2013). 3D printer selection: A decision-making evaluation and ranking model. *Virtual and Physical Prototyping*, *8*(3), 201–212.

Šafka, J., Ackermann, M., Bobek, J., Seidl, M., Habr, J., & Běhálek, L. (2016). Use of composite materials for FDM 3D print technology. *Materials Science Forum*, *862*, 174–181.

Salem Bala, A., & Bin Wahab, S. (2016). Elements and materials improve the FDM products: A review. *Advanced Engineering Forum*, *16*, 33–51.

Singh, R., & Singh, S. (2014). Development of nylon based FDM filament for rapid tooling application. *Journal of The Institution of Engineers (India): Series C*, *95*(2), 103–108.

Sood, A. K., Ohdar, R. K., & Mahapatra, S. S. (2010). Parametric appraisal of mechanical property of fused deposition modelling processed parts. *Materials & Design*, *31*(1), 287–295.

Stanek, M., Manas, D., Manas, M., Navratil, J., Kyas, K., Senkerik, V., & Skrobak, A. (2012). Comparison of different rapid prototyping methods. *International Journal Math Computer Simulation*, *6*, 550–557.

Stansburry, J. W., & Idacavage, M. J. (2016). 3D printing with polymers: Challenges among expanding options and opportunities. *Dental Materials*, *32*(1), 54–64.

Steffens, D., Alvarenga Rezende, R., Santi, B., Alencar de Sena Pereira, F. D., Inforçatti Neto, P., Lopes da Silva, J. V., & Pranke, P. (2016). 3D-printed PCL scaffolds for the cultivation of mesenchymal stem cells. *Journal of Applied Biomaterials & Functional Materials*, *14*(1), 19–25.

Stoof, D., & Pickering, K. (2018). Sustainable composite fused deposition modelling filament using recycled pre-consumer polypropylene. *Composites Part B: Engineering*, *135*, 110–118.

Studart, A. R. (2016). Additive manufacturing of biologically inspired materials. *Chemical Society Reviews*, *45*(2), 359–376.

Suárez, L., & Domínguez, M. (2020). Sustainability and environmental impact of fused deposition modelling (FDM) technologies. *The International Journal of Advanced Manufacturing Technology*, *106*(3-4), 1267–1279.

Tambrallimath, V., Keshavamurthy, R., Saravanbavan, D., Kumar, G. P., & Kumar, M. H. (2019). Synthesis and characterization of graphene filled PC-ABS filament for FDM applications. In: *AIP Conference Proceedings*, Vol. 2057(1), Surathkal: AIP Publishing LLC, 020039.

Torres, J., Cole, M., Owji, A., & DeMastry Gordon, A. P. (2016). An approach for mechanical property optimization of fused deposition modelling with polylactic acid via design of experiments. *Rapid Prototyping Journal*, *22*(2), 387–404.

Vaezi, M., Seitz, H., & Yang, S. (2013). A review on 3D micro-additive manufacturing technologies. *The International Journal of Advanced Manufacturing Technology*, *67*(5-8), 1721–1754.

Villalpando, L., Eiliat, H., & Urbanic, R. J. (2014). An optimization approach for components built by fused deposition modelling with parametric internal structures. *Procedia CIRP*, *17*, 800–805.

Wang, X., Jiang, M., Zhou, Z., Gou, J., & Hui, D. (2017). 3D printing of polymer matrix composites: A review and prospective. *Composites Part B: Engineering, 110*, 442–458.
Wang, X., Xu, S., Zhou, S., Xu, W., Leary, M., Choong, P., … Xie, Y. M. (2016). Topological design and additive manufacturing of porous metals for bone scaffolds and orthopaedic implants: A review. *Biomaterials, 83*, 127–141.
Wang, Y., Castles, F., & Grant, P. S. (2015). 3D printing of NiZn ferrite/ABS magnetic composites for electromagnetic devices. *MRS Online Proceedings Library Archive, 1788*, 29–35.
Watson, J. K., & Taminger, K. M. B. (2018). A decision-support model for selecting additive manufacturing versus subtractive manufacturing based on energy consumption. *Journal of Cleaner Production, 176*, 1316–1322.
Wong, K. V., & Hernandez, A. A. (2012). Review of additive manufacturing. *ISRN Mechanical Engineering, 2012*, 1–10.

CHAPTER

15

Out-of-autoclave as a sustainable composites manufacturing process for aerospace applications

Nadlene Razali[1,2], Muhd Ridzuan Mansor[1], Ghazali Omar[1,2], Syed Ahmad Faiz Syed Kamarulzaman[3], Mohd Hanafee Zin[3] and Nadia Razali[4]

[1]Faculty of Mechanical Engineering, Universiti Teknikal Malaysia Melaka, Melaka, Malaysia, [2]Aerospace Malaysia Innovation Centre (AMIC), Kompleks Teknologi 3, German Malaysian Institute, Jalan Ilmiah, Taman Universiti, Selangor, Malaysia, [3]Section of Environmental Engineering Technology, Universiti Kuala Lumpur—Malaysian Institute of Chemical and Bioengineering Technology, Melaka, Malaysia, [4]Centre for Advanced Research on Energy (CARe), Universiti Teknikal Malaysia Melaka, Melaka, Malaysia

15.1 Introduction

Nowadays, awareness of the need for sustainability in people's daily lives has grown tremendously. They are taking seriously the effects of modernization on the environment and the well-being of its inhabitants, as they are more knowledgeable of the negative impact of modernization on the planet, such as abrupt change in climate, depletion of the ozone layer due to greenhouse gases (GHGs), and water and air pollution. As a result, many solutions have been developed and implemented by industries, researchers, and government authorities to improve global sustainability in many aspects including socially and economically.

Design for Sustainability
DOI: https://doi.org/10.1016/B978-0-12-819482-9.00011-3

One of the solutions making a significant impact toward sustainability is the implementation of sustainable design or design for sustainability (DfS). The DfS concept strives to improve the environmental performance of products and processes by holistically covering the entire life cycle phases, from raw material extraction until its end-of-life. Furthermore, the DfS concept also tries to get the best of the environmental benefits, while at the same time supporting the products and processes with the economic objective of gaining the maximum profit and sustain the well-being of human life and other living creatures. Dieter (2000) has listed several focuses toward achieving the sustainable design objective for products and processes, such as minimal use of material and energy as well as having a low environmental impact and using sustainable materials (Dieter, 2000).

The aerospace industry has also been aggressively taking the initiative toward achieving similar sustainability goals in the development of its aircraft components and related manufacturing processes. Aircraft have been developed using advanced materials such as polymer composites to make them lighter and structurally safe to operate, and to reduce the fuel consumption and subsequently GHG emissions. Similar polymer composite materials also showed better performance of payload weight over cost ratio. In terms of the manufacturing of components, aircraft makers are implementing more energy-efficient processing methods to reduce the energy consumption during the production stage. At the present moment, the aerospace industry is using higher quantities of composite materials to replace existing metal-based components, and these composite components need to be manufactured with high quality while adhering to very stringent specific performance requirements set by the aviation authorities to ensure the highest safety of passengers (Linganiso & Anandjiwala, 2016). The most common manufacturing process is autoclave manufacturing (Centea, Grunenfelder, & Nutt, 2015). Autoclave manufacturing technology, which requires the optimum controlled levels of heat and pressure to parts and assemblies, is particularly suitable and largely employed for this purpose. In autoclave manufacturing, the quality of advanced composite material parts greatly depends on the thermal and pressure cycles imposed on the composite material during the curing process.

There are many aerospace industries still maintaining autoclaves as a leading tool to manufacture aerospace products and materials. This is because of the high investment that has been made and the autoclave can produce excellent, high-quality products. The autoclave method can also process different types of composite materials such as thermosets and thermoplastics with different types of designs, complexity, and shape contours. Most aerospace products and structures are from thermoset materials because the price of these materials is lower than thermoplastic. In the aerospace industry the production of quality aerospace products and structures in accordance with the required specifications is very strict. Composites prepreg and autoclave cure have traditionally been associated with standard aerospace manufacturing processes (Grunenfelder, Dills, Centea, & Nutt, 2016; Lee et al., 2018).

In addition, high quality requirements are very important to improve the efficiency and cost-effectiveness of the structure system as well as ensuring a reliable and consistent manufacturing process. The use of autoclave machines is robust, well understood, and has the stability for the production of high-quality structural products and composite components. With the autoclave method, composite laminates such as carbon fiber and thermoset resin have good compression as well as interfacial bonding and less voids. Voids are dangerous faults for aerospace products that can lead to product failure. Fig. 15.1 shows the autoclave in Composites Technology Research Malaysia (CTRM) situated in Melaka, which is one of the largest aerospace manufacturing companies in Malaysia.

However, despite the autoclave advantages in producing high-quality products, the design of the autoclave system is very complex and is expensive to maintain and operate. The complex design of the autoclave system includes mechanical engineering, electrical, process control, and sophisticated computer system control and instrumentation to ensure the product meets specifications and safety standards. The autoclave manufacturing process is needed for higher capital investment and

FIGURE 15.1 Autoclave at Composites Technology Research Malaysia in Batu Berendam, Melaka, Malaysia [Composites Technology Research Malaysia Sdn. Bhd. (CTRM) (Composites Technology Research Malaysia Sdn. Bhd. CTRM, 2020)].

the production cycle time takes longer, for example, the average time is over 6 hours per cure. Guilherme et al. (2017) reported that many large aircraft companies invested huge capital in autoclaves and automated production machines to meet their high demands, for around 80–120 aircraft per year (Guilherme et al., 2017). From the total production cost of an aircraft, about 10% of the costs are incurred during the curing and assembly stages of the manufactured composite parts. Due to this limitation, prime airframe manufacturers are exploring alternatives to the autoclave process to manufacture composite parts with lower cost, as well as reducing the energy consumption during production as occurs when using an autoclave process, such as the out-of-autoclave (OOA) process.

15.2 Out-of-autoclave as a sustainable manufacturing process

One notable solution to reducing the cost as well as energy consumption when using an autoclave process is to manufacture composite structural components using OOA technology. OOA technology enables composites to be produced using only vacuum pressure, hence eliminating the cost of purchasing and operating an autoclave (Bond, Griffith, & Hahn, 2008). Furthermore, the lower cost of manufacturing is also due to the use of low-temperature tooling in OOA technology, such as wood or syntactic foam (Ridgard, 2000). Improvement actions have also been conducted on OOA technology to obtained comparable composite quality, especially in terms of low void contents after curing as part of the autoclave process. OOA can be applied for many applications requiring low production volume but high-quality composite component such as wind turbine blades for renewable energy generation, marine vessels, unmanned aerial vehicles, general aviation and sports aircraft, as well as high-performance automotive vehicles.

Brosius (2014) stated that the OOA process offers an improved sustainability manufacturing path with cost savings in operation. From the perspective of DfS, because OOA only requires vacuum pressure during operation, very low energy consumption is required to produce the components compared to an autoclave process, hence generating less GHG emissions during manufacturing and throughout the component life cycle. Energy savings obtained through OOA technology are converted into valuable cost savings for the component manufacturers, and lower price for aviation passengers using the aircraft (Brosius, 2014). Ogale, Potluri, Rittenschober, and Schlimbach (2014) also stated that OOA is environmentally friendly compared to the autoclave process due to it operating based on a press process which requires generally low to no pressure systems for curing the composite laminate (Ogale et al., 2014).

Furthermore, Brosius (2014) also stated that OOA also offers economic advantages to produce composite components at lower cost by enabling a shorter process life cycle time and lower capital as well as tooling investment when compared to autoclave-based composite manufacturing processes. This, in return, increases the utilization of composite components in aircraft because of their cost-effectiveness over traditional engineering materials such as metals (Brosius, 2014).

Centea et al. (2015) summarized several reports comparing the environmental and economic differences between the OOA process and autoclave process to produce composite components (Centea et al., 2015). In their report, the OOA process in particular vacuum bag (VBO), which requires low pressure to operate, may reduce scrap rates by mitigating autoclave-induced effects such as honeycomb crush and face-sheet dimpling. OOA may also allow integrated manufacturing of large parts that would otherwise require expensive and time-consuming assembly (Centea et al., 2015). Centea et al. (2015) also summarized the study by Witik et al. (2012) comparing the environmental impact of OOA and autoclave composite manufacturing methods (Witik et al., 2012). They revealed that using life cycle assessment analysis, due to the lower energy consumption, the production of aircraft composite components using an OOA processing method was able to reduce the environmental impact categories such as GHG emissions, resource use, ecosystem quality, and human health compared to an autoclave processing method by 10%–20%. In a similar report, the findings from Teuscher, Witik, Cohades, and Michaud (2012) compared the production of a composite ribbed part was also included (Teuscher et al., 2012). The results obtained showed that OOA which required an oven-cure process as compared to an autoclave process had better environmental performance due to lower energy consumption during operation. Furthermore, the use of OOA may also solve technical constraints related to autoclave process component production.

Based on the above environmental and cost advantages, statistics showed that currently OOA technology already represents about 14% or approximately 1158 metric tonnes of aerospace composite shipments. Examples of aircraft components manufactured using OOA technology are Airbus A340 and A380 wing fixed leading edges, Airbus A340 and A380 keel beam ribs, Boeing 787 pressure bulkheads, Boeing 787 and Airbus A350 XWB trailing edge component packages, Gulfstream Aerospace Corp.'s G650 rudder, elevator and floor beams, Mitsubishi Aircraft Corp.'s MRJ regional jet's vertical stabilizer, wingtip extensions for Northrop Grumman's RQ-4B Global Hawk, pressurized cabin section of the Bombardier Aerospace Learjet 85, Diamond Aircraft and Cirrus Design general aviation airframes, and RTM's propeller blades for turboprop aircraft, turbofan engine fan blades and spacers, containment cases and stator vanes, and foldable wings, control surfaces and launch tubes, and containers for missiles and munitions (Composites World, 2014).

15.3 Types of out-of-autoclave molding process

Currently, there are several OOA molding processes which have been developed and implemented for producing composite aerospace components. Among the OOA processes are resin transfer molding (RTM), vacuum-assisted resin transfer molding (VARTM), same qualified resin transfer molding (SQRTM) developed by Radius Engineering Inc., vacuum bag only (VBO), controlled atmospheric pressure resin infusion (CAPRI) developed by Boeing, double vacuum bag infusion molding, multiaxial compression molding, and thermoplastic composite forming and welding (Composites World, 2014). Fig. 15.2 summarizes the various types of OOA process. Descriptions of several of the OOA processes are given in the subsequent section (Brosius, 2014).

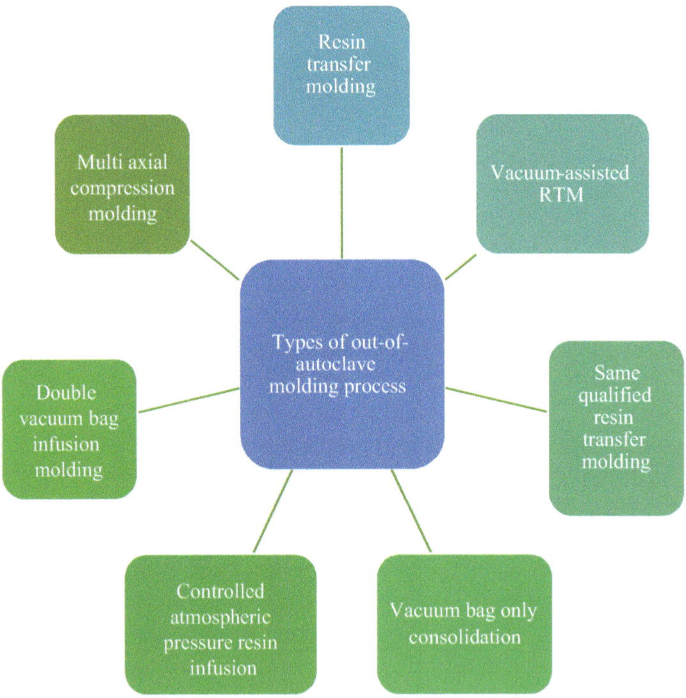

FIGURE 15.2 Various types of out-of-autoclave molding process.

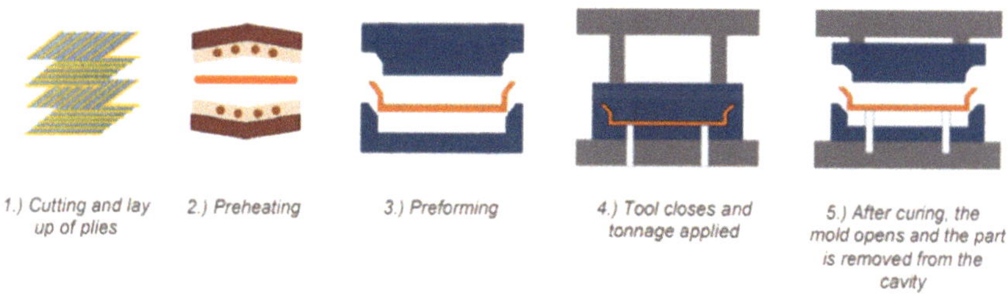

FIGURE 15.3 Prepreg compression molding process (Pasco & Kendall, 2018).

15.3.1 Hot press molding

One variation of the OOA process is a hot press molding, with added fast cured interdiffusion technology in order to counter the cost-effectiveness of producing the aerospace product. The materials are cured in 2 minutes, and it takes about 10 minutes to produce a seat pan product. Compression molding is a method of molding in which the molding material, generally preheated, is first placed in an open, heated mold cavity (female) or form (male) using a two-part mold system as shown in Fig. 15.3. The upper part of the mold is fix closed with the lower mold and compressed with appropriate pressure to ensure the material flows uniformly in the mold cavity (Wu, Centea, & Nutt, 2014). This method can fabricate components or aerospace structures with different lengths, thicknesses, and complexity. In addition, this method reduces energy consumption and has a shorter fabrication time. Thermoset composites are the most commonly used in compression molding.

In this process, for thermosetting matrices, normally an intermediate material called a molding compound is used, this is really a semicured composite. Sheet molding compound and bulk molding compound (BMC) are two widely used thermosetting-based molding compounds.

The hot press OOA process is also known as thermal press curing (TPC). TPC is a process that involves compressing a formed thermoset composite laminate between a heated mold (Özgül & Bektaş, 2019). The novelty of TPC, shown schematically in Fig. 15.4, is how it achieves uniform normal (i.e., hydrostatic) pressure and temperature distributions across the entire 3-D thermoset composite part surface.

15.3.2 Vacuum bag consolidation

In the aerospace industry, OOA means vacuum bag only (VBO) consolidation of carbon fiber prepreg with the product cured using an oven. As a result, this type of OOA molding method can save energy as well as produce products with low voids content.

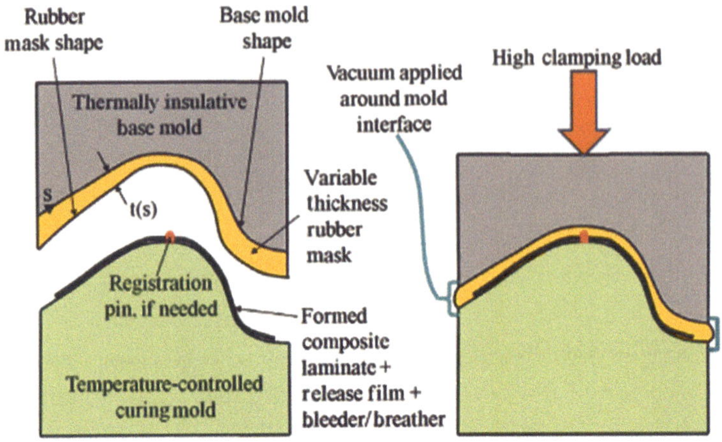

FIGURE 15.4 Schematic of the thermal press curing process for composite curing with a variable thickness rubber layer t(s) shown (left) and molds clamping the formed composite laminate (right) (Walczyk & Kuppers, 2012).

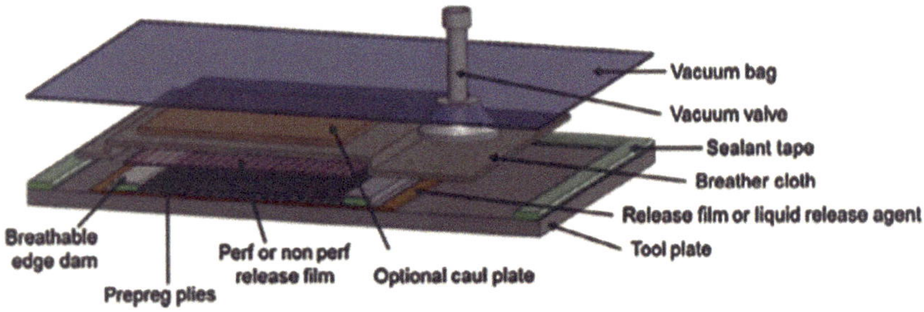

FIGURE 15.5 Vacuum bagging assembly for laminate manufacture (Centea et al., 2015).

In addition, VBO processing ensures not only a reduction in operating costs and energy savings, but also can shorten the time it takes to produce a product and reduce investment cost. The set-up system is cheaper than an autoclave. Fig. 15.5 shows the VBO OOA process.

A lot of research has been carried out on VBO to produce composite laminates (Centea, Grunenfelder, & Nutt, 2015; Levy, Kratz, & Hubert, 2015; Ma, Centea, Nilakantan, & Nutt, 2014). The VBO pre-pregs require a good hand lay-up process and every involved parameter must be set correctly. The pressure vacuum used is low at atmospheric pressure (101.3 kPa) to produce defect-free products without voids and bubbles between the resin and fiber, good mechanical compression, and even dimensional stability (Centea & Nutt, 2015; Hamill, Centea, & Nutt, 2015). Fig. 15.6 shows a flowchart of the VBO manufacturing process.

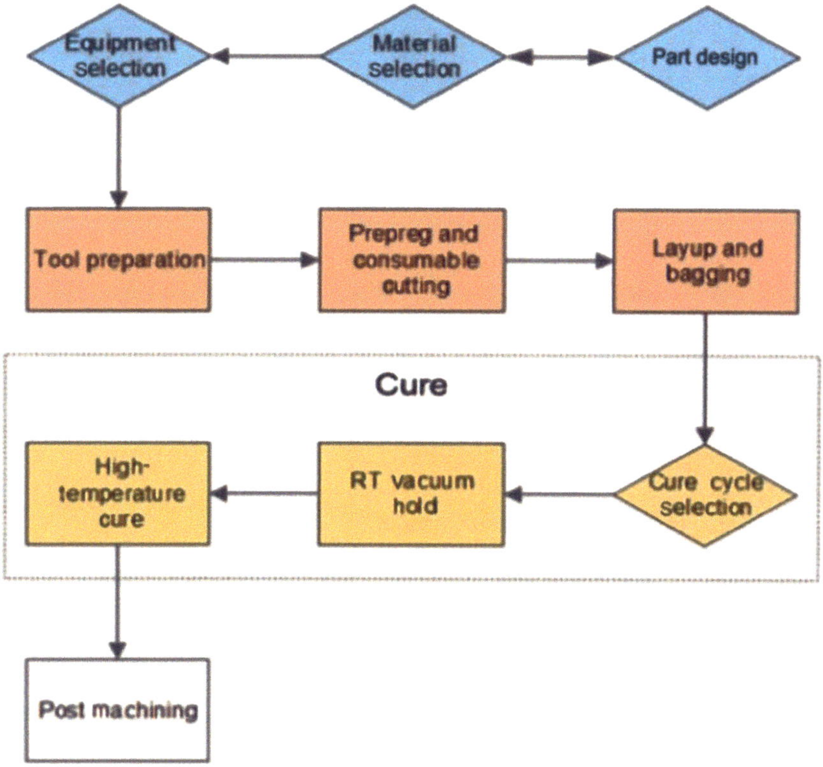

FIGURE 15.6 Flowchart of a traditional vacuum bag prepreg manufacturing process (Centea & Nutt, 2015).

15.3.3 Resin transfer molding

Resin transfer molding (RTM) is one of the OOA technologies used in the aerospace industry (Guitard & Trochu, 1996; Lin, Murphy, & Hahn, 2000; Sozer, Simacek, & Advani, 2012). Research has been done on the capabilities of RTM in order to improve the product quality and efficiency of the operation. As a result of a literature review, RTM has been shown to be capable of producing high-quality components, complex geometric shapes, dimensional accuracy, consistent with good mechanical properties (Ageyeva & Sibikin, 2019). RTM usually uses an aluminum closed mold. Fig. 15.7 shows the RTM process. The RTM process involves hand lay-up on the mold. Then, the mold is closed, heated, and placed under a vacuum in order for the resin to uniformly permeate with the fiber. Finally, the resin is injected into the mold and held until the resin is cured.

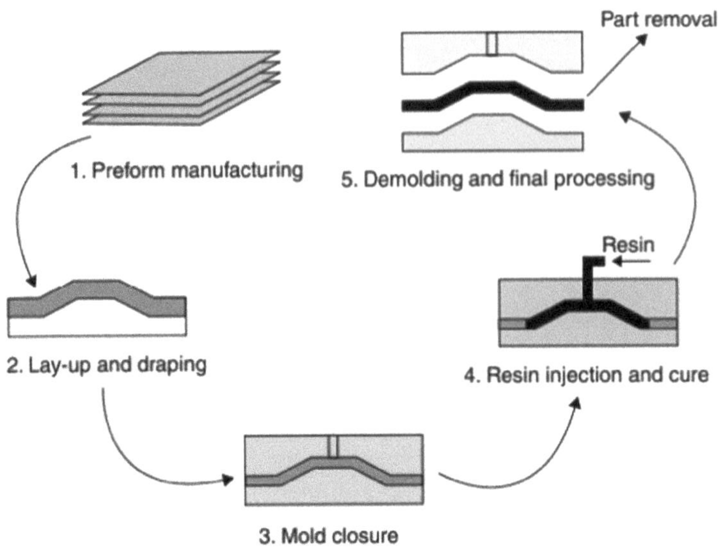

FIGURE 15.7 Resin transfer molding process (Advani & Sozer, 2011).

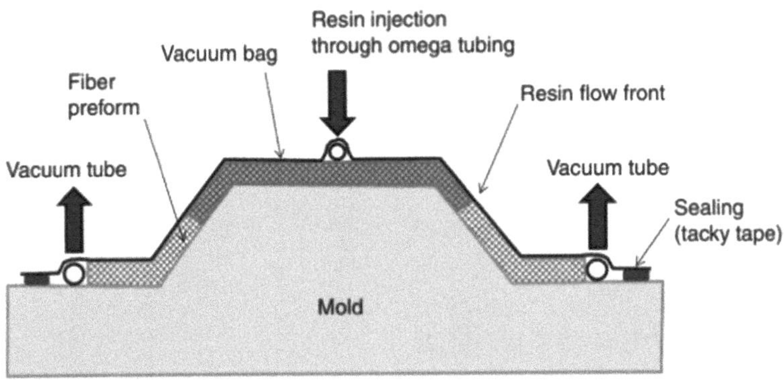

FIGURE 15.8 Vacuum-assisted resin transfer molding process (Advani & Sozer, 2011).

15.3.4 Vacuum-assisted resin transfer molding

Fig. 15.8 shows the VARTM process. VARTM is a simple process which includes a vacuum bag for removing trapped air during the product fabrication process (Sozer et al., 2012). This process consists of a side rigid mold and flexible vacuum plastic that will compress the resin or preform placed on the mold. The resin is placed on the mold surface under atmospheric pressure. However, VARTM needs manual lay-up labor in order to make the preformed material before being placed on the mold (Glancey, 2010). There has been some research done on VARTM, especially on carbon fiber reinforced polymer (Abusrea, Han, Arakawa, & Choi, 2018; Arulappan, Duraisamy, Adhikari, & Gururaja, 2014; Asadi, Abusrea, Arakawa, Colton, & Kalaitzidou,

2018). The fabrication process can produce high-quality products at minimum cost, and it is able to manufacture large, complex composite parts with good quality and high production rates (Hsiao & Heider, 2012).

15.4 Application of the out-of-autoclave process in manufacturing an aircraft composite seat pan

In this section, a case study demonstrating the application of OOA to produce aircraft seat pans is discussed. The study was focused to evaluate the effect of the resin film on the mechanical properties of an aircraft seat pan made from carbon fiber reinforced epoxy (CFRE) using an OOA hot press molding method. The aim of this study was to explore alternative ways to improve the cost efficiency of the OOA operation and fabrication of the composite seat pan component. Thermoset epoxy resin was used to formulate the composite laminate because of the matrix good structural properties, versatility, easy cure, and modest cure temperatures (Gopi, Balakrishnan, Sreekala, Pius, & Thomas, 2017). Similar factors also influenced aircraft composite designers to apply epoxy matrix in designing critical structures (Hamerton & Kratz, 2018).

15.4.1 Materials and methods

15.4.1.1 Materials

This work utilized epoxy carbon fiber prepreg. The transparent thermosetting epoxy system has been formulated for the production of fiber reinforced composite materials with flame resistance and self-extinguishing properties. Table 15.1 presents the properties of the materials.

15.4.1.2 Sample preparation

The samples were first fabricated using a hand lay-up method. After the lay-up process, samples were pressed by a hot press. However, optimization process parameter (pressure, temperature, holding time) of the samples was needed to identify first in order to obtained good cured samples. The first step in the optimization procedure was to vary the temperature of the process. It is crucial because the specifications of the material are received by the supplier is not same when it process with different environment condition. The laboratory environment was not similar to the supplier environment.

The second step was to set various pressures of the hot press at between 100–200 kg/cm^2 with the holding time again being held constant at 2 minutes. The resulting composite plates obtained from these

TABLE 15.1 Properties of carbon fiber reinforced epoxy prepreg.

Description	Properties
Chemical nature	Transparent flame-retardant thermosetting epoxy
Curing temperature range	Typically 120°C–160°C
Dynamic viscosity	Approximately 500 Poise @ 60°C, frequency 10 rad/s
T_g after curing (1) ASTM D 3418	130°C–135°C, after 120 minutes @ 110°C
	140°C–145°C, after 90 minutes @ 120°C
	150°C–155°C, after 40 minutes @ 130°C

DSC heating rate 20°C/min.

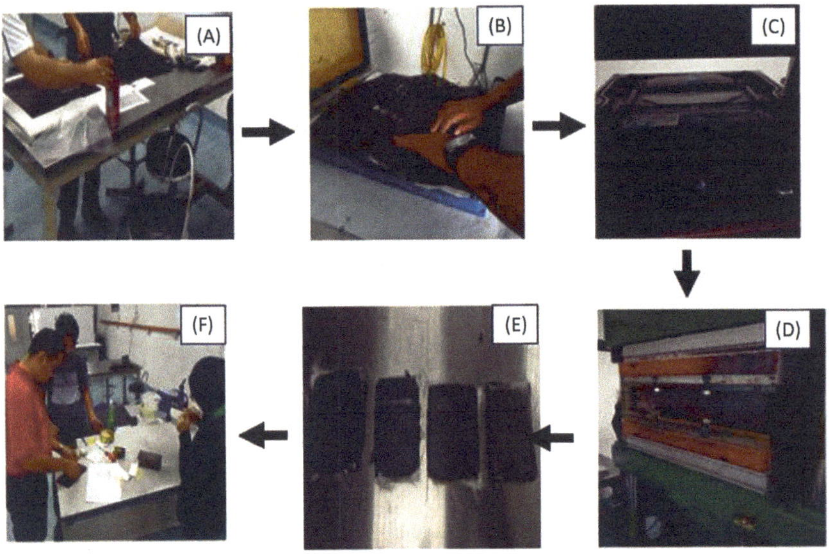

FIGURE 15.9 Fabrication process of the composite samples: (A) hand lay-up process, (B) de-form process, (C) the lay-up samples being placed in the mold, (D) hot compression process at 2 minutes, (E) composite sample of carbon fiber reinforced epoxy, and (F) the samples being cut according to the mechanical test requirements.

two optimization sequences possessed a nominal thickness of 3 mm as stated in ASTM D790 (Flexural Test) for the thickness of the sample. The optimum temperature to fabricate the composites sample must be set at 125°C and the optimum pressure applied is 200 kg/cm^2. The samples were fabricated with three different composites layers: (A) as-received samples; (B) the sample with a resin film; and (C) samples without a resin film. Fig. 15.9 shows the fabrication process of the seat pan composite samples.

15.4.1.3 Sample characterization

Material characterization and comparison were conducted on three different samples which were the as-received samples from the industry of CFRE, CFRE with resin film embedded, and CFRE without resin film embedded. The material characterization for all samples covered the flexural test, hardness test, and morphological analysis.

1. *Flexural test*

   Flexural tests were performed using a three-point bending setup according to ASTM D790 standard. The span length to the specimen thickness ratio is maintained at 16:1 as described in ASTM D790. At least five specimens were tested for each condition (A, B, C). The samples with dimensions of 15 × 1.27 × 0.3 cm (length × width × thickness) were tested using a ZWICK Z50 machine (1 mm/min). The flexural strength was then obtained using the expressions:

$$\sigma_f = \frac{3PL}{2bd^2}$$

   where L is the support span; b is the width of the specimen; d is the thickness; P is the maximum load; and m is the slope of the initial straight-line portion of the load–displacement curve.
2. *Microhardness*

   This test method covers the determination of the microindentation hardness of materials using Vickers indenters under a test force of 9.8 N and at least three indentation results were taken for each sample (Ahmad & Jamluddin, 2018).
3. *Scanning electron microscopy (SEM)*

   Morphological studies were performed in detail on the cross-section surface of the samples using a JOEL scanning electron microscope (SEM), at an accelerating voltage of 12 kV. The samples were coated with platinum to provide electrical conductivity, which did not significantly affect the resolution, thus allowing for high-quality results.

15.5 Results and discussion

15.5.1 Flexural test

The results obtained from the flexural test performed indicated that the fabricated sample with resin film has the highest flexural strength followed by the as-received sample and the sample without resin film, as shown in Table 15.2. This might be because of the extra resin layer to the composite structure improving the mechanical properties of the

TABLE 15.2 Flexural strength of carbon fiber reinforced epoxy seat pan.

Sample	Flexural strength (MPa)
A. As-received samples	721.53
B. Fabricated sample with the resin film	784.99
C. Fabricated sample without the resin film	533.33

composites. The fabricated sample without resin film had the lowest flexural strength due to the weakness of the composite structure.

15.5.2 Microhardness

A hardness test was performed to obtain the degree of hardness of surface deformation due to the curing and resin permeability effect. Generally, hardness properties are one of the most important properties of these materials. Changes in this property can be ascribed to the different types of material used as polymerization occurs during the curing process of composite material. The as-received sample hardness value was the highest compared to the other fabricated samples. One of the factors that influence the decrease in composite hardness is the depth of cured resins (Webo, Masu, & Maringa, 2018). Based on the observations in Fig. 15.10, the as-received sample resin is fully cured at the upper surface. The indentation marks of sample A were also found not to be very clear due to the hard surface of the sample.

For sample B, the indentation mark is clear on the surface, this might be because of the resin film layer before the fiber glass. It creates a clear plastic deformation of the sample surface because the extra resin effect covers the glass fiber layer. Sample C gives the lowest hardness due to the resin permeability problem. The resin does not fully cover the outer layer (glass fiber). The other possibility of the result of the hardness value might be due to different compositions of composite resins. Table 15.3 shows the hardness values for all the samples.

15.5.3 Morphology analysis

The morphology analysis was conducted on the lay-up cross-section of each sample for unpolished (as cut) fibers coated with platinum. Secondary electron (SE) mode was used in the SEM analysis to differentiate layers within the whole structure. The SEM morphology of a cross-section of the multiple layer structure is shown in Fig. 15.11 for composite samples. All images were taken at 50× magnification (at 0° tilting) under SE mode.

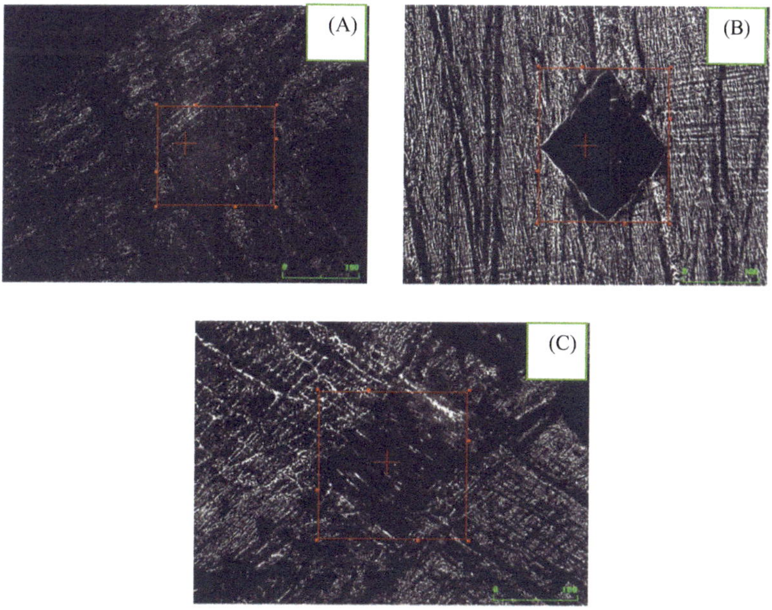

FIGURE 15.10 (A) Indentation mark of as-received samples, (B) indentation mark of fabrication sample with the resin film, and (C) indentation mark of fabrication sample without the resin film.

TABLE 15.3 Hardness properties of carbon fiber reinforced epoxy seat pan samples.

Sample	Hardness (Hv)
A. As-received samples	79.67
B. Fabricated sample with the resin film	60.34
C. Fabricated sample without the resin film	50.04

As observed in Fig. 15.11, the resin does not fully cover the carbon fiber and glass fiber layer based on the cross-section of the sample. Furthermore, the densification and smoothness were quite different between each layer (from the top to bottom of each sample) and also when compared between samples due to influences of processing variables such as temperature, dwell time, and applied pressure. From the results obtained, the as-received sample showed better resin permeability throughout the carbon fiber compared to the other samples, whereby a very few gaps between the laminate layers were observed in the sample. For the fabricated samples, gaps between the layers of each carbon fiber could be clearly seen, which indicates the wetting problem occurred due to incomplete resin permeability between the carbon and the glass fiber.

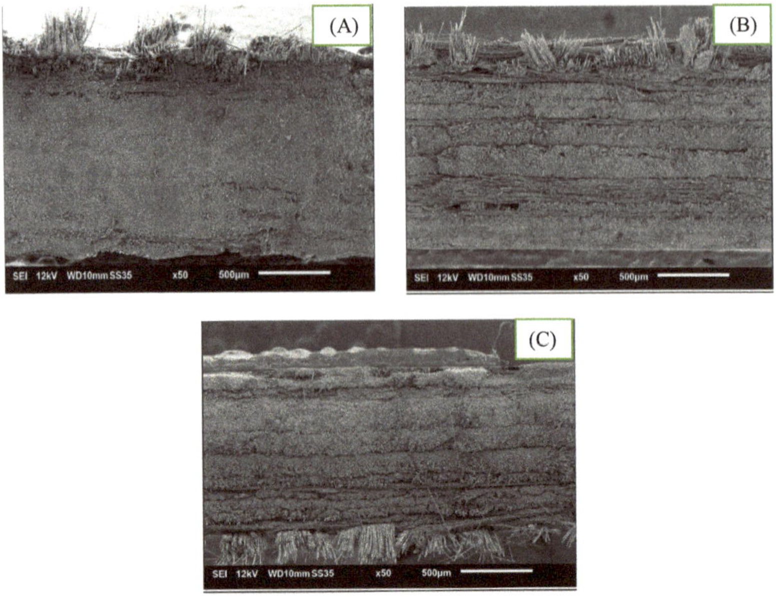

FIGURE 15.11 (A) Cross-section of the received sample, (B) cross-section of the fabricated sample with the resin film, and (C) cross-section of the fabricated sample without the resin film.

It can be concluded that the as-received seat pan samples have the best compactness and resin coverage on the carbon fiber, followed by the sample with the resin film and then the one without the resin film.

15.6 Conclusions

The OOA process is an attractive and viable processing solution in supporting the DfS goal, especially in energy conservation toward improved sustainability performance in aerospace applications. When compared to an autoclave process, OOA offer lower energy consumption and hence better environmental performance due to its processing method only involving vacuum pressure. Furthermore, the OOA process also offers economic advantages in terms of lower production life cycle time and cost of equipment and tooling compared to the autoclave process. Both environmental and economic features shown by the OOA process are highly aligned with the concept of DfS to produce sustainable products. The case study of OOA implementation for producing composite aircraft seat pans also showed that the process was able to give good end-products with acceptable structural properties. However, a similar case study also showed variation of structural performance obtained when different processing parameters were subject to the same composite product. This indicates that proper selection of processing parameters is crucial to producing high-quality composite components using the OOA process.

Acknowledgments

The authors gratefully acknowledge Aerospace Malaysia Innovation Center for funding this project through Industry-University research collaboration, RMK11 2017 P3.2 Development of localized sourcing and manufacturing capabilities of composite aircraft seats. The authors also thank S.M. Sapuan and M.R. Mansor for the tremendous support provided during the writing of this chapter.

References

Abusrea, M. R., Han, S.-w, Arakawa, K., & Choi, N.-s (2018). Bending strength of CFRP laminated adhesive joints fabricated by vacuum-assisted resin transfer molding. *Composites Part B, 156*, 8–16. Available from: https://doi.org/10.1016/j.compositesb.2018.08.041.

Advani, S. G., & Sozer, E. M. (2011). *Process modelling in composites manufacturing* (2nd ed.). CRC Press.

Ageyeva, T., & Sibikin, I. (2019). A review of thermoplastic resin transfer molding: Process modeling and simulation. *Polymers, 11*(1555), 1–28.

Ahmad, F. A. G., & Jamluddin, M. (2018). Hardness, tensile and microstructure assessment of carbon/glass hybrid composite laminate. *Journal of Mechanical Engineering, 15*(2), 91–105.

Arulappan, C., Duraisamy, A., Adhikari, D., & Gururaja, S. (2014). Investigations on pressure and thickness profiles in carbon fiber-reinforced polymers during vacuum assisted resin transfer molding. *Journal of Reinforced Plastics and Composites, 0*(0), 1–16. Available from: https://doi.org/10.1177/0731684414560220.

Asadi, A., Abusrea, M. R., Arakawa, K., Colton, J., & Kalaitzidou, K. (2018). A comparison of CFRP composite laminated joints fabricated with vacuum assisted resin transfer molding. *EXPRESS Polymer Letters, 12*(9), 781–789.

Bond, G. G., Griffith, J. M., & Hahn, G. L. (2008). Non-autoclave (prepreg) manufacturing technology. In *Proceedings of the 40th SAMPE fall technical conference*, Tennessee.

Brosius, D. (2014). *Out-of-autoclave manufacturing: The green solution*. Composites World. <https://www.compositesworld.com/articles/out-of-autoclave-manufacturing-the-green-solution>

Centea, T., Grunenfelder, L. K., & Nutt, S. R. (2015). A review of out-of-autoclave prepregs – Material properties, process phenomena and manufacturing considerations. *Composites: Part A, 70*, 132–154. Available from: https://doi.org/10.1016/j.compositesa.2014.09.029.

Centea, T., & Nutt, S. R. (2015). Manufacturing cost relationships for vacuum bag-only prepreg processing. *Journal of Composite Materials*, 1–17. Available from: https://doi.org/10.1177/0021998315602949.

Composites Technology Research Malaysia Sdn. Bhd. (CTRM). (2020). Facilities in CTRM. <https://www.ctrm.com.my/> Accessed 17.03.2020.

Composites World. (2014). *The market for OOA aerocomposites, 2013–2022*. Composites World. <https://www.compositesworld.com/articles/the-market-for-ooa-aerocomposites-2013-2022>

Dieter, G. E. (2000). *Engineering design: A materials and processing approach* (3rd ed). Singapore: McGraw-Hill.

Glancey, J. L. (2010). Vacuum-assisted resin transfer molding. *Innovations in Materials Manufacturing, Fabrication, and Environmental Safety*, 531–544.

Gopi, S., Balakrishnan, P., Sreekala, M. S., Pius, A., & Thomas, S. (2017). Green materials for aerospace industries. Biocomposites for high-performance applications. Elsevier Ltd. Available from: https://doi.org/10.1016/B978-0-08-100793-8.00011-9.

Grunenfelder, L. K., Dills, A., Centea, T., & Nutt, S. (2016). Effect of prepreg format on defect control in out-of-autoclave processing. *Composites Part A*. Available from: https://doi.org/10.1016/j.compositesa.2016.10.027.

Guilherme, F., Cassu, S. N., Diniz, M. F., de Faria Gonzales Leal, T. P., Sanches, N. B., & de Cassia Lazzarini Dutra, Rita (2017). Evaluation of out-of-autoclave (OOA) epoxy system. *Polimeros*, *27*(4), 353–361.

Guitard, D., & Trochu, F. (1996). Concurent method for permeability of measurement in resin transfer molding. *Polymer Composites*, *17*(1), 149–158.

Hamerton, I., & Kratz, J. (2018). *Chapter 9—The use of thermosets in modern aerospace applications. Thermosets* (2nd ed.). Elsevier Ltd. Available from: https://doi.org/10.1016/B978-0-08-101021-1.00009-5.

Hamill, L., Centea, T., & Nutt, S. (2015). Surface porosity during vacuum bag-only prepreg processing: Causes and mitigation strategies. *Composites Part A*, *75*, 1–10. Available from: https://doi.org/10.1016/j.compositesa.2015.04.009.

Hsiao, K., & Heider, D. (2012). Vacuum assisted resin transfer molding (VARTM) in polymer matrix composites. Manufacturing techniques for polymer matrix composites (PMCs). Woodhead Publishing Limited. Available from: https://doi.org/10.1533/9780857096258.3.310.

Lee, J., Ni, X., Daso, F., Xiao, X., King, D., Blanco, T., ...& Jose, S. (2018). *Advanced carbon fiber composite out-of-autoclave laminate manufacture via nanostructured out-of-oven conductive curing out-of-oven*, 1–10. <https://doi.org/10.1016/j.compscitech.2018.02.031>

Levy, A., Kratz, J., & Hubert, P. (2015). Air evacuation during vacuum bag only prepreg processing of honeycomb sandwich structures: In-plane air extraction prior to cure. *Composites: Part A*, *68*, 365–376. Available from: https://doi.org/10.1016/j.compositesa.2014.10.013.

Lin, M. Y., Murphy, M. J., & Hahn, H. T. (2000). Resin transfer molding process optimization. *Composites Part A: Applied Science and Manufacturing*, *31*. Available from: https://doi.org/10.1016/S1359-835X(99)00054-8.

Linganiso, L. Z., & Anandjiwala, R. D. (2016). Fibre-reinforced laminates in aerospace engineering. Advanced composite materials for aerospace engineering. Elsevier Ltd. Available from: https://doi.org/10.1016/B978-0-08-100037-3.00004-3.

Ma, Y., Centea, T., Nilakantan, G., & Nutt, S. (2014). Vacuum bag only processing of complex shapes: Effect of corner angle, material properties and processing conditions. In *Proceedings of the American society for composites 2014*, American Society of Composites 29th Technical Conference/16th US-Japan Conference on Composite Materials/ASTM D-30 MeetingAt: San Diego, CA, United States.

Ogale, A., Potluri, P., Rittenschober, B., & Schlimbach, J. (2014). Out-of-autoclave curing of composites for high temperature aerospace applications. In *Proceedings of SAMPE Seattle 2014, 2–5 June 2014, Seattle, WA*.

Özgül, F., & Bektaş, N. B. (2019). Investigation of the effects of additives on mechanical properties of E-glass reinforced thermoset composites. *Acta Physica Polonica A*, *135*(5). Available from: https://doi.org/10.12693/APhysPolA.135.915.

Pasco, C., & Kendall, K. (2018). Characterisation of the thermoset prepreg compression moulding process. In *Automotive composites conference & exhibition 2016*, SPE Automotive and Composites Divisions, Michigan, USA.

Ridgard, C. (2000). Advances in low temperature curing prepregs for aerospace structures. In *Proceedings of the 45th international SAMPE symposium, CA*.

Sozer, E. M., Simacek, P., & Advani, S. G. (2012). Resin transfer molding (RTM) in polymer matrix composites. Manufacturing techniques for polymer matrix composites (PMCs). Woodhead Publishing Limited. Available from: https://doi.org/10.1533/9780857096258.3.243.

Teuscher, R., Witik, R., Cohades, A., & Michaud, A. (2012). Out of autoclave processing of ribbed parts: Technical, economic and environmental assessment. In *Proceedings of the 11th international conference on flow processing in composite materials*.

Walczyk, D., & Kuppers, J. (2012). Thermal press curing of advanced thermoset composite laminate parts. *Composites Part A*, *43*(4), 635–646. Available from: https://doi.org/10.1016/j.compositesa.2011.12.008.

Webo, W., Masu, L., & Maringa, M. (2018). The impact toughness and hardness of treated and untreated sisal fibre-epoxy resin composites. *Advances in Materials Science and Engineering 2018*, 1–10.

Witik, R. A., Gaille, F., Teuscher, R., Ringwald, H., Michaud, V., & Månson, J. E. (2012). Economic and environmental assessment of alternative production methods for composite aircraft components. *Journal of Cleaner Production*, *29–30*, 91–102. Available from: https://doi.org/10.1016/j.jclepro.2012.02.028.

Wu, M. S., Centea, T., & Nutt, S. R. (2014). Compression molding of reused in-process waste–Effects of material and process factors. *Advanced Manufacturing: Polymer & Composites Science*, *4*(1), 1–12. Available from: https://doi.org/10.1080/20550340.2017.1411873.

CHAPTER

16

Life cycle analysis of fused filament fabrication: A review

M.T. Mastura[1,2], M.R. Alkahari[3,4] and A.K. Syahibudil Ikhwan[1,2]

[1]Faculty of Mechanical and Manufacturing Engineering Technology, Universiti Teknikal Malaysia Melaka, Melaka, Malaysia, [2]Centre for Advanced Research on Energy, Universiti Teknikal Malaysia Melaka, Melaka, Malaysia, [3]Faculty of Mechanical Engineering, Universiti Teknikal Malaysia Melaka, Melaka, Malaysia, [4]Advanced Manufacturing Centre, Universiti Teknikal Malaysia Melaka, Melaka, Malaysia

16.1 Introduction

Sustainability is not a new term in product design and manufacturing as understanding of its importance is rising concurrently with the new technologies. Every stage of technology development considers the sustainability elements to support the policies and law that have been created by governments in all countries across the world. Technology in manufacturing is improving daily and all parties that participate in technology development are carefully making decisions to help meet the sustainability requirements. Technology in additive manufacturing (AM) has become popular in recent years among researchers because of its flexibility and lower negative environmental impact compared with traditional manufacturing processes. Fused filament fabrication (FFF) is one of the technologies in AM that is widely found in many product designs and development processes. The operational principles of FFF that optimize material consumption would likely appeal to green manufacturers to fabricate their products with less waste. However, how "green" FFF is should be asked as the dominant materials used as a

Design for Sustainability
DOI: https://doi.org/10.1016/B978-0-12-819482-9.00020-4

feedstock for FFF are plastics. Therefore a study of the life cycle analysis (LCA) of FFF should be performed in order to evaluate the environmental impact of this technology.

In FFF, material or filament is extruded through a hot nozzle on the horizontal platform. The printer in FFF is used as a mechanism that creates a digital 3D model from computer-aided design (CAD) software to a real part that is fabricated from computer numerical control that moves the extrusion head according to the tool path. Therefore there is almost zero waste of materials in the post-processing of FFF. Moreover, materials that are used as filament in FFF can be varied according to previous studies (Ngo, Kashani, Imbalzano, Nguyen, & Hui, 2018). Thermoplastics are the prime materials for FFF. Plastic waste as a pollution issue that has never been solved in has become a major drawback of products fabricated using FFF. Therefore a lot of effort has been exerted by researchers to solve this issue by reducing the amount of polymer in FFF filament. Graphene (Kalsoom, Nesterenko, & Paull, 2016), natural fibers (Matsuzaki et al., 2016), synthetic fibers (Zhang, Ma, & Tan, 2017), metal, (Bourell et al., 2017), and ceramics (Hofstätter, Pedersen, Tosello, & Hansen, 2017) are among the materials that have been added to polymer filaments to reduce the volume of polymer and produce composite filaments for FFF. Furthermore, recycled polymers also have been studied as a filament to produce FFF parts. Hence, the flexibility of materials for FFF could be a factor contributing to the sustainable elements in FFF.

In this chapter, a review of the LCA of FFF is carried out. The fundamentals, principles, and operations of FFF are explained in the next section. LCA is explained later to gain an understanding of its framework. Earlier studies in which analyses have been performed also are discussed. Later, comparisons of FFF with other technologies in AM with regards to the environmental impact are discussed. Finally, conclusions about this study are made in the last section.

16.2 Principles of fused filament fabrication

AM is defined as the process that produces objects from 3D model CAD data through use of layer-by-layer fabrication techniques. AM is opposed to the subtractive manufacturing (SM) process such as machining where material is removed layer-by-layer. AM has been said to be able to revolutionize how we manufacture things in the future. It has competitive advantages over conventional processes due to its potential for rapid product development, controllable material properties, ability to accommodate customization, ability to fabricate complex designs, less tooling, simple setup, and others. There are many types and variations of AM processes available today. Some of these common processes are known as FFF/fused deposition modeling (FDM), selective laser sintering (SLS), selective laser melting, direct metal laser sintering, material jetting (MJ), stereolithography (SLA), digital light processing, electron beam melting, laminated object modeling, laser engineered net shaping (LENS), etc.

The process technology and market share for the commercialized technology is presented in Fig. 16.1. FFF is currently the most commonly used AM technology worldwide with 69% of the market, followed by SLA at 15%, SLS at 12%, and then MJ and metal sintering, both at 2%. This is influenced by the relatively low price and simplicity of the FFF system. Although there are many AM processes available, only a few are reflected in Fig. 16.1 as a significant number are either still under proprietary manufacturers or under research and development. Nevertheless, these processes have potential to be commercialized in the near future. Due to high number of variations in AM processes, many attempts have been made to classify these processes by their common characteristics. This classification is important as one of the methods for furthering understanding of AM and also to facilitate further analysis of AM products and processes.

One of the classical ways to classify the process is based on the fundamental or baseline technology for the respective AM processes. Under this classification, processes are classified based on whether they are under extrusion process, laser based, melting, sintering, etc. (Burns, 1993; Kruth, Leu, and Nakagawa 1998). Another classification was made based on the initial state of the raw material during AM processing. Hence, they are either solid-based processes, liquid-based processes, or powder-based processes (Chua & Leong, 2014).

One of the important classifications is given in ASTM F2792-12a (ASTM F2792-12a, 2012) where AM is classified into several main categories based

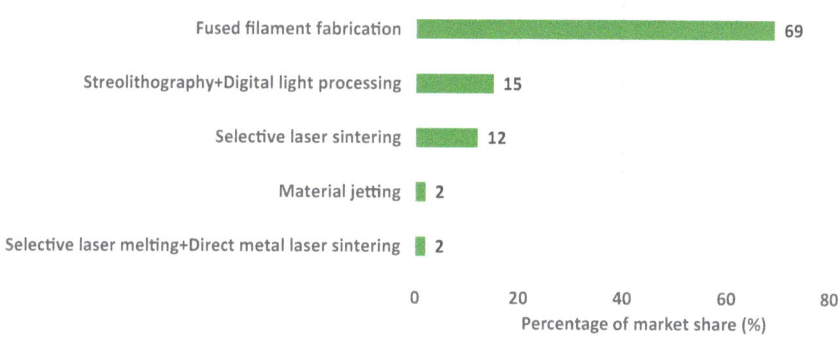

FIGURE 16.1 Market share of AM/3D printing technologies (Statista, 2018).

on the process for the technology user's reference. The standard defines seven main classification groups for AM, namely material extrusion, binder-jetting, MJ, powder bed fusion, vat photopolymerization, directed energy deposition, and sheet lamination. This classification is based on the mechanism and principle during processing to produce the layer-by-layer product which is the main characteristic of AM. FFF, also known as FDM, is classified under material extrusion as the semimolten material is extruded from a nozzle during processing. Binder-jetting involves a liquid bonding agent that is selectively deposited to join material powders which spread on the platform surface. MJ generally uses a special mechanism to produce droplets of material to be deposited layer-by-layer. This differs from powder bed fusion where thermal energy which might be from a laser, electron beam, or other is used to selectively fuse regions of the powder bed. In directed energy deposition, materials are melted as they are being deposited by focused thermal energy. Another group is vat photopolymerization. This group of AM is characterized by the application of liquid photopolymer to selectively cure a defined position based on the CAD data by light activation. The last group under this ASTM classification is sheet lamination. Sheet lamination refers to the process where sheets of materials are cut in the XY plane and then bonded together to produce a 3D structure.

A more recent and general classification was given by a joint international collaboration between the International Organization for Standardization (ISO) and ASTM International, where consensus for developing standard and classification was made (International Standard Organisation, 2015). In this cooperation, a structure for standard development was defined where a common structure has multiple levels and hierarchy of AM standards based on general standard, category standard, and specialized standard, which further refined base raw materials, process/equipment, or finished parts. Understanding the classification and standardization of AM is crucial as it opens up a generalized approach for technology use and further analysis.

Among the AM processes, FFF is regarded as the most versatile and cost-effective technique. Expiration of the patent for AM in 2009 (Scoot, 1992), specifically for FFF, opened up a new dimension in the academic and industry where there was a sudden increase in machine development related to this technology. As a result, there has been a drastic decrease in machine price and an increase in the development of open-source systems, especially for FFF.

FFF is an AM processes that is based on the extrusion of semimolten material from a nozzle. Generally, in FFF, the material is heated to a certain range of temperature. Commonly the temperature during processing ranges from 180°C to 300°C based on the type of material, filament size, system configuration, and other factors. The materials used in FFF are in the form of filaments. The most commonly used materials in FFF are thermoplastic materials such as acrylonitrile butadiene styrene (ABS) and polylactic acid (PLA). In pace with industry and academic needs, increasing variations of the materials are being developed in the form of filaments which allow FFF-based 3D printed parts to be developed easily. Some of these materials are TPU, nylon, wood, and composite materials. Recently, some of proprietary manufacturers also have developed their own materials in order to enhance the mechanical properties of the 3D printed materials. FFF basic principles are illustrated in Fig. 16.2.

During FFF processing, the filament material from the spool is pulled by a mechanism powered by a stepper motor which then drives the filament to the heating component, which is the hot end (liquifiers). The hot end has a heating element that is able to melt the filament and turn it to the semimolten stage. The semimolten filament is then extruded through the constricted nozzle as it may move in at least in three different axis directions. This movement is accomplished either through the movement of the table and/or nozzle depending on the configuration of the machine. As the movement in the XY plane occurs, materials are extruded and layered on the built platform continuously. Later the movement in the Z axis occurs, which allows materials to be added to the top of the previous layer. This process is repeated layer-by-layer based on the CAD data until a complete 3D component is fabricated.

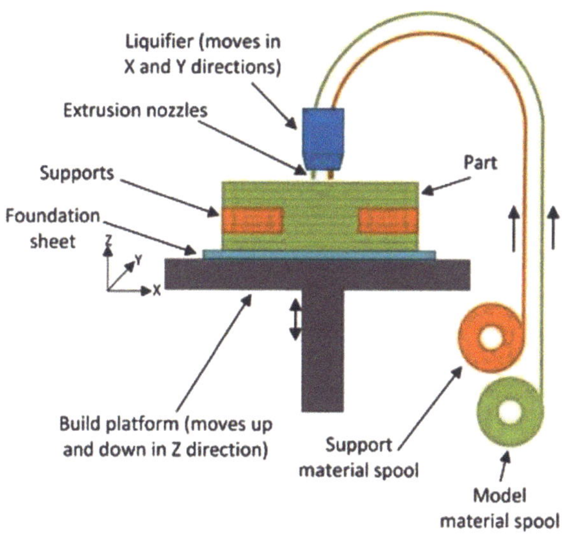

FIGURE 16.2 Schematic of fused filament fabrication (Gebisa & Lemu, 2018).

16.3 Life cycle analysis methodology

During product development, LCA is one of the most important activities that should be performed at an early stage of the process. Some designers include this activity in material selections or manufacturing process selections or conceptual design selections. Estimation of the life cycle of the product could reduce time, cost, and risk of the wrongly chosen design of the product. LCA is performed to evaluate the environmental impact that could occur during the lifetime of the product.

There are several terms used to describe the lifetime of the product in LCA, including cradle-to-grave, cradle-to-gate, and cradle-to-cradle. LCA of the product for cradle-to-grave means that the evaluation of the environmental impact of the product starts from extraction of the raw materials until disposal of the product is performed. Kafara, Süchting, Kemnitzer, Westermann, and Steinhilper (2017) performed cradle-to-grave LCA for conventional and AM of mold cores that are used in carbon fiber reinforced polymer production. They found that an AM process of mold core in carbon fiber reinforced polymer production exhibits lower environmental impact than a conventional manufacturing process. For cradle-to-gate, the assessment only includes partial life of the product which is from extraction of the raw materials to the factory gate; before the product is transported to customers. A study conducted by Bekker and Verlinden (2018) evaluated the environmental impact of wire and arc AM, green sand casting, and CNC milling from cradle-to-gate life cycle assessment. The results from this study are used to optimize the design with regards to the environmental impact of three manufacturing techniques.

Meanwhile, cradle-to-cradle assessment is for recyclable products that have a zero waste in a closed-loop lifetime.

Generally, results from the LCA would imply the consumption of energy and natural resources, level of toxicity, ozone depletion, and release of greenhouse gases. With regard to this, engineers decide what options they have to reduce the negative environmental impact and what the major contributions are of the impact. Engineers can optimize the design of a product by changing the shape, size, function, materials, or fabrication methods. Ma, Zhao, Pu, and Li (2018) performed material selections for automotive components by considering the life cycle assessment to evaluate the negative environmental impacts and selecting the optimal sustainable materials. In addition, Mastura, Sapuan, Mansor, and Nuraini (2016) also conducted a study in selecting materials for an automotive antiroll bar by considering the voice of environment at the early stage of selection and comparing the energy consumption and CO_2 footprint of the top two selected materials. On the other hand, Sun, Liu, Lu, Zhang, and Zhao (2017) conducted a study on cradle-to-grave LCA of a lightweight automotive engine hood design by comparing the LCA results of three different designs. In addition, manufacturers that perform LCA can select the best manufacturing techniques and change the process parameters to reduce the negative impacts and develop a green manufacturing process. Kek and Vinodh (2016) proposed an integration framework in selecting a sustainable manufacturing process that includes environmental performance assessment. They integrated the LCAs method with the analytic network process to calculate the impact score and compare the alternatives to select the best option. Gaha, Yannou, and Benamara (2016) proposed a new methodology that considered LCA and CAD tools in selecting a green manufacturing process. The aim was to select the most suitable manufacturing process based on CAD features that have lower negative environmental impacts.

Generally, a LCA framework includes four phases, which are goal and scope definition, inventory analysis, impact assessment, and interpretation phase (International Standard Organisation, 1997). Along with the LCA process, there are a few tools and software that commonly are used by researchers and engineers such as SimaPro, Gabi, and Eco-Indicator. These tools are used to help engineers to evaluate and estimate the environmental impact such as ozone depletion, energy consumption, greenhouse gas emission, toxicity, and eutrophication.

16.3.1 Goal and scope definition

In the framework of LCA, goal and scope definition is the initial phase that should not be underestimated by practitioners. During this first phase of LCA, activity in defining the aim and scope of the analysis should be performed in an exact approach to avoid misunderstanding and to ensure the validity of the final results. Here, practitioners have to exactly determine the intention of the application and carry out the study that should be stated in the goal of LCA. In addition, identification of the target audience also should be included in defining the goal of the LCA. The scope of the LCA defines the boundaries of the analysis in order to keep practitioners focused on the objective of the study and prevent unnecessary assessment that wastes time, cost, and energy. Generally, the scope defined by the practitioners should include the product system to be studied, functional unit, system boundary, data requirements, assumptions and limitations, and type and format of the report required for the study. Here, the practitioners can determine the beginning and end of the life cycle specifically. A functional unit is a basic essential in LCA where it is used as a reference unit. System boundaries would set the input and output of the system such as resources that are associated with the product system.

Yao, Hu, Di Maio, and Cucurachi (2019) in their study have shown the goal and scope in lab-scale LCA of 3D printing geo-polymer concrete. The goal in their study was to identify the environmental hotspots in the 3D printing technology system. A 0.307 kg 3D printing geo-polymer concrete product was set as the functional unit. The study only focused on the physical output of the technology and the scope was defined as a cradle-to-gate LCA study. On the other hand, Belculfine (2016) also performed a study on LCA and set a target for his study to compare the environmental impact of a plastic part fabricated by a desktop 3D printer in a distributed manufacturing model with a similar plastic part fabricated by injection molding in a conventional manufacturing model. In this case, the volume of ABS parts is used as the functional unit to enable comparison between the two systems. In another comparison study by Mami, Revéret, Fallaha, and Margni (2017), the closure of an aircraft door throughout the aircraft's lifetime (35 years) is set as a functional unit in a study to assess the environmental impacts throughout the entire life cycle of an aircraft doorstop made by 3D printing and by traditional machining. The goal of the study also determined the life cycle costs of the method and identified optimal options by applying the proposed eco-efficiency model and comparing the manufacturing techniques.

Generally, the goal and scope of the LCA are set up by the practitioner to ensure the conducted study is reasonable and useful to the associated target audience. The system boundaries that are identified in the early phase of study would be set up to limit the scope and ensure the focus area of the study. Moreover, functional unit is a key element of the LCA study, where it is used as a standard specification. Hence, the initial phase of a LCA study should be performed accurately to prevent misconception and for it to be achievable. This phase would be the main reference point in whole study process.

16.3.2 Inventory analysis

In the second phase of a LCA, data need to be collected and all the information regarding the system must be ready. Here, an inventory analysis is conducted following the first phase. Information about the input and output of the system is collected in the inventory analysis and calculation procedures are performed. An analysis study is used to identify the input of resources, products or semiproducts, and output of waste, products, or emissions and data are generated from databases with unit processes. The databases contain quantified physical elementary flows for the system and the practitioner selects the relevant inputs and outputs of the system such as material resources, air, water, land, electricity, and fuel. At the end of this phase, practitioners would obtain a list of data associated with the product system.

Meteyer, Xu, Perry, and Zhao (2014) performed a study to propose a model of energy and material consumption of a binder-jetting process. Since there is a lack of documentation of energy consumption and material flow data limits of binder-jetting technology, they presented a model that was validated by experimental data to assist practitioners of LCA in industry to understand the energy and material consumption during the binder-jetting process. Cerdas, Juraschek, Thiede, and Herrmann (2017) presented their study on LCA of 3D printed products in a distributed manufacturing system and took the frame of eyeglasses as a case study. They also listed inventory data for the cradle-to-grave of printed parts. The data were obtained from literature studies and databases of Granta Design Limited. In addition, Yeon, Rew, Kang, & Choi, 2015 presented an inventory analysis for the two spall repair methods that were used to compare the existing and proposed spall repair methods throughout the life cycle of the concrete.

Inventory analysis for AM could be challenging because not much data and information are readily available for practitioners to use in LCA. Common resources that are used by other traditional manufacturing processes could be used as a reference such as electricity, water, and plastic materials. It is assumed that an AM process such as 3D printing information is provided from the machine creator and should be stored in the data inventory for future reference. Inventory analysis includes data from the transportation of raw materials, generation of electricity, water, production, and origin of the raw materials and air. Some researchers have taken the initiative to propose a new model for the calculation procedure of the physical flow of the technology.

16.3.3 Impact assessment

Impact assessment in LCA is conducted to identify and record the physical flows and intervention of the product system toward the environment. According to ISO 14040 standard (The International Standards Organisation, 2006), there are several elements that are necessary in this third phase, which are selection of the impact categories, classification of the output from phase two by assigning them into appropriate impact categories, and characterization of the impact profile. Furthermore, normalization of the impact profile and grouping of the impact categories or weighting using weighting factors for each impact category. The results are used as decision support specifically in a comparison analysis study.

Impact assessment of 3D printing includes the calculations and scoring of the impact of the technology toward the environment. In most traditional manufacturing processes, greenhouse gases and NOx emissions are usually calculated in the environmental evaluation in decision-making. Kreiger, Mulder, Glover, and Pearce (2014) performed a study on a LCA of distributed recycling of post-consumer high-density polyethylene for 3D printing and found that the distributed process could lower greenhouse gas emissions. They also suggested that the application of open-source 3D printing could lower the environmental impact compared with the current system. In another study, Belculfine (2016) presented impact categories of the distributed manufacturing process that includes raw material transportation until the 3D printing process. The impact categories that were evaluated in his study were ozone depletion, global warming, smog, acidification, eutrophication, carcinogenics, respiratory effects, eco-toxicity, and fossil fuel depletion. In his study, a higher value of fossil fuel depletion was exhibited in the whole process. The lowest impact was carcinogenics. The results would help manufacturers in reducing the environmental impact of the 3D printing process by changing the process parameters or changing the method of transportation of the materials.

Since the prime materials for 3D printing parts are plastics, impact analysis toward the environment and human health should be identified. An overview of LCA of AM performed by Baumers et al. (2017) mentioned concerns about exposure to emissions that could harm the health of machine operators. They are exposed to ultrafine particles that are released during the hot extrusion of polymer materials that used a filament in 3D printing. Frăţilă and Rotaru (2017) addressed some of the ecological issues of additive manufacturing in their study. The ecological issues are related to the materials and energy consumption, health and safety, transport, and waste management. By sustaining the principle of sustainable AM, efficient materials and energy use and material waste management is essentially needed to demonstrate an environmentally friendly manufacturing process. Hence, the appropriate LCA should be performed to ensure the negative impacts toward the environment have been identified accurately.

16.3.4 Interpretation

Interpretation is the final phase in the LCA framework. In this phase, findings from the inventory analysis and impact analysis are concluded to meet the goal and scope that are set at the first phase of the LCA. All the information that was collected needs to be verified, checked, and evaluated to make it easy for the practitioner to explain the conclusion, limitations, and recommendations based on the goal and scope of the LCA study. Sensitivity analysis and uncertainty analysis are conducted at this phase to assist practitioners strengthen the conclusions. Moreover, a report generated at this phase could help decision makers in weighting and normalizing all the decision criteria that meet their goals. Therefore this phase requires critical thinking and good understanding from all related parties.

Mohammed et al. (2017) reported a study on the environmental impact of 3D printing that specifically focused on the carbon footprint. The study showed the impact of recyclable plastic filaments that reduce negative environmental impacts. The technology of 3D printing could reduce plastic waste by converting it into valuable products. Liu, Jiang, Zhang, Li, and Zhang (2016) proposed a framework that integrates LCA with CAD. This would help decision makers in determining the right geometry, materials, and process parameters, while keeping the sustainability elements for 3D printing technology. A study conducted by Böckin and Tillman (2019) showed that material selection and environmental impact are the major considerations in implementing AM as nickel-alloy filaments increased the impacts on the environment based on their sensitivity analysis results. Hence, sensitivity analysis is essentially carried out at the last phase of the LCA to identify factors that influence the results of the analysis.

16.4 Comparison of life cycle assessment of fused filament fabrication with other additive manufacturing processes

Over the past few decades, AM has evolved from a niche process for rapid prototyping to a legitimate manufacturing process for low volume with complicated parts production in comparison to conventional manufacturing process (i.e., injection molding). Being the most innovative, cost-effective technique, with less processing procedures than other AM processes (Huang, Leu, Mazumder, & Donmez, 2015; Kafara et al., 2017), the technological and economic benefits of FFF have been widely documented. However, environmental issues of FFF have been at the center of attention due to the importance of environmental protection. A consideration of the eco-design aspect in FFF in every design and fabrication process can reduce any potential environmental impacts. Eco-design may encompass efficiency, health and safety, remanufacturing, recycling, source reduction, waste minimization, and is usually associated with life cycle assessment (Glavič & Lukman, 2007).

According to Huang, Liu, and Mokasdar (2013), most of the chemicals used in AM machines are not really harmful to humans, except for a few such as propylene glycol (produced by FFF machines) and photopolymers (produced by LENS). Past studies have evaluated the environmental performance of the FFF process in particular. The environmental performance measurement of FFF takes into account several characteristics such as materials, energy consumed, waste, disposal, emissions, and ultrafine particles. In general, the FFF process has equally good environmental characteristics compared with other common AM processes like SL and LS. Nevertheless, FFF faces several other challenges related to energy consumption and emissions during production. Table 16.1 shows several environmental criteria of FFF compared with common AM technologies.

Mehrpouya et al. (2019) suggested that, through LCA, a reduction in energy and material consumption can be performed to make FFF more sustainable. According to these authors, there are several sustainability dimensions that can be considered to address the environmental issues of FFF. It can be investigated based on three dimensions, as follows:

1. *Resource consumption*—energy and material consumption represent the main resource consumption for FFF. This is due to the machine, subsystem, and material consumptions. Energy consumption may be higher, particularly when the FFF process involves a lengthy fabrication time at high temperature. Also, an optimal design would save material usage by up to 40%.
2. *Pollution control*—AM eliminates the use of harmful chemicals that can be harmful to the environment, for example, casting release compounds, cutting fluids, and forging lubricants. Also, FFF uses

TABLE 16.1 Environmental criteria of fused filament fabrication compared with common AM processes, adapted and adjusted from Huang et al. (2013).

Process	Fused filament fabrication	Stereolithography	Laser sintering
Raw material	Acrylonitrile butadiene styrene, polylactic acid	Liquid photopolymer	Nylon, metal, ceramic, wax
Energy consumed	Heat	UV laser beam	High-power laser beam
Laser used	No	Yes	Yes
Solid residues	Material chips, support materials	Small amount of resin, support materials	Material chips
Liquid residues	No	No	No
Disposal	Incineration, landfill, recycling	Incineration, landfill	Incineration, landfill, recycling
Emissions	CO_2, CO, SOx, PMc, NOx	CO_2	CO_2, CO, SOx

nontoxic materials such as ABS and PLA. Nonetheless, high emissions including gaseous and ultrafine particles can be emitted from thermoplastic materials during the FFF process (Jeon, Park, Kim, Park, & Yoon, 2019).

3. *Waste management*—AM produces 90% of material waste when to compared to conventional manufacturing processes (i.e., subtractive manufacturing). However, studies related to the optimization of material reduction of FFF remain limited (Kellens et al., 2018).

Numerous studies have investigated AM machines, particularly FFF, consuming high energy because they depend on the electrical consumption during the printing process (factors include part volume, layer thickness, process speed, etc.) and curing process (factors include type of materials, melting temperature, etc.) (Böckin & Tillman, 2019; Cerdas et al., 2017; Faludi et al., 2015; Suárez & Domínguez, 2020). Faludi et al. (2015) found that the energy consumption of the FFF process ranged between 23 and 346.4 kWh/kg. It is therefore necessary to optimize the process parameters to minimize the energy consumption of FFF devices. This parameter can be achieved by optimizing the design of the manufactured part by utilizing only the amount of materials needed for the product (Suárez & Domínguez, 2020). This can be done by reducing the material quantity, weight of the part, and also by simplifying the part shape, which in consequence, allows the overall footprint of the product to be reduced.

On the other hand, some studies have been conducted to assess the environmental impact of the FFF process. Previous studies by Bourhis et al. (2013) and Luo, Ji, Leu, & Caudill, 1999 used Eco-Indicator 95 to evaluate the environmental impact of several AM machines, including FFF devices, in which it allowed researchers to translate massive energy into a unit of environmental impact per kilogram expressed in mPts/kg. Table 16.2 shows the comparison of FFF with several common AM technologies on energy consumption and environmental impact perspectives.

The emission issues from FFF have also gained the attention of some researchers in recent years. The growing use of 3D printers using filaments, particularly ABS and PLA, in private and industrial spaces raises the question as to whether human exposure to emitted substances may cause adverse health effects. Several studies have demonstrated that the heating of thermoplastic filaments releases volatile organic chemicals and billions of airborne particles per minute. Studies also indicate that the potential inhalation and prolonged exposure of nanoparticles have a direct impact on consumers and have implications for health risks (Madl & Pinkerton, 2009; Suárez & Domínguez, 2020).

Ultrafine particle emission rates from FFF varied substantially depending on the make and model of the printer, the type of filament

TABLE 16.2 Comparison of fused filament fabrication with several common AM technologies on energy consumption and environmental impact, adapted from Bourhis et al. (2013).

Technology	Machines	Material	Energy consumption rate (kWh/kg)	Environmental impact (mPts/kg)
Fused filament fabrication	Fused deposition modeling (FDM) 1650	Acrylonitrile butadiene styrene	346.43	197.47
	FDM 2000		115.48	65.82
	FDM 8000		23.10	13.16
	FDM Quantum		202.09	115.19
Stereolithography (SLA)	SLA-250	Epoxy resin	32.48	18.51
	SLA-3000		41.41	23.60
	SLA-5000		20.70	11.80
Laser sintering	Sinterstation DTM 2000	Polyamide	40.01	22.81
	Sinterstation DTM 2500		29.77	16.97
	Vanguard HiQ		14.54	8.29
	EOSINT M250 Xtended	Metallic powder (bronze + Ni)	5.41	3.09
	EOSINT P760	Polyamide PA2200 Balance 1.0	36.50	20.81
		Polyamide PA2200 Speed 1.0	39.80	22.69
		Polyamide PA3200GF	26.30	14.99

material, nozzle and bed temperatures, the time of printing, and the volume of emissions (Azimi, Zhao, Pouzet, Crain, & Stephens, 2016; Stephens, Azimi, El, & Ramos, 2013). A critical study by Kwon et al. (2017) revealed that temperature was a primary factor influencing emission rates. Most of the particles emitted by FFF machines were nanoparticles emitted during the early stages of printing. The author revealed that the extruder temperature greatly affected particle emissions, regardless of the shape of the printed object, temperature conditions, or filament material. In their influential analysis on ultrafine particle emissions emitted from FFF, Kim et al. (2015) showed that high concentrations of emitted nanoparticles and aldehydes such as formaldehyde, acetaldehyde, and isovaleraldehyde are harmful to humans. The toxicity of the emissions is influenced by their size and morphology, in which small particles are more toxicities than larger ones.

Byrley, George, Boyes, and Rogers (2019) concluded that ABS filament produces higher particle emission rates than PLA. They reported that mean particle number concentration (PNC) emitted during FFF operation is 300,980 particles/cm^3 for ABS and 65,482 particles/cm^3 for PLA. However, inhalation of ultrafine particles from both materials is poisonous and may lead to detrimental health effects for operators. Table 16.3 shows the overall mean particle count diameters and mean PNC of FFF filaments.

Based on the findings by Kwon et al. (2017), the results suggest that preventive control measures are crucial for controlling the release of harmful ultrafine particles. The recommended preventive control measures when operating FFF machines are as follows:

1. The use of an enclosure with a high-efficiency filter can decrease the concentration of particles emitted during printing.
2. The use of the correct filter attached to the enclosure ventilation of the FFF machine can remove nanoparticles effectively.
3. The HEPA filter was the most effective filter for removing nanoparticles during FFF printing (removal effectiveness of 99.95% for ABS).
4. Use FFF machines only in well-ventilated indoor environments with open windows and local ventilation systems.
5. Preventive guidelines to reduce the risks of emitted pollutants for FFF need to be established by operators as a preventive control method.

TABLE 16.3 Overall mean particle count diameters (PCDs) and mean particle number concentration (PNC) of FFF filaments, adapted from Byrley et al. (2019).

Filament	Mean PCD (nm)	Mean PNC (#/cm^3)
ABS	53.0	–
PLA	36.7	–
ABS	39.3	–
PLA	56.7	13,900
ABS	47.4	654,000
PLA	44.7	641,000

(Continued)

TABLE 16.3 (Continued)

Filament	Mean PCD (nm)	Mean PNC (#/cm³)
ABS	66.0	126,975
PLA	29.6	115,173
ABS	10.5	533,800
PLA	14.0	93,608
ABS	48.0	68,175
PLA	42.2	21,635
ABS	–	347,286
PLA	–	1600
ABS	–	10,600
PLA	–	89,000
ABS	88.5	–
PLA	38.0	–
ABS	–	37,640
PLA	–	76,300
ABS	–	52,698
PLA	–	6736
ABS	32.6	1,740,000
PLA	108.1	49,000
ABS	–	18
PLA	–	–
ABS	48.5	287,827
PLA	40.4	82,590

16.5 Conclusions

This chapter analyzes the LCA of FFF, which is one of the most widely used techniques amongst the 3D printing methods. As there has been a sharp increase in the usage of FFF worldwide, there is a need to perform LCA analysis. This chapter presented detail analysis on the LCA of FFF and comparison with other AM techniques was made. Based on this study it can be concluded that more studies need to be conducted on reducing the emissions and energy consumption of FFF. The emissions are more prevalent when the most common materials for FFF is used (ABS). Comparison of FFF to other AM techniques on the LCA also found that more energy is consumed. Therefore further study of the life cycle assessment analysis on FFF with green materials needs to be conducted in order to reduce the emissions from material fabrication as FFF is the most widely used AM processes with the highest market share.

Acknowledgments

The authors would like to thank the Universiti Teknikal Malaysia Melaka for the financial support provided through the UTeM Short Term Grant (PJP/2020/FTKMP/PP/S01737).

References

ASTM F2792-12a, (2012). *Standard terminology for additive manufacturing technologies*, (Withdrawn 2015), ASTM International, West Conshohocken, PA, <www.astm.org>.

Azimi, P., Zhao, D., Pouzet, C., Crain, N. E., & Stephens, B. (2016). Emissions of ultrafine particles and volatile organic. *Environmental Science & Technology, 50*(3), 1260–1268.

Baumers, M., Duflou, J. R., Flanagan, W., Gutowski, T. G., Kellens, K., & Lifset, R. (2017). Charting the environmental dimensions of additive manufacturing and 3D printing. *Journal of Industrial Ecology, 21*, 9–14. Available from: https://doi.org/10.1111/jiec.12668.

Bekker, A. C. M., & Verlinden, J. C. (2018). Life cycle assessment of wire + arc additive manufacturing compared to green sand casting and CNC milling in stainless steel. *Journal of Cleaner Production, 177*, 438–447. Available from: https://doi.org/10.1016/j.jclepro.2017.12.148.

Belculfine, E. (2016). *The Environmental Impact of Desktop 3D Printing in a Distributed Manufacturing Model: Analyzing Spare Plastic Parts Fabricated by Home Users* (Doctoral dissertation).

Böckin, D., & Tillman, A. M. (2019). Environmental assessment of additive manufacturing in the automotive industry. *Journal of Cleaner Production, 226*, 977–987. Available from: https://doi.org/10.1016/j.jclepro.2019.04.086.

Bourell, D., Kruth, J. P., Leu, M., Levy, G., Rosen, D., Beese, A. M., & Clare, A. (2017). Materials for additive manufacturing. *CIRP Annals - Manufacturing Technology, 66*(2), 659–681. Available from: https://doi.org/10.1016/j.cirp.2017.05.009.

Bourhis, F. Le, Kerbrat, O., Hascoët, J., Mognol, P., Bourhis, F., Le., ... Mognol, P. (2013). Sustainable manufacturing: Evaluation and modeling of environmental impacts in additive manufacturing. *International Journal of Advanced Manufacturing Technology, 69*, 1927–1939. Available from: https://doi.org/10.1007/s00170-013-5151-2.

Burns, M. (1993). *Automated fabrication: improving productivity in manufacturing*. Englewood Cliffs: Prentice Hall.

Byrley, P., George, B. J., Boyes, W. K., & Rogers, K. (2019). Particle emissions from fused deposition modeling 3D printers: Evaluation and meta-analysis. *Science of the Total Environment, 655*, 395–407. Available from: https://doi.org/10.1016/j.scitotenv.2018.11.070.

Cerdas, F., Juraschek, M., Thiede, S., & Herrmann, C. (2017). Life cycle assessment of 3D printed products in a distributed manufacturing system. *Journal of Industrial Ecology, 21*, S80–S93. Available from: https://doi.org/10.1111/jiec.12618.

Chua, C., & Leong, K. (2014). *3D printing and additive manufacturing: Principles and Applications, Fifth Edition of Rapid Prototyping* (5th Edition). Singapore: World Scientific Publishing Company.

Faludi, J., Bayley, C., Bhogal, S., Iribarne, M., Campbell, R. I., & Bourell, D. (2015). Comparing environmental impacts of additive manufacturing vs. traditional machining via life-cycle assessment introduction. *Rapid Prototyping Journal, 21*(1), 1–52.

Frăţilă, D., & Rotaru, H. (2017). Additive manufacturing—A sustainable manufacturing route. *MATEC Web of Conferences, 94*, 03004. Available from: https://doi.org/10.1051/matecconf/20179403004.

Gaha, R., Yannou, B., & Benamara, A. (2016). Selection of a green manufacturing process based on CAD features. *International Journal of Advanced Manufacturing Technology, 87*(5–8), 1335–1343. Available from: https://doi.org/10.1007/s00170-015-7499-y.

Gębisa, A. W., & Lemu, H. G. (2018). Investigating effects of Fused-Deposition Modeling (FDM) processing parameters on flexural properties of ULTEM 9085 using designed experiment. *Materials, 11*, 500.

Glavič, P., & Lukman, R. (2007). Review of sustainability terms and their definitions. *Journal of Cleaner Production, 15*(18), 1875–1885. Available from: https://doi.org/10.1016/j.jclepro.2006.12.006.

Hofstätter, T., Pedersen, D. B., Tosello, G., & Hansen, H. N. (2017). State-of-the-art of fiber-reinforced polymers in additive manufacturing technologies. *Journal of Reinforced Plastics and Composites, 36*(15), 1061–1073. Available from: https://doi.org/10.1177/0731684417695648.

Huang, S. H., Liu, P., & Mokasdar, A. (2013). *Additive manufacturing and its societal impact: A literature review*, 1191–1203. Available from: https://doi.org/10.1007/s00170-012-4558-5.

Huang, Y., Leu, M. C., Mazumder, J., & Donmez, A. (2015). Additive manufacturing: Current state, future potential, gaps and needs, and recommendations. *Journal of Manufacturing Science and Engineering, 137*, 1–10. Available from: https://doi.org/10.1115/1.4028725.

International Standard Organisation. (1997). *Environmental assessment-Principles and framework*. ISO 14040. <https://www.iso.org/standard/23151.html>.

International Standard Organisation. (2015). *Additive manufacturing—General principles—Terminology*. ISO/ASTM 52900:2015.

Jeon, H., Park, J., Kim, S., Park, K., & Yoon, C. (2019). Effect of nozzle temperature on the emission rate of ultrafine particles during 3D printing. *International Journal of Indoor Environment and Health*. Available from: https://doi.org/10.1111/ina.12624.

Kafara, M., Süchting, M., Kemnitzer, J., Westermann, H. H., & Steinhilper, R. (2017). Comparative life cycle assessment of conventional and additive manufacturing in mold core making for CFRP production. *Procedia Manufacturing, 8*(October 2016), 223–230. Available from: https://doi.org/10.1016/j.promfg.2017.02.028.

Kalsoom, U., Nesterenko, P. N., & Paull, B. (2016). Recent developments in 3D printable composite materials. *RSC Adv., 6*(65), 60355–60371. Available from: https://doi.org/10.1039/C6RA11334F.

Kek, V., & Vinodh, S. (2016). LCA integrated ANP framework for selection of sustainable manufacturing processes. *Environmental Modeling and Assessment, 21*(4), 507–516. Available from: https://doi.org/10.1007/s10666-015-9490-2.

Kellens, K., Peng, T., Kellens, K., Tang, R., Chen, C., & Chen, G. (2018). Sustainability of additive manufacturing: An overview on its energy demand and environmental impact. *Additive Manufacturing, 21*, 694–704. Available from: https://doi.org/10.1016/j.addma.2018.04.022.

Kim, Y., Yoon, C., Ham, S., Park, J., Kim, S., Kwon, O., & Tsai, P.-J. (2015). Emissions of nanoparticles and gaseous material from 3D printer operation. *Environmental Science & Technology, 49*, 12044–12053. Available from: https://doi.org/10.1021/acs.est.5b02805.

Kreiger, M. A., Mulder, M. L., Glover, A. G., & Pearce, J. M. (2014). Life cycle analysis of distributed recycling of post-consumer high density polyethylene for 3-D printing filament. *Journal of Cleaner Production, 70*, 90–96. Available from: https://doi.org/10.1016/j.jclepro.2014.02.009.

Kruth, J. P., Leu, M. C., & Nakagawa, T. (1998). Progress in additive manufacturing and rapid prototyping. *Ann CIRP, 47*(2), 525–540.

Kwon, O., Yoon, C., Ham, S., Park, J., Lee, J., Yoo, D., & Kim, Y. (2017). Characterization and control of nanoparticle emission during 3D printing. *Environmental Science and Technology, 51*(18), 10357–10368. Available from: https://doi.org/10.1021/acs.est.7b01454.

Liu, Z., Jiang, Q., Zhang, Y., Li, T., & Zhang, H. C. (2016). Sustainability of 3D printing: A critical review and recommendations. ASME 2016 11th international manufacturing science and engineering conference, MSEC 2016. Available from:https://doi.org/10.1115/MSEC2016-8618.

Luo, Y., Ji, Z., Leu, M. C., & Caudill, R. (1999). Environmental performance analysis of solid freedom fabrication processes. In Proceedings of the 1999 IEEE international symposium on electronics and the environment (Cat. No. 99CH36357). Danvers: IEEE. (1–6).

Ma, F., Zhao, Y., Pu, Y., & Li, J. (2018). A comprehensive multi-criteria decision making model for sustainable material selection considering life cycle assessment method. *IEEE Access, 6*, 58338–58354. Available from: https://doi.org/10.1109/ACCESS.2018.2875038.

Madl, A. K., & Pinkerton, K. E. (2009). Health effects of inhaled engineered and incidental nanoparticles. *Critical Reviews in Toxicology, 39*(8), 629–658. Available from: https://doi.org/10.1080/10408440903133788.

Mami, F., Revéret, J. P., Fallaha, S., & Margni, M. (2017). Evaluating eco-efficiency of 3D printing in the aeronautic industry. *Journal of Industrial Ecology, 21*, S37–S48. Available from: https://doi.org/10.1111/jiec.12693.

Mastura, M. T., Sapuan, S. M., Mansor, M. R., & Nuraini, A. A. (2016). Environmentally conscious hybrid bio-composite material selection for automotive anti-roll bar. *The International Journal of Advanced Manufacturing Technology, 89*(5–8), 2203–2219. Available from: https://doi.org/10.1007/s00170-016-9217-9.

Matsuzaki, R., Ueda, M., Namiki, M., Jeong, T. K., Asahara, H., Horiguchi, K., ... Hirano, Y. (2016). Three-dimensional printing of continuous-fiber composites by in-nozzle impregnation. *Scientific Reports, 6*(December 2015), 1–7. Available from: https://doi.org/10.1038/srep23058.

Mehrpouya, M., Dehghanghadikolaei, A., Fotovvati, B., Vosooghnia, A., Emamian, S. S., & Gisario, A. (2019). The potential of additive manufacturing in the smart factory industrial 4.0: A review. *Applied Sciences (Switzerland), 9*(18). Available from: https://doi.org/10.3390/app9183865.

Meteyer, S., Xu, X., Perry, N., & Zhao, Y. F. (2014). Energy and material flow analysis of binder-jetting additive manufacturing processes. *Procedia CIRP, 15*, 19–25. Available from: https://doi.org/10.1016/j.procir.2014.06.030.

Mohammed, M. I., Mohan, M., Das, A., Johnson, M. D., Singh Badwal, P., McLean, D., & Gibson, I. (2017). A low carbon footprint approach to the reconstitution of plastics into 3D-printer filament for enhanced waste reduction. *KnE Engineering, 2*(2), 234. Available from: https://doi.org/10.18502/keg.v2i2.621.

Ngo, T. D., Kashani, A., Imbalzano, G., Nguyen, K. T. Q., & Hui, D. (2018). Additive manufacturing (3D printing): A review of materials, methods, applications and challenges. *Composites Part B: Engineering, 143*(December 2017), 172–196. Available from: https://doi.org/10.1016/j.compositesb.2018.02.012.

Scoot, C. (1992). U.S. Patent 5121329, Apparatus and method for creating three-dimensional objects. U.S. Class: 364/468. Stratasys, Inc., Minneapolis, MN (filed: 30 October 1989. date of patent: 9 June 1992).

Statista. (2018). Most used 3D printing technologies worldwide 2018. Accessed from: https://www.statista.com/statistics/756690/worldwide-most-used-3d-printing-technologies/

Stephens, B., Azimi, P., El, Z., & Ramos, T. (2013). Ultra fine particle emissions from desktop 3D printers. *Atmospheric Environment, 79*, 334–339. Available from: https://doi.org/10.1016/j.atmosenv.2013.06.050.

Suárez, L., & Domínguez, M. (2020). Sustainability and environmental impact of fused deposition modelling (FDM) technologies. *The International Journal of Advanced Manufacturing Technology, 106*(3–4), 1267–1279. Available from: https://doi.org/10.1007/s00170-019-04676-0.

Sun, X., Liu, J., Lu, B., Zhang, P., & Zhao, M. (2017). Life cycle assessment-based selection of a sustainable lightweight automotive engine hood design. *International Journal of Life Cycle Assessment, 22*(9), 1373–1383. Available from: https://doi.org/10.1007/s11367-016-1254-y.

The International Standards Organisation. (2006). Environmental management—Life cycle assessment—Principles and framework. Iso 14040, 2006, 1–28. Available from: https://doi.org/10.1136/bmj.332.7550.1107.

Yao, Y., Hu, M., Di Maio, F., & Cucurachi, S. (2019). Life cycle assessment of 3D printing geo-polymer concrete: An ex-ante study. *Journal of Industrial Ecology*. Available from: https://doi.org/10.1111/jiec.12930.

Yeon, J., Rew, Y., Kang, J., & Choi, K. (2015). *Life cycle assessment-based feasibility study of spall damage rehabilitation using 3D printing technology*. 54th ASC Annual International Conference Proceedings. Minneapolis.

Zhang F., Ma G., Tan Y. (2017). The nozzle structure design and analysis for continuous carbon fiber composite 3D printing. In 2017 7th International Conference on Advanced Design and Manufacturing Engineering (ICADME 2017). Atlantis Press.

CHAPTER

17

Application of lightweight materials toward design for sustainability in automotive component development

M. Fadzil[1,2], A.B. Abdullah[1], Z. Samad[1], F. Yusof[2,3] and Y.H.P. Manurung[4]

[1]School of Mechanical Engineering, Universiti Sains Malaysia (Engineering Campus), Penang, Malaysia, [2]Centre of Advanced Manufacturing and Material Processing (AMMP Centre), University of Malaya, Kuala Lumpur, Malaysia, [3]Department of Mechanical Engineering, Faculty of Engineering, University of Malaya, Kuala Lumpur, Malaysia, [4]Smart Manufacturing Research Institute (SMRI), Faculty of Mechanical Engineering Universiti Teknologi MARA, Selangor, Malaysia

17.1 Introduction

The automotive industry is one of most important economic sectors in the world, contributing over €2.75 trillion annually, corresponding to 3.65% of the global GDP (Saberi, 2018). The current estimated global vehicle population of 1.2 billion is predicted to rise to around 2 billion by 2040 (Sioshansi & Webb, 2019). The annual global sales of new vehicles are increased from an estimated 65.9 million units in 2005 to 95.1 million units in 2018, largely fueled by rising demand from developing countries (OICA, 2019). These vehicles are essential to the working of the global economy and serve the mobility needs of the world's citizen. However, this rapid

Design for Sustainability
DOI: https://doi.org/10.1016/B978-0-12-819482-9.00006-X

growth will result in significant increases in global fuel demands, raw materials utilization, and air pollution.

A report by the Intergovernmental Panel on Climate Change stated that 14% of 2010 global greenhouse gas emissions were from the transportation sector (IPCC, 2014). In the United States, transportation makes up the largest portion (29%) of total greenhouse gas emissions, with light-duty vehicles contributing 59% of transportation emissions (Office of Transportation and Air Quality, 2019). Research on vehicle exhaust emissions were started in the late 1950s as part of wider programs on the reduction of air pollution. These initiatives were first led by the United States under the 1968 Clean Air Act, and the vehicle emission limits of thc EC in Europe in 1970. This has led to astounding improvements in spark ignition engine technology and the use of three-way catalytic converters, reducing emission of exhaust pollutants by up to 90% (Smith & Davies, 1996). However, as basic greening technologies have already been implemented through advanced conventional technologies, future improvements in emission control require investments in other areas including lightweight technologies. Carbon dioxide regulations are still being created, with many countries enacting increasingly stringent laws to reduce emissions. Thus, to address the issue on carbon dioxide emissions, waste, fossil fuel dependence, rising energy prices, and other environmental concerns, the automotive industry should consider sustainability as an integral part of their business, and focus on lightweighting strategies for producing the next generation of vehicles.

17.2 Sustainability in automotive production

Nowadays, the automotive industry is aligning its production more closely to the principles of sustainability. The general concept in designing for sustainability is striking the right balance between economic objectives with social, ethical, and environmental considerations. Sustainability in the automotive industry has been discussed by various researchers (Mayyas, Qattawi, Omar, & Shan, 2012; Vaz, Rauen, & Lezana, 2017). Several tools have been developed to incorporate sustainability in manufacturing such as design for environment and life cycle assessment, design for resources and energy, and design for sustainability (Rosen & Kishawy, 2012). The majority of sustainability implementations in the automotive industry are achieved through incremental improvements to existing technologies rather than radical innovations that require major changes in skills and automotive operations (van den Hoed, 2007). Most of the current efforts on sustainability are centered on the immediate issue of environmental effects, due to the tightening global legislations on exhaust emissions. One such aspect is the focus on the energy efficiency of both the product and the production process.

17.3 Design for energy efficiency by lightweighting

A major source of environmental impact is the energy consumed by a product during its phase of usage. For passenger cars, 86% of the total lifetime energy use from the start of production up until end-of-life (disuse or scrapping) phases is expended on carrying their own weights and occupants (Watarai, 2006). Weight reduction is as effective as engine efficiency improvement and friction loss reduction as it has a ripple effect on fuel economy. A heavier vehicle not only requires high power for mobility, the power losses are also effectively proportional to the vehicle mass. Reducing weight has a larger proportional impact on the total vehicle energy load than improvements in aerodynamics or tire rolling resistance (Isenstadt et al., 2016).

Vehicle weight is an important factor in the design of an automobile as it has a direct effect on driving dynamics (handling, braking, ride, acceleration), agility, cargo capacity, and fuel consumption (Heuss, Müller, van Sintern, Starke, & Tschiesner, 2012). Vehicle mass exhibits a direct correlation with the fuel consumption rate. A reduction of 6%–8% in fuel consumption can be obtained by reducing the weight of a car by 10% (Joost, 2012). Reducing the vehicle weight by 340 kg can improve the fuel economy by 40% without compromising safety (Gutowski, Yang, Li, Dean, & Zhang, 2017). In terms of exhaust emissions, a reduction of 5% in vehicle body weight corresponds to a reduction of 1.3%–1.8% of CO_2 emissions, and increases to 2.7%–3.6% for a 10% weight reduction (Fontaras & Samaras, 2010). A 100 kg reduction in the weight of a car corresponds to a 9 g reduction of CO_2 per kilometer (Hirsch, 2011). The European regulations target average CO_2 emissions of 95 g/km in 2021 and a further 37.5% reduction by 2030, coupled with a financial penalty of €95 per g/km of exceedance (European Union, 2014, 2019). The automotive industry must therefore decrease the overall vehicle weight while being able to provide high levels of safety and comfort. However, vehicle weights have increased over the years due to consumer demands for improved interior comfort and advanced electronic systems for safety, navigation, and entertainment, as shown in Fig. 17.1. This trend is projected to further rise with the transition from internal combustion engines (ICEs) to electric drivetrains, which increases the total kerb weight of the vehicle by approximately 200 kg over their gasoline equivalents (Faria, Moura, Delgado, & De Almeida, 2012).

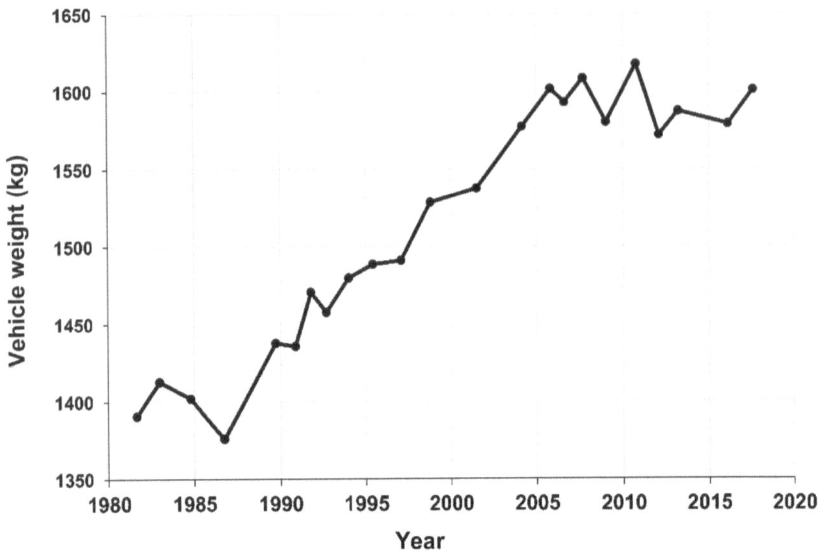

FIGURE 17.1 Average new vehicle weight (1980–2018) [Hula, French, Maguire, Bunker, and Rojeck (2019)]. *Source: Adapted from Hula, A., French, R., Maguire, A., Bunker, A., & Rojeck, T. (2019).* The 2018 EPA automotive trends report: Greenhouse gas emissions, fuel economy, and technology since 1975 (EPA-420-R-19-002, March 2019) *(No. EPA-420-R-19-002; p. 208). Office of Transportation and Air Quality.*

Lightweighting is one of the key strategies for meeting future regulation standards and to counterbalance the added weights of interior, safety, and electronic systems of modern vehicles. It is relevant for both CO_2-optimized ICEs, hybrid electric vehicles, as well as full battery electric vehicles. The weight reduction lowers the amount of energy needed to drive the vehicle, regardless of the efficiency and type of propulsion system. The benefit can be even larger for electric vehicles, as it reduces battery sizes and cost and/or increases range. It is predicted that the share of lightweight automotive components will increase from 29% in 2010 to 67% in 2030 (Heuss et al., 2012). However, the utilization of lightweight materials has so far been limited due to the high costs of potential lightweight solutions and consumer's limited willingness to pay for weight reduction. Nevertheless, manufacturers are expressing a high level of confidence in lightweighting. In the 2019 Wards Auto survey of the automotive engineering and design community, 24% of the respondents said lightweighting and use of lightweight structural materials was their primary efficiency strategy for meeting 2025 CAFE (Corporate Average Fuel Economy) standards, surpassing other responses of engine efficiency (14%) and electrification (19%) (Winter, 2019). Manufacturers need to evaluate the risk of lightweighting strategies considered since some strategies, although technically viable, pose risks in terms of functional equivalency and consumer acceptance. These risk factors can be categorized into three levels (Mascarin, Hannibal, Raghunathan, Ivanic, & Clark, 2016):

- Low-risk strategies: Readily implementable by automakers because there are both material and process familiarity and confidence in performance and economic;
- Medium-risk strategies: Technically viable processes, but less familiarity in high-volume production and/or less certainty of performance, quality, and cost relative to existing practices;
- High-risk strategies: Significant material or process cost reductions or significant changes in vehicle performance or features.

Other important issues which require resolutions before a new lightweight material can be specified include its effects on vehicle dynamics, durability, damageability, and repair and crash worthiness (Bhattacharya & Alpas, 2017). These attributes relate to the effect of metallurgical characteristics and the impact of the part design and manufacturing process on the material and product performance. Current lightweight materials in production include high-strength steels, aluminum alloys, magnesium, plastics, and composites. The design of lightweight structures should consider the need to design for strength (the material does not fail) and to design for stiffness (it does not elastically bend too much). These two construction principles may overlap, depending on the purpose of the lightweighted component.

17.4 Lightweight metal alloys in automotive applications

Usage of nonferrous metal alloys to create lightweight vehicles is not new. Lightweight metals, such as aluminum, have been used since the early 1930s in the construction of a niche group of cars built for speed. Being a mature technology, the benefit of metal alloys over nonmetal alternatives is their compatibility with existing production infrastructure. Nonmetal alternatives, such as composites, are costly to produce and require major changes to production processes. However, these lightweight metal alloys are expensive and may require different processing parameters as compared to steel.

A material-based approach of directly replacing steel with lightweight alternatives, while maintaining the same workpiece geometry, will certainly reduce the component's weight. However, the use of light metals must not decrease the properties and performance of the product. Thus, specific material properties, such as those shown in Table 17.1, should not be overlooked (Kleiner, Geiger, & Klaus, 2003).

TABLE 17.1 Specific material properties of selected lightweight metals (Kleiner et al., 2003).

Material properties	Units	Steel	Aluminum	Magnesium	Titanium
Density, (ρ)	kg/dm^3	7.83	2.8	1.74	4.5
Tensile modulus, (E)	GPa	210	70	45	110
Tensile strength, (R_m)	N/mm^2	300–1200	150–680	100–380	910–1190
Specific strength, (R_m/ρ)	10^6 N mm/kg	38–153	54–243	57–218	202–264
Specific stiffness, (E/ρ)	10^9 N mm/kg	26.8	25.0	25.9	24.4
Dent resistance, ($\sqrt{Rm}/\rho$)	10^6 N$^{1/2}$ mm^2/kg	4.4	9.3	11.2	7.7
Shell stiffness, ($\sqrt[3]{Em}/\rho$)	10^7 N$^{1/3}$ mm$^{7/3}$/kg	7.6	14.7	20.4	10.6

The current lightweight metallic alternatives to conventional automotive steel alloys are aluminum, magnesium, advanced high strength steel, titanium, and metal matrix composites (MMCs). Their applications are gradually spreading from luxury and specialty brands to mass-produced models. A high percentage of weight reduction using lightweight materials has been seen specifically for larger vehicles, such as trucks and pickups.

17.4.1 Advanced high-strength steels

The biggest change in recent years for the automotive industry has been the widespread use of advanced high-strength steel to replace traditional low-carbon steel. Advanced high-strength steel is considered a lightweight material as the part can be made thinner due to its increased strength, achieving 20%–30% weight savings in commercial vehicles (Mohrbacher, Spöttl, & Paegle, 2015). High-strength steel families, such as dual-phase (DP) and TRIP steels, are already in use in body-in-white (BIW) applications. The first generation of advanced high-strength steels (AHSS) includes the main low-alloyed steel grades of DP, TRIP, complex phase, and hot stamping steels. The usage of larger amounts of alloying elements gave rise to the second generation of AHSS, such as high-manganese steel, and the third-generation AHSS such as quenched and partitioned (Q&P) steels, and medium-Mn steels or TWIP. The third-generation AHSS provides the high strength required (over 1000 MPa) with significant ductility and elongation. These high-strength parts are mostly used for safety components where minimal deformation is expected such as in B-pillars or bumpers. Press-hardened steels are currently being developed to further improve the crash resistance and energy absorption. Fig. 17.2 summarizes the three generations of AHSS for automotive applications (Zhao & Jiang, 2018).

In body shell applications, the use of AHSS may not be suitable since there is a reduction in the thickness of the material, which is related to dent resistance and shell stiffness. Reducing the sheet thickness will decrease the part's stiffness, even if the Young's modulus of steel, between 190 and 225 GPa, is still much higher than that for light alloys. In this case, the use of aluminum and magnesium shows much better properties than steels as shells with the same weight have a higher wall thickness due to lower densities of the materials (Kleiner et al., 2003).

Low-density medium-Mn steel is a promising lightweight grade which utilizes 0.2–0.5 wt.% carbon, 2–8 wt.% manganese, and 5–8 wt.% aluminum. The addition of Al to Fe-C steels reduces both the density and Young's modulus. Every 1 wt.% addition of Al reduces the density by 1.3% and the Young's modulus by 2% (Chen, Rana, Haldar, & Ray, 2017). However, the utilization of Fe-Mn-Al-C steels in the automotive industry is still not widespread due to the limited

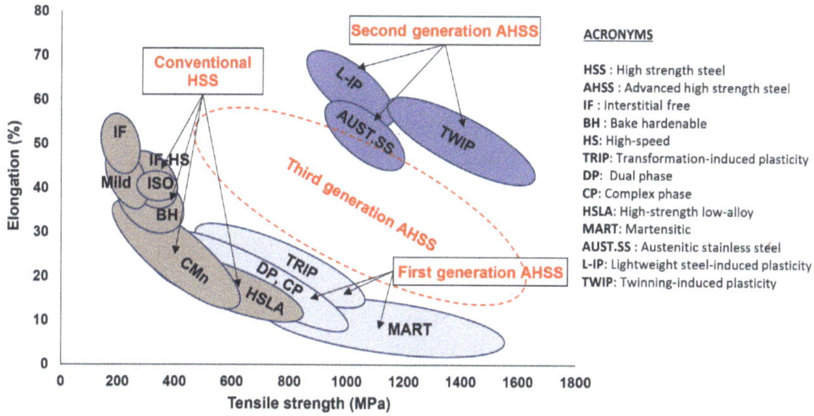

FIGURE 17.2 Tensile strength versus elongation (%) for various types of automotive steels (based on Zhao & Jiang, 2018).

knowledge on their application properties such as fatigue behavior, formability, weldability, and coatability. Thus, from the current technological point of view, the advantage of low-density steel has yet to compensate for the difficulties in producing and processing Fe-Mn-Al-C on an industrial scale.

17.4.2 Aluminum

Aluminum has favorable properties for automobile manufacturing such as corrosion resistance, light weight, high strength, and ease of fabrication. It is also readily available, has good recyclability, and can be easily incorporated into mass production. Aluminum offers weight savings of up to 40% over traditional steel structures, although in term of cost it can be 30% higher than steel (Gutowski et al., 2017). The highest volume of aluminum usage in automotive components has been in the conventionally casted parts such as the engine block, cylinder heads, and chassis assembly (Hirsch, 2004). Its usage has been extended to other automotive structures, such as the roof hood, outer body panels, and enclosures.

Aluminum alloy sheets were introduced as alternatives to steel sheets in high-volume automotive production in the mid-1980s. Automotive alloys used for body sheets include the Al-Cu 2000 series alloys (AA2008, AA2036), the Al-Mg 5000 series alloys (AA5053, AA5023, AA5182 and AA5052, AA5754), and the Al-Mg-Si 6000 series alloys (AA6016, AA6022 and AA6111) (Hirsch, 2004; Sakurai, 2008). The current predominant aluminum-based alloys for automotive body panels are the AA6000 series alloys as they are heat-treatable and their strengths controllable by heat treatment, although the work-hardenable AA5000 series is still being used in some applications, especially for complex inner parts due to its high formability. Several AA7000 series alloys have been produced that meet the strength, elongation, and corrosion requirements, but require warm forming technology for press operations (Long & Newman, 2017).

For outer panel application, the low Young's modulus of aluminum is improved by increasing the sheet thickness without sacrificing weight reduction. A thicker aluminum sheet with the same characteristics as a thinner steel sheet is still lighter than a steel sheet (Morita, 1998). Apart from a few critical components related to cabin safety (such as the roof rail, longitudinal front, and upper beams), it is now possible to make an entire BIW from aluminum, resulting in significant weight reduction and improved performance.

The development of high-strength aluminum alloys greater than 600 MPa and 8% total elongation are based on the high-strength 7000 series alloys that will be competitive with AHSS. Simulation studies have shown good formability for temperatures of 177°C–204°C. However, issues in forming such as the effectiveness of lubricants, effect of cooling, and springback during warm forming is yet to be addressed (Long & Newman, 2017).

17.4.3 Magnesium

Magnesium is the lightest of all structural metals (~1.8 g/cm^3), and is 33% lighter than aluminum (~2.7 g/cm^3), 75% lighter than steel (~7.2 g/cm^3) (Subic & ICSAT, 2012), and has a comparable strength to weight ratio to aluminum. A 20%–70% weight reduction with increased power performance can be obtained by replacing automotive steel components with magnesium alloys (Kumar, Sasanka, Ravindra, & Suman, 2015). The Superlight Car project by the EUC targets a reduction of 35% of BIW weight by using 7% of magnesium in sheet and cast components (Niu, Skszek, Fabischek, & Zak, 2014).

Although the use of magnesium in vehicles started in the 1930s, automotive applications of magnesium alloys are limited to a few components such as in the instrument panel beams, transfer cases, steering components, and radiator supports (Luo & Sachdev, 2012). It is used to produce engine blocks, gearboxes and clutch housing, oil pans, and wheel and engine cradles, where stress can be distributed, and high loads avoided. Magnesium interior components are also utilized where corrosion is not an issue such as in steering wheel armatures and steering column brackets (Luo, 2013).

Magnesium alloy components are usually cast rather than stamped as it has excellent high-pressure die castability. Compared to aluminum alloys, the lower heat capacity of magnesium alloy allows for reduced cycle time, and its high fluidity enables the casting of thinner, larger, and more detailed components. Wrought magnesium alloys are currently used to a very limited extent due to the lack of suitable alloying elements and the technological restrictions imposed by its hexagonal lattice crystal structure, which resists plastic deformation. Thus, only a limited number of body and chassis components are currently made of magnesium. However, thin-wall structural components can be fabricated using high-pressure die casting for components of variable thicknesses, ranging from 2 to 7 mm. This enables the fabrication of body components such as liftgates, front end structures, and instrument panels (Knight, 2015).

The major barrier to the applications of magnesium sheet components in vehicle structures is the limited formability of magnesium sheet and the production cost of the sheet itself. Wrought magnesium alloys are made by ingot metallurgy and subsequent hot rolling, making it a very expensive process to produce magnesium sheet products. New alloy systems being researched for improved formability and strengths are based on the Mg-Mn (ME and MJ Series) and the novel Mg-Zn (ZK, ZE, ZW series) alloys (Kumar & Sasanka, 2017). Low-temperature warm forming of Mg ZEK 100 has been shown to have better formability than AZ31B at blank temperatures above 230°C for the forming of magnesium inner door panels (Niu et al., 2014).

Magnesium is significantly more challenging and is not on the same level of processing familiarity in the industry as aluminum. Commercial usage of magnesium sheet for automotive applications is dependent on further reductions in material and process costs and improved production capability. The recent development of new age-hardenable low-cost magnesium alloy having excellent room-temperature formability comparable to medium-strength aluminum alloys is expected to overcome the long-standing issues in application of magnesium sheet metals for automotive applications (Bian et al., 2017).

17.4.4 Titanium

Titanium and its alloys have tensile strengths from 210 to 1380 MPa, which are equivalent to the range of strengths in steel alloys (Costa, Miranda, Quintino, & Yapp, 2007), but at approximately 60% of its density (4.5 g/cm^3 vs. 7.8 g/cm^3) (Sachdev, Kulkarni, Fang, Yang, & Girshov, 2012). Its strength is retained even at high temperatures and it has excellent corrosion resistance. Automotive applications of titanium fall into three broad categories, which are (1) reciprocating components (engine valves and connecting rods), (2) extreme temperatures and thermal gradients (turbochargers and exhaust systems), and (3) suspension components (springs). Using titanium in automotive applications not only benefits mass reduction, but also in obtaining more engine power and better ride and handling.

However, titanium is extremely expensive at a $18–110 per kg as compared to only $0.70–1.30 per kg for sheet steel and $2.20–11 per kg for sheet aluminum (Froes, Friedrich, Kiese, & Bergoint, 2004). Thus, its usage is limited primarily to low-volume or niche vehicles, or in unique and high-end applications. Developments in titanium alloys are reported for the primary metal production and powder metallurgy processing, advanced titanium aluminide alloys, and "meltless" processes for producing high-quality titanium powder for powder metallurgy processes (Sachdev et al., 2012). A summary of lightweight metal alloys for automotive applications is shown in Table 17.2.

TABLE 17.2 Summary of lightweight metal alloys for automotive applications [*Source*: From Taub & Luo (2015). Advanced lightweight materials and manufacturing processes for automotive applications. *MRS Bulletin*, 40, 1045–1054; Hurles & Froes (2002). Lowering the cost of titanium. *The AMPTIAC Newsletter*, 6(2), 10].

Material	Relative weight	Relative cost per part	Key advantages	Limitations	Automotive applications
Traditional Steel	100%	100%	• Established material • Good Formability • Widely available	• High density	Body structure, closures, and ancillary parts
AHSS	75%–85%	100%–150%	• Very high strength	• Poor formability • Low ductility	Safety cage, A- and B-pillars reinforcements
Aluminum	50%–60%	130%–200%	• Low density • Increased dent resistance	• Not easily weldable • Paint problems and premature corrosion	Cylinder heads, roof, and outer body panels
Magnesium	40%–45%	15%–250%	• Very low density for automotive metal	• Poor room temperature formability	Engine blocks, steering wheel armatures, gearbox housing
Titanium	60%	2900%	• Highest strength to weight ratio • Heat and corrosion resistant	• Very expensive	Engine valves, exhaust, springs

17.5 Sustainable design: tailor welded blanks

The car body frame, or BIW, forms the structure that supports the vehicle. It represents approximately 30% of the total body weight, making it the heaviest vehicle element (Hirsch, 2011). Typically, the BIW is made up of more than 300 panels, mainly assembled by spot welding. BIW structures are typically metal-based and produced from stamped metal sheets. Steel remains the predominant material used as the car body; thus, it has the highest weight-saving potential through increasing use of lightweight content.

Prior to the introduction of tailor welded blanks (TWBs), automobile parts were individually stamped based on the different thickness and material composition of each part and then welded to form a complete piece. ThyssenKrupp Steel AG first introduced the TWBs in the mid-1980s by laser welding two coated steel sheets to produce a part that was wider than the hot-dip galvanizing line could coat at the time. Since then, the manufacture of TWBs has progressed to include welding steels of different thicknesses, coatings, and material grades of carbon and stainless steel (Eisenmenger & Szymanski, 2012). Subsequent developments have also introduced the production of aluminum TWB (Das, 2000). The differences in composition and thickness of the TWB allow designers to select and strengthen critical areas without adding significant weight. This improved structural rigidity also eliminates the need for support for parts that require reinforcements, reducing the number of parts and the overall weight. It can also be used in energy absorption areas to achieve controlled collapse in an accident. The technology also improves material utilization, as the segments of the TWB can be nested to optimally fit onto the sheet of the stock material.

Presently, the thickness ratio often needed for automotive TWB ranges from 1.5:1 to 2:1 (Kunkel & Hovanski, 2016). They are used in the fabrication of side panels, wheel arches, floors, doors, and side members. For example, in the car door subassembly shown in Fig. 17.3, a thicker, harder material on the hinge side is connected to a thinner, lighter, and more malleable sheet metal for the rest of the door. Commercial applications of tailored blanks are in carbon steels, aluminum, and stainless steel, with thicknesses varying between 0.6–3.0 mm (Eisenmenger & Szymanski, 2012).

The benefits of tailored blanks include (Yan, 2005):

1. Improved raw material utilization and reduction of scrap: Nesting the various blank segments during the blanking process reduces the engineered scrap content;
2. Improvement of functional performance: Properties such as structural rigidity, corrosion resistance, and improved crash absorption can be obtained by combining blanks of different compositions and thicknesses to obtain properties tailored to performance requirements;
3. Reduction in manufacturing costs: Reduction in the number of stamping and assembling operations. For example, with traditional processing, the door panel and hinge reinforcement are punched and stamped separately, and then joined together. In TWB, the two sheets

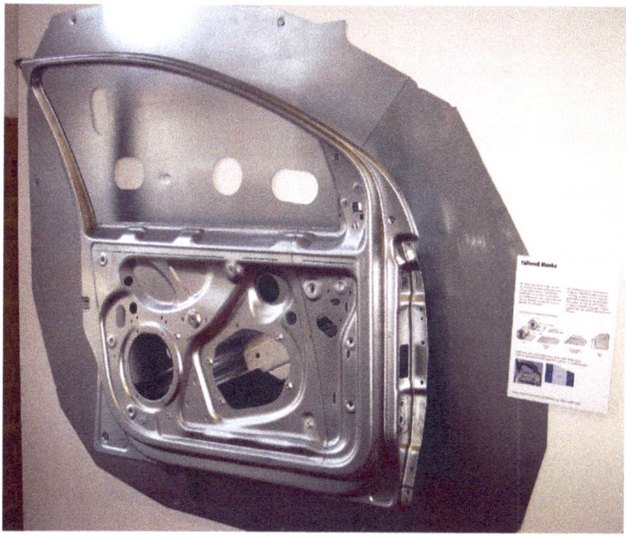

FIGURE 17.3 Car door panel fabricated from tailor welded blank Stahlkocher (2006). *Source: Courtesy of Stahlkocher. (2006).* Tailored Blank Hoesch museum. *Retrieved from <https://commons.wikimedia.org/wiki/File:Tailored_Blank_Hoesch_Museum.jpg>.*

are first welded and then stamped, consolidating the number of stamping operations;

4. Better finish: butt–joint weld is less visually noticeable and can be hidden more easily than other overlapping edge joints;
5. Lower operating costs: TWB reduces both the space occupied and inventory in the warehouse, reducing logistical and administrative costs.

However, the disadvantages of TWB are the higher cost of blanks and the increased cost of machinery. Very precise machineries are required for TWB as it requires very tight dimensional tolerances and high welding quality to withstand stamping operations without the loss of material performance.

17.6 Production and processing of tailor welded blanks

Traditional car manufacturing is extremely capital intensive and acts as a barrier for designing and deploying new lightweight materials and components. The processes involved in the production of tailor welded components are blanking, welding, and stamping, as shown in Fig. 17.4, to produce a TWB automotive door. Steel- and aluminum-based TWBs are already commercialized and used in automotive production, while

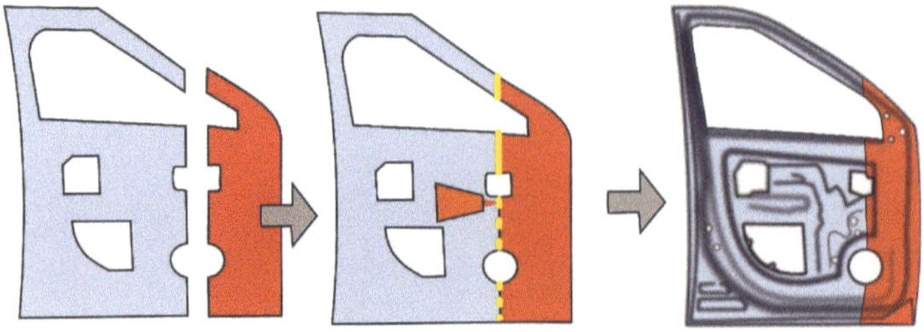

FIGURE 17.4 Process of TWB automotive door production Sun et al. (2018). *Source: Sun, G., Tan, D., Lv, X., Yan, X., Li, Q., & Huang, X. (2018). Multi-objective topology optimization of a vehicle door using multiple material tailor-welded blank (TWB) technology. Advances in Engineering Software, 124, 1–9.*

other lightweight and hybrid metal combinations are still being actively researched. Development items in TWB studies are the metal alloys of the blank, the butt-welding technology, and the press-forming processes involved. Other evaluations on post-production criteria of TWB such as surface quality, corrosion resistance, in-service performance, and end-of-life should also be taken into consideration.

TWB are usually welded in the butt–joint configuration using laser welding, electron beam welding (EBW), or friction stir welding (FSW). Tailored blanks are usually joined with a linear weld seam, and less often with nonlinear weld seams. Laser beam welding (LBW) has been the choice for tailor blank applications, initially using CO_2 lasers and recently using the more efficient neodymium-doped yttrium aluminum garnet (Nd:YAG) and fiber laser systems.

17.6.1 Laser beam welding

The major advantage of LBW is the low heat input, which minimizes welding stresses and reduces distortion. This is achieved by the high energy density of the beam producing a small fusion zone at high welding speeds. Laser welding reduces the heat-affected zone to a minimum, but the workpieces to be joined must have high-quality edges and may need special preparations. Laser welding has the advantage of high productivity and ease of automation, making it suitable for a mass production environment. Current industrial laser welding systems are based on carbon dioxide (CO_2), Nd:YAG, diode, and fiber laser technologies.

Welding lasers used in early TWB productions are in the 5–10 kW power range (Petring, 2004). High-power (>10 kW) CO_2 lasers have been used to weld the TWB for steel (Bayraktar, Moiron, & Kaplan, 2006). However, Nd:YAG systems have gained popularity over CO_2 laser systems due to their simplified process conditions, energy efficiency, and flexible installations. The new generation of fiber lasers (power levels from 100 to 1000 W) presents several benefits for industrial purposes such as high power with low beam divergence, flexible beam delivery, and low ownership costs. A ytterbium fiber laser with 1064–1070 nm wavelengths allows improved coupling into high-reflectivity materials. Studies have been conducted on the use of low-power fiber lasers for the welding of aluminum 7075 (200 W laser power). However, it was found that a CW mode laser had only 0.17 mm penetration depth (Salleh, Ishak, Romlay, & Aiman, 2016). In another study, a very low speed (2–3 mm/s) at 311 W power fiber laser, focusing slightly into the workpiece, achieved a penetration depth of 1.02 mm for the bead-on-plate welds of aluminum 7075-T6 (Paleocrassas & Tu, 2007).

17.6.2 Friction stir welding

FSW is a solid-state joining process developed by The Welding Institute in 1991 and since then has garnered significant interest in the R&D community (Smith & Lord, 2007). It uses a nonconsumable rotating tool to plastically deform and mix the workpiece materials at the joining interface as the tool travels along the welding path, as shown in Fig. 17.5. The process is considered a green technology and is energy-efficient and environmentally friendly. It is potentially useful for the joining of materials that are conventionally difficult to be joined by fusion welding processes. Past investigators have shown that FSW is

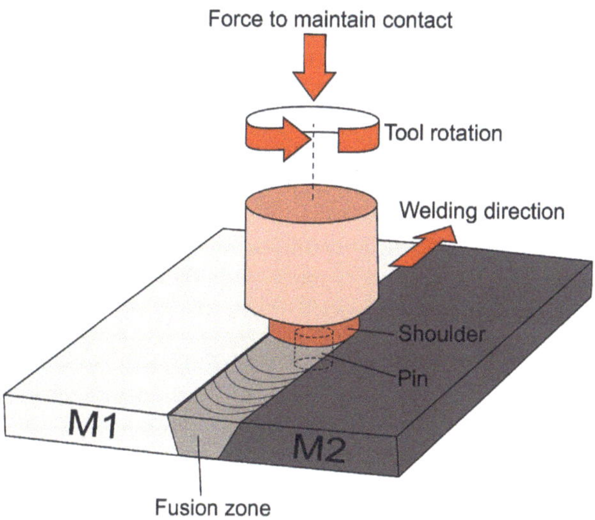

FIGURE 17.5 Friction stir welding process.

suitable for the welding of soft metal alloys, Al-based MMCs, and dissimilar metals. However, the process is associated with low welding speeds and high welding loads, which reduce the productivity and energy efficiency of the process (Arora, Mehta, De, & DebRoy, 2012).

FSWs of aluminum alloys are well established and are being used in production by several automotive companies. Current research and developments in FSW are in joining of hybrid aluminum alloys (Zhong, Wu, & Padhy, 2017), hard materials (steel and titanium), hybrid material joining (aluminum to steel or magnesium) (Coelho, Kostka, dos Santos, & Kaysser-Pyzalla, 2012; Kimapong & Watanabe, 2004), ultrasonic hybrid FSW processes (Zhong et al., 2017), auxiliary heating by plasma or laser (Tang & Shen, 2017), powder-assisted FSW (Inada, Fujii, Ji, Sun, & Morisada, 2010), and high-speed FSW processes (Hovanski et al., 2015).

17.6.3 Stamping and forming processes of TWB

In the forming process of TWB, it should be noted that welded parts deform differently than uniform materials due to differences in the mechanical properties and thicknesses of the welded materials. Formability of TWB can be evaluated using formability testing methods, such as the swift cup, limiting dome height, and forming limit curves.

Different thicknesses and strengths result in inhomogeneous springback. When forming tailored blanks, the following new problems arises (Kleiner et al., 2003):

1. Tribology: there will be a variation in the friction coefficients of TWB due to the differences in surface texture and coatings;
2. Mechanical properties: the forming behavior cannot directly be estimated from both single materials. In addition, the weld seam may have increased hardness, lower ductility, and different failure modes as compared to the parent materials;
3. Material flow: The difference in thicknesses may lead to wrinkling and cracks, especially in the weld seam. In addition, material in the flange would preferably flow in the milder portion of the blank. Thus, different tool concepts, such as variable clamping force, may be necessary for forming of tailored blanks.

The effect of temperatures in the range of 200°C–300°C has been reported to increase the ductility of aluminum, which can be exploited to increase the forming limits of aluminum sheet with standard stamping tools. Elevated temperature forming may benefit the mechanical properties and formability of TWB, as demonstrated by the hot forming quenching process for the sheet metal forming of friction stir welded TWBs (Ko, Kim, Ko, & Kim, 2014).

At very low strain rates, some metals can exhibit extremely high deformation behavior. Most aluminum alloys demonstrate superplastic behavior depending on the metallurgical structure (grain size about 10 μm), temperature (350°C–550°C), and strain rate (about 10^{-5}–10^{-2}/s) (Kleiner et al., 2003). This behavior can be exploited for superplastic forming (SPF) for the manufacturing of complex aerodynamic shapes. However, the process requires a long cycle time, thus limiting its applications for the automotive mass production. SPF can be applied to TWB to obtain more detailed, more complex, and lighter components. However local flow stress differences between the weld and the sheet material need to be considered. In a series of uniaxial tensile tests on TWB under SPF conditions, it was found that longitudinal-weld specimens exhibited significantly lower elongations of about 40%–60% compared to 220%–360% of the parent material. Therefore the flow stress and ductility incompatibilities should be taken into account during component design (Davies, Vetrano, Smith, & Pitman, 2002).

17.6.4 Magnesium tailor welded blanks

The commercialization of magnesium TWBs is still in its infancy, and their applications are mainly seen in prototypes and preindustrial trials. Magnesium alloy (e.g., AZ32) sheet products are available with thicknesses of 0.8–30 mm and widths of up to 1850 mm (Kulekci, 2008). Magnesium has limited formability at room temperature, which is further reduced for welded specimens with variations in the microstructure and thicknesses. However, most conventional forming processes can be performed when the material is heated to between 230°C–370°C. Similar to aluminum, laser welding can be used to join magnesium alloy sheets, although welding without additional filler wire is not possible due to the extensive evaporation of magnesium (Bräunig, Düring, Hartmann, & Viehweger, 2005). Weldability of magnesium alloy was reported to be significantly better using Nd:YAG as compared to CO_2 lasers due to its shorter wavelength (Cao, Jahazi, Immarigeon, & Wallace, 2006). A study on Nd:YAG laser-welded AZ32 magnesium alloy thin sheets (1.2 mm) has shown that the average tensile strength of 256 MPa obtained was within 96%–100% of the base material, but the ductility was significantly reduced (Sorgente, Palumbo, Scintilla, & Tricarico, 2013).

FSW has been made on AZ31, AM50, AM60, and AZ91 using lower feed rates as compared to aluminum. A study on FSW of AZ31 sheet (3.2 mm thickness) shows that the resultant strain profiles are nonsymmetric with respect to the welding line, due to the material flow during FSW. The best results were obtained for tool rotating speeds of 2000 rpm and an advancing velocity of 80 mm/min (Bruni, Buffa, Fratini, & Simoncini, 2010).

17.6.5 Titanium tailor welded blanks

The material properties, weldability, and forming behavior of titanium alloys are well documented in many studies. However, the extremely high cost of titanium has limited its commercial use in high-volume automotive productions. Nevertheless, ongoing research on the joinability and formability of titanium alloys has shown some insights into the possibility of using titanium alloys in TWB applications (Lin Prakash et al., 2018). A full titanium body supercar, the Icona Vulano Titanium, was made with 0.5 mm thick quarter panels.

Titanium has excellent properties such as high strength, low weight, and good corrosion resistance. However, it is very difficult to manufacture and process. Titanium-based joining is usually produced using EBW technology, which is similar to laser welding but takes place under a vacuum environment to avoid the formation of hard and brittle compounds. Titanium alloy sheets have poor drawability and high springback at room temperature, and joints cause additional difficulty in forming titanium welded sheets.

Numerical and experimental warm forming (temperature of 550°C) was investigated by Lai, Chan, and Chow (2007) for a Ti-6Al-4V laser-welded specimen with a thickness combination of 0.7 mm/1.0 mm. Forming tests have shown that failure was often found on the thinner section and at radial corner sides. Lin Prakash et al. (2018) in their investigation on the formability of EBW commercial pure titanium (CP Ti) of different thickness combinations (1.0 and 1.2 mm thicknesses), have shown that the elongations of the TWB were lower than the parent materials. The necking and subsequent failure occurred in the thinner 1.0 mm side, indicating good weld quality. It was also found that the rolling direction of the CP Ti sheets influences the elongation and ultimate tensile strengths, in which lower values were recorded for specimens with the welding line parallel to the rolling direction. Similarly, a formability study on the TWB of commercially pure titanium grade 2 welded to titanium alloy grade 5 (Ti-6Al-4V) has shown the drawability of the TWB deteriorated by 10%–16% as compared to the grade 5 base material. The weld dislocation in the direction of the grade 5 sheet was also observed (Winowiecka, Więckowski, & Zawadzki, 2013).

17.6.6 Aluminum tailor welded blanks

Aluminum car body applications (outer panels and interior components) use AA6000 series and AA5000 series in thicknesses ranging from 0.8 to 1.3 mm thick, which should possess good formability in combination with sufficient strength (Fridlyander et al., 2002). Aluminum TWBs are already in commercial production, with line performance capable of processing aluminum in the range of 0.9–3.0 mm thickness and a maximum blank length of 3600 mm (Farish, 2017). The joining processes in the production of aluminum TWB are usually made by laser welding or FSW.

LBW of aluminum alloys still suffers from seam imperfections such as notches and holes in the seam, which reduces the mechanical properties of the joint. Compared to steel, aluminum alloys have higher reflectivity and thermal conductivity, lower viscosity and surface tension, and higher hydrogen solubility in the molten state, making it difficult to obtain good laser-welded joints (Leong, Sabo, Sanders, & Spawr, 1997). This process instability is caused by reactions in the system laser beam, vapor cloud, and keyhole. It has been shown that vapor plume may refract part of the laser beam and reduce the beam intensity distribution on the workpiece. LBW of steel is more stable compared to aluminum because of the higher viscosity of steel. In addition, the behaviors of alloying elements such as silicon and magnesium reduce the viscosity, resulting in more irregularities in the weld seam. LBWs of aluminum alloys AA5182 and AA5754 have been made for autogenous welding of automotive body components. The results obtained for high-power CO_2 (4.2 kW laser power) and Nd:YAG laser welds have shown comparable tensile strengths with ultimate and yield strengths exceeding 90% of the parent material. Corresponding formability results showed a formability of 50%–70% of the parent material and fracture during dome tests occurring in the fusion zone, as there in an increase in hardness of 10% in the fusion zone (Leong et al., 1997). Similarly a high-power 2.0 kW YAG laser has been used to successfully produce full penetration butt joints for AA5083 (2.0 and 3.0 mm thicknesses) alloys (Okon, Dearden, Watkins, Sharp, & French, 2002). Low-power welding of AA7075 aluminum has been tested with 200 W laser power, achieving 50% penetration depth using the pulse wave mode, and only 20% penetration using the continuous wave (CW) mode (Salleh et al., 2016). A 300 W fiber laser was used in bead-on-plate welds of AA7075-T6, which recorded 430 MPa UTS, which was 75% of that of the parent material (Paleocrassas & Tu, 2007).

There are several challenges in forming aluminum TWB as compared to steel tailored blanks due to reduced formability of hardenable aluminum alloys. As the hardness and strength would decrease in the heat-affected zone surrounding the weld seam, consideration has to be made of the position and orientation of the weld seam in correlation to the drawing direction (Merklein & Geiger, 2002). This can be managed using nonlinear weld seams in which the position and shape of the weld beam can be adjusted to be suitable for forming directions. In addition, intrinsic tailored blanks are locally heat-treated aluminum blanks, in which the mechanical properties are locally changed by local heat treatment. This local change in mechanical properties is caused by the microstructural changes induced by local heat treatment such as laser heating (Merklein & Geiger, 2002). It was shown that local laser heat treatment increases the deep-drawability of AA6181-T4 sheets from 2.1 to 2.6, and reduces the punch force by 10%.

FSW of aluminum TWB overcomes the problems caused by the heat associated with the joining process. Mechanical properties of FSW are especially beneficial for AA6000 series aluminum. However, FSW seam starts and stop should be kept in low strain or removed for stamping. Commercial production of FSW aluminum TWB is already established, with the first dedicated production line in North America launched in 2014 (United States Department of Energy, 2016). Macro-performance analysis of nonuniform aluminum blanks includes formability (Kim et al., 2010), springback (Adnan, Abdullah, & Samad, 2017), fatigue (Garware, Kridli, & Mallick, 2010), and deep-drawability (Kesharwani, Panda, & Pal, 2015).

17.6.7 Hybrid tailor welded blanks

Studies into the joining and forming of dissimilar materials pave the way into their potential use as tailor welded components. The main problem of thermal joining for dissimilar materials is the formation of very hard intermetallic phases, which increases the brittleness of the joint (Schubert, Klassen, Zerner, Walz, & Sepold, 2001). This problem can be overcome by controlling the material diffusion in the formation of the intermetallic phases, which depends on process-related temperature–time cycles (Kreimeyer, Wagner, & Vollertsen, 2005). One such technique is by lowering the overall energy input into the joining, such as combining a high-power laser with high joining speeds. An intermetallic layer of less than 10 μm is considered acceptable for technical applications, which does not significantly deteriorate the ductility and tensile strength of the joint (Meco, Pardal, Ganguly, Williams, & McPherson, 2015). Usually, the processing window is quite small and depends on the increasing power of the laser beam. In the case of aluminum to steel LBW, the aluminum melts and brazes the steel (which remains unmelted) and has been demonstrated to produce steel-Al TWB, as shown in Fig. 17.6.

In contrast, the LBW of aluminum to magnesium is different as both materials have similar melting points, resulting in the molten metals flowing into each other. Cross-sectional evaluation of the weld has shown the expected increase of hardness at the weld zone but no intermetallic layers could be found (Schubert et al., 2001).

CO_2 LBW of titanium (thickness 0.8 mm) to aluminum (thickness 1.15 mm) in a butt–joint configuration has been demonstrated as having

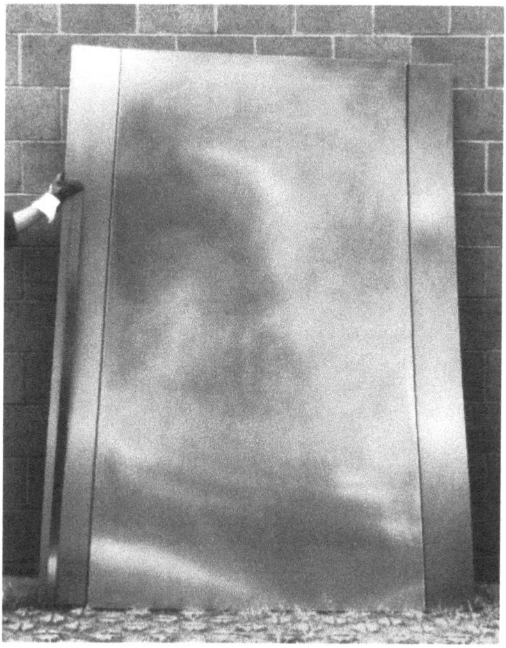

FIGURE 17.6 Tailored blank with outer parts of steel and an aluminum sheet in the middle. *Source: Courtesy of Schubert, E., Klassen, M., Zerner, I., Walz, C., & Sepold, G. (2001). Light-weight structures produced by laser beam joining for future applications in automobile and aerospace industry. Journal of Materials Processing Technology, 115, 2–8.*

minimal intermetallic phase layers of less than 2 μm. The laser beam power used was 4.5 kW at a joining speed of 100 mm/s. The resulting joint had a static tensile strength of around 200 MPa, which was equivalent to 80% of the aluminum base material. Failure occurs in the HAZ with the crack propagating from the HAZ of the Al side (Kreimeyer et al., 2005).

Solid-state joining processes, such as FSW, overcome the problems of poor weldability and intermetallic formation in the joining of different materials. FSW has been used to weld AA6016 (2 mm thick) to AZ91 and AM50 (2.6 mm thick) (Bergmann, Schürer, & Ritter, 2013). It was found that the welding speed has to be reduced compared to similar material welds, and the lowest welding condition (f = 19 mm/min, tool rotation = 700 rpm) gives the highest tensile strength. Similarly, AA5052P-0 (2 mm thickness) was joined to AZ31B-O (2 mm thickness) to obtain a joint efficiency of 72% with a maximum average tensile strength of about 143 MPa (welding speeds = 100 mm/min and tool speed = 1400 rpm) (Shigematsu, Kwon, & Saito, 2009).

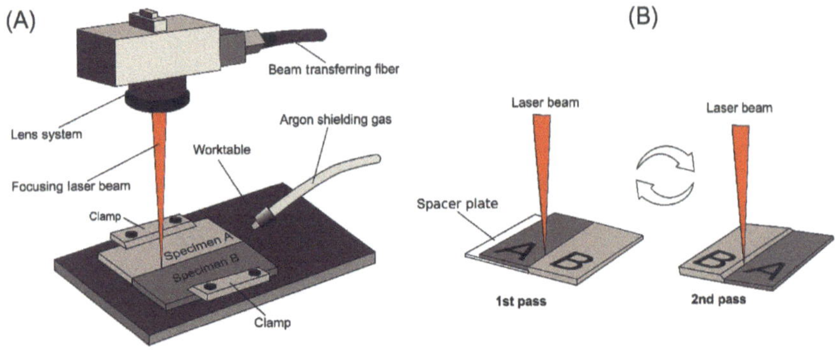

FIGURE 17.7 Experimental set up for double-sided fiber laser welding of dissimilar aluminum alloy showing (A) apparatus set up, and (B) successive double-sided welding.

17.7 Case study: low-power laser-welded dissimilar aluminum tailor welded blanks

In this study, a low-power fiber laser was used to join dissimilar AA5052-H32 (1.5 and 2.0 mm thickness) to AA6061-T6 (1.0 mm thickness) specimens in a butt weld configuration, similar to a TWB joining arrangement. The laser selected has a maximum power of 314 W CW mode and uses a scanner mirror to deliver the laser beam along a predetermined path. Initial penetration tests have shown that the maximum laser penetration is approximately 778 μm at 280 W laser power and welding speed of 20 mm/s. Hence, double-sided welding is proposed, in which the weld is sequentially carried out on both sides of the specimen joint. The experimental set up is shown in Fig. 17.7.

The highest tensile strength obtained for dissimilar 5052-H32 (1.5 mm) to 6061-T6 (1.0 mm) welded at 280 W laser power and 10 mm/s welding speed was 214 MPa, which is 93% of the lower strength parent material (5052-H32 at 230 MPa), as shown in Fig. 17.8. The micrograph in Fig. 17.9 shows the penetration depth of the laser beam on both sides of the joint, although several instances of porosities are detected, which would decrease the joining strength of the specimen.

Thus, there is potential in the use of low-power fiber lasers for the fabrication of dissimilar aluminum TWB. Hence, the combination of light material (aluminum), design (TWB), and low energy processing (low-power fiber laser) can be a potential strategy in increasing sustainability levels in automotive production.

17.8 Future trends and directions

With the increasing content of light materials in automotive vehicles, it is estimated that the use of high-strength steel will decline from 15%

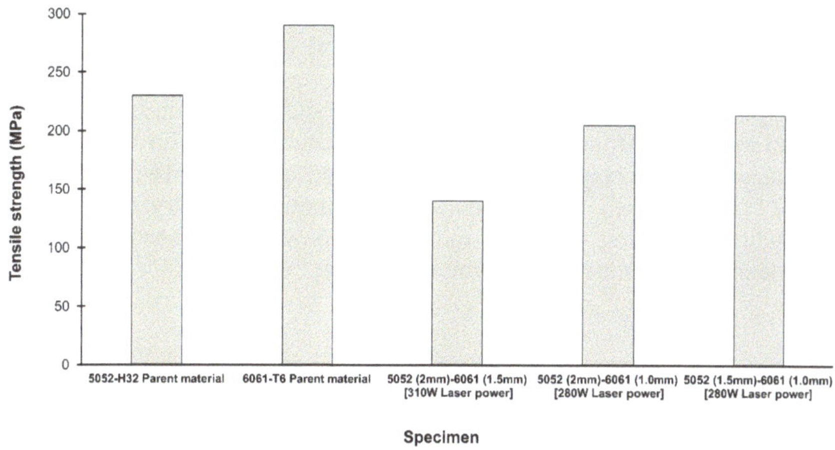

FIGURE 17.8 Tensile strength results for low-power fiber laser weld of dissimilar aluminum TWB.

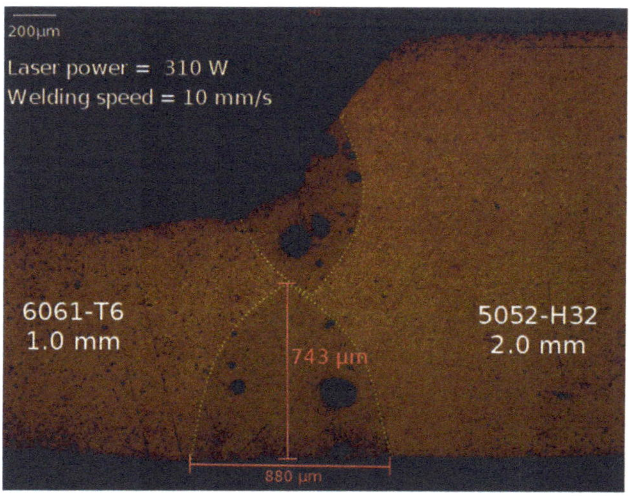

FIGURE 17.9 Micrograph of the cross-section of the laser-welded joint showing laser beam penetration.

of total vehicle weight in 2020 to 5% by 2040 (Smith, Spulber, Modi, & Fiorelli, 2017). The adoption of AHSS steel and aluminum will rise steadily, with increasing use in safety cage parts and components. Magnesium will also see more prominent usage in interior components, such as the instrument panel crossbeams. By 2030, it is also predicted that 96% of vehicle programs will consider aluminum for BIW applications. The future generation of vehicles will most likely be made up of multimaterial combinations, mixing various types of AHSS steel, aluminum alloys, magnesium, plastics, and other materials to achieve the weight, costs, and performance targets. The joining and forming methods for these multimaterial components and structures are important technologies to be developed in the future.

The production processes of light materials such as aluminum and magnesium are emission and energy intensive. Focusing only on their weight advantage without considering production and end-of-life issues may not be the best approach for a sustainable automotive industry. In the low-carbon lightweighting approach, manufacturers work together with suppliers to significantly reduce CO_2 emissions along the entire value chain.

17.9 Conclusions

Lightweighting strategies enable the automotive industry to address both economic fuel efficiency and environmental concerns about exhaust emissions. They play a key role in both the interim and long-term improvements in vehicle energy consumption. Design for sustainability is achieved by the use of lightweight materials in weight-saving designs, such as TWBs, to further enhance the solution not only in terms of the product itself but also in the manufacturing process involved. Lower manufacturing costs can be obtained by considering new technologies such as energy-efficient fiber lasers and FSW in the production process. For example, an efficient low-to-medium power fiber laser system can be set up with relatively low capital entry and utilization costs as compared with existing high-powered laser systems.

Commercially, the main competition for lightweighting is between aluminum and steel. The use of aluminum in automotive production is increasing, while steel is still dominant with its AHSS variation. Aluminum-based TWB is a readily available technology, and enhanced aluminum TWB and hybrid Al-steel will be seeing an entry into commercialization as it matures from the numerous ongoing studies. Other lightweight materials, particularly magnesium, will be seen in sheet metal applications and subsequently TWB applications.

Finally, the current production processes of these lightweight metal alloys have large environmental footprints. The assessment of the environmental impact during the entire life cycle should also be considered alongside efforts on lightweighting and cost-effective production processes to achieve greater levels of sustainability in the automotive industry.

Acknowledgment

The authors gratefully acknowledge the support of Universiti Sains Malaysia RUI grant (grant no: 1001/PMEKANIK/8014031) in the preparation of this work.

References

Adnan, M. F., Abdullah, A. B., & Samad, Z. (2017). Springback behavior of AA6061 with non-uniform thickness section using Taguchi Method. *The International Journal of Advanced Manufacturing Technology, 89*, 2041–2052.

Arora, A., Mehta, M., De, A., & DebRoy, T. (2012). Load bearing capacity of tool pin during friction stir welding. *The International Journal of Advanced Manufacturing Technology, 61*, 911–920. Available from: https://doi.org/10.1007/s00170-011-3759-7.

Bayraktar, E., Moiron, J., & Kaplan, D. (2006). Effect of welding conditions on the formability characteristics of thin sheet steels: Mechanical and metallurgical effects. *Journal of Materials Processing Technology, 175*, 20–26.

Bergmann, J. P., Schürer, R., & Ritter, K. (2013). Friction stir welding of tailored blanks of aluminum and magnesium alloys. *Key Engineering Materials, 549*, 492–499.

Bhattacharya, S., & Alpas, A. T. (2017). Tribology of aluminum and aluminum matrix composite materials for automotive components. In O. Faruk, J. Tjong, & M. Sain (Eds.), *Lightweight and sustainable materials for automotive applications* (1st ed., pp. 303–328). CRC Press.

Bian, M. Z., Sasaki, T. T., Suh, B. C., Nakata, T., Kamado, S., & Hono, K. (2017). A heat-treatable Mg–Al–Ca–Mn–Zn sheet alloy with good room temperature formability. *Scripta Materialia, 138*, 151–155.

Bräunig, S., Düring, M., Hartmann, H., & Viehweger, B. (2005). Magnesium sheets for industrial application. In K. U. Kainer (Ed.), *Magnesium* (pp. 955–961). Weinheim, FRG: Wiley-VCH Verlag GmbH & Co. KGaA.

Bruni, C., Buffa, G., Fratini, L., & Simoncini, M. (2010). Friction stir welding of magnesium alloys under different process parameters. *Materials Science Forum, 638–642*, 3954–3959.

Cao, X., Jahazi, M., Immarigeon, J. P., & Wallace, W. (2006). A review of laser welding techniques for magnesium alloys. *Journal of Materials Processing Technology, 171*, 188–204.

Chen, S., Rana, R., Haldar, A., & Ray, R. K. (2017). Current state of Fe-Mn-Al-C low density steels. *Progress in Materials Science, 89*, 345–391.

Coelho, R. S., Kostka, A., dos Santos, J. F., & Kaysser-Pyzalla, A. (2012). Friction-stir dissimilar welding of aluminium alloy to high strength steels: Mechanical properties and their relation to microstructure. *Materials Science and Engineering: A, 556*, 175–183.

Costa, A., Miranda, R., Quintino, L., & Yapp, D. (2007). Analysis of beam material interaction in welding of titanium with fiber lasers. *Materials and Manufacturing Processes, 22*, 798–803.

Das, S. (2000). Aluminum tailor welded blanks. *Advanced Materials and Processes, 157*, 41–42.

Davies, R. W., Vetrano, J. S., Smith, M. T., & Pitman, S. G. (2002). Mechanical properties of aluminum tailor welded blanks at superplastic temperatures. *Journal of Materials Processing Technology, 128*, 38–47.

Eisenmenger, M., & Szymanski, D. (2012). Tailor welding coils with the fiber-delivered laser. *Industrial Laser Solutions-for Manufacturing, 27*(6). Available from: https://www.industrial-lasers.com/welding/article/16485310/tailor-welding-coils-with-the-fiberdelivered-laser.

European Union (2014). Regulation (EU) No 333/2014 of the European Parliament and of the Council of 11 March 2014 amending Regulation (EC) No 443/2009 to define the modalities for reaching the 2020 target to reduce CO_2 emissions from new passenger cars. *Official Journal of the European Union, 103*, 15–21.

European Union (2019). Regulation (EU) 2019/631 of the European Parliament and of the Council of 17 April 2019 setting CO_2 emission performance standards for new passenger cars and for new light commercial vehicles, and repealing Regulations (EC) No 443/2009 and (EU) No 510/2011 (Text with EEA relevance.). *Official Journal of the European Union, 111*, 13–53.

Faria, R., Moura, P., Delgado, J., & De Almeida, A. T. (2012). A sustainability assessment of electric vehicles as a personal mobility system. *Energy Conversion and Management, 61*, 19–30.

Farish, M. (2017). *Tailor-made*. Retrieved 20.09.2019, from Automative Manufacturing Solutions website: <https://www.automotivemanufacturingsolutions.com/tailor-made/35724.article>.

Fontaras, G., & Samaras, Z. (2010). On the way to 130g CO_2/km—Estimating the future characteristics of the average European passenger car. *Energy Policy, 38*, 1826–1833.

Fridlyander, I. N., Sister, V. G., Grushko, O. E., Berstenev, V. V., Sheveleva, L. M., & Ivanova, L. A. (2002). Aluminum alloys: Promising materials in the automotive industry. *Metal Science and Heat Treatment, 44*, 365–370.

Froes, F. H., Friedrich, H., Kiese, J., & Bergoint, D. (2004). Titanium in the family automobile: The cost challenge. *JOM, 56*, 40–44.

Garware, M., Kridli, G. T., & Mallick, P. K. (2010). Tensile and fatigue behavior of friction-stir welded tailor-welded blank of aluminum alloy 5754. *Journal of Materials Engineering and Performance, 19*, 1161–1171.

Gutowski, V., Yang, W., Li, S., Dean, K., & Zhang, X. (2017). Lightweight nanocomposite materials. In O. Faruk, J. Tjong, & M. Sain (Eds.), *Lightweight and sustainable materials for automotive applications* (1st ed., pp. 277–302). CRC Press.

Heuss, R., Müller, N., van Sintern, W., Starke, A., & Tschiesner, A. (2012). *Lightweight, heavy impact*. McKinsey & Company.

Hirsch, J. (2004). Automotive trends in aluminium-The European perspective. *Materials Forum, 28*, 15–20.

Hirsch, J. (2011). Aluminium in innovative light-weight car design. *Materials Transactions, 52*, 818–824.

Hovanski, Y., Upadhyay, P., Carsley, J., Luzanski, T., Carlson, B., Eisenmenger, M., & Hartfield-Wunsch, S. (2015). High-speed friction-stir welding to enable aluminum tailor-welded blanks. *JOM, 67*, 1045–1053.

Hula, A., French, R., Maguire, A., Bunker, A., & Rojeck, T. (2019). *The 2018 EPA automotive trends report: Greenhouse gas emissions, fuel economy, and technology since 1975 (EPA-420-R-19-002, March 2019)* (No. EPA-420-R-19-002; p. 208). Office of Transportation and Air Quality.

Inada, K., Fujii, H., Ji, Y., Sun, Y., & Morisada, Y. (2010). Effect of gap on FSW joint formation and development of friction powder processing. *Science and Technology of Welding and Joining, 15*, 131–136.

IPCC. (2014). Summary for policymakers. In O. Edenhofer, R. Pichs-Madruga, Y. Sokona, E. Farahani, S. Kadner, K. Seyboth, & ... S. Schlömer (Eds.), *Climate change 2014: Mitigation of climate change. Contribution of working group III to the fifth assessment report of the intergovernmental panel on climate change*. Cambridge University Press.

Isenstadt, A., German, J., Bubna, P., Wiseman, M., Venkatakrishnan, U., Abbasov, L., ... Kolwich, G. (2016). *Lightweighting Technology Development and Trends in U.S. Passenger Vehicles; Working Paper 25* (pp. 1–24). Washington, DC: International Council on Clean Transportation.

Joost, W. J. (2012). Reducing vehicle weight and improving U.S. energy efficiency using integrated computational materials engineering. *JOM, 64*, 1032–1038.

Kesharwani, R. K., Panda, S. K., & Pal, S. K. (2015). Experimental investigations on formability of aluminum tailor friction stir welded blanks in deep drawing process. *Journal of Materials Engineering and Performance, 24*, 1038–1049.

Kim, D., Lee, W., Kim, J., Chung, K.-H., Kim, C., Okamoto, K., ... Chung, K. (2010). Macro-performance evaluation of friction stir welded automotive tailor-welded blank sheets: Part II—Formability. *International Journal of Solids and Structures, 47*, 1063–1081.

Kimapong, K., & Watanabe, T. (2004). Friction stir welding of aluminum alloy to steel. *Welding Journal, 83*, 277.

Kleiner, M., Geiger, M., & Klaus, A. (2003). Manufacturing of lightweight components by metal forming. *CIRP Annals-Manufacturing Technology, 52*, 521–542.

Knight, B. (2015). Magnesium alloys in the automotive market. *Presented at the AMM'S North American Automotive Metals Conference 2015, Michigan, USA*. Retrieved from <https://www.amm.com/events/download.ashx/document/speaker/7981/a0ID000000X0kJUMAZ/Presentation>

Ko, D.-H., Kim, J.-H., Ko, D.-C., & Kim, B.-M. (2014). Improvement of weldment properties by hot forming quenching of friction stir welded TWB sheet. *Advances in Mechanical Engineering, 2014*, 1–9.

Kreimeyer, M., Wagner, F., & Vollertsen, F. (2005). Laser processing of aluminum–titanium-tailored blanks. *Optics and Lasers in Engineering, 43*, 1021–1035.

Kulekci, M. K. (2008). Magnesium and its alloys applications in automotive industry. *The International Journal of Advanced Manufacturing Technology, 39*, 851–865.

Kumar, D. S., & Sasanka, C. T. (2017). *Magnesium and its alloys. Lightweight and sustainable materials for automotive applications* (pp. 329–368). CRC Press.

Kumar, D. S., Sasanka, C. T., Ravindra, K., & Suman, K. N. S. (2015). Magnesium and its alloys in automotive applications—A review. *American Journal of Materials Science and Technology*. Available from: https://doi.org/10.7726/ajmst.2015.1002.

Kunkel, G. A., & Hovanski, Y. (2016). From the lab to your driveway: Aluminum tailor-welded blanks. *Welding Journal, 95*, 36–39.

Lai, C. P., Chan, L. C., & Chow, C. L. (2007). Warm forming simulation of titanium tailor-welded blanks with experimental verification. *AIP Conference Proceedings, 908*, 1621–1626, Porto (Portugal): AIP.

Leong, K. H., Sabo, K. R., Sanders, P. G., & Spawr, W. J. (1997). Laser welding of aluminum alloys. *Lasers as Tools for Manufacturing II, 2993*, 37–44, International Society for Optics and Photonics.

Lin Prakash, P. S., Biswal, S. K., Roy, G. G., Jha, M. N., Mascarenhas, M., & Panda, S. K. (2018). Effect of orientation of weld line on formability of electron beam-welded dissimilar thickness titanium sheets. *Journal of Materials Engineering and Performance, 27*, 5913–5925.

Long, R., & Newman, J. (2017). Development of low-cost, high-strength automotive aluminum sheet—Arconic. Lightweight Materials 2016 Annual Report (pp. 72–78). Washington, DC: US Department of Energy.

Luo, A. A. (2013). Magnesium casting technology for structural applications. *Journal of Magnesium and Alloys, 1*, 2–22.

Luo, A. A., & Sachdev, A. K. (2012). *Applications of magnesium alloys in automotive engineering. Advances in wrought magnesium alloys* (pp. 393–426). Elsevier.

Mascarin, A., Hannibal, T., Raghunathan, A., Ivanic, Z., & Clark, M. (2016). *Vehicle lightweighting: Mass reduction spectrum analysis and process cost modeling* (No. INL/EXT--16-38001, 1363637). Available from: https://doi.org/10.2172/1363637.

Mayyas, A., Qattawi, A., Omar, M., & Shan, D. (2012). Design for sustainability in automotive industry: A comprehensive review. *Renewable and Sustainable Energy Reviews, 16*, 1845–1862.

Meco, S., Pardal, G., Ganguly, S., Williams, S., & McPherson, N. (2015). Application of laser in seam welding of dissimilar steel to aluminium joints for thick structural components. *Optics and Lasers in Engineering, 67*, 22–30.

Merklein, M., & Geiger, M. (2002). New materials and production technologies for innovative lightweight constructions. *Journal of Materials Processing Technology, 125*, 532–536.

Mohrbacher, H., Spöttl, M., & Paegle, J. (2015). Innovative manufacturing technology enabling light weighting with steel in commercial vehicles. *Advances in Manufacturing, 3*, 3–18.

Morita, A. (1998). *Aluminium alloys for automobile applications. Proceedings of the 6th International Conference on Aluminium Alloys.* (pp. 25–32). Toyohashi, Japan: The Japan Institute of Light Metals.

Niu, X. P., Skszek, T., Fabischek, M., & Zak, A. (2014). Low temperature warm forming of magnesium ZEK 100 sheets for automotive applications. *Materials Science Forum, 783–786*, 431–436.

OICA. (2019). World motor vehicles sales by country and type 2005-2018. *International organization of motor vehicle manufacturers.* Retrieved from <http://www.oica.net/wp-content/uploads/total_sales_2018.pdf>.

Office of Transportation and Air Quality. (2019). *Fast facts U.S. transportation sector greenhouse gas emissions 1990 –2017* (No. EPA-420-F-19-047; p. 5). United States Environmental Protection Agency.

Okon, P., Dearden, G., Watkins, K., Sharp, M., & French, P. (2002). Laser welding of aluminium alloy 5083. *International Congress on Applications of Lasers & Electro-Optics, 2002*, 158364. LIA.

Paleocrassas, A., & Tu, J. (2007). Low-speed laser welding of aluminum alloy 7075-T6 using a 300-W, single-mode, ytterbium fiber laser. *Welding Journal-New York, 86*, 179.

Petring, D. (2004). Laser applications in European automotive manufacturing: Historical review and recent trends. *Journal of the Japan Welding Society, 73*, 539–546.

Rosen, M. A., & Kishawy, H. A. (2012). Sustainable manufacturing and design: Concepts, practices and needs. *Sustainability, 4*, 154–174.

Saberi, B. (2018). The role of the automobile industry in the economy of developed countries. *International Robotics & Automation Journal, 4*, 179–180.

Sachdev, A. K., Kulkarni, K., Fang, Z. Z., Yang, R., & Girshov, V. (2012). Titanium for automotive applications: Challenges and opportunities in materials and processing. *JOM, 64*, 553–565.

Sakurai, T. (2008). The latest trends in aluminum alloy sheets for automotive body panels. *Kobelco Technology Review*, 22–28.

Salleh, M., Ishak, M., Romlay, F., & Aiman, M. (2016). A study on bead-on-plate welding of AA7075 using low power fiber laser. *Journal of Mechanical Engineering and Sciences, 10*, 2065–2075.

Schubert, E., Klassen, M., Zerner, I., Walz, C., & Sepold, G. (2001). Light-weight structures produced by laser beam joining for future applications in automobile and aerospace industry. *Journal of Materials Processing Technology, 115*, 2–8.

Shigematsu, I., Kwon, Y.-J., & Saito, N. (2009). Dissimilar friction stir welding for tailor-welded blanks of aluminum and magnesium alloys. *Materials Transactions, 50*, 197–203.

Sioshansi, F., & Webb, J. (2019). Transitioning from conventional to electric vehicles: The effect of cost and environmental drivers on peak oil demand. *Economic Analysis and Policy, 61*, 7–15.

Smith, A., & Davies, H. (1996). A review of the history of emission legislation, urban and national transport trends and their impact on transport emission. *Transactions on the Built Environment, 23*, 293–302.

Smith, B., Spulber, A., Modi, S., & Fiorelli, T. (2017). *Technology roadmaps: Intelligent mobility technology, materials and manufacturing processes, and light duty vehicle propulsion.*

Smith, I. J., & Lord, D. D. (2007). FSW patents-a stirring story. *SAE Transactions*, 772–780.

Sorgente, D., Palumbo, G., Scintilla, L. D., & Tricarico, L. (2013). Evaluation of the strain behaviour of butt joints on AZ31 magnesium alloy thin sheets welded by Nd:YAG laser. *The International Journal of Advanced Manufacturing Technology, 67*, 2753–2763.

Stahlkocher. (2006). *Tailored blank hoesch museum.* Retrieved from <https://commons.wikimedia.org/wiki/File:Tailored_Blank_Hoesch_Museum.jpg>

Subic, A., & ICSAT (Eds.), (2012). *Sustainable automotive technologies 2012: Proceedings of the 4th international conference.* Berlin: Springer.

Sun, G., Tan, D., Lv, X., Yan, X., Li, Q., & Huang, X. (2018). Multi-objective topology optimization of a vehicle door using multiple material tailor-welded blank (TWB) technology. *Advances in Engineering Software, 124*, 1–9.

Tang, J., & Shen, Y. (2017). Effects of preheating treatment on temperature distribution and material flow of aluminum alloy and steel friction stir welds. *Journal of Manufacturing Processes, 29*, 29–40.

United States Department of Energy. (2016). *Report on technology transfer and related technology partnering activities at the national laboratories and other facilities for fiscal year 2014* (p. 75). United States Department of Energy.

van den Hoed, R. (2007). Sources of radical technological innovation: The emergence of fuel cell technology in the automotive industry. *Journal of Cleaner Production, 15*, 1014–1021.

Vaz, C., Rauen, T., & Lezana, Á. (2017). Sustainability and innovation in the automotive sector: A structured content analysis. *Sustainability, 9*, 880.

Watarai, H. (2006). *Trend of research and development for magnesium alloys-reducing the weight of structural materials in motor vehicles* (pp. 84–97). NISTEP Science & Technology Foresight Center.

Winowiecka, J., Więckowski, W., & Zawadzki, M. (2013). Evaluation of drawability of tailor-welded blanks made of titanium alloys Grade 2 | | Grade 5. *Computational Materials Science, 77*, 108–113.

Winter, D. (2019). Auto engineers: Higher efficiency possible but U.S. buyers not interested. *WardsAuto*. Retrieved from <https://www.wardsauto.com/car-management-briefing-seminars/auto-engineers-higher-efficiency-possible-us-buyers-not-interested>.

Yan, Q. (2005). *Tailor-welded blanks and their production* (S. Deng, A. Matsunawa, Y. L. Yao, & M. Zhong, Eds.). Beijing, China.

Zhao, J., & Jiang, Z. (2018). Thermomechanical processing of advanced high strength steels. *Progress in Materials Science, 94*, 174–242.

Zhong, Y. B., Wu, C. S., & Padhy, G. K. (2017). Effect of ultrasonic vibration on welding load, temperature and material flow in friction stir welding. *Journal of Materials Processing Technology, 239*, 273–283.

Index

Note: Page numbers followed by "*f*" and "*t*" refer to figures and tables, respectively.

D

Q

R

FSC
www.fsc.org
MIX
Papier | Fördert
gute Waldnutzung
FSC® C083411

Kolibri 360 GmbH
Ferdinand-Jühlke-Straße 7
99095 Erfurt, Deutschland
produktsicherheit@kolibri360.de